Texts in Mathematics

Volume 4

Chapters in Probability

Texts in Mathematics Series Editor
Dov Gabbay   dov.gabbay@kcl.ac.uk

Chapters in Probability

Craig Smoryński

ISBN 978-1-84890-067-7

College Publications
Scientific Director: Dov Gabbay
Managing Director: Jane Spurr
Department of Computer Science
King's College London, Strand, London WC2R 2LS, UK

http://www.collegepublications.co.uk

Cover designed by Laraine Welch
Printed by Lightning Source, Milton Keynes, UK

Contents

0

Preface

Three individuals beside myself are responsible for this book. John Baldwin is the first, having arranged for me to teach the course it grew out of. This was a course on the History of Mathematics given to middle school teachers from the Chicago Public School system who wanted to learn more mathematics with an eye to teaching the more specialised topics of middle school mathematics. When I learned that Probability was now being taught in middle school, I decided to devote a week to the subject. This took place during the last week of the course.

Although I was writing up my notes throughout the course, I tend to be expansive after the initial presentation and, as I started the write up on Probability well after the completion of the course, there were no deadlines to keep the writing in check. My exposition grew and grew.

In the United States there is a constant shortage of teachers trained to teach mathematics, and those prepared to teach other subjects are pressed into service. Some of my students were such with weak mathematical backgrounds and, with them in mind, I automatically added the necessary explanations of concepts and definitions of terms. My friend Eckart Menzler-Trott recognised before I did that I was no longer writing up notes on the History of Probability, but a sort of history-based introduction to Elementary Probability Theory. He also demanded I include English translations of the important papers on the subject and got in touch with our mutual friend Jan von Plato, who donated to the project his copy of Ivo Schneider's German translations of most of these papers. The end result does not meet Eckart's original demands or justify von Plato's generosity, but the reader will find quite a number of excerpts from the original books and papers, along with passages from historic textbooks, remarks and opinions of various mathematical historians, and sundry other items I found quotable.

Regarding these excerpts, I note that those from English sources are fairly exact transcriptions, preserving spelling, punctuation, and (with one exception) notation. Departures from this practice are some elements of layout (e.g., linebreaks), the numbering of displayed formulæ, the deletion of some bibliographic references, and, alas, the use of the single modern "s" for the two

typographically distinct versions of the letter used in the older literature. I simply couldn't find the character in the available LATEX fonts. In the translated
passages I have been less demanding. And in some cases I have translated from
German translations. I think my translations are overall correct and serviceable,
if not literary masterpieces. The footnoted comparisons to other translations
of Jakob Bernoulli's comments on his famous theorem (Chapter 5, section 1,
below) ought to give a good indication of their overall reliability.

The final choice of topics might be explained as my answer to the question: what should the college-educated student coming out of a mathematics
appreciation course covering Probability know? Obviously he should know the
material up to and including binomial probability. For philosophical and cultural reasons, I deem it necessary for such a student to have knowledge of the
Law of Large Numbers and the Central Limit Theorem. And these bring one
to Statistics, a topic I do not go into deeply but could not entirely avoid.

The present book is, however, far too technical for a mathematics appreciation course, the middle or high school probability course, or the service course
for college freshmen, and goes into quite some technical detail in Chapters 5 and
6. Nonetheless, I think it not too technical for the average student of the quantitative sciences who has completed the basic Calculus course or, as textbooks
used to say, acquired the necessary mathematical maturity. One cannot discuss
even Elementary Probability to any depth without mentioning the Calculus.
With an eye to my middle school teachers who were innocent of knowledge of
this most important field of mathematics, I have relegated the essential use of
the Calculus to a few technical asides, and have added occasional explanatory
material. The mathematical appendix is, in fact, devoted to offering such explanations, as well as adding some supplementary technical material that didn't
require coverage in the main text. This includes brief discussions of differentiation, integration and limits in general, though for the most part the reader can
get by knowing that $\int_a^b f(x)dx$ is an abbreviation for the area under the curve
$y = f(x)$ over the interval $[a, b]$.

At the opposite extreme, the technically advanced reader will undoubtedly
be disappointed at what I leave out. There is for example no proof of the Central
Limit Theorem. I was sorely tempted to include more in the mathematical
appendix, but something had to be done to keep the book from topping 600
pages.

Two auxiliary features of this book are the use of postage stamps and
banknotes as illustrations and a longish appendix on the use of the calculator.
Aside from the fact that a number of writers have chosen to illustrate their texts
philatelically, thus making such practice fashionable, there is some pædagogical
value in it. Their use here gives some indication of what is available that the
prospective probability teacher might want to exhibit in class.

The inclusion of material on the calculator requires some explanation. First,
as to the choice of the *TI-83 Plus* as my standard for discussion, there are several reasons. A simple reason is that the *TI-83 Plus* was the standard where I
taught most recently and I am quite familiar with it. Another is the widespread
use of calculators from Texas Instruments in education in the United States.

Now there are essentially three families of educational calculators produced by Texas Instruments: the *TI-30*, the *TI-83* and company, and the *TI-89*. The *TI-30* is great for middle school, but, though I found it useful in statistical computations years ago, I now deem it inadequate for the purposes of the present book, where the ability to calculate directly with lists and to program the calculator are paramount. Also needed are, of course, the calculator's graphical capabilities and also useful are the built-in statistical functions of the *TI-83* or *TI-89*. Now, the *TI-89* is such an advanced calculator that I deem it axiomatic that anyone who purchases it will have all the experience necessary in list processing and programming on a calculator as not to require any additional instruction on my part on these matters. For the level at which this book is written, such experience cannot be assumed and the reader will be at the *TI-83* level (albeit using perhaps an older *TI-82* or a more modern *TI-84* with or without some of the words "Silver Millennium Edition" appended to its name).

The manual for the *TI-83 Plus* is fairly good and I do not attempt to replace it with my Appendix, but to supplement it. I assume the reader has familiarised himself with the calculator sufficiently so as to perform basic tasks such as graphing functions, setting the window, etc. I do not assume familiarity with programming or with non-numeric data types like strings, lists, and matrices. Incidentally, for those opting to use the more advanced *TI-89*, I note that although the syntax of its programming langauge differs from that of the *TI-83 Plus*, it is still a form of BASIC and one should have no difficulty translating the programs presented here to that calculator. I add that the final section of the appendix on the calculator contains additional material, on random number generation and calculator simulation of random processes, that was so computational in character I decided to place it in this appendix rather than in the main body of the text.

The overall structure of the book is simple: there are three parts. Part I discusses Elementary Probability Theory, the material one would see in the middle school, high school, or college level general education service course, albeit presented from a more advanced standpoint. Although I would encourage him to proceed further, the middle school teacher can probably stop here or skip ahead to Appendix A if he is not very familiar with the calculator. Some of the material in Chapter 4 on using the area under the Bell Curve to approximate binomial probability might seem irrelevant as regards teaching the aforesaid elementary courses: the calculator yields such probabilities quickly and accurately. This "normal" approximation to the binomial is part of the history of Probability, as well as part of the history of its teaching. It also provides motivational background for the Central Limit Theorem of Chapter 6.

Part II on the Law of Large Numbers is divided into two chapters, the first on Bernoulli's Theorem, and the second on its aftermath. Both chapters have their share of technical detail, and, more interesting, historical and other examples illustrating the issues involved.

Part III consists of five appendices. Appendix A is the instruction on the use of the calculator mentioned earlier. Appendix B is the mathematical appendix, the first couple of sections being devoted to filling in some mathematical background and the latter sections expanding the main text with supplementary material. Appendix C is a short essay on stamp collecting serving as an excuse to include a few probability-related stamps that didn't fit in elsewhere. Appendix D includes a few tables alluded to in the text and the last two sections of Appendix B. And Appendix E is a short annotated bibliography with suggestions for further reading.

Part I

Elementary Probability

1

The Error of Brother Lucas

It is traditional[1] to assign the birth of probability theory to Blaise Pascal (1623 - 1662) and his correspondence on the subject with Pierre de Fermat (*c.* 1601 - 1665) of 1654. This is nicely expressed in the opening lines of Hacking's book on the origins of the discipline:

> In 1865 Isaac Todhunter published *A History of the Mathematical Theory of Probability from the Time of Pascal to that of Laplace.* It remains an authoritative survey of nearly all work between 1654 and 1812. Its title is exactly right. There was hardly any history to record before Pascal, while after Laplace probability was so well understood that a page-by-page account of published work on the subject became almost impossible. Just six of the 618 pages of text in Todhunter's book discuss Pascal's predecessors.[2]

Freudenthal and Steiner state this a bit more forcefully:

> One can place the birth of the probability calculus sharply with the year 1654. Although one had known games of chance from time immemorial, the earlier traces of probability theoretic contemplation are either vague or historically insignificant. Two problems were the midwives. They are supposed to have been posed by the Chevalier de Méré to Pascal. Pascal corresponded on them in 1654 with Fermat.[3]

---

[1] *Cf.*, e.g., Isaac Todhunter (1820 - 1884), *A History of the Mathematical Theory of Probability; From the Time of Pascal to That of Laplace*, 1865; Oystein Ore (1899 - 1968), "Pascal and the invention of probability theory", *The American Mathematical Monthly* 67 (1960), pp. 409 - 419; Hans Freudenthal and Hans-Georg Steiner, "Aus der Geschichte der Wahrscheinlichkeitstheorie und der mathematischen Statistik", in: Heinrich Behnke, Günther Bertram and Robert Sauer, eds., *Grundzüge der Mathematik IV*, Vandenhoek & Ruprecht, Göttingen, 1966; or Ian Hacking, *The Emergence of Probability; A Philosophical Study of Early Ideas About Probability*, Cambridge University Press, Cambridge, 1975.

[2] Hacking, *op. cit.*, p. 1.

[3] Freudenthal and Steiner, *op.cit.*, p. 151.

Against this, M.G. Kendall writes in an interesting paper on Pascal's predecessors that

> The first article in this series by Dr F.N. David[4] (1955) has reviewed the development of dicing and gaming up to the time of Fermat and Pascal, who are popularly but erroneously supposed to have founded the calculus of probability.[5]

Again Hacking:

> It is true that we first find European calculations on chances in work like that of Pacioli [1494], a book famous as the origin of double-entry book-keeping. But what is notable is not that problems of chance occur in early works of arithmetic... but that these books were quite unable to solve the new problems. No one could solve them until about 1660, and then everyone could.[6]

The man referred to by Hacking is Fra Luca Pacioli (*c.* 1445 - 1517), aka Lucas de Burgo, an itinerant mathematician and Franciscan brother who taught arithmetic in various Italian cities and universities; and the work referred to is his *Summa de arithmetica, geometria, proportioni et proportionalita* (1494, 2nd edition 1523). This work departs from the tradition of practical arithmetics like the famous *Liber abbaci* of Leonardo of Pisa (*c.* 1170 - *c.* 1240) in that it was encyclopædic in coverage. The *Summa* is best known outside mathematics for offering the first printed account of double-entry bookkeeping, referred to by him as the "method of Venice", an appellation that suggests he learned it from Venetian merchants and did not invent it himself. The *Summa* is a compendium, not a work of great originality. Nor is it without error.[7] Nonetheless, it is an important work, influencing such 16th century mathematicians as Girolamo Cardano (1501 - 1576), Niccolò Tartaglia (1499 *or* 1500 - 1557), and Rafæl Bombelli (1526 - 1572). The book is important in the history of probability for two problems it raises.

Before mentioning the probabilistic problems of the *Summa*, I should digress to say a few words about Pacioli, whose name is not as familiar in the history of mathematics as the more substantial Newton (1642 - 1727) or Leibniz (1646 -

---

[4] F.N. David, "Studies in the history of probability and statistics I. Dicing and gaming (A note on the history of probability)", *Biometrika* 42 (1955), pp. 1 - 15. The journal *Biometrika* ran quite a few articles on the history of probability, of which David's was the first in the series and Kendall's the second. David incorporated material from these articles in the first three chapters of her very readable and informative *Games, Gods and Gambling; A History of Probability and Statistical Inference*, Charles Griffin & Co., Ltd., London, 1962 (reprinted by Dover Publications, Mineola (New York), 1998). Her assessment of credit to Pascal's predecessors is more generous.

[5] M.G. Kendall, "Studies in the history of probability and statistics II. The beginnings of a probability calculus", *Biometrika* 43 (1956), pp. 1 - 14; here, p. 1.

[6] Hacking, *op. cit.*, p. 5.

[7] *Cf.*, e.g., footnote 19, below, or Tartaglia's reference to one such error cited on page 10, below.

Two philatelic portraits of Luca Pacioli derived from the famous painting by Jacopo d' Barbari. The Italian stamp (above) was issued in 1994 on the occasion of the 500th anniversary of the publication of the *Summa*. The Sri Lankan stamp was issued the same year to celebrate 500 years of double-entry bookkeeping as reported on in the *Summa*. Note that in each case artistic license has been applied in interpreting Barbari's painting. (*Exercise.* Look up the original to determine the differences from the original.)

1716), or the more colourful Cardano or Tartaglia. Pacioli moved in high circles and is depicted as St. Peter the Martyr in an altarpiece painted by no less an artist than Piero della Francesca (*c.* 1415 - 1492), and he is the subject of a painting by Jacopo d' Barbari. He and the second scientific Leonardo, a certain da Vinci (1452 - 1519), were together in the service of Ludovico Sforza in Milan and, after Sforza was captured by the French, roomed together in Florence. Leonardo consulted Pacioli in mathematics and in return illustrated Pacioli's *Divina proportione*, published in 1509. This latter work consisted of three parts: the first, *Compendio de divina proportione*, concerned the divine proportion ϕ and other geometric topics; the second was a treatise on architecture; and the third was an Italian translation of a work *De corporibus regularibus* of Piero della Francesca.

One or both of Pacioli's probabilistic problems are reproduced in historical accounts. For example, in Ore we read

> A team plays ball [in] such [a way] that a total of 60 points is required to win the game, and each inning counts 10 points. The stakes are 10 ducats. By some incident they cannot finish the game and one side has 50 points and the other 20. One wants to know what share of the prize money belongs to each side. In this case I have found that opinions differ from one to another, but all seem to me insufficient in their arguments, but I shall state the truth and give the correct way.[8]

The second problem is similar, but involved three players:

---

[8] Ore, *op. cit.*, p. 414.

Three compete with the cross bow and the one who first obtains six first places wins; they stake 10 ducats among themselves. When the first has four best hits, the second three, and the third two, they do not want to continue and decide to divide the prize fairly. One asks what the share of each should be.[9]

Kendall offers the following account:

20. Fra Luca dal Borgo, or Paccioli, was an itinerant teacher of mathematics whose *Summa de Arithmetica, Geometria, Proportioni et Proportionalità*, published in 1494, was widely studied in Italy. He considers a simple version of what later became known as the problem of points[10]: *A* and *B*, playing a fair game (not dice, but *balla*, presumably a ball game) agree to continue until one has won six rounds; but the match has to stop when *A* has won five and *B* has won three. How should the stakes be divided?
21. Paccioli makes very heavy weather of this, but his solution amounts to saying that the stakes should be divided in the proportion 5 : 3.[11]

Gnedenko elucidates:

He suggested that the prize money be divided up proportionally between the number of wins. His solution to the first problem would be to give 5/8 of the prize money, or 13.75 ducats, to the winning team and 3/8 of the prize money, or 8.25 ducats, to the second. Accordingly, Pacioli solved the second problem by awarding 4 and 4/9 ducats to the first place player, 3 and 3/9 ducats to the second, and 2 and 2/9 ducats to the third.[12]

As the attentive reader might have discerned, there are a couple of discrepancies here. Ore cites the problem with a 5 : 2 ratio (and Hacking agrees), while Gnedenko and Kendall offer a 5 : 3 ratio. Moreover, according to Ore's reading of the problem, there are 10 ducats at stake, while the shares of the stake in Gnedenko's telling total up to 22 ducats. This sort of discrepancy between works on the history of mathematics is common.[13] Ideally, one would resolve such an issue by going to the original source. If one cannot do this, perhaps because of the unavailability of the source material, inability to read the source

---

[9] *Ibid.*

[10] It is also called the *division problem*. Pacioli's *Summa* offers the first printed account of the problem of points. There are older discussions in manuscript (as one might expect from his reference to the opinions of others). A good account of them and the social-historical background to the problem is given by Ivo Schneider, "The market place of games of chance in the fifteenth and sixteenth centuries", in: Cynthia Hay, ed., *Mathematics from Manuscript to Print, 1300 - 1600*, Oxford University Press, Oxford, 1988.

[11] Kendall, *op. cit.*, p. 7.

[12] Boris Vladimirovich Gnedenko (1912 - 1995) (translation by Igor A. Ushakov) *Theory of Probability*, 6th edition, CRC Press, 1998, p. 416.

[13] An even more serious discrepancy is coming up.

in the original language, lack of time or simple laziness, one could note that the abstract principle is the same in either case, declare the difference unimportant, and then (appropriately in discussing probability) flip a coin to decide whose figures to deal with. Pacioli's *Summa* is, however, readily accessible online at the University of Seville, and Hacking gives the page reference. One finds Pacioli's discussion on folio pages 197 and 197v and, if one checks, one does indeed find the figures given by Hacking and Ore. Of course, one needs to be able to translate Pacioli's Italian to verify that the second player's 20 points is cited as his score and not the lead the first player has over him... Fortunately, and most conveniently, the relevant passages of several early works on probability have been translated into modern German in a source book edited by Ivo Schneider[14]. The translation of Pacioli's text was done by Schneider and Ettore Casari with the assistance of Rudolf Haller and is in full agreement with the translations of the problem offered by Ore and Hacking. Thus, the correct rendering of Pacioli's problem has a total stake of 10 ducats, and they are playing for 60 points, having to quit when one player has 50 points and the other 20.

It appears that Kendall and Gnedenko derived their versions of the problem not from Pacioli, but from Tartaglia's account, which reads:

Brother Luca from Borgo set forth the following problem:
A company plays ball to 60 points for a whole game, whereby 10 points are to be awarded for individual plays. In all they stake 22 ducats. Due to certain circumstances, they cannot finish the whole game at a stage where one party has 50 points and the other 30. One asks, what share of the stake is each entitled to. In this problem the aforesaid Brother Lucas says he finds several proposed solutions in one direction or another, but that to him all the arguments appear insufficient and that the correct method and true is such that one can carry out the calculation in three ways.
The first claims that one must consider the maximum number of individual plays that can be made by one and the other player. One finds this to be 11, if in fact both have shown 50 points, and one sees, so he says, that the one with 50 has a share of 5/11 of the necessary individual plays and the one with 30 a share of 3/11.
He further says, however, that one party may take 5/11 of the mentioned 22 ducats and the other may take 3/11; this makes 8/11 altogether; furthermore he says that one must proceed as in business, where one says, if 22 ducats correspond to 8/11, what corresponds to 5/11 and 3/11. If one proceeds in this manner, one finds, that the party with 50 points will receive 13 3/4 ducats and the one with 30 points will receive 8 1/4 ducats.
His rule seems to me to be neither beautiful nor good. For, if by chance

[14] Ivo Schneider, ed., *Die Entwicklung der Wahrscheinlichkeitstheorie von der Anfängen bis 1933; Einführungen und Texte*, Wissenschaftliche Buchgesellschaft, Darmstadt, 1989.

one of the parties had 10 points and the other none and one were to proceed by his rule, it would transpire that the party with 10 points would take all and the other would get nothing at all, which would be completely without sense, that one with 10 might take the total.[15]

I already quoted Kendall as saying "Pacioli makes very heavy weather of this". I wouldn't really say he does this. There is one point that could be clearer. After stating the problem, Pacioli continues

> For this problem I have found varying proposals for solution, which go in one direction or another; all seem to me insufficient with respect to their arguments. However the truth is that which I will say together with the correct way.
>
> I say that you can proceed in three ways.
>
> First: You must determine how many individual plays in total can be made by the two parties; these are 11, namely then, if both have already shown 50 points. Now you see what share of all these individual plays the one with 50 points has: he has namely 5/11; and the one with 20 points has 2/11. Thus the one party can claim 5/11 of the stake for his share and the other party 2/11. All told that makes 7/11. Then 7/11 corresponds to the 10 ducats; what is the party with the share of 5/11 entitled to and what is the party with the share of 2/11 entitled to? Thus he with 50 points will receive 7 1/7 ducats and he with 20 points will receive 2 6/7 ducats.[16] Finished.
>
> Another way is similar: i.e, they can make all told 110 points. See what part of it the 50 points is; you will find, as above, 5/11 and for 20, respectively, 2/11; proceed as above!
>
> The third, very short way is, that you sum what the two parties together have: i.e., 50 and 20 make 70. And this is the divisor, by which 70 corresponds to 10 ducats. What are the party with 50 points and the party with 20 points entitled to? Etc.[17]

Pacioli's solution to the problem does take into account the fact that some players are closer to winning the prize than others. His solution is, thus, more sophisticated than the simple one of returning to each player the amount he staked. But is his solution indeed correct, or even fairer? We have already had a glimpse of the criticism by Tartaglia. Ore quotes Cardano:

---

[15] *Ibid.*, pp. 18 - 19.

[16] This is, to me, the mystery sentence, possibly earning the description "heavy work". Here, however, we can follow Tartaglia's lead and say the stake is to be divided, as in business practice, in the proportion of the amounts earned,

$$5/11 : 2/11,$$

i.e., 5 : 2, thus giving the first player 5/7 of the 10 ducats, i.e., 7 1/7 ducats. And the second receives the remaining 2 6/7 ducats.

[17] Schneider, *Entwicklung...*, *op.cit.*, pp. 11 - 12.

And there is an evident error in the determination of the shares in the game problem as even a child should recognize, while he (Paciuoli) criticizes others and praises his own excellent opinion.[18]

This quotation is from Cardano's *Practica arithmetice et mensurandi singulari*[19] (1539) about which Gnedenko reports

Pacioli's problem of the stack sharing before the game terminated was also interesting for Cardano. In his book *"Practice of General Arithmetics"* published in 1539, Cardano gave some critical notices about Pacioli's solution. He wrote that the suggestion to share the stack proportionally to the numbers of games won by each gambler did not take into account how many games each of them should win to be a winner. In his opinion, Cardano suggested that if s is the number of games which should be won and p and q are the numbers of games actually won by the first and second players, then the stack was to be shared as

$$[1 + 2 + \ldots + (s - q)] : [1 + 2 + \ldots + (s - p)].$$

We will show below that Cardano's suggestion in general is erroneous and leads to the correct results only in some particular cases.[20]

Cardano explains the rationale behind his choice. It is that the split should be according to the chances the individual players would have of winning were they to continue playing. If, for example, in playing for 10 points they break off when one player has 9 and the other 7, the first player needs only 1 point and the other needs 3. Let us say they are tossing a coin. The first player can win in 1, 2, or 3 tosses, while the second must win 3 in succession. Cardano reasons that the first player has 1 chance of winning in 1 toss, 2 in 2 tosses (the coin could land in his favour in on either toss), and 3 in 3 tosses, thus giving him $1 + 2 + 3 = 6$ chances in all. The second player has only 1 chance, hence the split should be $(1 + 2 + 3) : 1 = 6 : 1$. The idea that the split should be proportional to the odds is absolutely correct, but the calculation of these odds is fundamentally flawed.

Assuming the Kendall-Gnedenko transcription of Pacioli's first problem, the split according to Pacioli would be 5 : 3, i.e., the first player would get 5/8 of the prize, while, according to Cardano, the ratio would again be $(1 + 2 + 3) : 1 = 6 : 1$, i.e., the first player would get 6/7 of the prize. The "correct", i.e., the modern, accepted ratio is 7:1, i.e., the first player getting 7/8 of the prize. Cardano is silent on the problem of three players, but the extension is obvious.

[18] Ore, *op. cit.*, p. 414.

[19] The *Practica arithmetice* is basically a rewrite of the *Summa*, with numerous corrections to the latter. A number of its chapters refer directly to Pacioli in their titles, and, indeed, the last chapter is titled "Caput ultimum de erroribus Fratris Luce" [Final chapter on the errors of Brother Lucas]. The material of importance to the history of probability occurs in chapters 61 and 68 (the final chapter), of which the most relevant material is to be found in Schneider's *Entwicklung...*, *op. cit.*, pp. 15 - 17. The full Latin text can be found online at the University of Seville.

[20] Gnedenko, *op. cit.*, pp. 418 - 419.

As already noted, Tartaglia also got into the act. He criticises Pacioli in a passage variously described:

> Cardano's arch enemy, Tartaglia, feels himself on swaying ground when he deals with the division problem in his *General Trattato* (1556). The margin displays the warning *"Error di Fra Luca dal Borgo,"* and Tartaglia gives his own rule, but with the reservation: "Therefore I say that the resolution of such a question is judicial rather than mathematical, so that in whatever way the division is made there will be cause for litigation."[21]

> Tartaglia returned to the problem of stack sharing in his *"La Prima Rarte del General Trattato di Numerix et Misure"* [22] published in 1556. His approach was presented in Chapter 20 *"The Error of Brother Lucas from Borgo."* This critic is correct and has a strong basis: "This rule does not seem to me either beautiful or good. So, if one side has won 10 points and another none at all, then the stack should be given to the first gambler but this makes no sense."[23]

We find Kendall and Gnedenko presenting Tartaglia's method differently. According to Gnedenko, Tartaglia's solution was to start with the even distribution of the prize and make the deviation therefrom proportional to the difference in points won. For Pacioli's first problem, with the points collected according to Kendall and Gnedenko, this would give the first player

$$\frac{1}{2} + \frac{50 - 30}{60} = \frac{1}{2} + \frac{1}{3} = \frac{5}{6}$$

of the prize and the second

$$\frac{1}{2} + \frac{30 - 50}{60} = \frac{1}{2} - \frac{1}{3} = \frac{1}{6}$$

of the prize. This is all very nice, but consider the alternate reading of Pacioli's problem where the two players have 50 and 20 points, respectively. Gnedenko's reading of Tartaglia allots

$$\frac{1}{2} + \frac{50 - 20}{60} = \frac{1}{2} + \frac{1}{2} = 1,$$

i.e., all of the prize money to the first player. And, if the second player has only 10 points, the first player gets

$$\frac{1}{2} + \frac{50 - 10}{60} = \frac{1}{2} + \frac{4}{6} = \frac{7}{6}$$

of the prize money, and the second player not only gets nothing, but has to come up with an additional 1/6 of the prize to give the first—having a small

---

[21] Ore, *op. cit.*, p. 414.

[22] *Sic!* The correct title is *La prima parte del general trattato di numeri et misure* [*The first part of a general treatise on numbers and measures*].

[23] Gnedenko, *op. cit.*, p. 419.

chance of winning when they discontinue the game is worse than losing outright! Obviously, something is wrong.

Kendall also starts with an even distribution and then says of Tartaglia

He then argues that the difference between A's score (five) and B's score (three) being two, and this being one-third of the number of games needed to win (six), A should take one-third of B's share and the total stake should be divided in the ratio 2:1. Or so I interpret his rather prolix discussion. It would appear that if A has x and B y games in hand when the total number required to win is z, Tartaglia's rule requires that A takes a proportion $\frac{1}{2} + (x - y)/(2z)$ of the stake.[24]

So again we have a discrepancy. For, Gnedenko says Tartaglia would assign the first player 5/6 and the second player 1/6 of the stakes. Kendall's reading is more reasonable than Gnedenko's, but is it correct? Again we look to Schneider's translation. Tartaglia begins with the example of a score being 10 points to 0 in a bet in which each player has staked 22 ducats:

...if one by chance had 10 and the other 0, then the one with 10 had a sixth of the whole game; and therefore I say, that he in this case must receive a sixth of the ducats that they have per party staked; i.e., if one stakes 22 ducats per party, he must receive a sixth of the said 22 ducats, i.e., 3 2/3 ducats; this together with his own 22 ducats makes 25 2/3 ducats, and the other party may take the remainder, and this remainder is 18 1/3 ducats. If now one party had 50 and the other 30, one must subtract 30 from 50. There remains 20 and these 20 are a third of the whole game. Therefore one may (aside from his own share) also take a third of the money of the other party, and this third is 7 1/3 ducats, which, together with his own yields 29 1/3 ducats. The other party may take the remainder, namely 14 2/3 ducats. If one proceeds thus, nothing unreasonable as by the solution of Brother Lucas will result.[25]

The proportions agree with Kendall.

By way of a simple summary, note that for Tartaglia's example in which the first player won a single inning and the second none, the split would be

1 : 0 according to Pacioli,

$(1 + 2 + \ldots + 6) : (1 + 2 + \ldots + 5) = 21 : 15 = 7 : 5$ according to Cardano,

$(\frac{1}{2} + \frac{1}{6}) : (\frac{1}{2} - \frac{1}{6}) = \frac{4}{12} : \frac{2}{12} = 2 : 1$ according to Tartaglia (Gnedenko),

$(\frac{1}{2} + \frac{1}{12}) : (\frac{1}{2} - \frac{1}{12}) = \frac{7}{12} : \frac{5}{12} = 7 : 5$ according to Tartaglia (Kendall).

As percentages, these yield 100%, 58 1/3%, 66 2/3%, and 58 1/3%, respectively, of the prize going to the first player. The accepted solution is to give this player approximately 62.30% of the prize money.

---

[24] Kendall, *op.cit.*, p. 7.
[25] Schneider, *Entwicklung, op. cit.*, p. 19.

Two years after Tartaglia's *Trattato*, another work appeared, *Due Brevi e Facili Trattati, il Primo d' Arithmetica, l' Altro di Geometria* by Giovanni Francesco Peverone (1509 - 1559?), who very nearly got it right.

Peverone's near miss, however, appears to have had no impact and the problem of points, though frequently discussed, had to wait nearly a century until Pascal came up with a solution. Before we discuss this, however, it might be instructive to consider the attempts we have seen up to now.

The problem is to determine a function $f(x, y, z)$ defined for integers $0 \leq x < z, 0 \leq y < z$ which gives the proportion of the prize player A should receive in playing for z points should the game be interrupted when player A has won x points and player B has won y points. Adding two obvious calculations to those offered so far, we have the following collection:

Primogeniture:
$$f_{Pr}(x, y, z) = \begin{cases} 1, & x > y \\ \frac{1}{2}, & x = y \\ 0, & x < y \end{cases}$$

Equidistribution:
$$f_E(x, y, z) = \frac{1}{2}$$

Pacioli:
$$f_{Pa}(x, y, z) = \frac{x}{x+y}$$

Cardano:
$$f_C(x, y, z) + \frac{1 + 2 + \ldots + (z - y)}{(1 + \ldots + (z - y)) + (1 + \ldots + (z - x))}$$
$$= \frac{\dfrac{(z - y)(z - y + 1)}{2}}{\dfrac{(z - y)(z - y + 1)}{2} + \dfrac{(z - x)(z - x + 1)}{2}}$$
$$= \frac{(z - y)(z - y + 1)}{(z - y)(z - y + 1) + (z - x)(z - x + 1)}$$

Tartaglia:
$$f_{T,k}(x, y, z) = \frac{1}{2} + \frac{x - y}{kz},$$

where $k = 1$ according to Gnedenko and $k = 2$ according to Kendall. We have already given a good reason not to accept Tartaglia's function when $k = 1$, but we have not ruled out $k = 2$—nor, for that matter, have we ruled out $k = 3, 4, 5, \ldots$ or even a function, say $k(z) = z$. There are infinitely many choices of functions f. How do we decide which ones are fair and which are not?

The functions exhibited all satisfy some conditions of fairness and realism. All but $f_{T,1}$ satisfy the realistic constraint that

$$0 \le f(x, y, z) \le 1.$$

And they all satisfy

$$f(x, y, z) + f(y, x, z) = 1,$$

i.e., all the money goes back to the two players. They all satisfy the fairness requirement that the player with the higher score does no worse than his opponent,

$$x \ge y \Rightarrow f(x, y, z) \ge f(y, x, z),$$

in fact,

$$x \ge y \Rightarrow f(x, y, z) \ge \frac{1}{2}.$$

The equidistribution function is fair in that, as neither player has completed the task, no one has a claim to the whole prize and it can be returned to the players. (It is implicitly assumed that the game is fair: each player is as likely as the other to win a point and each stakes the same amount.) It lacks sophistication in that it is insensitive to the fact that one player might be ahead when the game is discontinued. That the question of how the prize should be divided is asked at all indicates that one wants such sophistication, and we thus delete f_E from our list of candidates.

Tartaglia's criticism of f_{Pa} is that, so long as $x < z$, player B still has a chance of winning and should not receive nothing. Another requirement might be monotonicity: for fixed y,

$$x_0 > x_1 \Rightarrow f(x_0, y, z) > f(x_1, y, z),$$

i.e., the closer player A is to winning the game, the greater his share. f_{Pa} does not satisfy this for $y = 0$, as already $f_{Pa}(1, 0, z) = 1$. We can also note that z is not taken into account by f_{Pa}: That A has won, say, 2 points to B's single point when the game is discontinued should mean less and less the larger z is. Yet f_{Pa} assigns the same share of the take regardless of how large z is. Thus we scratch f_{Pa} from the list of candidates for a fair distribution of the prize. We similarly delete f_{Pr}.

Thus, we are left with Cardano's and Tartaglia's solutions. Now, f is not needed when $x = z$, but we could make it a requirement that the formula for calculating f yield 1 when $x = z$ and $y < z$. f_C satisfies this, but

$$f_{T,k}(z, y, z) = \frac{1}{2} + \frac{z - y}{kz}$$

generally fails to have this property. Note that, for $y = 0$, we get

$$f_{T,k}(z, 0, z) = \frac{1}{2} + \frac{1}{k} = 1$$

only for $k = 2$. And for this choice of k,

$$f_{T,2}(z, 1, z) = \frac{1}{2} + \frac{z - 1}{2z} = \frac{2z - 1}{2z} < 1.$$

So, should we scratch Tartaglia's function from our list? Does this mean that we should accept Cardano's solution? Or, should we continue searching for more candidates?

1 Exercise. i. Let $x = 1, y = 0, z = 2$ and compare $f_C, f_{T,2}$ for these arguments. The modern solution gives player A 3/4 of the prize.
ii. Do the same for $x = 2, y = 0, z = 3$. The modern solution gives A 7/8 of the prize.

2 Exercise. Compare $f_C(1, 0, z)$ and $f_{T,2}(1, 0, z)$. What values do these functions give if one further assumes $z = 3$? The modern solution gives 11/16 of the prize to player A.

3 Exercise. Define the function

$$f_{C,2}(x, y, z) = \frac{z - y}{(z - y) + (z - x)} = \frac{z - y}{2z - x - y}$$

and discuss its possible use as a solution to the problem of points. Is it better or worse than f_C? What about

$$f_{C,3}(x, y, z) = \frac{(z - y)(z - y + 1)(z - y + 2)}{(z - y)(z - y + 1)(z - y + 2) + (z - x)(z - x + 1)(z - x + 2)} ?$$

Or

$$f_{C,4}(x, y, z) = \frac{1^2 + 2^2 + \ldots + (z - y)^2}{(1^2 + 2^2 + \ldots + (z - y)^2) + (1^2 + 2^2 + \ldots + (z - x)^2)} ?$$

Of the solutions thus far proposed[26], only that of Cardano has grappled with the fundamental issue: Given that A has won x points and B has won y, how likely is it that A would win z points if the game were to continue? And he got it wrong. The calculation of the probability in question is not trivial and only the most trivial such calculations had been performed by the mid-16th century.

There are two forms that probability theory assumes today, discrete and continuous. The continuous form is abstract and advanced. At best we can indicate it here by citing a simple example. If we draw a circle in a square and lay the drawing flat on the ground, the probability that a raindrop landing on the square also lands inside the circle would equal the ratio of the area of the circle to that of the square. Discrete probability is conceptually, if not always computationally, simpler. In its simplest manifestation one has a set S, called the *fundamental probability set*, or more simply the *sample space*. Its elements

---

[26] There are other attempts. I refer the reader to Edith Dudley Sylla, *The Art of Conjecturing, together with a Letter to a Friend on Sets in Court Tennis*, Johns Hopkins University Press, Baltimore, 2006, p. 68 for the 1491 attempt by Filippo Calandri. Sylla precedes Calandri's contribution with an informative discussion of Pacioli's solution (pp. 66 - 67) and the tradition it stems from (pp. 63 - 69).

are called (possible) *outcomes*.[27] The act of choosing an outcome at random is termed an *experiment*, and a designated set of outcomes is deemed an *event*. For example, if one tosses a coin twice, the possible outcomes are

head followed by head, usually written HH
head followed by tail, usually written HT
tail followed by head, usually written TH
tail followed by tail, usually written TT.

The sample space in this case would be the collection of all possible outcomes: $\{HH, HT, TH, TT\}$. An experiment would be performed by tossing a coin twice, and its outcome would be the result of the experiment. An example of an event might be "a single head turns up", i.e., the set $\{HT, TH\}$. Another way of looking at this might be to consider the outcomes as

no head occurs, $0H$
a single head occurs, $1H$
two heads occur, $2H$.

This would give the sample space $\{0H, 1H, 2H\}$. The event "a single head turns up" is now the set $\{1H\}$ and has only the single outcome. This description is less desirable as, for example, it does not allow us to consider "a head followed by a tail" or "a head comes first" as events. Also, there is a lack of symmetry among the outcomes, as $1H$ is twice as likely to occur than either outcome $0H, 2H$, as one can verify by repeating the experiment a number of times and recording the results. Thus, in the simplest case we assume outcomes to be fully analysed and *equally likely*.

With all of this we would define the probability of an event $E \subseteq S$ to be the ratio of the number of elements in E to the number in S, i.e., the proportion of elements of S that are in E. For example, if S consists of the collections of double tosses of a coin and E is the event "a single head turns up", then $E = \{HT, TH\}$ has two elements, while $S = \{HH, HT, TH, TT\}$ has four. The probability of E occurring, written $P(E)$ is thus

$$P(E) = \frac{\text{number of elements of } E}{\text{number of elements of } S} = \frac{2}{4} = \frac{1}{2}.$$

A pædagogical aside: I've never taught probability to middle school students, nor to high school students. However, I have taught it in courses on Finite Mathematics to college freshmen and can report that there are two types of students—those who believe it is obvious that the probability in question is $1/2$ and that it is thus a waste of time to set up the sample space and describe the event as a set of outcomes, and those more docile ones who do as they are told. In my experience the former students cannot solve any of the problems that occur later in the course, while the latter students have no difficulty determining the probabilities of the trickiest problems. It is important in teaching the subject to carefully go through the checklist: what is the sample space?

---

[27] In older accounts, one finds the words *cases* or *chances*.

what is the event? how many outcomes are in the sample space? how many in the event?

In the beginnings, the key concepts were not singled out and named, and in the initial gropings, it was not the probability of an event, but the *odds* that were often calculated. Thus, instead of their calculating the ratio

$$\frac{\text{number of elements of } E}{\text{number of elements of } S},$$

early probabilists calculated the ratio

$$\frac{\text{number of elements of } E}{\text{number of elements of } S \text{ not in } E}.$$

Either way, the difficulty would be in the determination of S and the counting of the numbers of elements of E and S. The initial examples were small enough that S and E could be determined by simple enumeration.

The first written book to discuss these matters was Cardano's *Liber de ludo aleæ*[28], possibly written around 1563 or 1564, but first published a century later in 1663 in his collected works. The book has very little probability in it and is, in fact, more of a handbook for gamblers. But it does include the rudiments of the basics of discrete probability.

Cardano's book went unpublished for a century and the history of probability next concerns itself with a minor manuscript of Galileo (1564 - 1642): *Sopra le scoperte dei dadi*[29] [*On the discoveries with dice*], believed written between 1613 and 1623, and published in his collected works in 1718 as *Considerazione sopra il guoco dei dadi* [*Thoughts about dice-games*]. This paper is a response to a question put to him, probably by his patron the Grand Duke of Tuscany:

> Now I, in order to oblige him who has ordered me to produce whatever occurs to me about the problem, will expound my ideas...[30]

The problem put forth was an apparent paradox in odds involved in tossing three dice:

> ...although 9 and 12 can be made up in as many ways as 10 and 11, and therefore they should be considered as being of equal utility to these, yet it is known that long observation has made dice-players consider 10 and 11 to be more advantageous than 9 and 12. And it is clear that 9 and 10 can be made up by an equal diversity of numbers (and this is also true of 12 and 11): since 9 is made up of 1.2.6, 1.3.5, 1.4.4, 2.2.5, 2.3.4, 3.3.3, which are six triple numbers, and 10 of 1.3.6,

---

[28] An English translation by S.H. Gould appears in Ore's biography of Cardano, *Cardano, the Gambling Scholar*, Princeton University Press, Princeton, 1953, and was reprinted as Gerolamo Cardano, *The Book of Games of Chance*, Holt, Rinehart & Winston, New York, 1961.

[29] An English translation by E.N. Thorne appears as an appendix to David, *Games, Gods...*, *op. cit.*, pp. 192 - 195.

[30] *Ibid.*, pp. 65 and 192.

1564   1642

Galileo Galilei

POSTE ITALIANE L. 70

Space exploration and astronomy are popular subjects among topical stamp collectors and the post offices of the world have obliged such collectors in full measure. Depicted on many of these stamps is Galileo, either directly via his portrait or indirectly via his telescope, one of his drawings, etc. His scientific interests outside astronomy are virtually ignored, although his law of the pendulum is referenced in one stamp of Ecuador from 1966. His native Italy has often commemorated him philatelically and given us a wide choice from which to choose a representative. My choice is the attractive portrait shown, one of a pair of stamps issued in 1964 in celebration of his 400th birthday. Additional philatelic images of Galileo can be found on pages 217 and 520 in other connexions.

1.4.5, 2.2.6, 2.3.5, 2.4.4, 3.3.4, and in no other ways, and these also are six combinations.[31]

That the Duke, who was not a mathematician, should raise the question shows that the intuition that the chance of winning is determined by the proportion of favourable to unfavourable outcomes was not limited to the mathematical specialist. It also illustrates the difficulty in counting the outcomes, or, put differently, in determining the *correct* sample space, i.e., one in which the various outcomes are equally likely. This is also a difficulty beginning students have. In the present case I explain the situation by suggesting the students imagine the game played with different coloured dice, which, in honour of Galileo, we may take to be the green, white, and red of the Italian flag. If we do this, we see that, for example, the sum $1 + 2 + 6$ yielding 9 is not a single possibility, but 6, as in the table below. Similarly 1.3.5 and 2.3.4 each represents 6 possibilities.

green	1	1	2	2	6	6
white	2	6	1	6	1	2
red	6	2	6	1	2	1

The combination 3.3.3 is uniquely obtainable, and the two combinations 1.4.4 and 2.2.5 each represents 3 possibilities, e.g., those for 1.4.4 listed in the next table. Therefore, as Galileo enumerates, we have yet another table from which

green	1	4	4
white	4	1	4
red	4	4	1

---

[31] *Ibid.*, p. 192.

combination	1.2.6	1.3.5	1.4.4	2.2.5	2.3.4	3.3.3
possibilities per combination	6	6	3	3	6	1

we determine that the total number of possibilities of obtaining a 9 is

$$6 + 6 + 3 + 3 + 6 + 1 = 25.$$

As for 10, we have the table below.

combination	1.3.6	1.4.5	2.2.6	2.3.5	2.4.4	3.3.4
possibilities per combination	6	6	3	6	3	3

But

$$6 + 6 + 3 + 6 + 3 + 3 = 27,$$

and there are two more possibilities for obtaining a 10 than for a 9.

Galileo in fact produced a table giving the combinations and the number of ways of producing the combinations for all sums from 3 to 10, noting that the numbers for the sums 11 to 18 simply reverse the list of totals.

One suspects Galileo would have had no trouble solving the problem of points had it crossed his path. As it is, the solution had to wait until the Chevalier de Méré brought the problem to Pascal, who quickly solved it and entered into a correspondence with Fermat on the subject. The correspondence is not complete, but what there is has been translated twice into English and is readily available[32] and is, for the most part, readable. A number of topics are discussed, and some letters are missing, so the correspondence does not form a smooth exposition. However, some of the individual parts are so clear and well-written that it is hard to improve upon them. For example, in Pascal's letter of 29 July 1654, we read

> This is the way I go about it to know the value of each of the shares when two gamblers play, for example, in three throws, and when each has put 32 pistoles at stake:
> Let us suppose that the first of them has *two* (points) and the other *one*. They now play one throw of which the chances are such that if the first wins, he will win the entire wager that is at stake, that is to say 64 pistoles. If the other wins, they will be *two* to *two* and in consequence,

---

[32] The first translation, by Vera Sanford, appears in David Eugene Smith, ed., *A Source Book in Mathematics*, Cambridge University Press, Cambridge, 1929, reprinted by Dover Publishing Company, New York, 1959. A second translation, by Maxine Merrington, appears as an appendix to David, *Games, Gods...*, *op. cit.* Additionally, David herself translates long passages from the letters in the main text of her book. Sanford's translation of one letter of the correspondence is also reproduced—and dissected—in Keith Devlin, *The Unfinished Game; Pascal, Fermat, and the Seventeenth-Century Letter that Made the World Modern* (Basic Books, New York, 2008), an entertaining and informative account of the matter.

if they wish to separate, it follows that each will take back his wager that is to say 32 pistoles.

Consider then, Monsieur, that if the first wins, 64 will belong to him. If he loses, 32 will belong to him. Then if they do not wish to play this point, and separate without doing it, the first should say "I am sure of 32 pistoles, for even a loss gives them to me. As for the 32 others, perhaps I will have them and perhaps you will have them, the risk is equal. Therefore let us divide the 32 pistoles in half, and give me the 32 of which I am certain besides." He will then have 48 pistoles and the other will have 16.

Now let us suppose that the first has *two* points and the other *none*, and that they are beginning to play for a point. The chances are such that if the first wins, he will win all of the wager, 64 pistoles. If the other wins, behold they have come back to the preceding case in which the first has *two* points and the other *one*.

But we have already shown that in this case 48 pistoles will belong to the one who has *two* points. Therefore if they do not wish to play this point, he should say, "If I win, I shall gain all, that is 64. If I lose, 48 will legitimately belong to me. Therefore give me the 48 that are certain to be mine, even if I lose, and let us divide the other 16 in half because there is as much chance that you will gain them as that I will." Thus he will have 48 and 8, which is 56 pistoles.

Let us now suppose that the first has but *one* point and the other *none*. You see, Monsieur, that if they begin a new throw, the chances are such that if the first wins, he will have *two* points to *none*, and dividing by the preceding case, 56 will belong to him. If he loses, they will be point for point, and 32 pistoles will belong to him. He should therefore say, "If you do not wish to play, give me the 32 pistoles of which I am certain, and let us divide the rest of the 56 in half. From 56 take 32, and 24 remains. Then divide 24 in half, you take 12 and I take 12 which with 32 will make 44...[33]

The solution described by Pascal is algebraic in nature. If we let $f(x, y, z)$ denote, as before, the fraction of the stake the first player is to receive if, in playing for a winning score of z points, they break off when the first player has x points and the second player has y points, then Pascal is essentially setting up the recursion:

$$f(z - 1, z - 1, z) = \frac{1}{2}$$

$$f(z - 1, y - 1, z) = f(z - 1, y, z) + \frac{1}{2}\left(1 - f(z - 1, y, z)\right)$$

$$= \frac{1}{2} + \frac{1}{2}f(z - 1, y, z)$$

$$f(z - (k + 1), y, z) = f(z - (k + 1), y + 1, z)$$

---

[33] Smith, *op.cit.*, pp. 548 - 549.

Philatelic portraits of Pascal (left) and Fermat (right). The portrait of Blaise Pascal on a French postage stamp was issued in 1944 in a set of stamps depicting famous Frenchmen of the 17th century. On the 300th anniversary of his death, in 1962, France issued another stamp in his honour, but I find it less attractive.

The stamp honouring Fermat was issued in 2001. The equation pasted all over the background represents Fermat's Last Theorem, the assertion that the equation has no nontrivial integral solutions for $n > 2$. The Last Theorem was finally proven in 1994 by Andrew Wiles.

$$+ \frac{1}{2}\big(f(z - k, y, z) - f(z - (k + 1), y + 1, z)\big)$$
$$= \frac{1}{2}\big(f(z - (k + 1), y + 1, z) + f(z - k, y, z)\big).$$

For $z = 3$, as Pascal demonstrated, the calculation by hand is simple enough. For larger values of z, a little automatic computation is called for. On the *TI-83 Plus*, we can use the following program[34] to generate a table of solutions in the form of a matrix:

```
PROGRAM:POINTS
:Disp "ENTER"
:Input "N=", N
:{N,N}→dim([A])
:1/2→[A](N,N)
:For(J,N−1,1,−1)
:1/2+1/2*[A](J+1,N)→[A](J,N)
:1−[A](J,N)→[A](N,J)
:End
:For(J,N−1,1,−1)
:For(K,N−1,1,−1)
```

---

[34] For those not familiar with matrices or with programming the calculator, I refer the reader to Appendix A at the end of the book for the explanations.

```
:1/2([A](J+1,K)+[A](J,K+1))→[A](J,K)
:End
:End
:DelVar J
:DelVar K
:DelVar N
:ClrHome
:[A]▶Frac .
```

The last line converting the elements of the matrix into fractions is not necessary, but is convenient for small values of N.

If we run this program, entering 3 when prompted for a value, we get the
matrix

$$\begin{bmatrix} 1/2 & 11/16 & 7/8 \\ 5/16 & 1/2 & 3/4 \\ 1/8 & 1/4 & 1/2 \end{bmatrix},$$

the (i, j)-th entry of which gives the fraction of the stakes going to the first
player if he has $j - 1$ wins to the second player's $i - 1$ wins[35]. Multiplying the
matrix by 64 pistoles yields Pascal's partial payoffs:

$$\begin{bmatrix} 32 & 44 & 56 \\ 20 & 32 & 48 \\ 8 & 16 & 32 \end{bmatrix}.$$

If we return to Pacioli's problem and enter 6 when prompted for a value of
N, we get the matrix

$$\begin{bmatrix} 1/2 & 319/512 & 191/256 & 219/256 & 15/16 & 63/64 \\ 193/512 & 1/2 & 163/256 & 99/128 & 57/64 & 31/32 \\ 65/256 & 93/256 & 1/2 & 21/32 & 13/16 & 15/16 \\ 37/256 & 29/128 & 11/32 & 1/2 & 11/16 & 7/8 \\ 1/16 & 7/64 & 3/16 & 5/16 & 1/2 & 3/4 \\ 1/64 & 1/32 & 1/16 & 1/8 & 1/4 & 1/2 \end{bmatrix}.$$

Recalling that the first player was assumed to have won 5 points and the second
player 2, we look at the entry in the 6-th column, 3-rd row and see the first
player deserves 15/16 of the money. Tartaglia's version of the problem gave the
second player 3 wins and the 6-th column, 4-th row grants the first player 7/8
of the money, as earlier announced without justification.

What we have done with our program, in fact, is that we have generated a
probability table, featuring the probability the first player would win the full
prize given the number of points 0 to $N - 1$ he has so far won and the number 0
to $N - 1$ his opponent has won. This connexion between the proposed division
of the stakes and the probability that a given player will win is presumably

---

[35] The rows and columns are numbered from 1 to N, while the number of points so
far won lies between 0 and $N - 1$, inclusive. Thus the description is always off by
1 from what one expects.

brought out more clearly in one of Fermat's missing letters and is described in Pascal's response to it.

Pascal illustrates Fermat's approach in his letter of 24 August by imagining two players tossing a two-sided die (for which we would substitute a coin) with sides labelled a and b. It is assumed the first player needs 2 points to win and the second player needs 3. If they were to continue playing, the game would be decided in 4 throws. (Why?) The fundamental probability set is thus the collection of sequences of 4 throws of the die, which we may represent by four-letter words in the restricted $\{a, b\}$-alphabet. There are 16 of these, of which those describing a win for the first player are

$aaaa$	$aaab$	$aabb$
	$aaba$	$abab$
	$abaa$	$abba$
	$baaa$	$bbaa$
		$baab$
		$baba$,

and those describing a loss are

$abbb$	$bbbb$
$babb$	
$bbab$	
$bbba$	.

Thus the first player has 11 chances to win, while the second has 5. The probability of the first player's win is then 11/16 and he should take 11/16 of the stake. Put differently, his odds of winning are 11 : 5 and the stake should be divided 11 : 5. If the reader runs the POINTS program with any number N greater than or equal to 4, and looks at the entry in row N − 2 (player 2 needs 3 wins) and column N − 1 (player 1 needs 2 wins), he will find 11/16.

Of particular interest here is the confusion subsequently expressed by Pascal about the sample space. Pascal had also been in contact with Gilles Personne de Roberval (1602 - 1675), who objected that, in our game with the two-faced die, one would not play all the games listed. For, once the first player wins two games or the second three, the play would stop. This would yield a sample space with the throws

aa	aba	$abba$
	baa	$baba$
		$bbaa$

favourable to the first player and the throws

bbb	$abbb$
	$babb$
	$bbab$

favourable to the second. However, this does not make the probability of the first player's winning equal to $6/10 = 3/4$, as the listed outcomes are not

equally likely. If we were to continue the play, player one would win the first
two throws 1/4 the time, they would come up with *aba* 1/8 the time, likewise
with *baa*, and each of the last three winning sequences for him would occur
1/16 of the time. Adding the probabilities of these outcomes yields

$$\frac{1}{4} + \frac{1}{8} + \frac{1}{8} + \frac{1}{16} + \frac{1}{16} + \frac{1}{16} = \frac{11}{16},$$

as before.

Pascal reports that he explained the correctness of Fermat's approach to
Roberval, but had an objection of his own:

> There is your method, when there are *two* players, whereupon you say
> that if there are more players, it will not be difficult to make the divi-
> sion by this method.
> 3. On this point, Monsieur, I tell you that this division for the two play-
> ers founded on combinations[36] is very equitable and good, but that if
> there are more than two players, it is not always just and I shall tell
> you the reason for this difference.
> . . .
> 4. Let us follow the same argument for *three* players and let us assume
> that the first lacks *one* point, the second *two*, and the third *two*. To
> make the division, following the same method of combinations, it is nec-
> essary to first discover in how many points the game may be decided
> as we did when there were two players. This will be in three points for
> they cannot play three throws without necessarily arriving at a deci-
> sion.
> It is now necessary to see how many ways three throws may be com-
> bined among three players and how many are favorable to the first,
> how many to the second, and how many to the third, and to follow this
> proportion in distributing the wager as we did in the hypothesis of the
> two gamblers.
> It is easy to see how many combinations there are in all. This is the
> third power of 3; that is to say, its cube, or 27. For if one throws three
> dice at a time (for it is necessary to throw three times), these dice
> having three faces each (since there are three players), one marked *a*
> favorable to the first, one marked *b* favorable to the second, and one
> marked *c* favorable to the third,—it is evident that these three dice
> thrown together can fall in 27 different ways. . . [37]

There then follows an enumeration of the 27 outcomes, *aaa, aab, . . .* and an
indication next to each which of the players the outcome favours. His table
lists

abb,   bab,   bba

as simultaneously favourable to the first *and* second players, and

---

[36] Today, the word "combination" has a slightly more restrictive technical meaning,
as we will discuss in Chapter 3, below.
[37] Smith, *op. cit.*, pp. 555 - 557.

$$acc, \quad cac, \quad cca$$

as favourable to both the first and third players. The result turns out to be a three-way split in the ratio 19 : 7 : 7. He claims that the overlapping cases should each count as only half a win for each of the two players and the split should be $16 : 5\frac{1}{2} : 5\frac{1}{2}$.

Pascal goes on to say that, using his method (i.e., the recursion equations, suitably modified for 3 players), one again obtains a $16 : 5\frac{1}{2} : 5\frac{1}{2}$ split, but if they quit playing as soon as one player gains the necessary number of points, the split will be 17 : 5 : 5.

In his reply, Fermat essentially noted that there was nothing wrong with his procedure and choice of sample space S, but that Pascal had erred in his delineation of the event E. While it is true that, for example, the outcome *abb* gives both the first and second players their needed points, the first player earned his first and the win is solely his. When the supposedly shared wins are reassigned according to who got his points first, the wins are in the proportion 17 : 5 : 5.

4 Exercise. Give the complete enumeration for Pascal's three player problem and verify Fermat's remarks.

5 Exercise. Solve Pacioli's second problem: Three compete with the cross bow, playing for 6 points. They break off playing when one player has 4, a second 3, and a third 2 points. In what proportions should the stakes be divided?

Conceptually, this last exercise is no worse than Pascal's example, and the reader should see how, in principle, to apply Fermat's method. However, once he starts enumerating the sample space, he will quickly see the difficulty—we are guaranteed a winner in 7 rounds, but $3^7 = 2187 =$ the number of elements in the sample space; this is a lot of sequences of a's, b's, and c's to write down and sort through. And Pascal's recursion is a nightmare: the evaluation of a share $f(w, x, y, z)$ (w, x, y points accumulated and z needed to win) depending on three other values,

$$f(w, x, y, z) = \frac{1}{3}\big(f(w+1, x, y, z) + f(w, x+1, y, z) + f(w, x, y+1, z)\big),$$

the number of subcomputations grows fairly large. As the calculator does not store 3-dimensional arrays, a generalisation of the POINTS program would have to treat a 3-dimensional $N \times N \times N$ matrix as, say, an $N^2 \times N$ matrix. Such a program is doable, but one has to exhibit some care in the bookkeeping. Probably the simplest thing to do without appeal to more theory is to write a program that generates all sequences of length 7, keeping track of which are wins for players one, two, and three. With appeal to further principles, it is easier to solve it by hand than to take the trouble to write a program. I postpone this to Chapter 3, where I discuss such principles.

What I would like to add at this point is the remark that Pacioli's, Cardano's, and Tartaglia's proposed solutions to the problem of points were not wrong; they were merely less sophisticated than the modern one. Today there are similar decision problems with similar *ad hoc* solutions that are widely accepted. From the world of sport there are the byzantine scoring systems of some Olympic sports that yield strange results, and in politics there are the American electoral college under which a candidate with fewer votes can win an election, and parliamentary systems under which a minority party can have its way on certain issues because a stronger party needs its coalition to retain power. The situation is quite similar to that of the problem of points: Various people have their individual desires on the resolutions of certain issues, but these desires are in conflict. The solution is sought by the devising of a scheme that will resolve the conflicts according to some (not necessarily explicitly formulated) criteria of fairness. Pacioli's solution met such a criterion, the player who was ahead got a larger stake, proportionally according to the distribution of points. Cardano and Tartaglia cited examples where Pacioli's distribution was paradoxical, not meeting some previously overlooked criterion of fairness, and proposed new solutions. As pointed out in the text, Tartaglia's solution is also paradoxical. The solution of Pascal and Fermat, making the distribution of the stake proportional to the probabilities the individual players would have of winning were the game to continue is the fairest solution anyone has offered. Yet it too can lead to paradoxical results, as we will discuss in the next chapter.

6 Project. Do a little research on Olympic scoring for, say, figure skating. Describe the rules. What are some of the paradoxical situations that can arise under it?

7 Project. Look up and write a report on Condorcet's Paradox and Arrow's Theorem. Is the latter really as significant as philosophers claim? Why or why not?

And, as long as I am ending this chapter with projects, here is a totally unrelated one, but one that came to mind in reading the Pascal-Fermat correspondence:

8 Project. (Report Cards) Scientists often make revealing comments about other scientists in their memoirs and correspondence. In writing to Fermat, Pascal had this to say about the Chevalier de Méré:

> I have no time to send you the proof of a difficult point which astonished M. so greatly, for he has ability but he is not a geometer[38] (which is, as you know, a great defect) and he does not even comprehend that a mathematical line is infinitely divisible and he is firmly convinced that

---

[38] At the time, the word "geometer" did not refer to a specialist in geometry, but to mathematicians in general.

it is composed of a finite number of points.[39] I have never been able to get him out of it. If you could do so, it would make him perfect.[40]

A few more comments are also from the history of probability. On 2 November 1910 the great Russian probability theorist Andrei Andreevich Markov (1856 - 1922) sent a postcard to Aleksandr Aleksandrovich Chuprov (1874 - 1926) criticising a passage in the latter's book:

> I note with astonishment that in the book of A.A. Chuprov, *Essays on the Theory of Statistics*, on page 195, P.A. Nekrasov, whose work in recent years represents an abuse of mathematics, is mentioned next to Chebyshev.[41]

Markov was not one to hold back. On 6 November he wrote Chuprov,

> Of course I was also surprised by your reference to Bruns whom I consider a negligible quantity.[42]

Markov also expressed his low opinion of Karl Pearson (1857 - 1936):

> The words, "the profound mathematical research of Pearson," in the new edition astonish me.[43]

---

[39] De Méré may have been behind the times on this, but he was certainly not alone. A brief discussion on the several views of the mathematical line can be found in the section, "The continuum from Zeno to Bradwardine", pp. 67 - 76, of my *History of Mathematics; A Supplement*, Springer-Verlag, New York, 2008.

[40] Smith, *op. cit.*, p. 552. This is taken from a letter from Pascal to Fermat dated 29 July 1654.

[41] Kh. O. Ondev, ed., (translation by Charles and Margaret Stein), *The Correspondence Between A.A. Markov and A.A. Chuprov on the Theory of Probability and Mathematical Statistics*, Springer-Verlag, New York, 1991, p. 3. Pafnuty Lvovich Chebyshev (1821 - 1894) was one of the great Russian mathematicians of the 19th century, whose work includes a fundamental result in the theory of probability. Nekrasov had claimed in 1898, without proof, to have generalised Chebyshev's result. In 1899 Markov showed Nekrasov to be in error. About Nekrasov, Oscar Sheynin later wrote

Pavel Alekseevich Nekrasov (1853 - 1924) was an outstanding mathematician who contributed to algebra, mathematical analysis and probability theory as well as to mechanics. However, around 1900 his works became unimaginably verbose and hardly understandable; he began connecting mathematics with religion and politics; and his arguments and general declarations often did not carry weight anymore sometimes becoming downright wrong and contradictory.

[42] *Ibid.*, p. 5. Ernst Heinrich Bruns (1848 - 1918) was a mathematician and astronomer who contributed to probability and statistics. In one of the delicious ironies of history, Markov's and Bruns's names are linked in the eponymously labelled *Markov-Bruns chains*.

[43] *Ibid.*, p. 51. Markov's correspondence with Chuprov begins just as the second edition of Chuprov's book was coming out. This opinion of Pearson, given in a letter of 4 December 1910, did not express his final judgement of Pearson, which, under Chuprov's influence, grew more favourable.

I cannot resist relating another favourite of mine. In his autobiography, the Nobel Prize winning physicist Emilio Segrè (1905 - 1989) mentions the outstanding 20th century mathematician John von Neumann (1903 - 1957):

> In the evening we occasionally played poker at the Staubs' house. We played in a very amateurish unsophisticated way. Sometimes John von Neumann joined us. I do not remember that he won particularly often, but he knew the odds of every card combination and of every move.[44]

I propose as a long-term project one start reading the correspondence, memoirs, and autobiographies of scientists, keeping a scratch pad handy to jot down any interesting comments on mathematicians or even other scientists, and collect them in a database labelled "Report Cards". One is bound to find something that can be used in class to liven up the occasional lesson.[45]

---

[44] Emilio Segrè, *A Mind Always in Motion; The Autobiography of Emilio Segrè*, University of California Press, Berkeley, 1993, p. 193.

[45] An alternative is to consult the various works of Howard Eves, who collected amusing anecdotes about mathematicians. His history textbook, *An Introduction to the History of Mathematics*, Holt, Rinehart, and Winston, New York, 1953 (currently in, I believe, its 6th edition) is peppered with such anecdotes.

2

Great Expectations

1 Mathematical Expectation

The correspondence between Fermat and Pascal went unpublished for a while. Mathematicians in France knew of it, but apparently were not aware of the details. The young Dutch diplomat and budding mathematician Christian Huyghens (or, Christiaan Huygens) (1629 - 1695) learned of it while visiting Paris, but could not get any specific information. Fermat was in far-off Toulouse, Pascal in a religious retreat not seeing visitors, and the Parisian mathematicians apparently knew little. Huyghens was intrigued and had to work the details out for himself. The result, although a short work, was another milestone in the history of probability. Written in Dutch, it was translated into Latin by Frans van Schooten (*c.* 1615 - 1660), who included it under the title *De ratiociniis in aleæ ludo* [*Calculating in games of chance*] in the fifth book of his *Exercitationes mathematicæ* [*Mathematical exercises*] (1657). Florence N. David offers the following assessment of Huyghens and his work:

> The scientist who first put forward in a systematic way the new propositions evoked by the problems set to Pascal and Fermat, who gave the rules and who first made definitive the idea of mathematical expectation was Christianus Huygens, Lord of Zelem and Zuylichem.[1]

> The treatise of Huygens *De Ratiociniis in Aleae Ludo* was, it is said, warmly received by contemporary mathematicians, and for nearly half a century it was the unique introduction to the theory of probability. It was not until the tremendous researches of the 1690 - 1710 period which resulted in *Essai d'Analyse sur les Jeux de Hasard* (Montmort, 1708), *Ars Conjectandi* (James Bernoulli, 1713), *Calcul des Chances à la statistique générale, à la statistique des décès et aux rentes viagères* (Nicholas Struyk, 1713), and *Doctrine of Chances* (Abraham de Moivre, 1718) that Huygens' work was superceded, and then not entirely since

---

[1] F.N. David, *Games, Gods and Gambling; A History of Probability and Statistical Inference*, Charles Griffin & Co., Ltd., London, 1962 (reprinted by Dover Publications, Mineola (New York), 1998), p. 110.

James Bernoulli[2] incorporated it in the *Ars Conjectandi*. Two English translations of it appeared during this period...[3]

Portraits of Christian Huyghens (right and below). Huyghens is most widely known for his invention of the pendulum clock, his discovery of the nature of Saturn's rings, and for his formulation of the wave theory of light. The Netherlands has honoured Huyghens on stamps several times. In 1928 the country issued the portrait to the right on one of four stamps celebrating famous Dutch scientists; in 1962 the pendulum clock appeared on a summer charity stamp; and in 1988, celebrating the 300th anniversary of England's Glorious Revolution, the Netherlands issued a stamp referencing Newton and Huyghens, the former via his prism and Huyghens via depictions of the pendulum clock and the rings of Saturn. The 25 guilder note pictured below was issued in 1955. Inside the large circle on the left is a second portrait of Huyghens in a watermark that does not show up here. The reference to Huyghen's discovery of the nature of Saturn's rings is clearly depicted. Also shown is "Hofwijck", the family estate, now a museum, where Huyghens spent the later years of his life.

Incidentally, the portraits in these reproductions are engraved copies of an engraving of... — for the full history see http://www.leidenuniv.nl/fsw/verduin.

De ratiociniis consisted of 14 demonstrated propositions and 5 problems for the reader. The first three propositions concern what is now called *mathematical expectation* or *expected value*:

[2] I.e., Jakob Bernoulli (1654 - 1705). See the inset on the entire family.
[3] David, *op. cit.*, pp. 115 - 116.

The Swiss stamp featuring Jakob Bernoulli was issued in 1994 in connexion with the International Congress of Mathematicians held that year in Zürich. Bernoulli's work on probability, specifically the Law of Large Numbers, which will be discussed in Chapter 5, is directly referred to by formula and graphical representation. The portrait, incidentally, is by Jakob's brother Nikolaus the Elder and was painted around 1687.

The Bernoulli, or Bernouilli, family is one of the great scientific dynasties and any work on the history of mathematics in the 17th and 18th centuries is bound to mention more than one member of the family. As some names are repeated and their names often appear translated, some confusion can arise. Thus I shall attempt to sort them out. A druggist named Jakob Bernoulli moved from Amsterdam to Basel. His son Nikolaus was an artist who had four sons: Jakob (1654 - 1705), Nikolaus ("the Elder"), Johann (1667 - 1748), and Hieronymus. Jakob and Johann are the famous feuding brothers, and are cited in the literature under various names. Jakob is also referred to as Jacques, James, and, as he was the first mathematical Bernoulli bearing the name, Jakob I. Johann is likewise variously referred to as Jean, John, and Johann I. Nikolaus was a painter, but he had a son likewise named Nikolaus (1687 - 1759) who was a doctor of law as well as a mathematician. This son is often referred to as Nikolaus I Bernoulli. Hieronymus was a druggist and had several children, none of whom, unfortunately, was a mathematician. Jakob I had a son Nikolaus ("the Younger") who, like his namesake uncle and grandfather, was an artist. Johann I was more fortunate in his offspring, siring three mathematicians: Nikolaus II (1695 - 1726), Daniel I (1700 - 1782), and Johann II (1710 - 1770). Johann II had three scientific sons, Johann III (1744 - 1807) who was a mathematician and astronomer, Daniel II (1757 - 1834) who was a medical doctor and assisted Daniel I, and Jakob II (1759 - 1789) who was another mathematician. This ended the mathematical chapter of the Bernoulli family history. Later Bernoullis distinguished themselves in other pursuits. In the next generation, for example, one finds Christoph Bernoulli (1782 - 1863), the son of Daniel II, being a professor of Natural History. The three most famous Bernoullis are Jakob I, Johann I, and Daniel I; those most important in the history of probability are Jakob I, Nikolaus I, and Daniel I.

I.   To have equal chances of getting a and b is worth $(a+b)/2$.

II. To have equal chances of getting a, b, or c is worth $(a+b+c)/3$.

III. To have p chances of obtaining a and q of obtaining b, chances being equal, is worth $(pa+qb)/p+q$.[4]

One will note that Huyghens does not express the expected value in terms of probabilities of events, but in terms of the chances of obtaining various payoffs. Thus, if one has a sample space S split into two non-overlapping events E_1 and E_2 and there are payoffs—a if E_1 occurs and b if E_2 occurs, Huyghens claims the expected value of a trial of the experiment in question is

$$\frac{|E_1|a + |E_2|b}{|S|},$$

where we write $|X|$ to denote the number of elements of a set X.[5] This last expression can be rewritten as

$$\frac{|E_1|}{|S|}a + \frac{|E_2|}{|S|}b,$$

i.e.,

$$P(E_1)a + P(E_2)b,$$

the form in which we would express it today.

The motivation for the definition of the concept comes from the intuitive idea that the probability of an event is the *relative frequency* of its occurrence in a large number of trials. If $p_i = P(E_i)$, and f_i is this frequency of occurrence in n trials, then

$$p_1a + p_2b \approx \frac{f_1}{n}a + \frac{f_2}{n}b = \frac{f_1a + f_2b}{n}$$

is the average payoff per trial over the long run, and, thus, what one should expect to win on average.

Propositions IV to IX are just instances of the problem of points, the first 4 for 2 players, and the next 2 for 3 players. For example,

IV. Suppose I play against an opponent as to who will win the first three games and that I have already won two and he one. I want to know what proportion of the stakes is due to me if we decide not to play the remaining games.[6]

Proposition X concerns a different problem:

---

[4] David, *op. cit.*, p. 116. Note the order of precedence: here, p and q are added before the division is performed. Your calculator will perform the division before the addition.

[5] The absolute value notation for the cardinality of a set is far from universal. One also sees $\#(X)$, $\mathrm{card}(X)$, etc.

[6] David, *op. cit.*, p. 116.

X. To find how many times one may wager to throw a six with one die.[7]

The problem was not new. Only one feeling extraordinarily lucky would bet even money that he could throw a 6 in a single throw of a die. If he wagered, say, 10 ducats against his opponent's 10 ducats, he could expect on average to win

$$\frac{1}{6} \cdot 10 + \frac{5}{6} \cdot (-10) = \frac{-4 \cdot 10}{6} = -6\frac{2}{3}$$

ducats, i.e., he would expect to lose 6 2/3 ducats. If he were allowed 2 throws, he could win on the first throw or lose the first and win the second. If we analyse the situation à la Fermat, and consider all 36 pairs of die tosses, we would find him winning with the throws

$$61, 62, 63, 64, 65, 66, 16, 26, 36, 46, 56,$$

i.e., 11 times out of 36, and thus losing 25 times out of 36. His expected value is

$$\frac{11}{36} \cdot 10 + \frac{25}{36} \cdot (-10) = \frac{110 - 250}{36} = \frac{-140}{36} = -3\frac{8}{9}.$$

He should thus expect to lose 3 8/9 ducats. The enumeration begins to be less pleasant with three tosses and one looks for shortcuts, as in Chapter 3. I shall leave it to the reader to consult that Chapter and work the result out for himself, noting only that 3 tosses gives him an expected value of $-1\ 31/54$, i.e., he should expect a loss, and for 4 throws his expected value is $115/324$ ducats, a small but positive amount. Thus, in the long run with 4 throws in which to obtain a 6, he should expect to come out ahead. Clearly he should demand at least 4 throws before agreeing to play the game.

Note that with 4 throws it is still not a fair game, for now the thrower has a slight advantage. Mathematically, we would call a game *fair* if neither player has a mathematical expectation of winning or losing, i.e., if the expected value of the game is 0 for each player. This concept allows one to turn even the single throw game into a fair one by adjusting the stakes. Suppose A wins if he can throw a 6 in a single toss of a die and B stakes 10 ducats saying A cannot do it. How much should A stake? This is a simple algebraic computation. If A puts up x ducats, his expected value is

$$V = \frac{1}{6} \cdot 10 + \frac{5}{6} \cdot (-x) = \frac{10 - 5x}{6}.$$

For a fair game, one must have $V = 0$, i.e., $10 - 5x = 0$, whence $x = 2$ ducats. And, indeed, if he plays the game repeatedly, in the long run he should break even.

It is important not to read too much into this. The expected value is a mathematical average. If the dice are fair, the coin is not weighted, etc., then the probability of an event is the relative frequency of its occurrence and the expected value is the average of the winnings over a large number of plays of the

---

[7] *Ibid.*, p. 116.

game. It says nothing about a particular instance and is not the determining factor in deciding whether or not to play. A pertinent example from Daniel Bernoulli (1700 - 1782) comes to mind:

> Somehow a very poor fellow obtains a lottery ticket that will yield with equal probability either nothing or twenty thousand ducats. Will this man evaluate his chance of winning at ten thousand ducats? Would he not be ill-advised to sell this lottery ticket for nine thousand ducats? To me it seems the answer is in the negative. On the other hand I am inclined to believe that a rich man would be ill-advised to refuse to buy the lottery ticket for nine thousand ducats.[8]

As regards the poor fellow, one cannot disagree with Bernoulli. The old adage, "A bird in the hand is worth two in the bush", applies. The 9000 ducats offered for the lottery ticket would make a sizable difference in the poor fellow's life, possibly more than a *possible* additional 11000 ducats should he, *by chance*, hold the winning ticket. If this were a repeatable event, he might be better off holding onto the ticket as, in the long run, he would average 1000 ducats each time. But clearly this is a once in a lifetime occurrence and a guarantee of 9000 ducats should not be passed up for what amounts to the flip of a coin.

I cannot agree with Bernoulli that the rich man is ill-advised to pass up the opportunity to purchase the ticket for 9000 ducats, unless he is purchasing it from the poor man as an act of disguised charity. For, the mathematical advantage does not apply. Casinos make money on smaller expected profits, but that is because they bet their customers a large, large number of times. While he mathematically expects a profit (i.e., that his expected value,

$$\frac{1}{2}(-9000) + \frac{1}{2}(11000) = \frac{2000}{2} = 1000,$$

is positive), his probability of winning is still only 1/2—and it would remain the same whether he paid too little (e.g., 9000), a fair price (10000), or too much (e.g., 15000).

[8] Daniel Bernoulli, *Specimen theoriæ novæ de mensura sortis* [*Exposition of a new theory of measurement of risks*] (1738). I have a thin monograph, *Daniel Bernouilli Specimen Theoriae Novae de Mensura Sortis* (Gregg Press, 1967), which contains translations into German and English. The German translation, by Alfred Pringsheim (1850 - 1941), accompanied by a short introduction by Ludwig Fisk, appeared in an earlier thin monograph, *Die Grundlage der modernen Wertlehre: Daniel Bernoulli, Versuch einer neuen Theorie der Wertbestimmung von Glücksfallen (Specimen Theoriae novae de Mensura Sortis)* [*The basis of the modern theory of values: Daniel Bernoulli, Attempt at a new theory of the determination of the value of chances*] (Verlag von Duncker & Humblot, Leipzig, 1896). The English translation is also a reprint: Louise Sommer, trans., "Exposition of a new theory on the measurement of risk", *Econometrica* 22 (1954), pp. 23 - 36. Following Fisk's introduction, the page numbering of the two translations in the joint monograph each begins with 21. The quotation is taken from page 24 of the English version, and corresponds to pp. 25 - 26 of the German.

2 The Petersburg Problem

Daniel Bernoulli's study of risk measurement was inspired by another question, this one posed by his cousin Nikolaus I Bernoulli (1687 - 1759), who included a variant of the problem in a letter to Pierre Rémond de Montmort (1678 - 1719), who published it in the second edition of *Essai d'Analyse sur le Jeux de Hazard* [*Attempt at the analysis of games of chance*] in 1713. The problem is known as the Petersburg Problem[9], as Daniel Bernoulli's paper was published in the *Papers of the Imperial Academy of Sciences in Petersburg*. Bernoulli repeats it as follows:

> *Peter tosses a coin and continues to do so until it should land "heads" when it comes to the ground. He agrees to give Paul one ducat if he gets "heads" on the very first throw, two ducats if he gets it on the second, four if on the third, eight if on the fourth, and so on, so that with each additional throw the number of ducats he must pay is doubled. Suppose we seek to determine the value of Paul's expectation.* My aforementioned cousin discussed this problem in a letter to me asking for my opinion. Although the standard calculation shows that the value of Paul's expectation is infinitely great, it has, he said, to be admitted that any fairly reasonable man would sell his chance, with great pleasure, for twenty ducats. The accepted method of calculation does, indeed value Paul's prospects at infinity though no one would be willing to purchase it at a moderately high price.[10]

The formal determination of the expected value in this case is a little tricky. The sample space in question is infinite,

$$S = \{H, TH, TTH, TTTH, \ldots\}, \tag{1}$$

and its outcomes are not equally likely. If we try to imitate Fermat by considering those sequences which continue after a head is obtained, our new sample space will consist of all infinite sequences of H's and T's, including the losing constant sequence $TTT\ldots$ Presumably this would make the outcomes equally likely, but we still run into the problem of dealing with infinite sets. In the present case, the old outcomes have become events: H is now the infinite set of all sequences beginning with H, TH is the set of all infinite sequences beginning with TH, etc.

What we can do is to pick a large number N and consider the space S_N of all sequences of H's and T's of length N, i.e., N-letter words in an alphabet restricted to the two letters H and T. If we do this, then by the methods of Chapter 3, we can find the probabilities of the event that the first H occurs on

the first throw to be $1/2$
the second throw to be $1/4$
the third throw to be $1/8$

---

[9] Also: St. Petersburg Problem and (St.) Petersburg Paradox.
[10] Bernoulli monograph, *op. cit.*, p. 31.

...
the N-th throw to be $1/2^N$.

Paul's expected value for a game of N throws is thus

$$V_N = \left(\frac{1}{2} \cdot 1 + \frac{1}{4} \cdot 2 + \frac{1}{8} \cdot 4 + \ldots + \frac{1}{2^N} \cdot 2^{N-1}\right) + \frac{1}{2^N} \cdot 0 \qquad (2)$$

$$= \left(\frac{1}{2} + \frac{1}{2} + \frac{1}{2} + \ldots + \frac{1}{2}\right) + 0 = \frac{N}{2},$$

where the last summand of (2) is the contribution of the losing sequence
$TT\ldots T$.

The expected value V of the original game could now be taken to be the
limiting value of the V_N's as N gets ever larger. There is no such limiting value
as the V_N's grow large without bound. This means V is either undefined or
infinite.

Alternatively, we can note that the probabilities of the events H, TH, TTH,
etc., once defined, remain fixed. For example,
for $N = 1$, $S_1 = \{H, T\}$, $H = \{H\}$, and $P(H) = 1/2$
for $N = 2$, $S_2 = \{HH, HT, TH, TT\}$, $H = \{HH, HT\}$, and $P(H) = 2/4 = 1/2$
for $N = 3$, $S_3 = \{HHH, HHT, HTH, HTT, THH, THT, TTH, TTT\}$,
$H = \{HHH, HHT, HTH, HTT\}$, and $P(H) = 4/8 = 1/2$
etc.
Thus we can go back to (1) and assign to the outcomes the probabilities

$$P(H) = 1/2, \quad P(TH) = 1/4, \quad P(TTH) = 1/8, \quad \ldots$$

and calculate the expected value directly:

$$V = \frac{1}{2} \cdot 1 + \frac{1}{4} \cdot 2 + \frac{1}{8} \cdot 4 + \ldots \qquad (3)$$

$$= \frac{1}{2} + \frac{1}{2} + \frac{1}{2} + \ldots = \infty.$$

I've gone through this rigmarole to emphasise that we are in unknown
territory. We can blindly apply the formula and determine that the expected
value is infinite. But do we know what this means? Ostensibly, the expected
value is the average amount one should expect to win if one were to play the
game a great many times. Obviously, if Paul played the game a lot of times,
he would still only win a finite amount each time. However, if he played it a
great number of times, the average winnings per game would exceed any finite
amount. Thus, it is to Paul's advantage to pay any amount to play the game.
But is this really the case?

It should come as no surprise to the reader to learn that no one agrees that
Paul should pay much to play Peter's game. The explanations offered and fair
stakes proposed vary and I don't know if there is a definitive solution to this
problem. We might compare the situation of the Petersburg Problem with that
of the problem of the division of points when Luca Pacioli first encountered

it. Before one had a convincing criterion by which to judge the adequacy of a solution—and the calculational power to determine the correct solution—, all one could do was propose a solution and state how it was preferable to previous proposals.

Here is a simple proposal for the solution to the Petersburg Problem: How many throws can Paul make before he expects to come up with heads? That is, what is the expected number of throws it takes to come up with heads? Well, half the time one would obtain H on the first throw, $1/4$ the time one would first obtain H on the second throw,... That is, the expected number of throws is given by the infinite sum,

$$\frac{1}{2} \cdot 1 + \frac{1}{4} \cdot 2 + \frac{1}{8} \cdot 3 + \frac{1}{16} \cdot 4 + \dots \tag{4}$$

Now this does have a finite sum and there are simple methods of finding it.

In a footnote[11] in his German translation of Bernoulli's paper, Pringsheim notes that

$$\frac{1}{(1-x)^2} = 1 + 2x + 3x^2 + \dots + nx^{n-1} + \dots, \tag{5}$$

whence, for $x = 1/2$,

$$2^2 = \frac{1}{(1/2)^2} = \frac{1}{(1-1/2)^2} = 1 + 2 \cdot \frac{1}{2} + 3 \cdot \frac{1}{4} + \dots,$$

and division by 2 yields

$$\frac{1}{2} \cdot 1 + \frac{1}{4} \cdot 2 + \frac{1}{8} \cdot 3 + \dots = 2. \tag{6}$$

Bernoulli, who in fact uses this sum himself as we will see shortly, would certainly have been familiar with such expansions of rational functions obtained by long division and would probably have recalled the rational function corresponding to the right hand side of (5). If not, he could easily have reversed the process: Let X be the infinite sum $1 + 2x + 3x^2 + \dots$ and observe

$$X - xX = (1 + 2x + 3x^2 + \dots) - (x + 2x^2 + 3x^3 + \dots)$$
$$= 1 + x + x^2 + x^3 + \dots = \frac{1}{1-x},$$

using the familiar formula for the sum of an infinite geometric progression. Dividing by $1 - x$ yields (5):

$$X = \frac{1}{(1-x)^2}.$$

I might also mention, for those who know the Calculus, that differentiating both sides of the equation,

---

[11] Bernoulli, *op. cit.*, p. 54.

$$\frac{1}{1-x} = 1 + x + x^2 + x^3 + \ldots,$$

will again yield (5). However, this approach is *ad hoc* and not as general as the preceding.

Today, with the emphasis on rigour, without having first established the legitimacy of the operations just performed on infinite series, an expositor would first find an expression for the partial sums of the series:

1 Lemma. *For $n \geq 1$,*

$$\frac{1}{2} \cdot 1 + \frac{1}{4} \cdot 2 + \ldots + \frac{n}{2^n} = \frac{2^{n+2} - 2n - 4}{2^{n+1}}. \tag{7}$$

Proof. By induction.
Basis. Observe,

$$\frac{2^{1+2} - 2 \cdot 1 - 4}{2^{1+1}} = \frac{8 - 2 - 4}{4} = \frac{2}{4} = \frac{1}{2}.$$

Induction step. Observe

$$\begin{aligned}
\frac{1}{2} \cdot 1 + \frac{1}{4} \cdot 2 + \ldots + \frac{n}{2^n} + \frac{n+1}{2^{n+1}} &= \frac{2^{n+2} - 2n - 4}{2^{n+1}} + \frac{n+1}{2^{n+1}} \\
&= \frac{2^{n+2} - n - 3}{2^{n+1}} \\
&= \frac{2}{2} \cdot \frac{2^{n+2} - n - 3}{2^{n+1}} \\
&= \frac{2^{n+3} - 2(n+1) - 4}{2^{n+2}}.
\end{aligned}$$ □

The rigorous expositor would then rewrite (7) as

$$\frac{1}{2} \cdot 1 + \frac{1}{4} \cdot 2 + \ldots + \frac{n}{2^n} = 2 - \frac{2n+4}{2^{n+1}} = 2 - \frac{n+2}{2^n}$$

and note that this differs from 2 by a mere $(n+2)/2^n$, a difference that gets very, very small as n gets very large. He would then conclude the limit of the partial sums to be 2. Finding the expression on the right-hand side of (7) is a little trickier than finding the sum of the infinite series, but there are techniques for doing this that one learns early in the study of the Calculus of Finite Differences.[12] It is not a matter of recalling one of infinitely many formulæ one has previously learned and memorised: I cannot overemphasise the fact that success in mathematics does not come from memorising and applying formulæ. This works for certain narrow classes of well-understood problems, but in general one has to use real thought. This is particularly true of applications of probability theory and is one of the reasons I was surprised to find the subject represented in a middle school textbook. I would think that students had been

---

[12] See the subsection on Δ, E, and Σ in the first section of Appendix B.

taught mathematics in elementary school as drill in basic algorithms, followed by memorisable formulæ and their applications, and that most middle school teachers, having studied general education and not mathematics education, should be ill-prepared to present the subject.

Anyway, getting back to the Petersburg Problem, we see that the expected number of throws Paul would have before the game is over and he collects his money is 2, which means he should realistically expect to win 2 ducats. Thus 2 ducats is a fair price to pay to play the game.

There is absolutely no justification for this solution. Why, for example, should we trust averaging the number of throws when we don't trust averaging the prizes awarded? The solution is formulaic, and students like and take comfort in formulæ. If presented this in class, students might just accept it unquestioningly, though some might feel a little uneasy about it, as if someone were trying to put something past them and they could not quite figure out what it was.

Unsurprisingly, there are other solutions to the problem. Daniel Bernoulli explained

> ...the determination of the *value* of an item must not be based on its *price*, but rather on the *utility* it yields. The price of the item is dependent only on the thing itself and is equal for everyone; the utility, however, is dependent on the particular circumstances of the person making the estimate. Thus there is no doubt that a gain of one thousand ducats is more significant to a pauper than to a rich man though both gain the same amount.[13]

Now, a fair price to enter the game is external and supposedly objective. Bernoulli proposes not to determine a fair price, but the price Paul should consider reasonable given his circumstances. This is not the mean of the values of the possible prizes, but the value of the mean of their utilities:

> If the utility of each possible profit expectation is multiplied by the number of ways in which it can occur, and we then divide the sum of these products by the total number of possible cases, a mean utility [moral expectation] will be obtained, and the profit which corresponds to this utility will equal the value of the risk in question.[14]

As for the measure of utility, he says

---

[13] Bernoulli, *op. cit.*, p. 24.

[14] *Ibid.*, p. 24. The entire passage is emphasised in the original. There is a footnote by the translator noting that "mean utility" is a literal translation of Bernoulli's "emolumentum medium". The bracketed "moral expectation", added by the translator, is an earlier, more widely used, looser translation. It was, for example, used by Isaac Todhunter in his *A History of the Mathematical Theory of Probability From the Time of Pascal to That of Laplace*, Cambridge University Press, Cambridge, 1865.

Now it is highly probable that *any increase in wealth, no matter how insignificant, will always result in an increase in utility which is inversely proportionate to the quantity of goods already possessed.*[15]

Bernoulli does not explain why he thinks this "highly probable". It is hardly obvious and indeed, although some mathematicians have accepted it without reservation, others have criticised it unequivocally. Mathematically, it bears a resemblance to the later Weber-Fechner Law in physiology. From the 1820's on, Ernst Heinrich Weber (1795 - 1878) introduced and applied quantitative methods in physiology and noted that one's ability to perceive an increase in the intensity of a stimulus diminishes as the intensity increases. This is readily verified by the simple experiment of lighting a match in a dark room and doing so in a well-lit room. The change in intensity of lighting is the same, but one will notice it in the first case and not in the second. Weber claimed, in fact, that there is a constant c such that a change ΔI in intensity I of a stimulus is just noticeable if

$$\frac{\Delta I}{I} = c.$$

Gustav Theodor Fechner[16] (1801 - 1887) expanded this in his *Elemente der Psychophysik* [*Elements of psychophysics*] (1860), claiming the change ΔR of the intensity of response R by an organism satisfies

$$\Delta R = c \cdot \frac{\Delta I}{I},$$

for some constant c, and derived from this the relation

$$R = C \ln I, \quad C \text{ a constant.}$$

As one learns in labs in introductory psychology courses, this is not mere numerical speculation, but an experimentally verifiable law.

What has all of this to do with Bernoulli? Well, insofar as the decision on whether or not to place a bet on what one considers a fair wager is psychological, dependent on the *perceived* value of the gain (ΔI, where I is one's current wealth), Bernoulli's notion of "moral expectation" becomes plausible. It might well model what the player perceives to be his expectation. The argument is not unassailable; it is reasoning by analogy and argues, at best, for plausibility.

I should probably add that the Weber-Fechner Law came well after Bernoulli's paper and I am not attempting here to reproduce his reasoning. I am merely suggesting that there is some reason to take Bernoulli's suggestion seriously.

---

[15] Bernoulli, *op. cit.*, p. 25.

[16] Fechner was a physicist of broad interest who pioneered the application of statistical techniques in general science. His most important work in probability and statistics was his *Kollektivmasslehre* [*Theory of measure of collectives*] (1892), a book that would inspire work in probability by Heinrich Bruns, Felix Hausdorff and, especially, Richard von Mises.

By the statement quoted, Bernoulli claims the difference in utility is inversely proportional to present wealth. He also states that it is directly proportional to the difference in wealth, thus yielding the same fundamental relation as the later Weber-Fechner Law,

$$\Delta y = b \frac{\Delta x}{x}, \tag{8}$$

where x = wealth, y = gain in utility, and b is the constant of proportionality. Bernoulli, being a mathematician, insisted this be true no matter how small Δx should be (Recall the phrase "no matter how insignificant".) and (8) became

$$dy = b \frac{dx}{x}, \tag{9}$$

a simple differential equation, the solution to which, as one learns in the Calculus, is

$$y = c + b \ln x, \tag{10}$$

for some constant c. If α is one's initial wealth, then for $x = \alpha$, one has $y = 0$ and one concludes $c = -b \ln \alpha$. Plugging this into (10) and simplifying, one finally gets

$$y = b \ln \frac{x}{\alpha}.$$

We are now in position to calculate Paul's expected utility:

$$U = \frac{1}{2} b \ln \frac{\alpha + 1}{\alpha} + \frac{1}{4} b \ln \frac{\alpha + 2}{\alpha} + \ldots + \frac{1}{2^{k+1}} b \ln \frac{\alpha + 2^k}{\alpha} + \ldots$$
$$= b \ln(\alpha + 1)^{1/2} + b \ln(\alpha + 2)^{1/4} + \ldots + b \ln(\alpha + 2^k)^{1/2^{k+1}} + \ldots$$
$$- b(\ln \alpha) \left(\frac{1}{2} + \frac{1}{4} + \frac{1}{8} + \ldots \right)$$
$$= b \ln \left[(\alpha + 1)^{1/2} (\alpha + 2)^{1/4} \cdots (\alpha + 2^k)^{1/2^{k+1}} \cdots \right] - b \ln \alpha,$$

using the sum of the geometric progression $1 = \frac{1}{2} + \frac{1}{4} + \frac{1}{8} + \ldots$ The expression U is rather ugly and we cannot evaluate it without knowing b. However, we don't need to evaluate it, for Bernoulli now asks what number D of ducats added to α will yield this utility. The utility of $\alpha + D$ is $b \ln((\alpha + D)/\alpha)$, whence

$$b \ln(\alpha + D) = b \ln \left[(\alpha + 1)^{1/2} (\alpha + 2)^{1/4} \cdots (\alpha + 2^k)^{1/2^{k+1}} \cdots \right],$$

i.e.,

$$\alpha + D = (\alpha + 1)^{1/2} (\alpha + 2)^{1/4} \cdots (\alpha + 2^k)^{1/2^{k+1}} \cdots \tag{11}$$

Ignoring the fact that the derivation implicitly assumed α not to be 0, Bernoulli now assumes Paul to have no money at all, i.e., $\alpha = 0$, and concludes the value to Paul in this case to be

$$D = \sqrt{1} \sqrt[4]{2} \sqrt[8]{4} \sqrt[16]{8} \cdots$$

$$= 2^0 \cdot 2^{1/2^2} \cdot 2^{2/2^3} \cdot 2^{3/2^4} \cdots$$
$$= 2^\beta,$$

where

$$\beta = \frac{1}{2^2} + \frac{2}{2^3} + \frac{3}{2^4} + \cdots$$
$$= \frac{1}{2}\left(\frac{1}{2} + \frac{2}{4} + \frac{3}{8} + \cdots\right)$$
$$= \frac{1}{2} \cdot 2, \text{ by our previous calculation,}$$
$$= 1,$$

and $D = 2^1 = 2$ ducats.

Bernoulli also calculates D for a few other choices of α:

> If he owned ten ducats his opportunity would be worth approximately three ducats; it would be worth approximately four if his wealth were one hundred, and six if he possessed one thousand. From this we can easily see what a tremendous fortune a man must own for it to make sense for him to purchase Paul's opportunity for twenty ducats.[17]

2 Project. The German translation gives the estimate $4\,1/3$ ducats for D for a man with 100 ducats to his name, and Todhunter agrees with this. In theory it should be a simple enough matter to decide if the English translation is indeed wrong, assuming Daniel Bernoulli made no arithmetic mistake. One simply calculates U for $\alpha = 100$ to sufficient accuracy to decide whether D is closer to 4 or $4\,1/3$. Indeed, since each factor of $\alpha + D$ in (11) is greater than 1, one will verify that Pringsheim and Todhunter are correct as soon as a product of its factors exceeds, say, 4.3. The computation, involving multiplication and root-taking, is labour intensive and one might prefer to automate the process. On the *TI-83 Plus* try entering the two expressions

Y$_1$=(A+2^(X−1))^(1/2^X)
Y$_2$=prod(seq(Y$_1$(X),X,1,N))−A

in the equation editor, and then storing 100 in the variable A. You can then successively store, say $5, 10, 15, \ldots$ in N and evaluate Y$_2$(0). [Note: Y$_2$ is a constant, so it can be evaluated at any value.] You should get the results of the table, below.

N	5	10	15	20	25	30
D=Y$_2$(0)	$-11.275\ldots$	$3.608\ldots$	$4.354\ldots$	$4.387\ldots$	$4.389\ldots$	$4.389\ldots$

This suggests $4\,1/3$ as closer to the correct value. However, experience with round-off errors tells us we cannot always trust our calculators when heavy calculation is involved. If we performed all our calculations on paper and always

---

[17] Bernoulli, *op. cit.*, p. 32.

rounded down, then a number like 4.354 for 15 factors would certainly tell us D was greater than 4 1/3. Verify that D is indeed greater than 4.3. Is it less than 4.4?[18] Does this exercise show that Pringsheim's translation is correct, or that he quietly corrected an error by Bernoulli? Would your answer be affected if I added that Pringsheim was not silent in another matter, one in which Bernoulli definitely made an error?

After presenting his results to the Imperial Academy in Petersburg, Daniel Bernoulli wrote to his cousin Nikolaus I Bernoulli to report on them. Nikolaus responded positively and forwarded him a letter from Gabriel Cramer (1704 - 1752), a Swiss mathematician best known for Cramer's Rule for solving systems of linear equations by means of determinants, though he himself did not originate the rule. Cramer agreed with Bernoulli in principle, if not in detail. His words are cited by the latter:

> The paradox consists in the infinite sum which A must pay to B. This seems absurd since no reasonable man would be willing to pay 20 ducats as equivalent. You [i.e., Nikolaus I Bernoulli] ask for an explanation of the discrepancy between the mathematical calculation and the vulgar evaluation. I believe that it results from the fact that, *in their theory*, mathematicians evaluate money in proportion to its quantity while, *in practice*, people with common sense evaluate money in proportion to the utility they can obtain from it.[19]

Cramer makes two independent proposals about the utility of money. The first is that there is a limit beyond which all amounts of money seem the same. He suggests 10 or 20 million ducats—for simplicity's sake he takes $2^{24} = 16777216$ ducats as offering maximum utility. With this he evaluates the expected value as

$$C = \frac{1}{2} \cdot 1 + \frac{1}{4} \cdot 2 + \ldots + \frac{1}{2^{24}} \cdot 2^{23} + 2^{24} \left(\frac{1}{2^{25}} + \frac{1}{2^{26}} + \ldots \right)$$
$$= \underbrace{\frac{1}{2} + \frac{1}{2} + \ldots + \frac{1}{2}}_{24} + \frac{2^{24}}{2^{25}} \left(1 + \frac{1}{2} + \frac{1}{4} + \ldots \right)$$
$$= 12 + \frac{1}{2} \cdot 2 = 13.$$

Thus, he calculates the "moral expectation" of the game for Paul to be 13 ducats.

A second solution offers an even more modest expectation:

> The equivalent can turn out to be smaller yet if we adopt some alternative hypothesis on the moral value of wealth... If, for example, we

---

[18] Pringsheim's analysis, carried out in a footnote spread out on German pages 53 - 54 of Bernoulli, *op. cit.*, shows that the infinite product of (11) does indeed converge and in fact is less than $2 + 2\alpha$, whence $D < 2 + \alpha$.

[19] *Ibid.*, p. 33.

suppose the moral value of goods to be directly proportionate to the
square root of their mathematical quantities,... my psychic expectation
becomes

$$\frac{1}{2}\sqrt{1} + \frac{1}{4}\sqrt{2} + \frac{1}{8}\sqrt{4} + \frac{1}{16}\sqrt{8} + \ldots = \frac{1}{2 - \sqrt{2}}. \tag{12}$$

However this magnitude is not the equivalent we seek, for this equivalent need not be equal to any moral expectation but rather should be of
such a magnitude that the pain caused by its loss is equal to the moral
expectation of the pleasure I hope to derive from any gain. Therefore,
the equivalent must, on our hypothesis, amount to

$$\left(\frac{1}{2-\sqrt{2}}\right)^2 = \left(\frac{1}{6-4\sqrt{2}}\right) = 2.9\cdots,$$

which is consequently less than 3, truly a trifling amount, but nevertheless, I believe, closer than is 13 to the vulgar evaluation.[20]

In plain English, the moral expectation or mean utility is given by the geometric
progression (12) and the number D of ducats yielding this utility is its square.

Daniel Bernoulli and Gabriel Cramer were only the first of a long line of
mathematicians who tried their hands at solving the Petersburg Problem. We
cannot survey them all. Two of them, however, are particularly worthy of our
attention and I am compelled to discuss them here. They are Georges Louis
Leclerc, Comte de Buffon (1707 - 1788), the celebrated natural historian, and
Jean le Rond d'Alembert (1717 - 1783), scientific editor of and contributor to
Diderot's *Encyclopédie* and noted mathematician. Buffon learned of the Petersburg Problem from Cramer in 1730 and around 1760 composed his *Essai
d'Arithmétique Morale*, which 103 page essay was published in the fourth volume of his *Supplément à l'Histoire Naturelle* in 1777. However, d'Alembert
published first, so I will begin with him.

Todhunter introduces d'Alembert as follows:

> D'ALEMBERT was born in 1717 and died in 1783. This great mathe-
> matician is known in the history of the Theory of Probability for his
> opposition to the opinions generally received; his high reputation in
> science, philosophy, and literature have secured an amount of attention
> for his paradoxes and errors which they would not have gained if they
> had proceeded from a less distinguished writer.[21]

D'Alembert wrote many of the mathematical articles of the *Encyclopédie, ou
Dictionnaire raisonné des sciences, des arts, et des metiers* (1751 - 1766),
though, oddly enough, not that on probability, which presumably was written
by Denis Diderot (1713 - 1784). Nonetheless, d'Alembert did discuss probability in various articles in the *Encyclopédie*. In the article "Croix ou Pile"

---

[20] *Ibid.*, pp. 34 - 35.
[21] Todhunter, *op. cit.*, p. 258. I glean most of my information on d'Alembert from
Chapter XIII (pp. 258 - 293) of Todhunter.

Buffon (left) and d'Alembert (right) on two French semipostals issued in 1949 and 1959, respectively, the former in celebration of famous Frenchmen of the 18th century.

[Cross or pile[22]] (1754), he made an egregious error. In considering the problem of throwing heads in two tosses of a coin, he rejected the sample space $\{HH, HT, TH, TT\}$ and its concomitant probability of $3/4$ in favour of the space $\{H, TH, TT\}$ by which he concluded the probability to be $2/3$. Likewise, for three throws of a coin he favoured $\{H, TH, TTH, TTT\}$ and $3/4$ as the probability of obtaining heads. In the same article he considers the Petersburg Problem and decides that the mathematical expectation is indeed infinite in the sense that it exceeds any finite bound.

In an article "Gageure" in a volume of the *Encyclopédie* published in 1757, he comments on the error, but still has doubts about the accepted theory. His most pertinent comments on probability occur in some of the articles of his *Opuscules mathématiques*, a collection of his mathematical essays published in 8 volumes from 1761 to 1780. In *Reflexions sur le calcul des Probabilités* published in the second volume, he has begun to change his mind about the Petersburg Problem, citing it as an example where the rule of expectation seems to fail. As a simpler example, he says, suppose Peter plays with James on the condition that a coin be tossed 100 times, and if a head appears on the last trial but not before, James will give 2^{100} crowns to Peter. The expected value to Peter is

$$0 \cdot \frac{2^{100} - 1}{2^{100}} + 2^{100} \cdot \frac{1}{2^{100}} = 1,$$

and, thus, Peter should pay James 1 crown at the beginning of the game for the opportunity to play. Not so, says d'Alembert, because the probability of

---

[22] One side of old coins bore crosses, whence "cross" came to mean the side of a coin bearing the cross; the opposite side was called pile. Hence "cross or pile" meant "head or tail".

Augustus de Morgan (*A Budget of Paradoxes II*, 2nd. edition, Open Court Publishing Company, Chicago, 1915, p. 4.) tells of Diderot annoying his colleagues in the St. Petersburg Academy with his constant atheistic pronouncements. The great Leonhard Euler (1707 -1783), assigned the task of shutting him up, declared to Diderot

$$\text{Sir, } (a + b^n)/n = x, \text{ whence God exists; answer!}$$

Being totally ignorant of mathematics, the embarrassed Diderot could do nothing but slink off to Paris... The story is undoubtedly not true, for, while he may not have been a creative mathematician, Diderot was certainly not ignorant of the subject. It was he, apparently, who wrote the unsigned article on probability in the *Encyclopédie*.

This article was reprinted in the new *Encyclopédie Méthodique*, where it was accompanied by a second essay on the subject by Marie Jean Antoine Nicholas Caritat, Marquis de Condorcet (1743 - 1794). Condorcet was one of a number of savants who perished in the Reign of Terror following the French Revolution. Though tragic, his end does have a certain amusement value: While attempting to flee the Terror, he showed up wounded and hungry at an inn and ordered an omelette. When asked how many eggs he wanted in his omelette, he responded with the extravagant guess of a dozen. Suspicion was aroused, he was arrested and escorted to prison, where he took his own life.

The stamp of Diderot was issued in France in 1958 in a set of semipostals honouring famous Frenchmen; that depicting Condorcet was issued in 1989 in a set commemorating the 200th anniversary of the French Revolution.

winning is so small that Peter will *certainly*, though not *necessarily*, lose.[23] And, indeed, this makes good, if not mathematically acceptable sense.

D'Alembert held several mathematically heretical views. One was that a very small probability ought to be regarded as 0. Another was that, if in tossing a coin one obtains tails three times in succession, the coin is more likely to come up heads on the fourth throw, and more likely still on a fifth throw should the result be tails four times in succession.[24] He applies this line of reasoning to the Petersburg Problem in an essay *Sur le calcul des Probabilités* in the fourth

---

[23] Todhunter, *op. cit.*, p. 262.

[24] *Ibid.*, p. 263.

volume of his *Opuscules*.[25] Here he suggests that the probability of a head failing to appear before the n-th throw is not $1/2^n$ as in the ordinary theory, but should be taken to be

$$\frac{1}{2^n\left(1+\beta n^2\right)}, \quad \beta \text{ a constant.}$$

As alternatives he proposes

$$\frac{1}{2^{n+\alpha n}} \quad \text{or} \quad \frac{1}{2^{n+\alpha(n-1)}}, \quad \alpha \text{ a constant,}$$

and even a fraction with the bizarre denominator

$$2^n\left(1+\frac{B}{(K-n)^{q/2}}\right),$$

where B, K are constants and q is an odd integer. In short, where Bernoulli and Cramer obtained a finite expectation by replacing the values by utilities and considering the sum,

$$\frac{1}{2}u_1 + \frac{1}{4}u_2 + \frac{1}{8}u_3 + \ldots + \frac{1}{2^n}u_n + \ldots,$$

d'Alembert replaces the probabilities to obtain a convergent series,

$$1p_1 + 2p_2 + 2^2 p_3 + \ldots + 2^n p_{n+1} + \ldots \tag{13}$$

He wouldn't be the last to try this. In 1769, Nicolas de Beguelin (1714 - 1789) published a memoir, "Sur l'usage du principe de la raison suffisante dans le calcul des probabilités" in the volume of *Histoire de l'Acad... Berlin* intended for 1767. One of his 6 solutions to the Petersburg Problem was just such a sum with $p_n = 1/((n-1)! + 1)$, for which choice the resulting series converges to approximately $2\,1/2$.[26] No justification for the choices of the sequences $p_1, p_2, \ldots$ is reported on by Todhunter in any of these cases, and, indeed, if d'Alembert could motivate any of his choices, why would he have produced the other two? Any positive number can be represented as a sum of the form (13), so without some justification of one's particular choice of $p_1, p_2, \ldots$, the value obtained, e.g., Beguelin's 2 1/2, is entirely without significance. I am inclined to say the same about the approach via utility functions, though I do consider Bernoulli's choice to be plausible enough not to reject it out of hand (which is not to say that I find it all that convincing).

Buffon took a more pragmatic approach.[27] One of the things he did was to perform an experiment. Describing Buffon's *Essai d'Arithmétique Morale*, Todhunter reports

---

[25] *Ibid.*, p. 275.
[26] *Ibid.*, pp. 331 - 332. (For those unfamiliar with the notation, $n!$ is carefully defined on page 64, below.)
[27] For Buffon, *cf.* pp. 344 - 349 of Todhunter.

The 18th section contains the details of an experiment made by Buffon respecting the *Petersburg Problem*. He says he played the game 2084 times by getting a child to toss a coin in the air. These 2084 games he says produced 10057 crowns. There were 1061 games which produced one crown, 494 which produced two crowns, and so on. The results are given in De Morgan's *Formal Logic*, page 185, together with those obtained by a repetition of the experiment. See also *Cambridge Philosophical Transactions*, Vol. IX, page 122.[28]

Augustus de Morgan (1806 - 1871) reports that Buffon played the game 2048 times, not 2084 as Todhunter said, and presents a table listing Buffon's results, those obtained "by a young pupil of mine, for his own satisfaction", and the values predicted by theory. These results are, as one would expect, in close agreement. De Morgan's pupil's experiment was a bit more disastrous for Peter, producing 53238 crowns, or ducats, or—de Morgan doesn't report which currency was involved. Curiously, although the throws agree well with theory, they don't agree too well on a fair price for the game. Buffon's experiment suggests an expected value of

$$\frac{10057}{2048} = 4.9106\ldots \approx 4.91 \text{ crowns,}$$

while de Morgan's pupil suggests

$$\frac{53238}{2048} = 25.995\ldots \approx 26 \text{ crowns.}$$

3 Project. Repeat the experiment yourself: Play the Petersburg game 2048 times, keeping a record of which throw the first head occurs on. What is the total number of crowns produced? What is the average per game? As tossing a coin and keeping records can quickly become boring, you might want to automate the process via computer simulation. On the *TI-83 Plus*, one will find a randInt(item on the PRB submenu of the menu accessed by the MATH button. Entering randInt(0,1) will randomly choose a 0 or 1, which we may take to represent H or T, respectively. The following program will play one game of Petersburg, or whatever you want to call it:

```
PROGRAM:PETE
:0→I              set up a counter I
:2→C              create a variable C for the coin
:Repeat C=0       repeat until heads shows up
:I+1→I            count the number of the toss
:randInt(0,1)→C   toss the coin
:End              end the loop
:DelVar C         we no longer need C.
```

---

[28] *Ibid.*, p. 346. I found two editions of de Morgan's book online, the first edition published in London in 1847 by Taylor and Walton, and a reprint published in 1926 in London by Open Court. The latter was typeset anew with more modern typography. There is less content per page and the page numbering differs, though the table of contents with the original page numbers is reprinted exactly. The page corresponding to page 185 in the first edition is page 213 in this one.

Write a program[29] BUFFON that will call PETE 2048 times producing a list telling how many times the first head occurred on the first throw (i.e., $l = 1$ after running PETE), the second throw (i.e., $l = 2$ after running PETE), etc. The hitch is that there is no *a priori* upper bound to the dimension of the list. Buffon's child never went beyond the 9th toss, while de Morgan's pupil did so 6 times, once to 16 tosses of the coin. I fancy one could safely choose a list of dimension 25, filling it with 0's before making the first call to PETE. After each run, one would simply increment the l-th element of this list. One could first test if l is not greater than the dimension of the list and, should it be, either display JACKPOT! on the screen and abort the program, or redimension the list. Once you have the completed list, multiply it by

$$\{1, 2, 4, \ldots, 2^{d-1}\},$$

where $d \geq 25$ is the dimension of your list, add the results, and divide by 2048 and compare your estimate of a fair price for the Petersburg game with Buffon's and de Morgan's. Carry out the experiment several times. [You might want to do this on the computer. I don't think I built any particular inefficiency into my version of BUFFON, but with 2048 iterations of a program with a variable number of steps it took my calculator just over 5 minutes. I came up, by the way, with a total of 17451 crowns, averaging approximately 8.52 crowns per game.]

Buffon had more to say. Like d'Alembert he felt that a very small probability could be treated as 0. To determine how small a probability should be to be sufficiently small to be treated as 0, he consulted the mortality tables and noted that, although the probability that a man aged 56 would die in the course of a day was 1/10000, such a man actually considers the probability of his imminent death to be 0. Thus, any probability smaller than 1/10000 could be taken to be 0.[30] Applying this to the Petersburg Problem, we get the expected value,

$$\frac{1}{2} \cdot 1 + \frac{1}{4} \cdot 2 + \ldots + \frac{1}{2^{13}} \cdot 2^{12} = \frac{13}{2} = 6.5.$$

Buffon did not stop here. He obtained an even lower estimate of 5 by considering

i.   the fact that there is only a finite amount of money, all of France possessing fewer than 2^{29} crowns;

ii.   the relative value of money, which he, like Bernoulli, took to be inversely proportional to the better's current wealth;

iii. the mortality of the players[31]; and

iv.   the identification of any probability less than 1/10000 with 0.

---

[29] For those not familiar with writing such programs, I refer to Appendix A, where I present my version of the program.

[30] Todhunter, *op. cit.*, p. 344.

[31] The name "Methuselah" pops up in one of Pringsheim's longer footnotes (pp. 46 - 52) to his translation of Bernoulli.

Todhunter, in recounting this[32], does not reveal how Buffon's calculation proceeds, but the important thing is not the calculation or the value, both of which can be disputed, but the enumeration of conditions. What is relevant here is that constraints of realism resolve the apparent paradox of the Petersburg Problem: the value of the game is infinite, but no one would pay much to play it.

Pure mathematicians are not generally concerned with such constraints as those raised in Buffon's considerations i, iii, and iv, and modern discussions tend to rely on utility functions à la Bernoulli, Cramer, or ii, or else they embedd the problem in a lot of deep theory that cannot be described here.[33] What can be described is the applied mathematician's simple, and most convincing, solution based on the first consideration, the finiteness of the amount of money in the world. I follow Shrisha Rao's presentation[34].

Suppose Peter's wealth is M ducats (or crowns, or dollars, or euros) and Paul is to put up m ducats for a fair game. Now, that game can go on for n tosses so long as $2^{n-1} \leq M$. After that, Peter cannot pay up. The expectation for n tosses being $n/2$, Paul should not put up more than $n/2$ ducats, i.e.,

$$m \leq \frac{n}{2},$$

whence $n \geq 2m$ and we have

$$M \geq 2^{n-1} \geq 2^{2m-1}.$$

Rao, being an American, works in dollars and notes that Paul should not put up \$10 unless Peter's fortune is at least

$$2^{2 \cdot 10 - 1} = 2^{19} = 526288$$

dollars. And, to put up \$50, which is well less than infinity, he should demand that Peter show him

$$2^{2 \cdot 50 - 1} = 2^{99} = 6.338 \times 10^{29}$$

dollars, "which is much greater than the total amount of money in the world". As for Buffon's 2^{29} crowns, note that, from

$$2^{29} \geq 2^{2m-1},$$

---

[32] Todhunter, *op. cit.*, pp. 345 - 346.

[33] The celebrated 20th century re-opening of the investigation is given by William Feller's (1906 - 1970) generalisation of the notion of fairness and his proof that his asymptotic solution (for which see his *An Introduction to Probability Theory and Its Applications*, 2nd edition, John Wiley & Sons, Inc., New York, 1957, pp. 235 - 239.) is fair. "Asymptotic" here means that he offers a value $E_n = \log_2 n$ such that as n grows large without bound, the probability that the total gain of playing the Petersburg game n times is "close" to nE_n tends to 1. This, of course, offers no advice on what a fair value is for Paul who is to play the game only once.

[34] Shrisha Rao, "A note on the St. Petersburg paradox", *Elemente der Mathematik* 56 (2001), pp. 102 - 104.

follows $2m - 1 \leq 29$, i.e., $m \leq 15$: Paul should not pay more than 15 crowns to play the game.

Another common sense solution to the problem is the consideration already discussed in section 1 with respect to the poor man and the rich man:

> It has been well remarked by Buffon, that the science of probabilities never professed to make the condition of a gambler the same as if he did not play; it only indicates the events of which we have most reason to expect the recurrence. Condorcet took away everything appearing paradoxical from the result, by an observation he made in a memoir on this subject in 1784. "It may often happen," says he, "that a reasonable man A will refuse to give B a sum b for the chance n of gaining a, although a be greater than $\frac{b}{n}$; and the reason may be, because A has not the opportunity of repeating the venture often enough to repair the loss which may accrue to him in a single trial, and because the sum ventured may be so great that its loss would occasion him an inconvenience, not at all counterbalanced by the advantages he could derive from his contingent gain." These are motives for inducing A to refrain from venturing, but cannot be made elements of the calculation as between him and a speculator B on the opposite event. No underwriter diminishes, or ought to diminish, his premium, on account of the small fortune of the party whose indemnity he guarantees.[35]

3 Pascal's Wager

There is another paradoxical bit of reasoning associated with the notion of expected value that goes by the name *Pascal's Wager* although when not dressed up in probabilistic language it goes back a number of centuries to a pious Christian named Arnobius (*fl. c.* 300 A.D.). It concerns the question of belief in God:

> But Christ himself does not prove what he promises. It is true. For, as I have said, there cannot be any absolute proof of future events. Therefore since it is a condition of future events that they cannot be grasped or comprehended by any efforts of anticipation, is it not more reasonable, out of two alternatives that are uncertain and that are hanging in doubtful expectation, to give credence to the one that gives some hope rather than to the one that offers none at all? For in the former case there is no danger if, as is said to threaten, it becomes empty and void; while in the latter case the danger is greatest, that is, the loss of salvation, if when the time comes it is found that it was not a falsehood.[36]

[35] John Lubbock and John Drinkwater-Bethune, *On Probability*, Baldwin & Chadock, London, 1830, p. 48.
[36] Quoted by David Eugene Smith, editor of: Augustus de Morgan, *A Budget of Paradoxes*, 2 volumes, 2nd edition, Open Court Publishing Company, Chicago, 1915, vol. II, p. 74.

Arnobius's argument has a gentleness and good will that deters one from harsh criticism. There may be some self-deception wrought by his evident belief that his is the only God in town[37], but there doesn't appear to be any conscious attempt to deceive.

Pascal also appears to have had the best of intentions when he dressed the argument up a bit. He is not as clear as Arnobius:

> Let us then examine this point, and say, "God is, or He is not." But to which side shall we incline? Reason can decide nothing here. There is an infinite chaos which separates us. A game is being played at the extremity of this infinite distance where heads or tails will turn up. What will you wager? According to reason, you can do neither the one thing nor the other; according to reason, you can defend neither of the propositions.
>
> Do not then reprove for error those who have made a choice; for you know nothing about it. "No, but I blame them for having made, not this choice, but a choice; for again both he who chooses heads and he who chooses tails are equally at fault, they are both in the wrong. The true course is not to wager at all."
>
> Yes; but you must wager. It is not optional. You are embarked. Which will you choose then? Let us see. Since you must choose, let us see which interests you least. You have two things to lose, the true and the good; and two things to stake, your reason and your will, your knowledge and your happiness; and your nature has two things to shun, error and misery. Your reason is no more shocked in choosing one rather than the other, since you must of necessity choose. This is one point settled. But your happiness? Let us weigh the gain and the loss in wagering that God is. Let us estimate these two chances. If you gain, you gain all; if you lose, you lose nothing. Wager, then, without hesitation that He is.— "That is very fine. Yes, I must wager; but I may perhaps wager too much."—Let us see. Since there is an equal risk of gain and of loss, if you had only to gain two lives, instead of one, you might still wager. But if there were three lives to gain, you would have to play (since you are under the necessity of playing), and you would be imprudent, when you are forced to play, not to chance your life to gain three at a game where there is an equal risk of loss and gain. But there is an eternity of life and happiness. And this being so, if there were an infinity of chances, of which one only would be for you, you would still be right in wagering one to win two, and you would act stupidly, being obliged to play, by refusing to stake one life against three at a game in which out of an infinity of chances there is one for you, if there were an infinity of an infinitely happy life to gain. But there is here an infinity of an infinitely happy life to gain, a chance of gain against a finite number of chances of loss, and what you stake is finite. It is all divided; wherever

---

[37] If you believe in his Christian God and the true Gods are those petty Olympians, then you are in grave danger. More on this later.

the infinite is and there is not an infinity of chances of loss against that of gain, there is no time to hesitate, you must give all.[38]

Pascal continue's the argument a bit. I am reminded of Lessing's fragment of *Faust*. In illustrating to his fellow German writers the dramatic possibilities of the legend, he took an amusing anecdote about Faust choosing a devil based on his speed and elaborated it past drama into the realm of somnolence. Pascal's wager was of some importance in philosophy and decision theory, and a thoughtful and careful examination of it is given by Ian Hacking[39]. The wager is not important in the history of probability in the way that the Petersburg Problem was, nor does it serve as a good motivating example of fundamental principles as do the problems to be discussed in the next chapter. It is, however, in its modern formulation, a good example of the misapplication of probabilistic concepts.

The usual presentation of Pascal's wager goes like this: God either exists or he doesn't: we may assume the probability of his existence to be $1/2$ (or, indeed, any non-zero value). Believing in him will involve a certain finite cost (F) (time and energy spent on religious observation, tithes to the church, etc.); not believing in him will cost you nothing in this life (assuming no danger of persecution!). If God exists and you believe, the rewards are infinite ($+\infty$), while if you don't believe, the punishment will be infinite ($-\infty$). If God doesn't exist, there will be neither reward nor punishment. So if you choose to believe in God your expected value is

$$E = \frac{1}{2}(+\infty) + \frac{1}{2}(-F) = +\infty, \tag{14}$$

while, if you choose not to believe, your expected value is

$$E = \frac{1}{2}(-\infty) + \frac{1}{2}(0) = -\infty. \tag{15}$$

Thus, you should choose to believe.

Laplace excoriates Pascal on this:

There are things so extraordinary that nothing can balance their improbability. But this, by the effect of a dominant opinion, can be weakened to the point of appearing inferior to the probability of the testimonies; and when this opinion changes an absurd statement admitted unanimously in the century which has given it birth offers to the following centuries only a new proof of the extreme influence of the general opinion upon the more enlightened minds. Two great men of the century of Louis XIV.—Racine and Pascal—are striking examples.

. . .

---

[38] Blaise Pascal, *Thoughts*, section 233 "Infinite–nothing", taken from: Monroe C. Beardsley, ed., *The European Philosophers from Descartes to Nietzsche*, The Modern Library, New York, 1960, pp. 116 - 117.

[39] Ian Hacking, *The Emergence of Probability*, Cambridge University Press, Cambridge, 1975, pp. 63 - 72.

There comes up naturally at this point the discussion of a famous argument of Pascal, that Craig[40], an English mathematician, has produced under a geometric form. Witnesses declare that they have it from Divinity that *in conforming* to a certain thing one will enjoy not one or two but an infinity of happy lives. However feeble the probability of the proofs may be, provided that it be not infinitely small, it is clear that the advantage of those who conform to the prescribed thing is infinite since it is the product of this probability and an infinite good; one ought not to hesitate then to procure for oneself this advantage.

This argument is based upon the infinite number of happy lives promised in the name of the Divinity by the witnesses; it is necessary then to prescribe them, precisely because they exaggerate their promises beyond all limits, a consequence which is repugnant to good sense. Also calculus teaches us that this exaggeration itself enfeebles the probability of their testimony to the point of rendering it infinitely small or zero. Indeed this case is similar to that of a witness who should announce the drawing of the highest number from an urn filled with a great number of numbers, one of which has been drawn and who would have a great interest in announcing the drawing of this number. One has already seen how much this interest enfeebles his testimony. In evaluating only at $\frac{1}{2}$ the probability that if the witness deceives he will choose the largest number, calculus gives the probability of his announcement as smaller than a fraction whose numerator is unity and whose denominator is unity plus the half of the product of the number of the numbers by the probability of falsehood considered *à priori* or independently of the announcement. In order to compare this case to that of the argument of Pascal it is sufficient to represent by the numbers in the urn all the possible numbers of happy lives which the number of these numbers renders infinite; and to observe that if the witnesses deceive they have the greatest interest, in order to accredit their falsehood, in promising an eternity of happiness. The expression of the probability of their testimony becomes then infinitely small. Multiplying it by the infinite number of happy lives promised, infinity would disappear from the product which expresses the advantage resultant from this promise which destroys the argument of Pascal.[41]

---

[40] Craig, or Craige, a contemporary of Newton's, wrote a much criticised little book on the weakening of the value of testimony over the generations. The first part of his book calculated that evidence supporting Christianity would expire around 3150 A.D. and, on the basis of a passage from scripture, that this was the first possible date for the return of Christ. The second part of his book was an elaboration of Pascal's argument. Craig's work is available in English translation: Richard Nash, *John Craige's Mathematical Principles of Christian Theology*, Southern Illinois University Press, Carbondale and Edwardsville, 1991.

[41] Frederick Wilson Truscott and Frederick Lincoln Emory (translators), Pierre Simon de Laplace, *A Philosophical Essay on Probabilities*, John Wiley & Sons, London, 1902, pp. 119 - 122.

Laplace included his discussion, as one might guess, in a chapter on the probability of testimony in legal arguments.

During the Second World War, physicists in Los Alamos faced their own Pascalian wager. Edward Teller, father of the American hydrogen bomb, raised a terrifying possibility:

> Teller found something else as well, or thought he did, and with his usual pellmell facility he scattered it before them. There are many other thermonuclear reactions besides the D + D reactions[42]. Bethe had examined a number of them methodically when looking for those that energized massive stars. Now Teller offered several which a fission bomb or a Super[43] might inadvertently trigger. He proposed to the assembled luminaries the possibility that their bombs might ignite the earth's oceans or its atmosphere and burn up the world.[44]

Exactly who made the suggestion is unclear:

> It happened when Teller was explaining his calculations of the heat yield of a nuclear blast and why it would trigger fusion in heavy hydrogen. In the midst of the explanation it apparently was Oppenheimer who saw the apocalypse.
> Would an atomic bomb instantly destroy the world?
> The awesome possibility was implied in the heat buildup that Teller had correctly calculated. The intensity of the heat was such that Oppenheimer and some of the other scientists feared it could ignite the heavy hydrogen that naturally occurs in seawater, or start a reaction in the nitrogen that constitutes eighty percent of the atmospheric envelope surrounding the earth. The oceans and the heavens would catch fire. There would be nothing left.
> The extent to which this fear was well founded is a matter of disagreement even among the physicists who were there. No one there considered the mathematical possibility higher than about one in three million. That would be a safe bet in any other enterprise, but such odds would be disturbingly low in the face of the consequences.[45]

Arthur Holly Compton described how he heard about the theory from J. Robert Oppenheimer:

> Was there really any chance that an atomic bomb would trigger the explosion of the nitrogen in the atmosphere or the hydrogen in the ocean? This would be the ultimate catastrophe. Better to accept the slavery of the Nazis than to run a chance of drawing the final curtain

---

[42] D = deuterium, or heavy hydrogen. The D + D reaction is the fusion of two deuterium atoms into a helium atom.

[43] I.e., the hydrogen bomb.

[44] Richard Rhodes, *The Making of the Atomic Bomb*, Simon and Shuster, New York, 1986, p. 418.

[45] Stanley A. Blumberg and Gwinn Owens, *Energy and Conflict; The Life and Times of Edward Teller*, G.P. Putnam's Sons, New York, 1978, p. 117.

on mankind!
We agreed there could be only one answer. Oppenheimer's team must
go ahead with their calculations.[46]

Rhodes continues, informing us that Hans Bethe redid Teller's calculations,
found Teller had made some unjustified assumptions, and convinced Teller and
everyone else that the world's end was "extremely unlikely". In an official report
this read, "The impossibility of igniting the atmosphere was thus assured by
science and common sense".[47]

Applying the numerical version of Pascal's argument, we have the alternatives of annihilation of mankind $(-\infty)$ and the consequences of winning the
war through nuclear bombs (F). The expectation, should they continue with
the programme and explode a nuclear device, would be

$$E = p(-\infty) + (1 - p)(F) = -\infty,$$

where p is the probability of igniting the oceans or the atmosphere. A Pascalian
would deem it imperative to halt the bomb project. Pascal's advice was not
followed and the rest is history. Evidently the similarity of Pascal's Wager to
the Petersburg Problem is greater than one might have imagined. In each case,
the determining factor is the imagined size of p, not the size of E.

In proper rhetorical fashion I should say that this is a "twice-told tale" and
has been repeated, but "twice-told" suggests only a single repetition. The fear
that scientists would destroy the world, followed by their pressing on in the face
of implausibility, and nothing happening is at least a thrice-told tale. Later in
the century the fear was of recombinant DNA research and, as I write, the
concern is that the Large Hadron Collider in Switzerland will create a black
hole that will swallow the Earth. Indeed, I am informed by one web site that
this has already happened—the hole has been created, is growing in size, and
will destroy the world on 21 December 2012, the exact date the Mayans deny
their ancestors claimed for the destruction of the Earth. I have my doubts about
this as, around 1699, John Craig proved the Second Coming of Christ will not
occur until after 3150 A.D. and there would be no point to Jesus's returning
after the Annihilation. In any event, what is relevant here is that those who are
in the best position to weigh the risks, the scientists themselves, always seem
to consider decisive the size of p, not that of E.

As I write, Pascal's argument is offered as the main justification for believing
in apocalyptic man-made *global warming*, a fashionable pseudo-scientific cause
not subscribed to by many scientists, but enthusiastically embraced by many
in Hollywood and others with no scientific education. As I am not interested
in controversy, I will not discuss this here, but refer the reader to Google,
suggesting an online search on "Pascal's Wager" + "global warming".

Let us return to Pascal's argument for believing in the existence of God.
We have Laplace not accepting its validity on the grounds that the probability
of infinite disaster is so low it can be ignored, the nuclear physicists rejecting a

---

[46] Rhodes, *op. cit.*, p. 419.
[47] *Ibid.*

similar wager on the same grounds. We can give a somewhat more sound yet elementary reason for rejecting the argument. Pascal offers only two alternatives: God exists and is a Pascalian God or he doesn't exist. How about a feminist refinement: God exists and is male, God exists and is female, or God does not exist? Let us assume with Pascal that the male God is the attention-starved demon who punishes those who do not believe in him, and let us further assume that the female God is a militant feminist who punishes those who insultingly consider Her to be male. The expectation of one who worships the traditional masculine deity would be

$$E = p_1(+\infty) + p_2(-\infty) + p_3(-F),$$

where p_1, p_2, p_3 are the respective probabilities of a male God, a female Goddess, and no god, and F is the finite cost of worship during one's life. The form is indeterminate and Pascal's argument settles nothing.

Reasonings about God tend to be fraught with emotion rather than rationality, as we shall see again in Chapter 5, sections 5 and 6, below, in de Moivre's reaction to some remarks of Nikolaus Bernoulli. So I suppose I'd best protect myself by declaring neutrality on the issue. I take no side on the outcome of the debate: I neither endorse the existence of any God nor deny it; I simply claim that the *form* of argument Pascal's Wager takes is totally bogus. Atheists and agnostics should accept this readily, atheists possibly for the wrong reason— they believe Pascal's conclusion false—, but theists might be too emotionally committed to the truth of the conclusion to see the fallacy of the argument. For the benefit of those at either extreme, let me offer a variant of Pascal's Wager free of emotional baggage: I should stay in bed today. For, if I get up I might go out and die horribly in a traffic accident $(-\infty)$, but if I stay in bed that will not happen. My expectation, should I get out of bed is

$$E = p_1(-\infty) + p_2(F_1) = -\infty,$$

while my expectation, should I stay in bed is

$$E = 0(-\infty) + 1(F_2) = F_2 = \text{finite}.$$

So, whether F_1, F_2 are good or bad, staying in bed all day offers me the better expectation.

The absurdity of Pascal's Wager is absolutely clear in this case. Without even trying, one quickly thinks of alternative cases: if I stay in bed, the boiler of the water heater could explode, destroying me and half the house[48]. Or, a meteorite could come crashing through the roof and crush me and my bed. There are even more probable domestic disasters, but they do not quite so quickly come to mind and, in any event, are too numerous to be listed here. If Pascal's Wager does not seem so immediately absurd in its original context,

---

[48] You will never guess who was recently visited by a gas company crew who observed everything in the house that was no longer up to code and related the various dangers inherent under such conditions.

it is only because one comes to it with preconceptions limiting the number of alternatives one will perceive: There is either Pascal's God or there is none. Citing the God's of Mount Olympus or a militant feminist to whom male chauvinism is the ultimate sin, though logically impeccable, may merely seem frivolous, and failure to consider such will obscure the fallacy of Pascal's Wager.

Ignoring all of this, there is still more to object to in Pascal's Wager. Ask yourself: What is the meaning of the numbers $\frac{1}{2}$ in (14) and (15)? People have mused over probability and chance since time immemorial, if archæological discoveries of gaming devices are any indication. And theorising about probability goes back at least as far as Cardano in the 16th century. Yet the question of what probability is was still a hot topic for philosophical discussion in the early decades of the 20th century. The question doesn't need to be addressed in most applications, where the answers given by pointing to the ratio of favourable to all cases or to relative frequency of occurrence are serviceable. But what does it mean to say that the probability that God exists is some number p?

I would rule out the relative frequency interpretation here. It would seem to require God occasionally to exist—He or She (or They) exists and passes judgment when some people die, and doesn't exist when others die. But surely God exists or doesn't exist; the relative frequency is 1 or 0, yielding probabilities 1 or 0. The true believer, for whom the probability is 1 and doesn't need convincing, will accept the argument; the atheist, for whom the probability is 0, will claim the argument shows he should not believe; and the agnostic, unable to choose between the two probabilities, will be left undecided.

As for the enumeration of equally likely cases, how is that to be done? There is no evidence for the existence of God, only the testimony of those who died centuries ago. So, once we've listed some alternatives, we have no objective means of assigning probabilities. In the face of total ignorance, beginning with Jakob Bernoulli, there was the *principle of insufficient reason* asserting that, there being no reason to assume any outcome more likely than another, they should be considered equally likely. Pascal used the principle implicitly in assigning $\frac{1}{2}$ to the probability of God's existence. But the principle of insufficient reason is easily criticised.[49] The trichotomy

Pascal's God exists, the Olympian God's exist, no God exists,

assigns $\frac{1}{3}$ to Pascal's God according to the principle. Suppose we suddenly realise that the Olympian Gods, should they exist, might be wearing modern clothing instead of the traditional ancient garb. We now have four "equally likely" cases and Pascal's God has been reduced in probability to $\frac{1}{4}$, while the Olympians have been bumped up to $\frac{1}{2} = \frac{1}{4}$ (ancient garb) $+ \frac{1}{4}$ (modern dress). We can, of course, increase the probability of the existence of Pascal's God by considering the various possibilities for His sartorial preferences...

The reader who might consider this criticism of the use here of the principle of insufficient reason less than convincing on account of its artificiality

---

[49] John Maynard Keynes devotes an entire chapter, "The Principle of Indifference", of his *A Treatise on Probability* (Macmillan and Co., London, 1921) to criticising this principle. I cite him on this matter on page 250, below.

The 20th century saw several attempts at explaining what probability was. Applications were outstripping the old standby of the ratio of favourable to all cases. Leading theorists and theories included Richard von Mises (1883 - 1953) and his attempt to base probability on relative frequencies, John Maynard Keynes (1883 - 1946) and his attempt to incorporate probability into logic, Bruno de Finetti (1906 - 1985) and his theory of subjective probability, and Andrei Nikolaevich Kolmogorov (1903 - 1987) and his axiomatic approach using measure theory.

Keynes is, of course, best known as an economist and one of the most influential men of the 20th century. Before he became an economist, however, he studied mathematics and philosophy. His *A Treatise on Probability* (1921) was a mathematically informed philosophical treatise that was quite influential in the philosophy of probability for a while, but did not prove as valuable in the long run as other approaches.

The first stamp pictured here features Keynes (right) and fellow economist Joseph Schumpeter whose economic theories rivalled those of Keynes. The stamp is one of a set of stamps issued by Portugal in 2000 honouring 20th century achievements. Kolmogorov is depicted on another stamp of the series (see p. 317, below).

Keynes was also chosen for an appearance on one of the stamps issued by Belgium for much the same purpose. The Belgian stamps offer amusing caricatures rather than straight portraits and I almost decided for the sake of decorum on presenting only the more staid Portuguese stamp here; but I like the Belgian stamp too much not to share it with you.

is reminded of the arbitrariness of Pascal's own dichotomy. The principle is usually deemed acceptable in those situations where there is some underlying symmetry to the cases. Dice, for example, have fairly identical faces, of equal area, and, unless they have been specially weighted, their centres of gravity are at the geometric centres of the dice—meaning there is no greater stability or instability to any one of the sides. In Pascal's Wager there is no enumeration of all the cases and no symmetry in sight.

Philosophers tend to take Pascal's Wager more seriously than do mathematicians, but not because of the questionable calculation. I refer the reader to the chapter "The Great Decision" of Hacking's book[50] for such a serious dis-

---

[50] Hacking, *op. cit.*, pp. 63 - 72.

cussion. Mathematicians tend to dismiss the argument out of hand or to mock
it. Noting that choosing to believe for the sake of infinite gain is not likely to
be smiled upon by a deity who takes motives into account, de Morgan wrote

> I wonder whether Pascal's curious imagination ever presented to him
> in sleep his convert, in a future state, shaken out of a red-hot dice-box
> upon a red-hot hazard-table[51], as perhaps he might have been, if Dante
> had been the later of the two.[52]

Laplace went so far as to show by calculation that the expected value was finite,
and de Morgan, likening the Wager to an annuity, concluded that "the present
value of an eternity is *not* infinitely great".[53]

---

[51] I.e., a table on which to roll the dice. The word "hazard" was often used in describing games of chance, i.e., gambling.

[52] De Morgan, *Budget*, vol. 2, p. 73.

[53] Karl Pearson, *The History of Statistics in the 17th & 18th Centuries against the
changing background of intellectual, scientific and religious thought*, Charles Griffin
& Co., Ltd., London, 1978, pp. 677 - 679.

Some Classic Problems

1 Pacioli's Second Problem

Recall the problem: The first of three players to win 6 points wins the game, but the players have to break off when one player we will call A has 4 points, a second player we call B has 3 points and a third we call C has 2 points. How do we divide the stakes?

Fermat offered a very simple method. We note that in any case the game would be decided in 7 more rounds, so we consider as our sample space the collection of all sequences of length 7 of possible wins of points, which sequences can be represented by 7-letter words in the alphabet consisting of letters A, B, C. We have but to enumerate all such words and count which represent wins for players A, B, C, respectively, to determine the probabilities the different players have of winning, and therewith their fair proportions of the stakes. The problem is that there are 2187 such words to enumerate. Even if we already had the 2187 words listed, counting the numbers of wins for each player would be a tedious task likely to give rise to error. Is there a better way?

The answer, of course, is "yes". There are methods of determining the number of elements in an event other than by direct enumeration. One breaks the event into component parts, determines their numbers, and the rule relating the number of elements of the event to the numbers of elements of its component parts.

The simplest example is the *Addition Principle*:

1 Theorem (Addition Principle). *If two sets X, Y have no overlap, the number of elements in their union is the sum of the number of elements of the two sets.*

In other words, if X has m elements and Y has n elements, and there is no overlap, then $X \cup Y$ has $m + n$ elements. More generally, if one has k sets $X_0, \ldots, X_{k-1}$, no two of which have an element in common, and if X_i has n_i elements then the union of the sets, $X_0 \cup X_1 \cup \ldots \cup X_{k-1}$ has $n_0 + \ldots + n_{k-1}$ elements.

For the problem at hand, in considering the event,

E: A wins,

we can first consider the separate events,

E_i: A wins in exactly i rounds,

where the numbers i range from 2 to 7, then count the number of outcomes in each sub-event E_i, and add them up. The six subtasks of counting the elements of $E_2, \ldots, E_7$ are more easily performed than the full task of counting the number of elements of E itself. These subtasks are, however, not trivial and require another counting principle—and some elaboration thereof.

Our second counting principle is one we have already made implicit use of.

2 Theorem (Sequential Principle). *Suppose we have a sequence of k tasks $T_0, T_1, \ldots, T_{k-1}$ to perform and suppose task T_i can be performed in n_i ways. Then we can perform all k tasks in $n_0 \cdot n_1 \cdots n_{k-1}$ ways.*

For example, if each of two tasks is to toss a coin, each toss being performable in 2 ways, the total number of ways of performing 2 tosses is $2 \cdot 2 = 4$, as we already know from the enumeration HH, HT, TH, TT. Pascal's discussion of Fermat's method of solving instances of the problem of points gives us two further examples. First, he considered the words of length 4 in an alphabet of two letters, a, b. Making such a word is a sequence of repetitions of the task of choosing one of the two letters, which can be done in two ways, and the overall repetition can thus be done in $2 \cdot 2 \cdot 2 \cdot 2 = 2^4 = 16$ ways. The three player version of his problem considered three-letter words in the $\{a, b, c\}$ alphabet, of which there are $3 \cdot 3 \cdot 3 = 3^3 = 27$. And our current sample space of 7-letter words formed from the letters A, B, C has $3^7 = 2187$ elements.

The tasks $T_2, T_3, \ldots, T_{k-1}$ may depend on the results of earlier tasks: T_2 on T_1, T_3 on T_1 and $T_2, \ldots$ What is important is that the numbers $n_2, n_3, \ldots, n_{k-1}$ not change. For example, suppose we want to know how many ways a set of, say, 5 elements an be ordered. We can think of the task of ordering the set as a sequence of the 5 subtasks:

choose an element of the set to be first
choose an element of the remaining 4 elements to be second
choose an element of the remaining 3 elements to be third
choose an element of the remaining 2 elements to be fourth
choose the remaining element to be last.

The successive subtasks are to choose an element from a set that varies with the results of the earlier subtasks, but whatever the set, the second subtask can be performed in 4 ways, the third in 3, the fourth in 2, and the fifth in 1 way. As the first subtask can be performed in 5 ways, the number of ways of performing the combined task is $5 \cdot 4 \cdot 3 \cdot 2 \cdot 1 = 120$.

The ordering problem is a special case of the more general *permutation* problem, which is to count how many ways one can choose and order a k-element subset from an n-element set, where, to make the problem non-trivial, one assumes $k \leq n$. This problem occurs frequently and there are special notations for it's solution:

$_nP_k$ or P_k^n = the number of permutations of n things taken k at a time.

Sometimes one writes r instead of k. Indeed, if one looks at the PRB submenu of the MATH menu on the *TI-83 Plus*, one will see $_nP_r$ as the second item on the list. It is an infix operator on the calculator: To evaluate, say, $_5P_3$, one enters

   5 $_nP_r$ 3

and the calculator will respond instantly with 60.

Before using the calculator one should, of course, apply the Sequential Principle to make the theoretical calculation. Choosing an ordered set of k elements from an n element set is a sequence of k tasks:

   choose the first element   (possible in n ways)
   choose the second element   (possible in $n-1$ ways)
   choose the third element   (possible in $n-2$ ways)
   $\vdots$
   choose the k-th element   (possible in $n-k+1$ ways).

Thus

$$_nP_k = n(n-1)\cdots(n-k+1).$$

And we can check this against our calculator: $5 \cdot 4 \cdot 3 = 60$.

One can illustrate permutations by considering a club electing officers, say, a president, vice president, secretary, and treasurer. If there are 10 members in the club, the number of possible ways of choosing the officers is

$$_{10}P_4 = 10 \cdot 9 \cdot 8 \cdot 7 = 5040.$$

For many tasks, the order is not important. If, for example, our club of 10 merely wishes to choose 4 members to form a committee to study an issue or to attend a convention and the members of the committee are to have equal standing, 5040 is too large a number. If members A, B, C, D are chosen, they are multiply listed among the 5040 permutations as $ABCD, ABDC, ACBD$, etc. In all each committee is listed as many times as there are orderings of its members, i.e., each committee is listed $_4P_4 = 4 \cdot 3 \cdot 2 \cdot 1 = 24$ times. The total number of different committees is thus $5040/24 = 210$.

Permutation-like subcollections taken without regard to order are called *combinations*, and the number of combinations of n things taken k at a time is variously written[1]

$$\binom{n}{k}, \quad _nC_k, \quad \text{or} \quad C_k^n.$$

Again, one often replaces "k" by the letter "r" and one can calculate $_nC_r$ on the calculator using the infix $_nC_r$ operator which is found next to the $_nP_r$ operator in the PRB submenu of the MATH menu. Thus, to calculate $_{10}C_4$ one would enter

---

[1] My preference is for the first of these, but it is sometimes more convenient in typesetting to use the second.

$$10 \; _nC_r \; 4$$

on the calculator and it would respond immediately with 210.

A formula for $_nC_k$ is readily obtained by recalculating $_nP_r$. Note that the task of choosing a permutation of n things taken k at a time can be broken down into two smaller tasks: First choose a combination of n things taken k at a time, which can be done in $_nC_k$ ways, and then order that set of k elements, which can be done in $_kP_k$ ways. The Sequential Principle yields

$$_nP_k = {}_nC_k \cdot {}_kP_k,$$

whence

$$_nC_k = \frac{{}_nP_k}{{}_kP_k} = \frac{n(n-1)\cdots(n-k+1)}{k(k-1)\cdots 1}. \tag{1}$$

A nicer formula is obtained by multiplying numerator and denominator by $(n-k)(n-k-1)\cdots 1$ and using the *factorial*, $j! = j(j-1)\cdots 1$:

$$_nC_k = \frac{n(n-1)\cdots(n-k+1)\cdot(n-k)\cdots 1}{k(k-1)\cdots 1 \cdot (n-k)\cdots 1} = \frac{n!}{k!(n-k)!}. \tag{2}$$

The numbers $_nC_k$ are called the *binomial coefficients* because of their occurrence in the *Binomial Theorem*,

$$(x+y)^n = {}_nC_0\, x^n + {}_nC_1\, x^{n-1}y + \ldots + {}_nC_k\, x^{n-k}y^k + \ldots + {}_nC_n\, y^n,$$

usually written

$$(x+y)^n = \binom{n}{0}x^n + \binom{n}{1}x^{n-1}y + \ldots + \binom{n}{k}x^{n-k}y^k + \ldots + \binom{n}{n}y^n, \tag{3}$$

or

$$(x+y)^n = \binom{n}{n}x^n + \binom{n}{n-1}x^{n-1}y + \ldots + \binom{n}{k}x^k y^{n-k} + \ldots + \binom{n}{0}y^n. \tag{4}$$

The equivalence of (3) and (4) follows from the identity

$$\binom{n}{k} = \binom{n}{n-k},$$

which may be established algebraically by appeal to (2), or combinatorially by noting that the act of choosing a k-element subset of an n-element set is the same as that of choosing the $n-k$ elements that are *not* supposed to belong to the subset.

The Binomial Theorem can be established combinatorially by writing

$$(x+y)^n = \underbrace{(x+y)}_{1}\, \underbrace{(x+y)}_{2} \cdots \underbrace{(x+y)}_{n}$$

and noting that the product consists of a huge sum of products of x's and y's taken from the factors numbered 1 to n. When like terms are combined, the

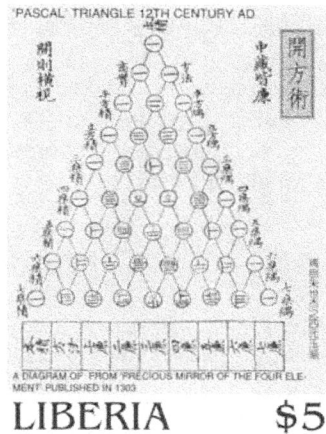

LIBERIA $5

I, for one, cannot imagine the United States Postal Service issuing a stamp celebrating the Binomial Theorem. Yet here it is, explicitly, if not quite correctly, stated on a North Korean stamp of 1993 commemorating Newton's 350th birthday. The errors are subtle: The binomial coefficient should appear as C_k^n and not C_n^k. And the attribution of credit is wrong: The Finite Binomial Theorem was known to Chinese scholars centuries earlier, as indicated on the Liberian stamp to the right depicting an illustration from the *Sìyuán yùjiàn* [*Precious mirror of the four elements*] of Zhū Shìjié (Pinyin, older Wade-Giles transliteration: Chu Shih-chieh) (*fl.* 1280 - 1303). The result was also known to mediæval Arab scholars and, reportedly, to Hindu scholars as well, before it was eventually rediscovered and popularised by Pascal, who referred to the triangular display of the binomial coefficients as the "arithmetical triangle". Today one refers to "Pascal's triangle". Newton's contribution to the Binomial Theorem was its generalisation of the formula to arbitrary rational exponents,

$$(1 + x)^q = \sum_{k=0}^{\infty} \binom{q}{k} x^k,$$

where

$$\binom{q}{0} = 1, \quad \binom{q}{k} = \frac{q(q-1)\cdots(q-k+1)}{k(k-1)\cdots 1} \text{ for } k > 0,$$

and $|x| < 1$. Of course, for probability it is the finite form of the Theorem, with positive integral exponents, that is of importance.

Curiously, the Liberian stamp displays an error as well: Zhū's table contains a typographical error in the 8th row, where the entries for $_7C_3$ and $_7C_4$ should both be 35. As one can see from the enlargement (inset, right), the first of these is given as 34, the horizontal bars tallying 10's and the vertical 1's.

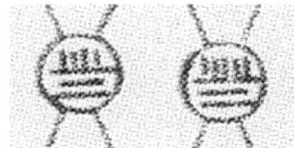

coefficient of $x^k y^{n-k}$ is the number of those products containing k x's and $n-k$ y's. How many are there? Well, it is a matter of choosing the k factors from the n factors to take the x from, and letting all the other factors contribute a y. This can be done in $\binom{n}{k}$ ways.

The Binomial Theorem generalises to a Multinomial Theorem, e.g., a Trinomial Theorem:

$$(x + y + z)^n = \binom{n}{n,0,0} x^n + \binom{n}{n-1,1,0} x^{n-1} y + \binom{n}{n-1,0,1} x^{n-1} z + \cdots,$$

where each product $x^i y^j z^k$, for non-negative integers i, j, k with $i + j + k = n$, is accompanied by a *trinomial coefficient*, which I write[2] as

$$\binom{n}{i, j, k}.$$

The determination of the trinomial coefficient is actually fairly easy: Again write the product,

$$(x + y + z)^n = (x + y + z)(x + y + z) \cdots (x + y + z).$$

We need to determine the number of ways we can choose i factors from which to take the x's, j from which to take the y's, and k from which to take the z's. This can be broken into a sequence of tasks:

  choose i factors from the n factors for the x's
  choose j factors from the remaining $n - i$ factors for the y's
  choose the remaining $k = n - i - j$ factors for the z's.

The first task can be performed in $_nC_i$ ways, the second in $_{n-i}C_j$ ways, and the third in only 1 $(= \ _{n-i-j}C_k)$ way. Thus, by the Sequential Principle,

$$\binom{n}{i, j, k} = {}_nC_k \cdot {}_{n-i}C_j \cdot {}_{n-i-j}C_k$$

$$= \binom{n}{i}\binom{n-i}{j}\binom{n-i-j}{k}$$

$$= \frac{n(n-1)\cdots(n-i+1)}{i!} \cdot \frac{(n-i)(n-i-1)\cdots(n-i-j+1)}{j!}$$

$$\cdot \frac{(n-i-j)\cdots 1}{k!} \text{ by (1)}$$

$$= \frac{n!}{i!j!k!}, \tag{5}$$

3 Exercise. Expand $(x + y + z)^3$ and verify that the coefficients of the terms are given by (5).

---

[2] I confess not to know what, if anything, is the standard notation and have merely chosen an obvious generalisation of the usual notation for binomial coefficients.

We now have everything we need to solve Pacioli's second problem.

The solution to Pacioli's problem chiefly consists in counting how many ways each player can win. To carry out this count, we will use both the Addition Principle and the multiplicative Sequential Principle, eventually employing trinomial coefficients. We start with player A.

A win for player A can occur well before all 7 rounds are played. He could win in as few as two rounds, or in three, or in four, or... Let E^A denote the event "A wins" and let us partition E^A into the subevents,

E_i^A : A wins in exactly i rounds,

and count the number of elements in each subevent.

A needs two points to win, whence for him to win in 2 rounds, he must win both. Thus an outcome in E_2^A is just a word $AAL_1L_2L_3L_4L_5$, where the letters L_i can be any of A, B, C. The initial A's can be chosen in only one way, and each of the L_i's in 3 ways. Thus the Sequential Principle tells us

$$|E_2^A| = 1 \cdot 1 \cdot 3 \cdot 3 \cdot 3 \cdot 3 \cdot 3 = 3^5 = 243.$$

To win in exactly three rounds, only one of the first two rounds can go to A and he must win the third. A typical outcome in E_3^A thus looks like

$$ALAL_1L_2L_3L_4 \quad \text{or} \quad LAAL_1L_2L_3L_4.$$

Now L_1, L_2, L_3, L_4 can be any of A, B, C and can thus be chosen in 3^4 ways. The letter L must be one of B, C and can be chosen in 2 ways. A can, of course, be chosen in only one way, but we have 2 choices of position among the first two letters of the word in which to place A. The total number of such words is, thus

$$|E_3^A| = 3^4 \cdot 2 \cdot 2 = 324.$$

To win in exactly 4 rounds, the 4th letter of the word must be an A and the last 3 can be any of A, B, C. Now the first 3 must include a single A and 2 from B, C. Again, choosing the A can be done in only one way, but it can go in any of 3 positions. And each of the other two positions can be filled in either of 2 ways. Thus,

$$|E_4^A| = 3^3 \cdot 3 \cdot 2^2 = 3^4 \cdot 4 = 324.$$

So far each case has been treated slightly differently; a unifying pattern begins to emerge in discussing a win in 5 rounds. A winning word has 4 letters, followed by an A, followed by 2 letters. The last 2 letters can be arbitrary and each can be chosen in 3 ways, thus yielding a factor of 3^2 in $|E_5^A|$. Now, in choosing the first 4 letters, we must choose 1 place for the A, and some places for B, and some for C. We cannot choose 3 places for B or the win would go to him. Thus, we must enumerate the possibilities. These are:

$1A, 2B, 1C$
$1A, 1B, 2C$
$1A, 0B, 3C.$

We use the trinomial coefficients to determine how many ways each of these tasks can be performed:

$$1A, 2B, 1C : \quad \frac{4!}{1!2!1!} = \frac{4 \cdot 3 \cdot 2 \cdot 1}{1 \cdot 2 \cdot 1 \cdot 1} = 4 \cdot 3 = 12$$

$$1A, 1B, 2C : \quad \frac{4!}{1!1!2!} = \frac{4 \cdot 3 \cdot 2 \cdot 1}{1 \cdot 1 \cdot 2 \cdot 1} = 12$$

$$1A, 0B, 3C : \quad \frac{4!}{1!0!3!} = \frac{4 \cdot 3 \cdot 2 \cdot 1}{1 \cdot 1 \cdot 3 \cdot 2 \cdot 1} = 4.$$

There are thus $12 + 12 + 4 = 28$ ways of choosing the first 4 letters and

$$\left| E_5^A \right| = 28 \cdot 3^2 = 252.$$

For 6 rounds, the 6th letter must be A, the 7th is arbitrary, and the first 5 letters break down into one of the possibilities:

$$1A, 2B, 2C : \quad \frac{5!}{1!2!2!} = \frac{5 \cdot 4 \cdot 3 \cdot 2 \cdot 1}{1 \cdot 2 \cdot 1 \cdot 2 \cdot 1} = 5 \cdot 2 \cdot 3 = 30$$

$$1A, 1B, 3C : \quad \frac{5!}{1!1!3!} = \frac{5 \cdot 4 \cdot 3 \cdot 2 \cdot 1}{1 \cdot 1 \cdot 3 \cdot 2 \cdot 1} = 20.$$

Thus

$$\left| E_5^A \right| = (30 + 20) \cdot 3 = 50 \cdot 3 = 150.$$

And, for 7 rounds, the last must go to A, leaving us with the following possibilities for 6 rounds:

$$1A, 2B, 3C : \quad \frac{6!}{1!2!3!} = \frac{6 \cdot 5 \cdot 4 \cdot 3 \cdot 2 \cdot 1}{1 \cdot 2 \cdot 1 \cdot 3 \cdot 2 \cdot 1} = 60.$$

Thus

$$\begin{aligned}
\left| E^A \right| &= \left| E_2^A \right| + \left| E_3^A \right| + \left| E_4^A \right| + \left| E_5^A \right| + \left| E_6^A \right| + \left| E_7^A \right| \\
&= 243 + 324 + 324 + 252 + 150 + 60 \\
&= 1353.
\end{aligned}$$

A's fraction of the stake is thus

$$\frac{1353}{2187} \approx .6186556927 \approx 62\%.$$

The shares going to B and C are determined similarly. B can win in $3, 4, 5, 6$ or 7 rounds.

$$\left| E_3^B \right| = 1 \cdot 3^4 = 81,$$

since B must win the first 3 rounds and the last 4 can go to any of the three players. For 4 rounds the possibilities for the first 3 rounds are

$$2B, 1A : \quad \frac{3!}{2!1!} = \frac{3 \cdot 2 \cdot 1}{2 \cdot 1 \cdot 1} = 3$$

$$2B, 1C: \quad \frac{3!}{2!1!} = 3.$$

Thus

$$\left| E_4^B \right| = (3+3) \cdot 3^3 = 6 \cdot 3^3 = 162.$$

For 5 rounds, the possibilities for the first 4 rounds are

$$2B, 1A, 1C: \quad \frac{4!}{2!1!1!} = \frac{4 \cdot 3 \cdot 2 \cdot 1}{2 \cdot 1 \cdot 1 \cdot 1} = 12$$

$$2B, 0A, 2C: \quad \frac{4!}{2!0!2!} = \frac{4 \cdot 3 \cdot 2 \cdot 1}{2 \cdot 1 \cdot 1 \cdot 2 \cdot 1} = 6.$$

Thus

$$\left| E_5^B \right| = (12 + 6) \cdot 3^2 = 18 \cdot 3^2 = 162.$$

For 6 rounds, we have

$$2B, 1A, 2C: \quad \frac{5!}{2!1!2!} = \frac{5 \cdot 4 \cdot 3 \cdot 2 \cdot 1}{2 \cdot 1 \cdot 1 \cdot 2 \cdot 1} = 30$$

$$2B, 0A, 3C: \quad \frac{5!}{2!0!3!} = \frac{5 \cdot 4 \cdot 3 \cdot 2 \cdot 1}{2 \cdot 1 \cdot 1 \cdot 3 \cdot 2 \cdot 1} = 10.$$

Thus

$$\left| E_6^B \right| = 40 \cdot 3 = 120.$$

And, finally, for 7 rounds we must have

$$2B, 1A, 3C: \quad \frac{6!}{2!1!3!} = \frac{6 \cdot 5 \cdot 4 \cdot 3 \cdot 2 \cdot 1}{2 \cdot 1 \cdot 1 \cdot 3 \cdot 2 \cdot 1} = 60.$$

The final reckoning for B is thus

$$\left| E^B \right| = 81 + 162 + 162 + 120 + 60 = 585,$$

and B's fraction of the stakes is

$$\frac{585}{2187} \approx .2674897119 \approx 27\%.$$

For C, we can, of course, cheat and simply subtract,

$$\left| E^C \right| = 2187 - \left| E^A \right| - \left| E^B \right| = 2187 - 1353 - 585 = 249,$$

and conclude C's fraction of the stakes to be

$$\frac{249}{2187} \approx .1138545953 \approx 11\%.$$

4 Exercise. Calculate $\left| E^C \right|$ directly by the method applied to the other players. Note that C can win in $4, 5, 6$ or 7 rounds.

2 Urn Problems

Other than to refer to repositories of the ashes of cremated family members, one doesn't use the word "urn" too much these days—until one takes a probability course. Urn problems form a proud tradition in probability going all the way back to Christian Huyghens. I've only read the statements of the 14 propositions and 5 problems of his *De ratiociniis in aleæ ludo*, so I cannot definitively state that the word "urn" (or, rather, its Latin equivalent) does not occur in this work. I can state that it does not occur in the English translations of these propositions and problems. Nonetheless, 2 of the 5 problems are what today would be classified as urn problems.

The basic urn problem consists of an urn containing several balls of various colours. One draws a fixed number of balls from the urn in succession and asks for the probabilities of various compositions of the draws. The examples from Huyghens are his second and fourth problems:

> 2. Three players A, B, C take twelve balls, eight of which are black and four white. They play on the following conditions; they are to draw blindfold, and the first player to draw a white ball wins. A is to have the first turn, B the next, C the next, then A again, and so on. Determine the chances of the players.

> 4. Twelve balls are taken, eight of which are black and four are white. A plays with B and undertakes in drawing seven balls blindfold to obtain three white balls. Compare the chances of A and B.[3]

Both of these problems are already quite sophisticated. The basic urn problem is more simply stated:

> An urn contains 12 balls, 8 of which are black and 4 of which are white. 7 balls are drawn in succession from the urn. What is the probability that exactly 3 of them are white?

But for an ambiguity, all three problems are readily solved by the methods of the preceding section. That is, each problem has two readings, each reading of which can be solved by these methods.

As noted in Chapter 2, *De ratiociniis* was the only introduction to the theory of probability for half a century. Among those who followed at the end of this period was Jakob Bernoulli with his *Ars conjectandi*, the first part of which incorporated *De ratiociniis*. Where Huyghens had given the answers to his problems without showing his work, Bernoulli gave full solutions and, with respect to problems 2 and 4, noted the variant readings. For example, for problem 2 he offered three solutions under the assumptions

  i. the ball is replaced after each draw;
  ii. the ball is not replaced after a draw and there is a common pool of

---

[3] Isaac Todhunter, *A History of the Mathematical Theory of Probability From the Time of Pascal to That of Laplace*, Chelsea Publishing Company, New York, 1965, p. 25.

12 balls; and

iii. the ball is not replaced after a draw and each player has his own pool of 12 balls to draw from.

Nowadays one would rule out the third interpretation by stating explicitly that the players take turns drawing the balls from a single urn in which there are 8 black and 4 white balls. And one would state explicitly which interpretation i or ii is meant by declaring the balls to be drawn *with replacement* or *without replacement*, respectively.

The three problems listed can all be solved by applying the counting tools of the preceding section. Take, for example, the basic problem of determining the probability of drawing exactly 3 white balls in 7 draws from an urn containing 8 black and 4 white balls. Drawing without replacement, the sample space S consists of all combinations of 12 things taken 7 at a time, whence

$$|S| = \binom{12}{7} = \frac{12!}{7!5!} = \frac{12 \cdot 11 \cdot 10 \cdot 9 \cdot 8}{5 \cdot 4 \cdot 3 \cdot 2 \cdot 1} = 792.$$

The event

E_3: exactly 3 of the balls are white

occurs if one chooses 3 white balls from the 4 white balls and 4 black balls from the 8. Thus,

$$|E_3| = \binom{4}{3}\binom{8}{4} = 4 \cdot 70 = 280$$

and

$$P(E_3) = \frac{280}{792} = \frac{35}{99} = .\overline{35},$$

or just over $1/3$.

Alternatively, since we imagine the balls drawn in succession, we can take as our sample space the set S of all permutations of 12 objects taken 7 at a time. Thus

$$|S| = {}_{12}P_7 = 12 \cdot 11 \cdot 10 \cdot 9 \cdot 8 \cdot 7 \cdot 6$$

and the event is

E_3: those permutations containing exactly 3 white balls.

We can choose an element of E_3 by first choosing which of the three positions in the permutations are white (possible in ${}_7C_3$ ways), then choosing the white balls (possible in ${}_4P_3$ ways), and then choosing the black balls (possible in ${}_8P_4$) ways. Thus

$$|E_3| = {}_7C_3 \cdot {}_4P_3 \cdot {}_8P_4$$
$$= \frac{7 \cdot 6 \cdot 5}{3 \cdot 2 \cdot 1} \cdot 4 \cdot 3 \cdot 2 \cdot 8 \cdot 7 \cdot 6 \cdot 5$$
$$= 7 \cdot 5 \cdot 4 \cdot 3 \cdot 2 \cdot 8 \cdot 7 \cdot 6 \cdot 5$$

and

$$P(E_3) = \frac{7 \cdot 5 \cdot 4 \cdot 3 \cdot 2 \cdot 8 \cdot 7 \cdot 6 \cdot 5}{12 \cdot 11 \cdot 10 \cdot 9 \cdot 8 \cdot 7 \cdot 6}$$

$$= \frac{4 \cdot 3}{12} \cdot \frac{5 \cdot 2}{10} \cdot \frac{8 \cdot 7 \cdot 6}{8 \cdot 7 \cdot 6} \cdot \frac{7 \cdot 5}{11 \cdot 9} = \frac{35}{99},$$

as before.

Drawing with replacement we can take as our sample space S the collection of all sequences of 7 balls from our urn of 12. In this case

$$|S| = 12^7$$

and the event we are interested in is

E_3: those sequences containing exactly 3 white balls.

Now, we can choose an element in E_3 by first choosing which 3 draws we will be choosing white balls on (possible in $_7C_3$ ways), then choosing the white balls (possible in 4^3 ways), and finally choosing the 4 black balls (possible in 8^4 ways). Thus

$$|E_3| = \binom{7}{3} \cdot 4^3 \cdot 8^4 = \frac{7 \cdot 6 \cdot 5}{3 \cdot 2 \cdot 1} \cdot 4^3 8^4 = 35 \cdot 4^3 \cdot 8^4$$

and

$$P(E_3) = \frac{35 \cdot 4^3 8^4}{12^7} = \frac{35 \cdot 4^3 \cdot 4^4 \cdot 2^4}{3^7 \cdot 4^7} = \frac{35 \cdot 2^4}{3^7}$$

$$= \frac{35 \cdot 16}{2187} = \frac{560}{2187} \approx .256,$$

or just over $1/4$.

One can similarly define E_4, E_5, E_6, E_7 and obtain

$$P(E_3) + P(E_4) + P(E_5) + P(E_6) + P(E_7) \tag{6}$$

as the solution to problem 4, i.e., the second problem cited from Huyghens. In practice, one would cut down on the arithmetic by noting that one obtains 1 when adding $P(E_0), P(E_1), P(E_2)$ to (6), whence (6) is equal to

$$1 - P(E_0) - P(E_1) - P(E_2).$$

Problem 2 can also be solved using the tools of the first section of this chapter, but I think it will be easier if we introduce a few new tools. The first is a probabilistic Sequential Principle replacing the counting Sequential Principle. The key new concept is that of conditional probability, which will be introduced three sections from now, in section 5.

3 The Birthday Problem

Before introducing conditional probability, I should digress to discuss a couple of famous problems that conceptually, if not historically, belong at this place in our development. One is the Birthday Problem, a relatively late arrival on the probability scene that added nothing to the theory. Problems in mathematics usually become famous because of their difficulties. Either a solution is so complex that only the expert is first able to unravel the intricacies involved, or the solution, like that of the problem of points or even the Petersburg Problem, requires new ideas. The Birthday Problem, like Bertrand's Box Paradox with which we shall end this chapter, is famous for a different reason. The problem can be a bit tricky for the neophyte, but its solution is not very involved and requires nothing new either conceptually or by way of technique. It, like the promised Box Paradox, is famous for its pædagogical use, being a good example of solving a problem by considering the complementary event—and, like the Box Paradox, it is almost paradoxical in nature.

The Birthday Problem first saw the light of day in a paper[4] written by Richard von Mises (1883 - 1953) in the late 1930's. The problem was popularised in the 1950's by William Feller (1906 - 1970), whose textbook presentation of the problem[5], however, lacks flair. Ideally the presentation of the problem is as follows:

> The probability instructor walks into the classroom, notices there are, say, 30 students, and announces that probably at least two of them share a birthday. The students, finding this nonintuitive, express their doubts. So the instructor surveys the birthdays of his students and, sure enough, two of them are identical. The students gasp in astonishment.

Actually, this outcome is not guaranteed, but there is a better than even chance the instructor is right.

I suppose at this point a good expositor would pause to explain why it is so surprising that it should be probable there be a match in birthdays. I have tried several times, but in each draft, after observing that the probability of matching a given birthday in one trial is $1/365$, my second step is so questionable in nature as to render the surprise implausible rather than natural. So I shall simply state that, before one starts thinking about the problem, it just *seems* unlikely that there will be a match in so small a sample.

When one does start thinking about it, if one goes about it directly, one quickly runs into complications. This is one of those problems where it is easier to count the number of outcomes in the complementary event than in the event itself. The general form of the complementary problem is easy to state. I quote Feller[6]

> We consider random samples of size r *with replacement* taken from a population of the n elements $a_1, \ldots, a_n$. We are interested in the event

---

[4] Richard von Mises, "Ueber Aufteilungs- und Besetzungs-Wahrscheinlichkeiten", *Revue de la Faculté des Sciences de l'Université d'Istanbul*, N.S. vol. 4 (1938 - 1939), pp. 145 - 163.

[5] William Feller, *An Introduction to Probability Theory and Its Applications*, 2nd ed., John Wiley & Sons, Inc., New York, 1957, pp. 31 - 32.

[6] *Ibid.*, p. 30. Note that he writes $(n)_r$ for ${}_nP_r$. I have taken the liberty in this passage of renumbering the formulæ to match the numbering of the present chapter.

A that in such a sample $(a_{j_1}, \ldots, a_{j_r})$ no element appears twice, that is, that our sample could have been obtained also by sampling without replacement. The last theorem[7] shows that there exist n^r different samples in all, of which $(n)_r$ satisfy the stipulated condition. Assuming that all arrangements have equal probability, we conclude that *the probability of no repetition in our sample is*

$$p = \frac{(n)_r}{n^r} = \frac{n(n-1)\cdots(n-r+1)}{n^r}. \tag{7}$$

He follows this observation with several examples, including the Birthday Problem:

(*d*) *Birthdays.* The birthdays of *r* people form a sample of size *r* from the population of all days in the year. The years are not of equal length, and we know that the birth rates are not quite constant throughout the year. However, in a first approximation, we may take a random selection of people as equivalent to a random selection of birthdays and consider the year as consisting of 365 days.

With these conventions we can interpret equation (7) to the effect that *the probability, p, that all r birthdays are different is*

$$p = \frac{(365)_r}{365^r} = \left(1 - \frac{1}{365}\right)\left(1 - \frac{2}{365}\right)\cdots\left(1 - \frac{r-1}{365}\right). \tag{8}$$

Again the numerical consequences are astounding. Thus for $r = 23$ people we have $p < \frac{1}{2}$, that is, *for 23 people the probability that at least two people have a common birthday exceeds $\frac{1}{2}$.*

Formula (8) looks forbidding, but it is easy to derive good numerical approximations to *p*. If *r* is small, we can neglect all cross products and have in a crude approximation

$$p \approx 1 - \frac{1 + 2 + \ldots + (r-1)}{365} = 1 - \frac{r(r-1)}{730}. \tag{9}$$

For $r = 10$ the correct value is $p = 0.883\ldots$; equation (9) gives the approximation 0.877.

For larger *r* we obtain a much better approximation by passing to logarithms. For small positive *x* we have $\log(1-x) \approx -x$, and thus from (8)

$$\log p \approx -\frac{1 + 2 + \ldots + (r-1)}{365} = -\frac{r(r-1)}{730}. \tag{10}$$

For $r = 30$ this leads to the approximation 0.3037 whereas the correct value is $p = 0.294$. For $r \leq 40$ the error in (10) is less than 0.08.[8]

---

[7] This was a theorem counting the total numbers, n^r and ${}_nP_r$, of samples with and without replacement, respectively.

[8] William Feller, *op. cit.*, pp. 31 - 32. The logarithm denoted "log" by Feller is the natural logarithm, nowadays more commonly denoted by "ln", and not the logarithm to base 10, for which "log" is usually reserved.

I like the approximations (9) and (10), but stress that the former is only accurate for small r. For $r = 23$, (9) yields $p = .3068$, whence a probability of at least 2 birthdays in common of .6932, which is simply wrong, and for $r = 30$ the formula yields $p = -.19178$, which is not only wrong but absurd. The values for r are large enough, however, to apply (10). The estimate for $r = 23$ using (10) is .4999982478, i.e., a probability of approximately .5000017522 for a match. This is an approximation guaranteed by Feller to be within .08 of the actual probability of success, which probability is thus guaranteed to lie between .4200017522 and .5800017522, i.e., it is not a good enough approximation to verify his claim that "for 23 people the probability that at least two people have a common birthday exceeds $\frac{1}{2}$". To verify this we must calculate the probabilities more exactly. Nowadays the product is not so forbidding. We have but to enter

$(365 \ _nP_r \ 23)/365\char`\^23$

in our calculators to obtain .4927027657, and thus

$$1 - .4927 \approx .5073$$

as the sought for probability.

Actually, even in Feller's day, the calculation would not have been all that formidable. Tables of logarithms of factorials go back to the 18th century at least and one has

$$\log \frac{_nP_r}{n^r} = \log \ _nP_r - r \log n$$

$$= \log \frac{n!}{(n-r)!} - r \log n$$

$$= \log n! - \log(n-r)! - r \log n.$$

Thus, using such a table[9], for $n = 365$ and $r = 23$ we have

$$\log p = \log 365! - \log 342! - 23 \log 365$$

$$\approx 778.3997452 - 719.7744243 - 23(2.56229)$$

$$\approx -.3073491,$$

whence my calculator tells me $p = .4927775341$ and the sought for probability of $1 - .4927775341 = .5072224659$ is indeed greater than $1/2$.

4 *Treize* (13)

Kindred to the Birthday Problem and much discussed in the history of probability is the game of *treize* ("thirteen"), also called *rencontre* ("coincidence"). It

---

[9] Karl Pearson included a table of logarithms of $n!$ for $n = 1$ to 2000 in his famous statistical tables earlier in the century. Indeed, I quote from a book of tables (Herbert Arkin and Raymond R. Colton, *Tables for Statisticians*, 2nd ed., Barnes & Noble, Inc., New York, 1963) which reproduces the table from the third (1930) edition of Pearson's collection.

makes its first appearance in the literature in the *Essai d'Analyse sur les Jeux de Hazard* of Pierre Rémond de Montmort (1678 - 1719). Todhunter offers the following description of the work:

> There are two editions of Montmort's work; the first appeared in 1708; the second is sometimes said to have appeared in 1713, but the date 1714 is on the title page of my copy, which appears to have been a present to 'sGravesande from the author. Both editions are in quarto; the first contains 189 pages with a preface of XXIV pages, and the second contains 414 pages with a preface and advertisement of XLII pages. The increased bulk of the second edition arises, partly from the introduction of a treatise on combinations which occupies pages 1 - 72, and partly from the addition of a series of letters which passed between Montmort and Nicholas Bernoulli with one letter from John Bernoulli. The name of Montmort does not appear on the title page or in the work, except once on page 338, where it is used with respect to a place.[10]

An excellent chapter on Montmort is given as Chapter 14, "Pierre Rémond de Montmort and the 'Essai d'Analyse'", in F.N. David's history of probability[11]. I quote a little:

> After his marriage he settled down on his country estate and set himself to work on the theory of probability. Quite why he chose this topic is not known, for he was no gambler. It was known in France and known to Montmort, if only from the éloge of James Bernoulli, that James had, at the time of his death, left the manuscript of a book[12] on the subject... He himself says
>
>> Several of my friends have urged me for a long time to see if I could not determine by algebra what is the advantage of the Banker in the game of Pharaoh. I would not have dared undertake this research... if the success of M. Bernoulli had not incited me for more than two years to try to calculate the different chances in this game... This gave me the idea of getting to the bottom of this matter and the desire to make up to the Public in some fashion for the loss that it has sustained in being deprived of the excellent work of M. Bernoulli....[13]

---

[10] Todhunter, *op. cit.*, p. 79. The edition of Montmort available online at Bielefeld bears the date 1713 on the title page.

[11] F.N. David, *Games, Gods and Gambling; A History of Probability and Statistical Inference*, Charles Griffin & Co., Ltd., London, 1962 (reprinted by Dover Publications, Mineola (New York), 1998, pp. 140 - 160 of the Dover edition.

[12] I will have much to say about Jakob Bernoulli's posthumous book in Chapters 4 and 5.

[13] It is perhaps a superficial observation, but I am inclined to note the parallel with Huyghens and his reproduction of the work of Fermat and Pascal. David makes the footnoted remark at this point, "Since he had published a number of the scientific works of others at his own expense it is a little curious that he did not try to publish *l'excellent ouvrage de M. Bernoulli.*"

The results of his researches were published in the *Essai d' Analyse sur les Jeux de Hasard* printed in Paris in 1708. This, the first edition, often passed over because of the greater length of the second edition and the incorporation of the Montmort-Bernoulli correspondence, is worthy of comment. As I have noted, one does not know quite what started Montmort off, but it is known that he wrote it in the comparative isolation of his country estate and it represents therefore the attempts of a competent mathematician, given the outline of the problems discussed by James Bernoulli, to solve them using the latest mathematical techniques, and to extend the field of application of these techniques to card games. This first edition is Montmort himself speaking; the second edition, while much more comprehensive, is probably a mixture of Montmort and Nicholas Bernoulli. The fact that he wrote at all is probably a fortunate one for the probability calculus. That a nobleman of France and an ex-canon of Notre-Dame should find such problems worthy of speculation and not an impious study would give the subject a certain cachet and air of respectability which left lesser mortals free to work undisturbed.[14]

For half a century, the only book available on probability had been that of Huyghens, which presented 14 worked out problems and offered 5 more for the reader. Montmort's was the first of a new batch that offered general solutions to more general problems, and analysed various popular games of chance. One of these, first discussed by Montmort, is *treize*, a precursor to the modern childrens' card game *snap*. I quote Todhunter:

160. Montmort next discusses the game of *Treize*; this discusson occupies pages 130 - 143 [15]. The problem involved is one of considerable interest, which has maintained a permanent place in works on the Theory of Probability.
The following is the problem considered by Montmort.
Suppose that we have thirteen cards numbered 1, 2, 3 ... up to 13; and that these cards are thrown promiscuously into a bag. The cards are then drawn out singly; required the chance that, once at least, the number on a card shall coincide with the number expressing the order in which it is drawn.
. . .
162. The game which Montmort calls *Treize* has sometimes been called *Rencontre*. The problem which is here introduced for the first time has been generalised and discussed by the following writers: DeMoivre, *Doctrine of Chances*, pages 109 - 117. Euler, *Hist. de l'Acad. . . . Berlin*,

---

[14] David, *op. cit.*, pp. 142 - 143.
[15] Todhunter is referring here and below to the second edition of the *Essai*. Montmort's discusion, along with an extract of a letter from Nikolaus Bernoulli to Montmort giving an alternate proof, can be found in English translation in H.A. David and A.W.F. Edwards, eds., *Annotated Readings in the History of Statistics*, Springer-Verlag New York, Inc., New York, 2001, pp. 19 - 31.

for 1751. Lambert, *Nouveaux Mémoires de l'Acad. . . . Berlin*, for 1771. Laplace, *Théorie . . . des Prob.*, pages 217 - 225. Michaelis, *Mémoire sur la probabilité du jeu de rencontre*, Berlin, 1846.[16]

The reason for interest in the game is made wonderfully clear by Feller:

> *Matches (coincidences).* The following problem with many variants and a surprising solution goes back to Montmort (1708). It has been generalized by Laplace and many other authors.
>
> Two equivalent decks of N different cards each are put into random order and matched against each other. If a card occupies the same place in both decks, we speak of a *match (coincidence* or *rencontre).* Matches may occur at any of the N places and at several places simultaneously. This experiment may be described in more amusing forms. For example, the two decks may be represented by a set of N letters and their envelopes, and a capricious secretary may perform the random matching. Alternatively we may imagine the hats in a checkroom mixed and distributed at random to the guests. A match occurs if a person gets his own hat. It is instructive to venture guesses as to how the probability of a match depends on N: How does the probability of a match of hats in a diner with 8 guests compare with the corresponding probability at a gathering of 10,000 people? It seems surprising that the probability is practically independent of N and roughly $\frac{2}{3}$.[17]

The solution to the problem is not difficult, though Montmort's own solution may have been messy, for he did not include it in either edition of his book. Todhunter reports[18]

> 161. In his first edition Montmort did not give any demonstrations of his results; but in his second edition he gives two demonstrations which he had received from Nicolas Bernoulli; see his pages 301, 302. We will take the first of these demonstrations.
>
> Let $a, b, c, d, e, \ldots$ denote the cards, n in number. Then the number of possible cases is $n!$ [19] The number of cases in which a is first is $(n-1)!$ The number of cases in which b is second, but a not first, is $(n-1)!-(n-2)!$. The number of cases in which c is third, but a not first nor b second, is $(n-1)!-(n-2)!-\{(n-2)!-(n-3)!\}$, that is $(n-1)!-2(n-2)!+(n-3)!$. The number of cases in which d is fourth, but neither a, b, nor c in its proper place is $(n-1)!-2(n-2)!+(n-3)!-\{(n-2)!-2(n-3)!+(n-4)!\}$,

---

[16] Todhunter, *op. cit.*, pp. 91 - 93. An English translation of Bernoulli's letter, sandwiched between Montmort's observations and a facsimile reproduction of de Moivre's treatment of the problem can be found in H.A. David and A.W.F. Edwards, eds., *Annotated Readings in the History of Statistics*, Springer-Verlag New York, Inc., New York, 2001.

[17] Feller, *op. cit.*, p. 90

[18] Todhunter, *op. cit.*, pp. 91 - 93.

[19] I have replaced Todhunter's typographically inconvenient archaic notation for the factorial by the modern, easy to typeset use of the exclamation mark. *Cf.* footnote 29 of page 131, below, for more information.

that is $(n-1)!-3(n-2)!+3(n-3)!+(n-4)!$ And generally the number of cases in which the mth card is in its proper place, while none of its predecessors is in its proper place, is[20]

$$(n-1)! - (m-1)(n-2)! + \frac{(m-1)(m-2)}{1.2}(n-3)!$$
$$- \frac{(m-1)(m-2)(m-3)}{3!}(n-4)! + \ldots\ldots + (-1)^{m-1}(n-m)!$$

We may supply a step here in the process of Nicolas Bernoulli by shewing the truth of this result by induction. Let $\psi(m,n)$ denote the number of cases in which the mth card is the first that occurs in its right place; we have to trace the connexion between $\psi(m,n)$ and $\psi(m+1,n)$. The number of cases in which the $(m+1)^{th}$ card is in its right place while none of the cards between b and the mth card, both inclusive, is in its right place is $\psi(m,n)$. From this number we must reject all those cases in which a is in its right place, and thus we shall obtain $\psi(m+1,n)$. The cases to be rejected are in number $\psi(m,n-1)$. Thus

$$\psi(m+1,n) = \psi(m,n) - \psi(m,n-1).$$

Hence we can shew that the form assigned by Nicolas Bernoulli to $\psi(m,n)$ is universally true.

Thus if a person undertakes that the mth card shall be the first that is in its right place, the number of cases favourable to him is $\psi(m,n)$, and therefore his chance is $\dfrac{\psi(m,n)}{n!}$.

If he undertakes that at least one card shall be in its right place we obtain the number of favourable cases by summing $\psi(m,n)$ for all values of m from 1 to n both inclusive: the chance is found by dividing this sum by $n!$

Hence we shall obtain for the chance that at least one card is in its right place,

$$1 - \frac{1}{2} + \frac{1}{3!} - \frac{1}{4!} + \ldots + \frac{(-1)^{n-1}}{n!}.$$

We may observe that if we subtract the last expression from unity we obtain the chance that no card is in its right place. Hence if $\phi(n)$ denote the number of cases in which no card is in its right place, we obtain

$$\phi(n) = n!\left\{\frac{1}{2} - \frac{1}{3!} + \frac{1}{4!} + \ldots - \frac{(-1)^{n-1}}{n!}\right\}.$$

Modern expositions proceed slightly differently, first proving a generalisation of the Addition Principle covering overlapping sets. For two sets, X,Y, this is easy: $|X|+|Y|$ counts all the elements in X and in Y and adds them together, each element in the intersection being counted twice. Hence

---

[20] The figure "1.2" would now be written "$1\cdot2$" or simply "2!".

$$|X| + |Y| = |X \cup Y| + |X \cap Y|,$$

i.e.,

$$|X \cup Y| = |X| + |Y| - |X \cap Y|.$$

For three sets,

$$
\begin{aligned}
|X \cup Y \cup Z| &= |X \cup Y| + |Z| - |(X \cup Y) \cap Z| \\
&= |X| + |Y| - |X \cap Y| + |Z| - |(X \cap Z) \cup (Y \cap Z)| \\
&= |X| + |Y| + |Z| - |X \cap Y| - \big(|X \cap Z| + |Y \cap Z| - |X \cap Y \cap Z|\big) \\
&= |X| + |Y| + |Z| - \big(|X \cap Y| + |X \cap Z| + Y \cap Z|\big) + |X \cap Y \cap Z|.
\end{aligned}
$$

In general, $|X_0 \cup X_1 \cup \ldots \cup X_{n-1}|$ will equal the sum of the cardinalities of the X_i's, minus the cardinalities of the pairwise intersections of X_i's and X_j's (for $i \neq j$), plus the sum of the cardinalities of the triple intersections of X_i's, X_j's, and X_k's (i, j, k all distinct), minus...

To give a formal statement and proof of the general result, let us agree on some notation. We are given a sequence $X_0, X_1, \ldots, X_{n-1}$ of sets and, for $i < j < k < \ldots$, we define

$$
\begin{aligned}
c_i &= |X_i| \\
c_{ij} &= |X_i \cap X_j| \\
c_{ijk} &= |X_i \cap X_j \cap X_k|
\end{aligned}
$$

etc., and

$$
S_1 = \sum_{i<n} c_i
$$

$$
S_2 = \sum_{i<j<n} c_{ij}
$$

$$
\vdots
$$

$$
S_n = c_{01\ldots n-1}.
$$

The following theorem often goes by the name, *The Inclusion-Exclusion Principle*.

5 Theorem (Addition Principle for Overlapping Sets). *Let $X_0, \ldots, X_{n-1}$ be $n \geq 2$ finite sets.*

$$|X_0 \cup \ldots \cup X_{n-1}| = S_1 - S_2 + S_3 - \ldots \pm S_n.$$

Proof. By induction on $n \geq 2$.
The basis has already been proven:

$$|X_0 \cup X_1| = |X_0| + |X_1| - |X_0 \cap X_1| = c_0 + c_1 - c_{01} = S_1 - S_2.$$

Assume the result true for any collection of $n \geq 2$ sets and let $n + 1$ sets $X_0, \ldots, X_{n-1}, X_n$ be given.

$$|X_0 \cup \ldots \cup X_{n-1} \cup X_n| = |X_0 \cup \ldots \cup X_{n-1}| + |X_n| - |(X_0 \cup \ldots \cup X_{n-1}) \cap X_n|. \quad (11)$$

By induction hypothesis,

$$|X_0 \cup \ldots \cup X_{n-1}| = \sum c_i - \sum c_{ij} + \sum c_{ijk} - \ldots \quad (12)$$

all $i < j < k < \ldots < n$. But

$$(X_0 \cup \ldots \cup X_{n-1}) \cap X_n = (X_0 \cap X_n) \cup \ldots \cup (X_{n-1} \cap X_n)$$

is also a union of n sets and the induction hypothesis applies:

$$|(X_0 \cap X_n) \cup \ldots \cup (X_{n-1} \cap X_n)| = \sum c_{in} - \sum c_{ijn} + \sum c_{ijkn} - \ldots \quad (13)$$

and combining (11) - (13) yields

$$
\begin{aligned}
|X_0 \cup \ldots \cup X_n| &= \left(\sum_{i<n} c_i - \sum_{i<j<n} c_{ij} + \ldots \right) + c_n - \left(\sum c_{in} - \sum c_{ijn} + \ldots \right) \\
&= \sum_{i<n+1} c_i - \sum_{i<j<n+1} c_{ij} + \sum_{i<j<k<n+1} c_{ijk} - \ldots \\
&= S_1 - S_2 + S_3 - \ldots,
\end{aligned}
$$

as was to be shown.   □

In Feller's textbook[21] we read a different proof. Given $X_0, \ldots, X_{n-1}$, consider an element x of $X_0 \cup \ldots \cup X_{n-1}$ and the sum

$$S_1 - S_2 + S_3 - \ldots \pm S_n.$$

This sum counts inclusions and exclusions. If x occurs in m of the sets $X_0, \ldots, X_{n-1}$, it is counted m times in S_1, $\binom{m}{2}$ times in adding up S_2, $\binom{m}{3}$ times in adding up S_3, etc. Thus it is counted

$$m - \binom{m}{2} + \binom{m}{3} - \ldots \pm \binom{m}{m} = \binom{m}{1} - \binom{m}{2} + \binom{m}{3} - \ldots \pm \binom{m}{m}$$

times in forming the sum $S_1 - S_2 + S_3 - \ldots \pm S_n$. But

$$
\begin{aligned}
\binom{m}{1} - \binom{m}{2} &+ \binom{m}{3} - \ldots \pm \binom{m}{m} \\
&= 1 - \binom{m}{0} + \binom{m}{1} - \binom{m}{2} + \binom{m}{3} - \ldots \pm \binom{m}{m} \\
&= 1 - \left(\binom{m}{0} - \binom{m}{1} + \binom{m}{2} - \binom{m}{3} + \ldots \mp \binom{m}{m} \right) \\
&= 1 - (1-1)^m = 1 - 0 = 1.
\end{aligned}
$$

---

[21] Feller, *op. cit.*, pp. 89 - 90.

Thus $S_1 - S_2 + S_3 - \ldots \pm S_n$ contains a net contribution of 1 for each element of the union, and the cardinality of the union equals $S_1 - S_2 + S_3 - \ldots \pm S_n$.

Before applying this to the problem of *treize*, it might be a good idea to rephrase what we have done in probabilistic terms. First we consider the Addition Principle for Probabilities where there is no overlap. To make a formal statement of the Principle, I pause first to note that probability theory predates set theory, and has its own terminology. Thus, if two events A, B are *disjoint*, i.e., if they have no elements in common when considered as sets of outcomes, one does not say they are disjoint, but that they *mutually exclude* each other, or that they are *mutually exclusive*. A collection of events $A_0, A_1, \ldots, A_{k-1}$ is *mutually exclusive* if each pair A_i, A_j for $i \neq j$ is mutually exclusive, i.e., if each pair is disjoint.

6 Theorem (Addition Principle for Probabilities). *Let the events $A_0, A_1, \ldots, A_{n-1}$ be mutually exclusive. Then*

$$P(A_0 \cup A_1 \cup \ldots \cup A_{n-1}) = P(A_0) + P(A_1) + \ldots + P(A_{n-1}). \qquad (14)$$

Again, one may find pre-set-theoretic notation used and in place of $A_0 \cup A_1 \cup \ldots \cup A_{n-1}$ one may find A_0 *or* A_1 *or* $\ldots$ *or* A_{n-1}.

When the events overlap, the sum on the right in (14) is too large and, as with the Addition Principle for Overlapping Sets, has to be adjusted. To this end, define, for $0 < i < j < k < \ldots \leq n$

$$p_i = P(A_i) \qquad\qquad S_1' = \sum_i p_i$$

$$p_{ij} = P(A_i \& A_j) \qquad\qquad S_2' = \sum_{i<j} p_{ij}$$

$$p_{ijk} = P(A_i \& A_j \& A_k) \qquad\qquad S_3' = \sum_{i<j<k} p_{ijk}$$

$$\vdots$$

7 Theorem (Additive Principle for Probabilities; General Case). *Let events $A_0, A_1, \ldots, A_{n-1}$ be given. Then*

$$P(A_0 \cup A_1 \cup \ldots \cup A_{n-1}) = S_1' - S_2' + S_3' - \ldots \pm S_n'. \qquad (15)$$

Theorem 7 follows readily from Theorem 5 by noting that, for $X_i = A_i$ and $c_i, c_{ij}, \ldots, S_i$ as before, if N denotes the cardinality of the sample space, then

$$p_i = \frac{c_i}{N}, \quad p_{ij} = \frac{c_{ij}}{N}, \quad p_{ijk} = \frac{c_{ijk}}{N}, \quad \ldots$$

$$S_1' = \frac{S_1}{N}, \quad S_2' = \frac{S_2}{N}, \quad \ldots$$

and

$$P(A_0 \cup A_1 \cup \ldots \cup A_{n-1}) = \frac{|X_0 \cup X_1 \cup \ldots \cup X_{n-1}|}{N}.$$

And (14), and thus Theorem 6, follows from (15) by noting that $S_2' = S_3' = \ldots = S_n' = 0$ when the events are mutually exclusive.

Getting back to the problem at hand, the game of *treize*, we can apply either Theorem 5 or Theorem 7. Since we went through all the trouble to prove it, let us use the latter. Let A_i be the event that the $(i+1)$-th cards of the two decks match. Keeping the first deck in its natural order, we take the sample space to be the set of all reshufflings of the second deck, i.e., the set of all permutations of the n cards of the second deck—of which there are $N = {}_nP_n = n!$ such. Thus, the cardinality of the sample space is $n!$.

Now p_i is the probability of a match in the $(i+1)$-th card. How many such permutations are there? Well, in choosing a permutation of $\{1, \ldots, n\}$, there is only one choice of where to place $i+1$ in the order, and ${}_{n-1}P_{n-1} = (n-1)!$ choices for the other $n-1$ cards. Thus

$$p_i = \frac{(n-1)!}{n!}.$$

Similarly, in calculating p_{ij}, the positions of cards $i+1, j+1$ are determined and there are ${}_{n-2}P_{n-2} = (n-2)!$ arrangements of the remaining cards. Thus

$$p_{ij} = \frac{(n-2)!}{n!}.$$

In general,

$$p_{i_0 \ldots i_{k-1}} = \frac{(n-k)!}{n!}.$$

Now these values are independent of the choices of i, j, etc., whence each S_k' is the sum of the constants $(n-k)!/k!$. There are ${}_nC_k$ such $i_0 < i_1 < \ldots < i_{k-1} < n$. Thus

$$S_k' = \frac{n!}{k!(n-k)!} \cdot \frac{(n-k)!}{n!} = \frac{1}{k!},$$

and

$$P(\text{at least one match}) = P(A_0 \cup A_1 \cup \ldots \cup A_{n-1})$$
$$= 1 - \frac{1}{2!} + \frac{1}{3!} - \ldots \pm \frac{1}{n!},$$

as promised.

After presenting this proof, Feller gives a small table of values to demonstrate how close these values are to each other for the various values of n. Rather than print his table here, I offer the following program for producing one's own table in the form of two lists $L_1 = \{1, 2, \ldots, n\}$ and L_2 being the list of probabilities.

```
PROGRAM:TRZTABLE
:Disp "PICK A NUMBER"
:Input "N=",N
:seq(X,X,1,N)→L₁
```

```
:N→dim(L₂)
:For(I,1,N)
:I→dim(L₃)
:1→L₃(1)
:For(J,2,I)
:−L₃(J−1)/J→L₃(J)
:End
:sum(L₃)→L₂(I)
:End
:SetUpEditor L₁,L₂
:DelVar I
:DelVar J
:DelVar N
:ClrList L₃ .
```

If you run this program and then open the List Editor, you will see a table of probabilities rapidly converging to .6321205588.

The actual limit is easily calculated by one who has taken a course in the Calculus, wherein is developed the power series expansion for e^x:

$$e^x = 1 + \frac{x}{1!} + \frac{x^2}{2!} + \frac{x^3}{3!} + \dots$$

Plugging -1 in for x, this specialises to

$$e^{-1} = 1 - 1 + \frac{1}{2!} - \frac{1}{3!} + \frac{1}{4!} - \dots,$$

whence

$$e^{-1} - 1 = -1 + \frac{1}{2!} - \frac{1}{3!} + \frac{1}{4!} - \dots,$$

and

$$1 - \frac{1}{e} = 1 - \frac{1}{2!} + \frac{1}{3!} - \frac{1}{4!} + \dots$$

$$= \lim_{n \to \infty} \left(1 - \frac{1}{2!} + \frac{1}{3!} - \dots \pm \frac{1}{n!} \right).$$

For those who have not taken a course in the Calculus, I note that $e \approx$ 2.718281828459045 is the base of the natural logarithm ln and there is the LN/e^x button evaluating the logarithmic and exponential functions for this base. And, for those not wanting to have to enter[22] e^1 to obtain e, the division button doubles as a button to recall this special constant, which will figure prominently in the next chapter.

---

[22] More accurately put, instead of entering $e^\wedge(1)$ by pressing e^x, 1, and), one would enter e by pressing the e button.

5 Conditional Probability

Suppose we have an urn containing the customary 8 black and 4 white balls, and we ask for the probability that the second ball is white when we draw two balls without replacement. Our sample space S consists of permutations of 12 things taken 2 at a time, whence

$$|S| = 12 \cdot 11 = 132.$$

The event E we are looking for can be represented

$$E = WW \cup BW,$$

i.e., the union of the events: white followed by white and black followed by white. Clearly

$$|WW| = 4 \cdot 3 = 12, \quad |BW| = 8 \cdot 4 = 32,$$

whence $|E| = 12 + 32 = 44$ and

$$P(E) = \frac{44}{132} = \frac{1}{3}.$$

[More simply, we could argue that if we choose the second ball first, we find the probability of drawing a white ball to be $4/12 = 1/3$. While some may find this obvious, others may view the argument with suspicion.]

Now suppose we know that the first ball drawn is white. What is the probability of the second ball also being white? This is called the *conditional probability*: the probability of the second ball being white *given that* the first ball was white. If we write

W_1: the first ball is white
W_2: the second ball is white,

we would write

$P(W_2|W_1)$: the probability of W_2 given that W_1 occurs.

And we would determine it by declaring W_1 to be our new sample space:

$$W_1 = \{\text{permutations in } S \text{ for which the first ball is white}\}.$$

And our event $E = W_2|W_1$ would be those permutations in W_1 in which the second ball is white. Now

$$|W_1| = 4 \cdot 11 = 44,$$

since there are 4 ways of choosing a white ball and 11 of choosing any other ball after it. And

$$|E| = {}_4P_2 = 4 \cdot 3 = 12,$$

whence

$$P(E) = \frac{12}{44} = \frac{3}{11}.$$

Notice that E is just $W_1 \cap W_2$, whence

$$P(W_2|W_1) = P(E) = \frac{|W_1 \cap W_2|}{|W_1|} = \frac{|W_1 \cap W_2|/|S|}{|W_1|/|S|} = \frac{P(W_1 \cap W_2)}{P(W_1)}.$$

In general, for events A, B, we have

$$P(B|A) = \frac{P(A \cap B)}{P(A)}, \tag{16}$$

whence

$$P(A \cap B) = P(A) \cdot P(B|A). \tag{17}$$

Formula (17) is useful in that $P(A)$ and $P(B|A)$ may be easier to compute directly than $P(A \cap B)$. In our urn problem, with $A = W_1, B = W_2$, we can reason as follows:

$$P(A) = P(W_1) = \frac{4}{12} = \frac{1}{3}$$

since in making our first choice there are 4 white balls out of 12 to choose from, and

$$P(B|A) = P(W_2|W_1) = \frac{3}{11},$$

since, having removed one white ball from the urn, there are 3 white balls out of 11 in our second choice. This yields

$$P(\text{both balls are white}) = P(A \cap B) = \frac{1}{3} \cdot \frac{3}{11} = \frac{1}{11}.$$

For multiple draws from an urn, we can generalise (17):

$$P(A \cap B \cap C) = P(A \cap B) \cdot P(C|A \cap B)$$
$$= P(A) \cdot P(B|A) \cdot P(C|A \cap B)$$
$$P(A \cap B \cap C \cap D) = P(A \cap B \cap C) \cdot P(D|A \cap B \cap C)$$
$$= P(A) \cdot P(B|A) \cdot P(C|A \cap B) \cdot P(D|A \cap B \cap C),$$

etc. Thinking of events as occurrences instead of as sets of outcomes, we might write & in place of $\cap$ and state the general principle as follows.

8 Theorem (Sequential Principle for Probabilities). *If we have a sequence $T_0, T_1, \ldots, T_{k-1}$ of tasks to perform in succession and $A_0, A_1, \ldots, A_{k-1}$ are events that can occur as the result of performing the respective tasks, then the probability of the successive occurrences of $A_0, A_1, \ldots, A_{k-1}$ is*

$$P(A_0 \& A_1 \& \ldots \& A_{k-1}) =$$
$$P(A_0) \cdot P(A_1|A_0) \cdot P(A_2|A_0 \& A_1) \cdots P(A_{k-1}|A_0 \& \ldots \& A_{k-2}).$$

9 Example. Let an urn contain 8 black and 4 white balls. Three balls are drawn in succession without replacement. Let

W_i: a white ball is drawn on the i-th draw
B_i: a black ball is drawn on the i-th draw.

We want to know, say, the probability of drawing white, then black, then white again, i.e., $P(W_1 \& B_2 \& W_3)$. By the Sequential Principle,

$$P(W_1 \& B_2 \& W_3) = P(W_1)P(B_2|W_1)P(W_3|W_1 \& B_2).$$

Now, on the first draw the urn contains 4 white balls out of 12, whence

$$P(W_1) = \frac{4}{12} = \frac{1}{3}.$$

If a white ball is chosen, there are now 11 balls in the urn, of which 3 are white and 8 are black. This means

$$P(B_2|W_1) = \frac{8}{11}.$$

Finally, if a black ball has been chosen on the second try, there are now 3 white and 7 black balls in the urn, 10 balls in all. Thus,

$$P(W_2|W_1 \& B_2) = \frac{3}{10}.$$

Putting it all together,

$$P(W_1 \& B_2 \& W_3) = \frac{1}{3} \cdot \frac{8}{11} \cdot \frac{3}{10} = \frac{4}{11 \cdot 5} = \frac{4}{55}.$$

This example was straightforward. More complicated examples are easily stated:

10 Example. An urn contains 8 black and 4 white balls. Three balls are drawn in succession without replacement. What are the probabilities of

   i. drawing white balls on the first and third draws;
  ii. drawing exactly 2 white balls; or
 iii. drawing at least 2 white balls?

Here it might be simpler if we go back to our notation of spelling words in the $\{W, B\}$-alphabet. Example 9 asked for the probability of spelling WBW (we can imagine, instead of balls, drawing lettered tiles without replacement). Example 10.i asks for words W_W, where either a W or a B can fill in the blank, thus it asks for words WBW or WWW. Example 10.ii asks for words with exactly 2 W's occurring in them, thus WWB, WBW, or BWW; and 10.iii asks for 2 or 3 W's, thus WWB, WBW, BWW, or WWW.

These probabilities are not too difficult to find. The events of Example 10 decompose into unions of non-overlapping sequential events and the probabilistic version of the Addition Principle allows us to add the individual probabilities:

  10.i: $P(W\_W) = P(WBW) + P(WWW)$
  10.ii: $P(2W) = P(WWB) + P(WBW) + P(BWW)$
  10.iii: $P(\geq 2W) = P(2W) + P(WWW),$

using the obvious abbreviations.

We can simplify the presentation of the solution of problems like this graphically by using a tree representation of the sample space. For Example 10, say, we can represent the first choice B or W by a simple branching from a starting node to a B and a W node; similarly, we branch to B and W from each of these nodes; and once again from the resulting nodes. If we place the various conditional probabilities next to the connecting lines, and multiply across each path, we should get the tree depicted in *Figure 1*, above. Multiplying across a

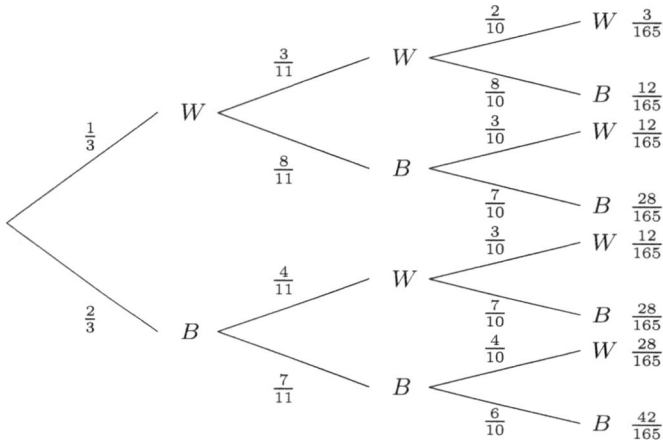

Fig. 1.

path is simply applying the Sequential Principle. For example,

$$P(WWW) = P(W)P(WW|W)P(WWW|WW) = \frac{1}{3} \cdot \frac{3}{11} \cdot \frac{2}{10} = \frac{3}{165}.$$

I haven't reduced these fractions to lowest terms, but have written them all with their least common denominator as the various parts of Example 10 require us, by the Addition Principle, to add some of these:

10.i: $P(W\_W) = P(WWW) + P(WBW) = \frac{3}{165} + \frac{12}{165} = \frac{15}{165} = \frac{1}{11}$

10.ii: $P(2W) = P(WWB) + P(WBW) + P(BWW)$
$$= \frac{12}{165} + \frac{12}{165} + \frac{12}{165} = \frac{36}{165} = \frac{12}{55}$$

10.iii: $P(\geq 2W) = P(WWW) + P(WWB) + P(WBW) + P(BWW)$
$$= \frac{3}{165} + \frac{12}{165} + \frac{12}{165} + \frac{12}{165} = \frac{39}{165} = \frac{13}{55}.$$

We, of course, do not need to draw the trees, but they organise the information clearly and are a handy way of guaranteeing we don't overlook any possibilities. The only drawback to the use of trees is they can grow quite large.[23] We do not always need to fill in the entire tree. For example, with

---

[23] And they are a pain to typeset.

problem 2 of Huyghens we can stop each branch as soon as a white ball is drawn. We thus get a tree as in *Figure 2*, below, which I present in a flattened form to save space[24]. Determining the probabilities that the various players

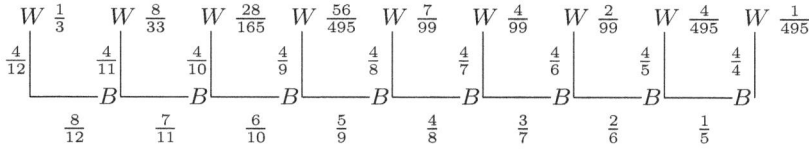

$$W\ \tfrac{1}{3} \qquad W\ \tfrac{8}{33} \qquad W\ \tfrac{28}{165} \qquad W\ \tfrac{56}{495} \qquad W\ \tfrac{7}{99} \qquad W\ \tfrac{4}{99} \qquad W\ \tfrac{2}{99} \qquad W\ \tfrac{4}{495} \qquad W\ \tfrac{1}{495}$$

$$\tfrac{4}{12}\quad B\ \tfrac{8}{12} \quad \tfrac{4}{11}\quad B\ \tfrac{7}{11} \quad \tfrac{4}{10}\quad B\ \tfrac{6}{10} \quad \tfrac{4}{9}\quad B\ \tfrac{5}{9} \quad \tfrac{4}{8}\quad B\ \tfrac{4}{8} \quad \tfrac{4}{7}\quad B\ \tfrac{3}{7} \quad \tfrac{4}{6}\quad B\ \tfrac{2}{6} \quad \tfrac{4}{5}\quad B\ \tfrac{1}{5} \quad \tfrac{4}{4}\quad B$$

Fig. 2.

will win is now a simple matter of adding the appropriate probabilities. A wins if the white ball occurs on draw 1, 4, or 7, whence

$$P(A) = \frac{1}{3} + \frac{56}{495} + \frac{2}{99} = \frac{165 + 56 + 10}{495} = \frac{231}{495} = \frac{7}{15} = .4\overline{6}.$$

Player B wins on draws 2, 5, or 8 and C wins on 3, 6, or 9:

$$P(B) = \frac{8}{33} + \frac{7}{99} + \frac{4}{495} = \frac{120 + 35 + 4}{495} = \frac{159}{495} = \frac{53}{165} = .3\overline{21}$$

$$P(C) = \frac{28}{165} + \frac{4}{99} + \frac{1}{495} = \frac{84 + 20 + 1}{495} = \frac{105}{495} = \frac{7}{33} = .\overline{21}.$$

Were we to allow replacement of the balls, the tree would be infinite as there is no limit to the number of black balls that could be drawn before a white ball appears. But the individual probabilities would be simpler: one would always draw a white ball with probability 1/3 and a black one with probability 2/3. The probability of drawing k black balls and then a white one is thus

$$\left(\frac{2}{3}\right)^k \frac{1}{3}.$$

A wins if the white ball appears on one of the draws 1, 4, 7, 10, ... and thus

$$P(A) = \frac{1}{3} + \left(\frac{2}{3}\right)^3 \frac{1}{3} + \left(\frac{2}{3}\right)^6 \frac{1}{3} + \left(\frac{2}{3}\right)^9 \frac{1}{3} + \cdots$$

$$= \frac{1}{3}\left(1 + \frac{8}{27} + \left(\frac{8}{27}\right)^2 + \left(\frac{8}{27}\right)^3 + \cdots\right)$$

$$= \frac{1}{3} \cdot \frac{1}{1 - 8/27} = \frac{1}{3} \cdot \frac{27}{27 - 8} = \frac{9}{19}.$$

And similarly for B and C:

$$P(B) = \frac{2}{3} \cdot \frac{1}{3} + \left(\frac{2}{3}\right)^4 \cdot \frac{1}{3} + \left(\frac{2}{3}\right)^7 \cdot \frac{1}{3} + \cdots$$

---

[24] And to make the typesetting easier.

$$= \frac{2}{3}P(A) = \frac{2}{3} \cdot \frac{9}{19} = \frac{6}{19}$$

$$P(C) = \left(\frac{2}{3}\right)^2 \cdot \frac{1}{3} + \left(\frac{2}{3}\right)^5 \cdot \frac{1}{3} + \left(\frac{2}{3}\right)^8 \cdot \frac{1}{3} + \cdots$$

$$= \frac{4}{9}P(A) = \frac{4}{9} \cdot \frac{9}{19} = \frac{4}{19}.$$

6 A Problem of Cardano

In his *Practica arithmetice* Girolamo Cardano stated the following problem:

17. A poor man went every day to the house of a rich man in order to play for a gold piece. When the poor man lost his gold piece he left the game; but when he won he would play again. Each would stake as much as the poor man possessed up to 4 plays and thereupon they would stop. Thus, for example, the rich man would stake one gold piece in the first play; if he won the game would end for the day; if he lost, the poor man possessed 2 gold pieces, whence the rich man would stake 2 gold pieces in the second play; if he won the playing would end at this point; if he lost the poor man possessed 4 gold pieces, whence the rich man also staked 4 gold pieces; correspondingly he staked 8 in the fourth play. Thus, if the rich man won, the poor man lost the 7 already won and one of his own gold pieces; if he had won, then he would have carried away 16 gold pieces, namely the 15 won thereby. It is asked: If the playing in this fashion were continued over several months under the assumption that luck and skill at play are equally distributed, who plays under more favourable conditions and by what percentage?[25]

Nowadays this is a fairly straightforward problem. The rich man has more opportunities for winning, but when he loses he can potentially lose more. We balance this out by calculating the expected values of each of the players. I refer to *Figure 3*, below. Here P denotes the event that the poor man wins and

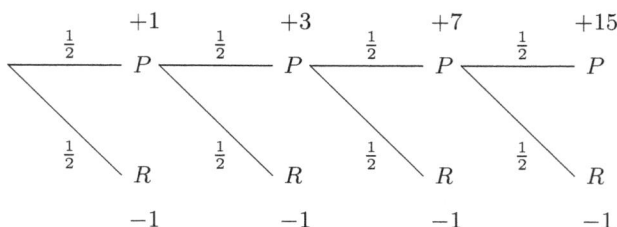

Fig. 3.

[25] Ivo Schneider, ed., *Die Entwicklung der Wahrscheinlichkeitstheorie von den Anfängen bis 1933; Einführungen und Texte*, Wissenschaftliche Buchgesellschaft, Darmstadt, 1989, p. 488.

R that the rich man does. The numbers above the P's denote the poor man's net win up to and including the given point, and the numbers (-1) below the R's his loss. His expected value is

$$(-1)P(R) + (-1)P(PR) + (-1)P(PPR) + (-1)P(PPPR) + 15P(PPPP)$$
$$= -1 \cdot \frac{1}{2} - 1 \cdot \frac{1}{4} - 1 \cdot \frac{1}{8} - 1 \cdot \frac{1}{16} + 15 \cdot \frac{1}{16}$$
$$= -\frac{8}{16} - \frac{4}{16} - \frac{2}{16} - \frac{1}{16} + \frac{15}{16} = 0.$$

Likewise, simply changing the signs of the amounts won, the above equations show the expected value of the rich man also to be 0. The game is evidently a fair one, allowing no advantage to either player.

We may view this problem as a nice example of the use of trees or of the concept of expected value, or as an introduction to the problems to which it is a precursor, an example of which will be given in the next section. But it also serves another purpose, as a further illustration of the truth of Hacking's remark (p. 4, above) that prior to Pascal and Fermat those who considered them "were quite unable to solve the new problems". For, indeed, Cardano cannot do it:

> The answer is clear. The progression from 4 is 10; thus the rich man would only need to stake 10 gold pieces; but he loses 15. Thus the rich man plays under less favourable conditions than the poor man, and because 5 is half of 10, it follows that his conditions are 50 per cent worse. By a continuation the poor man will therefore win much, so that in a year he will win 182 gold pieces, half the stakes.[26]

I don't know what to say to this. The sum of the arithmetic progression $1+2+3+4$ is indeed 10. It looks like he is miscalculating the chances of winning, as he did with the problem of points elsewhere in the *Practica arithmetice*. Since the rich man wins the day when the poor man loses a play, he has 1 chance if there is only one play, 2 if two plays are made, etc., so a total of 10 chances to win. The poor man must win all 4 plays to go home a winner, i.e., he has only 1 chance. Thus the odds in favour of the rich man winning the day are 10 to 1 and he should put up 10 times as much as the poor man to make things fair. But he might lose 15 gold pieces... But this calculation of the odds is wrong. The odds in favour of the rich man winning the day are 15 to 1, not 10 to 1. He should put up 15 gold pieces to the poor man's 1 piece.

If we accept Cardano's odds as 10 to 1, the poor man would have the expectation on a given day of

$$E = \frac{1}{11} \cdot 15 - \frac{10}{11} \cdot 1 = \frac{5}{11},$$

and thus should expect in a year's time to have won

$$365 \cdot \frac{5}{11} = 165 \frac{10}{11},$$

---

[26] *Ibid.*

gold pieces, not 182 such. The notion of expected value had not yet emerged in Cardano's day and we should not necessarily expect him to come up with the same final figure we would, even granting him his calculation of the odds. But I wonder how he comes up with the figure 182. It is approximately half of 365, the number of days in a year and he does mention that 5 is 50 per cent of 10, but I fail to see any connexion.

7 Gambler's Ruin

Cardano's problem, with the doubling of the rich man's stakes, bears a family resemblance to the later Petersburg Problem. But the problem it most closely resembles is the *Gambler's Ruin* insofar as the rich man and poor man quit their play each day when the poor man runs out of money. In the Gambler's Ruin, however, they don't arbitrarily stop after four games, and they play until one or other of the players runs out of money—or, in the formulation of Huyghens, counters[27]. The fifth problem of the *Ratiociniis* reads as follows:

> (5) *A* and *B* take each twelve counters and play with three dice on this condition, that if eleven is thrown *A* gives a counter to *B*, and if fourteen is thrown *B* gives a counter to *A*; and he wins the game who first obtains all the counters. Shew that *A*'s chance is to *B*'s as 244140625 is to 282429536481.[28]

This is a formidable problem and Huyghens gives no clue as to how he solved it. However, Jakob Bernoulli solves it in detail, offering three solutions:

> **Problem**. *A* and *B* each have 12 coins and play each other under the following conditions: If 11 eyes are thrown, then *A* gives a coin to *B*; should however 14 eyes be thrown, then *A* receives a coin from *B*. The one who first possesses all the coins has won the game. One finds 244 140 625 : 282 429 536 481 for the ratio of the chances of *A* and *B*.
>
> S o l u t i o n. The first thing to notice is that, of the 216 different possible throws with three dice, 15 throws with 14 eyes and 27 throws with 11 eyes can be found. Consequently there are 15 cases in which *A* receives a coin from *B* and 27 cases in which *B* gets one from *A*; with these there remain 174 cases in which each keeps his number of coins and therewith the current chances.

---

[27] Counters are coins without monetary value. Unlike coins, which would have been made of gold or silver, they were commonly made of copper. Today we would refer to them as "chips" or, more specifically, as "poker chips". In the loose translation of Bernoulli's *Ars conjectandi* that I will shortly be quoting from, the word *Münze* [coin] is used.

[28] Todhunter, *op. cit.*, p. 25.

The second thing to think about is that these 174 unsuccessful cases in which the chances of the players remain unalterred need not, by Theorem III, Corollary 4, to be considered; one has thus only 42 cases with three dice, of which 15 bring A and 27 bring B a coin.[29]

I interrupt Bernoulli at this point to note that today we might rigorously justify this practice of ignoring the neutral cases by noting

$$\frac{P(A)}{P(B)} = \frac{15/216}{27/216} = \frac{15}{27} = \frac{15/42}{27/42} = \frac{P(A|A \text{ or } B)}{P(B|A \text{ or } B)}.$$

So, in effect, he is considering a game in which one of A and B will win, A with probability $15/42$ and B with probability $27/42$. The basic concepts of probability theory were still emerging and had not yet crystalised. Thus he speaks not of the "probability" of A's winning or the "odds" that A will win, but of the "chances", meaning the number of cases favourable to A or B, respectively. Hence the simple step of removing a common factor now requires justification:

The third thing to remember about this is that by Theorem III, Corollary 2 the numbers of the cases 42, 15, and 27 can be replaced by the smallest relatively prime numbers 14, 5, and 9. In order to give the solution for the general case, however, I put the letters a, b, and c in their place in turn.

Following these remarks, in order to solve the problem before us, I now proceed in such a manner that I ask in order what chances the players have if each of them has one coin, two, three, four, ... coins, and, indeed, I go so far till I can conclude through induction the chances of the players if each possesses 12 coins.

If each player has only one coin, then obviously the ratio of the chances of A to those of B is as b to c.

If each player has two coins, then the toss causes either to come into possession of 3 coins or that only one remains to him. If 3 coins come into his possession, then he has b cases in which he will receive all four coins and thus win the stake, and c cases in which two coins remain, he is thus returned to his original chances, which will be denoted by z; this gives thus $\frac{b+cz}{a}$. If A holds only one coin, then he has b cases in which he will again have two coins and therewith return to the chances z, and c cases for the loss of the entire game; this makes $\frac{bz}{a}$ all told. For this, however, that A after the first toss has three coins, there are b cases and that he has only one coin, c cases; A has thus initially b cases for $\frac{b+cz}{a}$ and c cases for $\frac{bz}{a}$. It follows that the equation for his chances z is:

$$z = \frac{b^2 + 2bcz}{a^2},$$

whence follows

[29] R. Haussner, trans., *Wahrscheinlichkeitsrechnung (Ars conjectandi)*, Leipzig, 1899, p. 71.

$$z = \frac{b^2}{a^2 - 2bc} = \frac{b^2}{b^2 + c^2};$$

for B it consequently follows the remaining $\frac{c^2}{b^2+c^2}$ chances. The ratio of the two chances is therefore as b^2 to c^2.[30]

I doubt one wants to continue with this line of reasoning. The next case, in which both players have 3 coins is worked out in detail and he finally concludes the ratio of the chances to be b^3 to c^3. He then concludes by *Baconian* induction[31]—not mathematical induction—that, when each player has n coins to begin with, the odds of A's winning are b^n to c^n, in particular, for the values $b = 5, c = 9$, the odds are

$$b^{12} : c^{12} = 5^{12} : 9^{12} = 244\,140\,625 : 282\,429\,536\,481,$$

as Huyghens had declared.

Bernoulli doesn't stop here:

> One can also find this result without calculation as follows. If A has won all but one coin, he has b chances for the win, and B has, if he has won all but one coin c chances for the win. If A has won all but two coins, he has b chances to have all but one, i.e., to get the b previous ones; he thus has $b \cdot b = b^2$ cases for the win; B has, if he should find himself in the same situation, c^2 cases for the win. There are thus for each coin which the players lack to win, b cases for A and c cases for B in which they can succeed to the chances of the preceding case. If now at the beginning of the game each player has 12 coins and consequently lacks even so many coins to win, then the twelve powers of b and c yield the ratio of their chances, exactly as would have been found above.
>
> Whoever, however, for whom this calculation is not evident enough and who also does not see the inductive inference as sufficient proof, can reach the goal through a similar, abbreviated procedure, as was used by Huygens by Proposition XI, namely in which, as soon as he immediately went over to the case of six and from this to that of twelve coins, without considering all the intermediate cases. But here too no further calculation is required; because the calculation which has been carried through for the case where each player possesses two coins, remains valid if in place of one coin an arbitrary number, e.g., n of coins and in place of two coins $2n$ coins are substituted, if only instead

---

[30] *Ibid.*, pp. 71 - 72.

[31] One of the earliest writers to formulate the "scientific method", Francis Bacon (1561 - 1626) put a lot of emphasis on *induction*, whereby on making many observations, if one notices the same thing all the time, one declares it to be a general law. For philosophy the standard examples concern the rising of the sun and the colour of swans. The sun having always risen in the morning, we conclude by induction that it will rise again tomorrow. That this line of reasoning is not valid is illustrated by the swans. Until now every swan I have seen is white. Now suddenly I see a black swan...

of the numbers b and c of cases in which one of the two players wins or loses a coin, the number of cases in which every n coins can be won or lost is substituted. From here one then fully rigorously deduces that the ratio of the chances which the player has if they possess $2n$ coins equals the square of the ratio which is obtained if each player has n coins. As the ratio of chances in the case of three coins was found to equal $b^3 : c^3$, thus in the case of 6 coins it equals $b^6 : c^6$ and consequently again in the case of 12 coins equals $b^{12} : c^{12}$, as was also concluded by induction.[32]

Bernoulli's inductive solution essentially argued that for any number n of coins initially possessed by each player, an algebraic calculation would show the odds to be $b^n : c^n$. He verified this for $n = 1, 2, 3$, and claimed it held for general n.

His second proof is a mathematical induction: He proves by induction on n that the number of chances for A to win given that he has all but n coins is b^n, and the number for B under these circumstances is c^n, whence the ratio is $b^n : c^n$. The conclusion about the ratio is correct, but the rest is wrong: Suppose each player started with 12 coins and A now has 23 and B only 1. A has b chances to win the full game *in one step* by assumption. But he has other chances, such as losing the coin and then proceeding to win two.

11 Exercise. Is Bernoulli's third argument correct?

Bernoulli finishes his discussion of the problem by stating a generalisation: If A and B start with m and n coins, respectively, then the ratio of the chances of A's winning to those of B's winning is $m : n$ if $b = c$ and

$$\left(b^n c^m - b^{m+n}\right) : \left(c^{m+n} - b^n c^m\right),$$

otherwise. He declines to give the proof, declaring it to require a laborious calculation which he leaves to the reader.

Jakob Bernoulli died in 1705 and the *Ars conjectandi* was first published in 1713. Before this publication, in 1711, Abraham de Moivre (1667 - 1754) published *De mensura sortis, seu de probabilitate eventuum in ludis a casu fortuito pendentibus*, which, like Daniel Bernoulli's *Specimen theoriæ novæ de mensura sortis*, is often referred to simply as *De mensura sortis*. De Moivre's work was the first publication of the general formula. The demonstration appeared again in his great work on probability *The Doctrine of Chances* in 1718. About this demonstration Todhunter says

> Problem VII is the fifth of those proposed by Huygens for solution. We have already stated that De Moivre generalises the problem in the same way as James Bernoulli, and the result, with a demonstration, was first published in the *De Mensura Sortis*...De Moivre's demonstration is very ingenious, but not quite complete. For he finds the *ratio* of the chance that A will ruin B to the chance that B will ruin A; then he

---

[32] Haussner, *op. cit.*, pp. 74 - 75.

assumes in effect that in the long run one or other of the players *must* be ruined.[33]

It may be instructive to see what de Moivre has to say about this problem in *The Doctrine of Chances*. Below is the relevant passage from the third (1756) edition:

PROBLEM VII

Two Gamesters A *and* B, *each having* 12 *Counters, play with three Dice, on condition that if* 11 *Points come up,* B *shall give one Counter to* A*; if* 14 *Points come up,* A *shall give one Counter to* B*; and that he shall be the winner who shall soonest get all the Counters of his Adversary: what is the Probability that each of them has of winning?*

SOLUTION.

Let the number of Counters which each of them has be $= p$; and let a and b be the number of Chances they have respectively for getting a Counter, each cast of the Dice: which being supposed, I say that the Probabilities of winning are respectively as a^p to b^p; now because in this case $p = 12$, and that, by the preceding Lemma, $a = 27$, and $b = 15$, it follows that the Probabilities of winning are respectively as 27^{12} to 15^{12}, or as 9^{12} to 5^{12} or as 282429536481 to 244140625: which is the proportion assigned by *Huygens* in this particular case, but without any Demonstration.

Or more generally:

Let p be the number of the Counters of A , and q the number of the Counters of B; and let the proportion of the Chances be as a to b. I say that the Probabilities of winning will be respectively as $a^q \times \overline{a^p - b^p}$ to $b^p \times \overline{a^q - b^q}$; and consequently the Probabilities themselves will be

$$\frac{a^q \times \overline{a^p - b^p}}{a^{p+q} - b^{p+q}} = R, \text{ and } \frac{b^p \times \overline{a^q - b^q}}{a^{p+q} - b^{p+q}} = S.$$

DEMONSTRATION.

Let it be supposed that A has the counters E, F, G, H, &c. whose number is p, and that B has the Counters I, K, L, &c.[34] whose number is q : moreover, let it be supposed that the Counters are the thing played for, and that the value of each Counter is to the value of the following as a to b, in such manner as that E, F, G, H, I, K, L be in geometric Progression; this being supposed, A and B in every circumstance of their Play may lay down two such Counters as may be proportional to the number of Chances each has to get a single Counter; for in the beginning of the Play, A may lay down the Counter H, which is the

---

[33] Todhunter, *op. cit.*, p. 147.

[34] Note that J is skipped. In those days it was not yet separated from I.

lowest of his Counters, and B the Counter I, which is his highest; but H, I::a, b, therefore A and B play upon equal terms. If A beats B, then A may lay down the Counter I which he has just got of his adversary, and B the Counter K; but I, K::a, b, therefore A and B still play upon equal terms. But if A lose the first time, then A may lay down the Counter G, and B the Counter H, which he just now got of his adversary; but G, H::a, b, and therefore they still play upon equal terms as before: So that, as long as they play together, they play without advantage or disadvantage. Now the Value of the Expectation which A has of getting all the Counters of B is the product of the Sum he expects to win, and of the probability of obtaining it, and the same holds also in respect to B : but the Expectations of A and B are supposed equal, and therefore the Probabilities which they have respectively of winning, are reciprocally proportional to the Sums they expect to win, that is, are directly proportional to the Sums they are possessed of. Whence the Probability which A has of winning all the Counters of B, is to the Probability which B has of winning all the Counters of A, as the Sum of the terms, E, F, G, H, whose number is p, to the Sum of the terms I, K, L, whose number is q, that is as $a^q \times \overline{a^p - b^p}$ to $b^p \times \overline{a^q - b^q}$; as will easily appear if those terms, which are in geometric Progression, are actually summed up by the known Methods. Now the Probabilities of winning are not influenced by the Supposition here made of each Counter being to the following in proportion of a to b; and therefore when those Counters are supposed of equal value, or rather of no value, but serving only to mark the number of Stakes won or lost on either side, the Probabilities of winning will be the same as we have assigned.

COROLLARY 1.

If we suppose both a and b in a ratio of equality, the expressions whereby the Probabilities of winning are determined will be reduced to the proportion of p to q : which will easily apppear if those expressions are both divided by $a - b$.

COROLLARY 2.

If A and B play together for a Guinea a Game, and A has but one single Guinea to lose, but B any number, be it never so large; if A in each Game has the Chance of 2 to 1, he is more likely to win all the Stock of B than to lose his single Guinea; and just as likely, if the Stock of B were infinite.[35]

Augustus de Morgan calls de Moivre's demonstration "a method of the most striking ingenuity" [36], as indeed it is. De Moivre noticed that the probability in question depended in no way on the nature or value of the counters. He could thus arbitrarily decide to assign values to the counters, say

[35] Abraham de Moivre, *The Doctrine of Chances*, 3rd edition, 1756, pp. 51 - 53.
[36] Augustus de Morgan, *An Essay on Probabilities, And on Their Applications for Life Contingencies and Insurance Offices*, Longman, Orme, Brown, Green & Longmans, and John Taylor, London, 1838, Appendix the First, page i.

$c_1, \ldots, c_p, c_{p+1}, \ldots, c_{p+q}$ in such a way that the ratios of successive values is a/b:

$$\frac{c_i}{c_{i+1}} = \frac{a}{b}.$$

He further assumes that initially the first p belong to A and the last q to B, and that on any given play A stakes the last counter in the sequence belonging to him and B the first such belonging to him. Thus at any stage of play there is some number k such that A has the counters $c_1, \ldots, c_k$ and B the counters $c_{k+1}, \ldots, c_{p+q}$.

The probability that A wins on any individual play, given that either A or B wins, is $a/(a+b)$ and the probability that he loses is $b/(a+b)$. If he is staking c_k and B is staking c_{k+1}, the expected value for A is

$$\begin{aligned}
E &= \frac{a}{a+b} \cdot c_{k+1} - \frac{b}{a+b} \cdot c_k \\
&= \frac{a}{a+b} \cdot c_{k+1} - \frac{b}{a+b} \cdot \frac{a}{b} \cdot c_{k+1} \\
&= \frac{a}{a+b} \cdot c_{k+1} - \frac{a}{a+b} \cdot c_{k+1} = 0.
\end{aligned}$$

In short, his expectation for any play is 0 so long as A and B each have at least one coin to stake. His expectation for the whole game is thus also 0. Now, if $m : n$ are the odds that A will win the whole game (i.e., if $m : n$ is the ratio of the chances), then his mathematical expectation is

$$\begin{aligned}
0 &= \frac{m}{m+n} (c_{p+1} + \ldots + c_{p+q}) + \frac{n}{m+n} (-c_1 - \ldots - c_p) \\
&= \frac{m}{m+n} (c_{p+q} + \ldots + c_{p+1}) - \frac{n}{m+n} (c_p + \ldots + c_1) \\
&= \frac{m}{m+n} \cdot c_{p+q} \left(1 + \frac{a}{b} + \ldots + \left(\frac{a}{b}\right)^{q-1} \right) \\
&\quad - \frac{n}{m+n} \cdot c_{p+q} \left(\left(\frac{a}{b}\right)^q + \ldots + \left(\frac{a}{b}\right)^{q+p-1} \right),
\end{aligned}$$

whence

$$m \left(1 + \frac{a}{b} + \ldots + \left(\frac{a}{b}\right)^{q-1} \right) = n \left(\frac{a}{b}\right)^q \left(1 + \ldots + \left(\frac{a}{b}\right)^{p-1} \right). \tag{18}$$

Now, if $a = b$, (18) reduces to $mq = np$, i.e., $m/n = p/q$ as announced in Corollary 2. If $a \neq b$, we sum the geometric progressions in (18):

$$m \frac{(a/b)^q - 1}{a/b - 1} = n \left(\frac{a}{b}\right)^q \frac{(a/b)^p - 1}{a/b - 1},$$

i.e.,

$$m \left(\left(\frac{a}{b}\right)^q - 1 \right) = n \left(\frac{a}{b}\right)^q \left(\left(\frac{a}{b}\right)^p - 1 \right),$$

and multiplying by b^{p+q} yields

$$mb^p(a^q - b^q) = na^q(a^p - b^p) \qquad (19)$$

as claimed.

De Moivre's proof is not perfect. Since we do not allow division by 0, we would not divide the two sides of (19) by $a - b$ to obtain (18) when $a = b$. De Morgan criticises the proof and presents a different proof in an appendix to his own exposition of probability. In the fifth chapter of his book he presented Huyghens's Problem V and some rules for dealing with similar problems. This appendix begins as follows:

> Though the first part of the following reasoning is of a mathematical character, I have been induced to insert it by the consideration that the results of page 109. have never yet been introduced into an elementary work, nor even proved to the mathematician except either by incomplete or complicated trains of reasoning. Such being the case, perhaps even a well-informed mathematician might be excused for doubting some of the results of chapter V., and I have therefore digested the following demonstration, that no one who bears such a character may be able to weaken the evidence for the *necessity* of the pernicious results of gambling which that chapter is intended to afford.
> De Moivre was the first who gave a solution of the following problem, and by a method of the most striking ingenuity. But his demonstration has the defect of *assuming* that one or other of the players *must* be ruined in the long run. Laplace and Ampère,—the former in his *Théorie*, &c., the latter in a tract entitled, *Considérations sur la Théorie Mathématique du Jeu*, Lyons, 1802.—have also solved the problem: both solutions are of the highest order of difficulty, and cannot be rendered elementary. If my memory be correct, I have seen references to other solutions.
> The problem is as follows:—Two players, A and B, the first possessed of m times and the second of n times his stake, play at a game so constituted that it is a to b that A shall win any one game; required the probability which each has of ruining the other, if the game be indefinitely continued.[37]

There follows de Morgan's proof[38], which is perhaps a bit overly complicated. In his history, Todhunter avoids one unnecessary complication:

> It will be convenient to give the modern form of solution of the problem. Let u_x denote A's chance of winning all his adversary's counters when he has himself x counters. In the next game A must either win or lose a counter; his chances for these two contingencies are $\frac{a}{a+b}$ and $\frac{b}{a+b}$ respectively: and then his chances of winning all his adversary's counters are u_{x+1} and u_{x-1} respectively. Hence
> $$u_x = \frac{a}{a+b}u_{x+1} + \frac{b}{a+b}u_{x-1}.$$

---
[37] De Morgan, *op.cit.*, pp. i - ii.
[38] Perhaps one should say, his exposition of Laplace's proof.

One should say a few words about Laplace. Charles Coulston Gillispie's massive account of Laplace in the *Dictionary of Scientific Biography* begins with the words, "Laplace was among the most influential scientists in all history". The account of Laplace and his work in this *Dictionary* runs 29 sections covering 112 pages, not counting a 17 page bibliography. In writing this account, Gillispie enlisted the aid of Brian G. Marsden, C.A. Whitney, Ivo Schneider, Robert Fox, Stephen M. Stigler, and Ivor Grattan-Guinness.

Laplace had numerous publications in the fields of physics and mathematics, much of the content of which he collected and incorporated in two major works, each separately capable of securing for him a reputation as one of the world's great scientists. The first of these was the *Traité de mécanique céleste*, usually referred to as the *Mécanique céleste* or *Méchanique céleste*. This was published in five volumes, the first four from 1799 to 1805, and the fifth in instalments from 1822 to 1825. The *Mécanique céleste* was Laplace's most important work and translations into German and English were immediately undertaken. The most important of these was the 4 volume English translation, *Mécanique céleste by the Marquis de Laplace, Translated With a Commentary* (Boston, 1829 - 1839) by the American Nathaniel Bowditch (1773 - 1838), who printed these volumes at his own expense. Laplace's second major work was, of course, the *Théorie analytique des probabilités* (1812) which he accompanied with a popular exposition, the *Essai philosophique sur les probabilités* (1814).

At the mention of Laplace, de Morgan appended the following footnote. It is of no particular significance here. However, it makes a good addition to our "Report Card" of Project 8 of Chapter 1:

> The solution of Laplace gives results for the most part in precisely the same form as those of De Moivre, but according to Laplace's usual custom, no predecessor is mentioned. Though generally aware that Laplace (and too many others, particularly among French writers) was much given to this unworthy species of suppression, I had not any idea of the extent to which it was carried until I compared his solution of the problem of the duration of play, with that of De Moivre. Having been instrumental (in my mathematical treatise on Probabilities, in the *Encyclopedia Metropolitana*) in attributing to Laplace more than his due, having been misled by the suppressions aforesaid, I feel bound to take this opportunity of requesting any reader of that article to consider every thing there given to Laplace as meaning simply that it is to be found in his work, in which, as in the *Mécanique Céleste*, there is enough originating from himself to make any reader wonder that one who could so well afford to state what he had taken from others, should have set an example so dangerous to his own claims.

This equation is thus obtained in the manner exemplified by Huyghens in his fourteenth proposition...
The equation in Finite Differences may be solved in the ordinary way; thus we shall obtain

$$u_x = C_1 + C_2 \left(\frac{b}{a}\right)^x,$$

where C_1 and C_2 are arbitrary constants. To determine these constants we observe that A's chance is zero when he has no counters, and that it is unity when he has all the counters. Thus u_x is equal to 0 when x is 0, and is equal to 1 when x is $m+n$. Hence we have

$$0 = C_1 + C_2, \quad 1 = C_1 + C_2 \left(\frac{b}{a}\right)^{m+n};$$

therefore

$$C_1 = -C_2 = \frac{a^{m+n}}{a^{m+n} - b^{m+n}}.$$

Hence

$$u_x = \frac{a^{m+n} - a^{m+n-x}b^x}{a^{m+n} - b^{m+n}}.$$

To determine A's chance at the beginning of the game we must put $x = m$; thus we obtain

$$u_m = \frac{a^n(a^m - b^m)}{a^{m+n} - b^{m+n}}.$$

In precisely the same manner we may find B's chance at any stage of the game; and his chance at the beginning of the game will be[39]

$$\frac{b^m(a^n - b^n)}{a^{m+n} - b^{m+n}}.$$

Some words of explanation are in order. First note that whereas de Moivre used the word "chances" to denote the outcomes favourable to a given player, Todhunter uses the word to mean probability.[40] Thus, the chances of de Moivre would be a and b, while those of Todhunter are $\frac{a}{a+b}$ and $\frac{b}{a+b}$, respectively. Since it is the ratio (in lowest terms) that is sought, the same ratio,

$$\frac{a}{b} = \frac{a/(a+b)}{b/(a+b)},$$

will be obtained.

Todhunter's notation u_x thus denotes the conditional probability in the modern sense that A will win given that he has x counters and he finds it by breaking the task into two steps: perform one play of the game, and proceed from there to the finish. We can represent this graphically as in *Figure 4*, below.

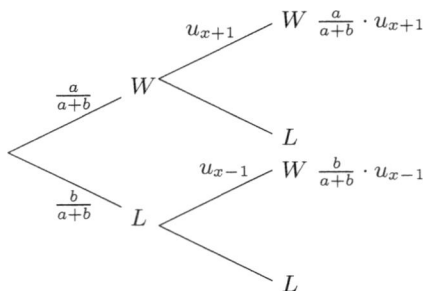

Fig. 4.

Here I let W denote a win for A and L a loss, the first occurrence referring to the individual play of the game and the second referring to the full game. Adding the winning probabilities on the right yields

$$u_x = \frac{a}{a+b} \cdot u_{x+1} + \frac{b}{a+b} \cdot u_{x-1}.$$

Shifting indices this reads

$$u_{x+1} = \frac{a}{a+b} \cdot u_{x+2} + \frac{b}{a+b} \cdot u_x. \tag{20}$$

To this we apply a couple of algebraic transformations:

$$0 = \frac{a}{a+b} \cdot u_{x+2} - u_{x+1} + \frac{b}{a+b} \cdot u_x$$

$$= u_{x+2} - \frac{a+b}{a} \cdot u_{x+1} + \frac{b}{a} \cdot u_x$$

$$= u_{x+2} - \left(1 + \frac{b}{a}\right) u_{x+1} + \frac{b}{a} \cdot u_x. \tag{21}$$

Now, if we introduce the operator E satisfying $E(f)(x) = f(x+1)$, (21) reads

$$E^2 u_x - \left(1 + \frac{b}{a}\right) E u_x + \frac{b}{a} u_x = 0,$$

i.e.,

$$\left(E^2 - \left(1 + \frac{b}{a}\right) E + \frac{b}{a}\right) u_x = 0. \tag{22}$$

Replacing E by the polynomial variable X yields a polynomial

$$X^2 - \left(1 + \frac{b}{a}\right) X + \frac{b}{a}$$

---

[39] Todhunter, *op. cit.*, pp. 62 - 63.

[40] And the translator of Bernoulli used the words "Hoffnungen" or "hope", which I have translated as "chances", and "Fälle" or "cases".

with roots 1 and b/a. In the Calculus of Finite Differences it is shown then that the solutions to (22) are all of the forms,

$$u_x = C_1 \cdot 1^x + C_2 \cdot \left(\frac{b}{a}\right)^x, \tag{23}$$

for $a \neq b$, or

$$u_x = C_1 \left(\frac{b}{a}\right)^x + C_2 x \left(\frac{b}{a}\right)^x = C_1 + C_2 x, \tag{24}$$

for $a = b$, in either case the bases of the exponentiation are the roots of the polynomial.

Since the game is lost when A has 0 counters, his probability of winning then is 0 and $u_0 = 0$. Likewise, $u_{m+n} = 1$. Plugging these values into (23) and (24), respectively, and solving for C_1, C_2 will yield the sought after probabilities and, taking their ratios, will solve the problem in these cases.

My rather detailed discussion of Todhunter's exposition of the proof does not yet address the issue raised by de Morgan—that de Moivre's proof rests on the assumption that one of the gamblers will win the game. Nor, of course, does it explain how de Morgan got around this, proving that one of the gamblers must be ruined and thus gambling is pernicious. The assumption is a subtle matter, easily overlooked. It is the point (cf. page 98, above) where we go from $m : n$ being the odds that A wins to the assumption that his loss on not winning is $c_1 + \ldots + c_p$. There is a third possibility, that A and B will keep swapping counters and the game will go on forever with neither player losing all of his counters. This assumption is not made in de Morgan's proof or in our exposition of Todhunter's version. De Morgan, in fact, goes back to Huyghens's formulation without Bernoulli's simplifying assumption. He considers the possibility that individual plays can be draws (as the 174 unsuccessful cases which Bernoulli dismissed). De Morgan chooses δ to represent these unsuccessful cases and replaces the recurrence relation (20) by

$$u_x = \frac{a}{a+b+\delta} u_{x+1} + \frac{b}{a+b+\delta} u_{x-1} + \frac{\delta}{a+b+\delta} u_x,$$

and shows that u_x again satisfies (23) or (24) according as $a \neq b$ or $a = b$. Thus, u_x is the same regardless of the value of δ.

12 Exercise. Verify this last assertion.

Referring now to the probabilities at the outset that A will win the entire game,

$$\frac{a^n(a^m - b^m)}{a^{m+n} - b^{m+n}},$$

and that B will win the entire game,

$$\frac{b^m(a^n - b^n)}{a^{m+n} - b^{m+n}},$$

de Morgan notes that their sum is 1:

The application of finite difference techniques as discussed here by Todhunter and de Morgan was developed mainly by Lagrange and Laplace. Joseph Louis Lagrange (1736 - 1813) was a slightly older contemporary of Laplace. The two of them were the leading French mathematicians of their day, at a time of French supremacy in science.

Lagrange gave a nice solution to the Duration of Play problem, a more general form of Gambler's Ruin than we consider here, and generalised de Moivre's work on the binomial distribution (to be discussed in the next chapter) to the multinomial distribution. His work in probability, however competent, broke no new ground and he was overshadowed by Laplace. Lagrange merits an entire chapter in the encyclopædic coverage by Todhunter, but rates only a few brief mentions in less exhaustive histories of probability.

The stamp pictured was issued by France in 1958 in a set of 4 stamps honouring French scientists.

> The sum of these two chances is unity, from which it appears that *one or other must be ruined in the long run*. These results agree entirely with those of De Moivre.[41]

To borrow the language of d'Alembert in one of his responses to the Petersburg Problem, one of the players will *certainly*, but not *necessarily* be ruined. It is possible for the two players to continue alternately gaining and losing a counter, but this is so highly improbable as to have probability 0.

There is another problem not raised by de Morgan, and possibly not recognised by him: All of these proofs depend on the unproven assumption that the probability in question exists. The proofs proceed by assuming the probability exists and then solving for it. The problem is that one is dealing with potentially infinite games, and strange things can happen when the infinite is involved. For example, we sum the geometric progression by solving for it: Let

$$S = a + ar + ar^2 + \dots$$

By noting

$$S = a + ar + ar^2 + \dots$$
$$rS = \quad ar + ar^2 + \dots$$

we obtain

$$(1 - r)S = S - rS = a$$

and

$$S = \frac{a}{1 - r}.$$

[41] De Morgan, *op. cit.*, p. iv.

But what happens if $a = 1$ and $r = -1$ or 2? We have

$$1 - 1 + 1 - 1 + 1 - \ldots = \frac{1}{1 - (-1)} = \frac{1}{2} \qquad (25)$$

$$1 + 2 + 4 + 8 + 16 + \ldots = \frac{1}{1 - 2} = -1. \qquad (26)$$

Similarly, we can find $1 - 2 + 3 - 4 + 5 - \ldots$ by letting x denote this sum and solving for it:

$$x = 1 - 2 + 3 - 4 + 5 - \ldots$$
$$0 + x = 0 + 1 - 2 + 3 - 4 + \ldots,$$

whence

$$2x = x + (0 + x) = 1 - 1 + 1 - 1 + 1 - \ldots = \frac{1}{2},$$

from which it follows that $x = 1/4$. Jakob Bernoulli discussed (25) and Gottfried Wilhelm Leibniz accepted it. And Euler claimed the validity of (26), something which Leibniz never would have done. Today we reject all three of these sums in the basic Calculus course.[42] Thus, when dealing with numbers attempting to find the measures of infinite sets, in the present case the probabilities of events, we cannot merely assume a number exists and solve for it. We must either first prove the existence of the number and then solve for it, or show in some direct way that the number found does indeed have whatever properties one is claiming for it. I have not read Laplace's or Ampère's proofs alluded to by de Morgan that one or the other of the players *must* be ruined in the long run, but it is conceivable that the higher orders of difficulty of these proofs arise from their possibly being more constructive and thus not resting on an existence assumption. This is the case with a more modern proof based on Markov chains.

When we read Todhunter's sentence, "It will be convenient to give the modern form of solution of the problem", we should remind ourselves that he was writing a century and a half ago. How "modern" is the proof he presented? Well, in his classic textbook of the 1950's, William Feller repeats it in Chapter XIV, "Random Walk and Ruin Problems", but he also comments on the existence problem:

> Strictly speaking, the probability of ruin is defined in a sample space of infinitely prolonged games, but we can work with the sample space of n trials. The probability of ruin in less than n trials increases with n and therefore has a limit. We call this *limit* "the probability of ruin." All probabilities in this chapter may be interpreted in this way without reference to infinite sample spaces.[43]

However, Feller also presents, in a later chapter, the treatment via Markov chains. The discussion of Markov chains is interesting in itself, and the present section being quite long already, I postpone their discussion to Appendix A.

---

[42] In higher mathematics we learn that there are senses in which these sums are correct. But the sums are not correct under the usual notion of summation.

[43] Feller, *op. cit.*, p. 313.

Andre Marie Ampère (1775 - 1836) is pictured here on a French stamp issued in 1936 to commemorate the centennial of his death. Two of his greatest influences were Buffon's *Histoire naturelle* and the *Encyclopédie*, of which, as already noted, d'Alembert was the scientific editor. Chronologically his main scientific interests passed from mathematics (1800 - 1814) through chemistry (1808 - 1815) to the work for which he is most famous, electrodynamics (1820 - 1827). In addition, all through his life he was keenly interested in philosophy and metaphysical questions. His tract on probability cited by de Morgan (page 99, above) was his first mathematical paper.

8 Bertrand's Box Paradox

Often, one wants to reverse the conditionality of the situation and find $P(A|B)$ where $P(B|A)$ is known. Such a question may arise, for example, in discussing causality[44]. A particularly famous example, often quoted under some slight change in configuration, is *Bertrand's Box Paradox*, first presented by Joseph Louis François Bertrand (1822 - 1900) in his textbook *Calcul des Probabilités*[45] (1889). The book opens with a long philosophical preface of 50 pages numbered in Roman numerals before presenting the table of contents and beginning the text proper. This text itself begins with the basic definition of probability for a sample space of equally likely outcomes and follows this immediately with the Box Paradox, offering it as an example of an incorrectly enumerated sample space, i.e., as yielding a sample space of outcomes that are not equally likely:

> 2. Three caskets[46] are of identical appearance. Each has two drawers, each drawer hides a medallion. The medallions of the first casket are gold; each of the second casket are silver; and the third casket contains one gold medallion and one silver medallion.
>
> One chooses a casket: what is the probability of finding within the drawers one gold piece and one silver piece?
>
> Three cases are possible and impartial since the three caskets are of identical appearance.

---

[44] While it is true that the Theory of Probability got its start in gambling, it was furthered by questions from other areas—annuities and insurance, errors of observation in astronomy, etc.

[45] The second edition, "conformable to the first", was published by Gauthier-Villars, Paris, 1907, and is available online. I refer to it on the assumption that it differs inessentially from the first edition.

[46] The French word is *coffret* and means something like small box, ornamental box, jewelry box (*coffret à bijou*), etc. The word "casket" is a commonly used single word translation of it, though in American usage "casket" is more likely to bring to mind a coffin—appropriately in line with the funerary meaning of "urn".

One case alone is favourable. The probability is 1/3.

The casket is chosen. One opens a drawer. Whatever medallion which is found therein, only two cases remain possible. The drawer which remains closed can contain a medallion the metal of which is or is not different from that of the first. Of these two cases, one alone is favourable to the casket which has two different pieces. The probability of having put the hand on this casket is thus 1/2.

Now how can we believe that opening a drawer suffices to change the probability and raise it from 1/3 to 1/2?

Perhaps the reasoning is not sound. Indeed, it is not.

After opening the first drawer two cases remain possible. Of the two cases, only one is favourable, this is true, but the second case does not have the same likelihood [as the first].

If the piece which one has in view is gold, the other is perhaps of silver, but one will have an advantage wagering it is of gold.

Suppose, for the sake of clarification, that instead of three caskets one has three hundred. One hundred contain two gold medallions, one hundred two silver medallions, and one hundred one gold and one silver medallion. In each casket one opens a drawer and sees consequently three hundred medallions. One hundred of them are gold and one hundred silver, this is certain; the other one hundred are in question, concerning the caskets from which the pieces are not the same: chance regulates the numbers.

One must expect, through opening the three hundred drawers, to see fewer than two hundred gold pieces: the principal part of the probability for which belongs to that of the hundred caskets for which the other piece is gold, and is thus greater than 1/2.[47]

Bertrand is here merely citing the caskets, nowadays usually replaced by desks with drawers, as a means of producing a sample space with outcomes of unequal likelihood. It is far too early in the book, 50 pages of preface notwithstanding, to treat conditional probability and actually solve the problem: Given that one chooses a gold coin (the modern day evolved form of the medallion) on one's first draw, what is the probability the second coin is silver, i.e., that the gold coin came from desk number III? Bertrand only first shows how to solve the problem in chapter VII on the probability of causes.

My French is nonexistent, so I cannot say with certainty that I did not merely miss it in glancing over this chapter, but I did not find a return to the Box Problem. I do find a simpler version on pp. 139 - 140:

Two urns, for example, are of identical aspect: the one contains one white ball and one black ball; the other ten black balls and one white. One chooses one of the urns, in fact one obtains a ball, it is white; what is the probability of having chosen the first urn?

Bertrand does not solve this problem directly but goes on first to handle the general case.

---

[47] Bertrand, pp. 2 - 3.

The tree diagram for this problem is depicted in the left half of *Figure 5*, below. One wants to fill in the probabilities of the tree of the right half of

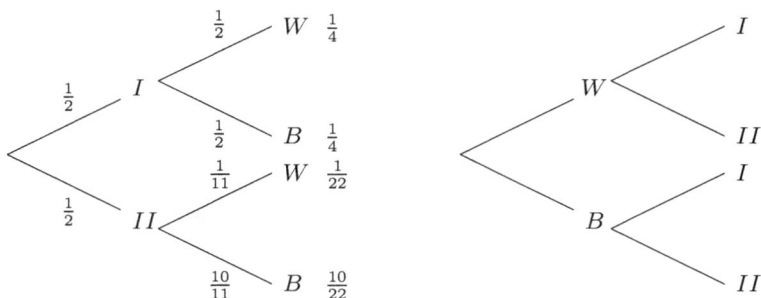

Fig. 5.

Figure 5. In particular, one wants the probability $P(I|W)$ that is to be placed above the WI line segment. Now the probability $P(W\&I)$ of the path WI is identical to the probability of the path IW of the first tree, which is $1/4$. And the probability of the first segment leading to the W of the uppermost branch is $P(W)$, which we can read off the first tree as being

$$\frac{1}{4} + \frac{1}{22} = \frac{11}{44} + \frac{2}{44} = \frac{13}{44}.$$

The uppermost path of our second tree now looks like the following and we

have but to solve for x:

$$\frac{13}{44}x = \frac{1}{4}$$
$$x = \frac{1}{4} \cdot \frac{44}{13} = \frac{11}{13}.$$

13 Exercise. Returning to Bertrand's problem, find the probability that the second medallion in the casket is silver given that the first one is gold, i.e., find the conditional probability that the third casket was chosen given that the drawer opened after choosing a casket contained a gold medallion.

The popularity of Bertrand's Box Paradox is a bit of a puzzle. He did not present it as a paradox, but merely as an example of the importance of equal likelihood of the outcomes in one's sample space if one is going to compute probabilities directly. Neither the notion of conditional probability nor that of inverting the order of the conditionality originated with him, and even the salient features of the problem were hardly novel. From half a century earlier, for example, Augustus de Morgan gave the following:

PROBLEM. A white ball is drawn, and from one or other of the following urns:

(3 white, 4 black)   (2 white, 7 black):

but before the drawing was made, it was three to one that the drawer should go to the first urn, and not to the second. What is the chance that it was the first urn from which the drawing was made?

We may immediately reduce the preceding to the case where all the antecedent circumstances are equally probable, by introducing urns enough of the first kind to make it 3 to 1 that the drawing is made from one or other of them. Let us suppose the urns to be as follows:

(3 white, 4 black)   (3 white, 4 black)   (3 white, 4 black)

(2 white, 7 black):

these urns being equally probable, the hypothesis of the problem exists. If we number the urns A_1, A_2, A_3, B, the chances which they severally give to the observed event are $\frac{3}{7}, \frac{3}{7}, \frac{3}{7}$, and $\frac{2}{9}$, the numerators of which, reduced to a common denominator, are 27, 27, 27, and 14. Consequently, the probability that A_1 was chosen, is $\frac{27}{95}$; and the same for A_2 and A_3. Therefore, the chance that one or other of the three, A_1, A_2, and A_3, was chosen, is $\frac{81}{95}$; which is the probability of the ball having been drawn from the urn (3 white, 4 black,) in the first statement of the problem.[48]

De Morgan's approach is not the one we take today, but the method is irreproachable. In fact, it was a not uncommon trick of the era, used to reduce problems with outcomes of unequal probabilities to problems in which all the outcomes were equally likely.

Perhaps, had de Morgan not so quickly added the extra urns and instead first found the numerators 27 and 14 reduced to a common denominator incorrectly yielding $\frac{27}{41}$ as the probability of the ball's having come from the first urn, modern textbooks would feature de Morgan's problem. But they don't; one often finds Bertrand's problem, mildly disguised through the transformations of the caskets into desks, the medallions into coins, and even the two drawers into single drawers containing two coins apiece. And, though Bertrand himself may not be named, it is his example that is cited and that is sometimes used to introduce conditional probability.

Bertrand's near paradoxical treatment notwithstanding, there is another reason why his example is so widely presented and de Morgan's is not: it is simpler. Indeed, it is almost as simple as it gets. But it can be simplified even further:

---

[48] De Morgan, *op.cit.*, pp. 58 - 59.

14 Exercise. i.[49] Suppose you meet my brother Peter and he says, "I have two children, and at least one of them is a girl." What would you judge to be the probability that my brother has a boy and a girl?

ii. Suppose Peter tells you that, in fact, the older child is a girl. What is the probability now?

---

[49] Keith Devlin, *The Unfinished Game; Pascal, Fermat, and the Seventeenth-Century Letter that Made the World Modern*, Basic Books, New York, 2008, pp. 154 - 155.

4

Binomial Probability

A particularly common kind of sequential probability, one that is presented in elementary courses on the subject, goes by the name *binomial probability*, this name deriving from the fact that the probabilities in question are terms in a binomial expansion. Binomial coefficients made their appearance in probability already in the Pascal-Fermat correspondence, and any number of mathematicians discussed them in connexion with counting—determining the number of combinations of n things taken k at a time, and using such determinations to compute probabilities. Binomial probability, as we know it today, however, makes its debut with Jakob Bernoulli in his posthumously published *Ars conjectandi*.

The *Ars conjectandi* was first published in Latin in 1713. It is a work in four parts. It has been translated into modern languages and such translations can be found easily online, but they are not all complete. From Google Books I obtained a French translation of the first part[1], the first three chapters of the second part in English[2], and what at first sight appeared to be the third and fourth parts in German[3], but which proved to be the complete work consisting of a single PDF file of the two-volume German translation in which the files for the two volumes were assembled in the wrong order.

---

[1] L.G.F. Vastel, trans., *L' Art de Conjecturer*, Caen, 1801.

[2] Francis Maseres, *The Doctrine of Permutations and Combinations, Being an Essential and Fundamental Part of the Doctrine of Chances*, London, 1795. I have truncated the title. Maseres (1731 - 1824), incidentally, has a certain notoriety in the history of mathematics as one of the last great opponents of negative numbers. On a more positive note, he was a populariser of science and published a number of classical works at his own expense.

[3] R. Haussner, trans., *Wahrscheinlichkeitsrechnung (Ars conjectandi)*, Leipzig, 1899. Having mentioned the rôle of Maseres in the popularisation and dissemination of knowledge in the preceding footnote, I should point out that Robert Karl Hermann Haussner's translation appeared as two volumes in *Ostwalds Klassiker*, a series of reprints of classical works in the exact sciences founded in 1859 by Wilhelm Ostwald (1853 - 1932), who would go on to win the Nobel prize in chemistry in 1909.

The first part of Bernoulli's book is an annotated version of *De ratiociniis* of Huyghens in which each passage from Huyghens is followed by a long discussion. It is in such discussive remarks to Proposition XII that we find binomial probability. Proposition XII, like Proposition XI before it, is a generalisation of Proposition X, which we discussed in Chapter 2 (page 33, above). The statements of these problems are as follows:

XI. To find how many times one should wager to throw 2 sixes with 2 dice.

XII. To find the number of dice with which one may wager to throw 2 sixes at the first throw.[4]

Propositions X, XI, and XII are all slightly complicated variations on a simpler theme. It is common in the history of mathematics for the initial problems considered not to be of the simplest variety and their solutions not to follow what we should now consider to be the simplest, most direct paths. In a modern theory-based course, one would first ask a question like: a coin is tossed 6 times; what is the probability of coming up heads exactly 3 times? That is, the prototypical problem of this sort repeats an experiment (tossing a coin, throwing a die, drawing a ball from an urn with replacement, etc.) with fixed probability p of *success* on a given *trial*[5], n times and asks for the probability of *exactly* k successes. In gambling, however, such a problem would not arise: one normally wins if there are *at least* m successes for some m and the question would be more complicated: what is the probability of at least m successes in n trials; or, what must n be for the probability of at least m successes to be $1/2$ or better?

In discussing Proposition XII, Bernoulli quotes Huyghens's presentation of the proposition and its solution, and follows up with a couple of explanatory comments. Both the presentation and the follow-up take about half a page. He then follows this with 7 1/2 pages[6] of generalisation. He considers dice with a faces, b favourable and $c = a - b$ unfavourable. (As with the two- and three-faced dice of the Pascal-Fermat correspondence, these dice may take some effort of imagination and one might prefer to think of drawing balls with replacement from an urn containing b white balls and c black ones.) Bernoulli, in solving the problem, considers the complementary problem of failing to make m successes in n trials. So he states the probability of failing on a given throw to be c/a and that of succeeding to be $(a - c)/a$. He then considers the first few cases, explicitly calculating the probabilities of overall failure for several small values of m, n. He follows this with a small table of such probabilities and the general formula,

[4] F.N. David, *Games, Gods and Gambling; A History of Probability and Statistical Inference*, Charles Griffin & Co., Ltd., London, 1962 (reprinted by Dover Publications, Mineola (New York), 1998), p. 116.
[5] Each instance of the experiment, or trial, is nowadays often called a *Bernoulli trial*.
[6] I refer to the German edition. The exact lengths vary from edition to edition, but the ratios ought to be roughly the same.

$$\frac{1}{a^n}\left(c^n + \binom{n}{1}bc^{n-1} + \binom{n}{2}b^2c^{n-2} + \ldots + \binom{n}{m-1}b^{m-1}c^{n-m+1}\right), \quad (1)$$

for the probability of scoring fewer than m successes in n trials.[7] The sought after probability is obtained by subtracting (1) from 1. Equivalently, it is given by

$$\frac{1}{a^n}\left(\binom{n}{m}b^mc^{n-m} + \binom{n}{m+1}b^{m+1}c^{n-m-1} + \ldots + \binom{n}{n-1}b^{n-1}c + b^n\right). \quad (2)$$

One reason for the preference for (1) over (2) is that, for $m = 2$ and $n > 4$, (1) will have fewer terms than (2) and will thus involve less work in computation. Another reason is that, for the problem at hand, where $m = 2$, (1) can uniformly be written as

$$\frac{1}{a^n}\left(c^n + nbc^{n-1}\right) = \frac{c^{n-1}}{a^n}(c + nb), \quad (3)$$

for all values of n, and the problem reduces to solving the equation,

$$\frac{c^{n-1}}{a^n}(c + nb) = \frac{1}{2}, \quad (4)$$

in n, where $a = 6, b = 1, c = 5$. Using the little table produced below, Bernoulli verified that $n = 10$ is the smallest positive integer satisfying

$$a^n = 6^n > (10 + 2n)5^{n-1} = (2c + 2nb)c^{n-1},$$

and thus that 10 is the minimal number of dice giving an even or better than even chance of winning. Presumably he made the same calculations and com-

a	$=$	6	c	$=$	5
a^3	$=$	216	c^2	$=$	25
a^9	$=$	$10\,077\,696$	c^4	$=$	125
a^{10}	$=$	$60\,466\,176$	c^8	$=$	$390\,625$
			c^9	$=$	$1\,953\,125$

$a^9 = 10\,077\,696 < 10\,937\,500 = 28 \cdot 390625 = (2c + 18b)c^8,$
$a^{10} = 60\,466\,176 > 58\,593\,750 = 30 \cdot 1953125 = (2c + 20b)c^9.$

parisons for $n = 2, 3, \ldots, 8$ before arriving at 9 and 10..

Bernoulli also gave a geometric construction based on a "logarithmic line", a term which in those days also referred to the graph of the exponential function. We can, of course, use our graphing calculators today in any number of ways

---

[7] A note in the German edition reports two liberties taken in the translation. Some of Bernoulli's calculations were omitted as being unnecessary, and the modern notation for binomial coefficients was used in place of the multiplicative expressions. The French edition keeps to Bernoulli's notation.

to solve (4), or, more accurately, since there is no integral solution, to solve for a real number x satisfying

$$\frac{5^{x-1}}{6^x}(5+x) = \frac{1}{2}. \tag{5}$$

Exponential functions tend to escape the calculator screen all too quickly, so one can take logarithms,

$$(x-1)\ln 5 - x\ln 6 + \ln(5+x) = -\ln 2,$$

and simplify:

$$x\ln 5/6 + \ln(5+x) + \ln 2/5 = 0.$$

If one now enters

   $Y_1 = X \ln(5/6) + \ln(5+X) + \ln(2/5)$

in the equation editor of the *TI-83 Plus* and graphs the functon in the standard window ($\mathsf{Xmin} = -10$, $\mathsf{Xmax} = 10$, $\mathsf{Xscl} = 1$, $\mathsf{Ymin} = -10$, $\mathsf{Ymax} = 10$, $\mathsf{Yscl} = 1$, $\mathsf{Xres} = 1$), one sees quickly that between $x = 9$ and $x = 10$ the graph is very close to the x-axis. That it actually does cross the axis and where it does is not clear with this window. One can hope for the best and press the TRACE button and follow the curve using the right arrow button and read the coordinates at the bottom of the screen to note a change in sign. If one does so, one sees that Y_1 decreases and finally changes sign between

$$x = 9.5744681, \text{ where } y = .01734858$$

and

$$x = 9.787234, \text{ where } y = -.0069502.$$

This is, of course, good enough to determine that the solution to (5) lies between $x = 9$ and $x = 10$. One can interpolate between these values of x or use the CALC button to determine more precisely the solution to (5), but such precision is unnecessary. If one were not convinced that the curve crossed the axis somewhere in the standard window, one could first change the window setting and regraph the equation. To make the axes visible, setting Xmin and Ymin to -1 is a good choice. Also, the function seems to be bounded above by 1, so choosing Ymax to be, say, 2 will give us greater vertical separation and possibly make obvious where the curve crosses the axis by visual examination, without having to resort to the TRACE button. As for Xmax, we already know that 10 is large enough, but if we weren't sure, we could up the value to, say, 15. If we choose these settings and graph the function, we see indeed that the curve crosses the x-axis somewhere between 9 and 10. This is even more clearly visible if we set Ymax equal to 1, and more clearly still if we set Xmin to be 9, Xmax to be 10, Ymin to be $-.1$, and Ymax to be .1: the curve crosses the x-axis somewhere between 9.7 and 9.8, though, again, this is far more information than we need. It is difficult, however, when one has such a powerful ~~toy~~ tool as a graphing calculator to stop exploring after one has done all that is strictly necessary to solve a problem.

Returning to probability, if we reflect on what we have done, we will realise that there are two issues confronting us here. First, how do we justify formulæ (1) and (2)? Second, can we use these formulæ to calculate these probabilities without a lot of work? And, if so, how?

The justification is fairly simple. To make matters easier on ourselves, let us write p for the probability of success on a given trial ($= (a-c)/a$ in Bernoulli's notation) and $q = 1-p$ for the probability of failure ($= c/a$ in Bernoulli's notation). Now the probability of having exactly k successes in n trials is

$$\binom{n}{k} p^k q^{n-k} = \binom{n}{k} p^k (1-p)^{n-k}.$$

To see this, let us represent the possible sequences of trials as paths through a tree, with probabilities assigned as in Chapter 3. *Figure 1*, below, illustrates this for three throws of a die in which success on a given throw means coming

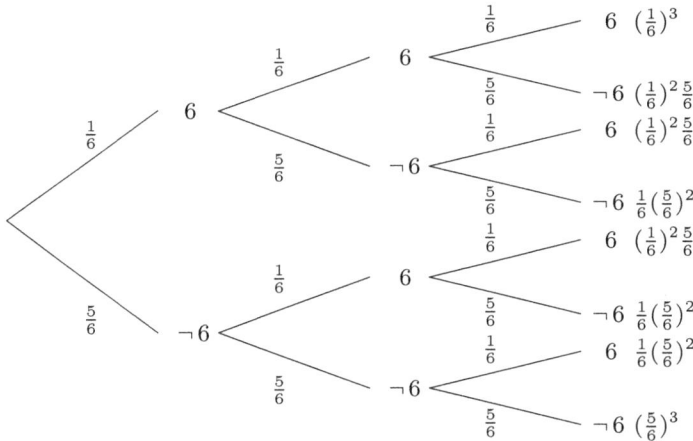

Fig. 1.

up with a 6. If we seek the probability of obtaining, say, exactly two 6's, we add the probabilities of those paths containing two 6's and a single non-6. The probabilities of such paths are the same—a product of two factors of $1/6$ and one of $5/6$. And the number of such paths is $_3C_2$, the number of choices of 2 throws out of 3 to come up 6. The overall probability is thus

$$\underbrace{\left(\frac{1}{6}\right)^2 \frac{5}{6} + \ldots + \left(\frac{1}{6}\right)^2 \frac{5}{6}}_{_3C_2} = \binom{3}{2}\left(\frac{1}{6}\right)^2 \frac{5}{6} = \frac{5}{72}.$$

In like manner, we see the winning paths for exactly k successes in n trials with probability p of success on a given trial all to have probability $p^k(1-p)^{n-k}$ and there to be $_nC_k$ such paths. This yields

$$\binom{n}{k} p^k (1-p)^{n-k}$$

as the overall probability of exactly k successes in n trials.

From this, (1) and (2) follow easily via the Addition Principle for Probabilities. For example, the probability of coming up with at least two 6's in three throws is the sum of the probabilities of obtaining exactly two 6's and exactly three 6's:

$$\binom{3}{2}\left(\frac{1}{6}\right)^2 \frac{5}{6} + \binom{3}{3}\left(\frac{1}{6}\right)^3 \left(\frac{5}{6}\right)^0 = \frac{5}{72} + \frac{1}{216} = \frac{15}{216} + \frac{1}{216} = \frac{2}{27} = \overline{.074}.$$

The issue of calculation is a much larger one. In theory it is really quite simple. If the probability of achieving a favourable outcome on an individual trial is p, and one repeats the experiment n times, to find the probability of *exactly* k successes, one calculates

$$A(n, p, k) = \binom{n}{k} p^k (1-p)^{n-k};$$

for *at least* m successes, one calculates

$$B(n, p, m) =$$

$$\binom{n}{m} p^m (1-p)^{n-m} + \binom{n}{m+1} p^{m+1}(1-p)^{n-m-1} + \ldots + \binom{n}{n} p^n; \quad (6)$$

and for *at most* m successes, one calculates

$$C(n, p, m) = \binom{n}{0}(1-p)^n + \binom{n}{1}p(1-p)^{n-1} + \ldots + \binom{n}{m}p^m (1-p)^{n-m}.$$

Each of these involves a lot of multiplications and, if one is going to do any amount of work with binomial probability—as statisticians, actuaries, and other applied mathematicians do—or one problem with a very large value for n, one will want a short cut. In former times this meant tables. Today this means previously programmed functions on one's calculator or in one's computer software. But tables of and programs for what? Values of A? of B? of C? Or, does one need all three?

From the standpoint of pure mathematics, $A(n, p, k)$ would seem the fundamental notion from which the others derive:

$$B(n, p, m) = A(n, p, m) + A(n, p, m+1) + \ldots + A(n, p, n)$$
$$C(n, p, m) = A(n, p, 0) + A(n, p, 1) + \ldots + A(n, p, m).$$

However, A can also be recovered from B or C:

$$A(n, p, k) = B(n, p, k) - B(n, p, k+1)$$
$$= C(n, p, k) - C(n, p, k-1),$$

and B, C can be derived from each other:

$$B(n, p, m) = 1 - C(n, p, m - 1)$$
$$C(n, p, m) = 1 - B(n, p, m + 1).$$

Any of the three can be taken as the primitive notion. Indeed, Todhunter notes about Abraham de Moivre's *Doctrine of Chances*:

> To find the chance that an event may happen *just r* times, De Moivre directs us to subtract the chance that it will happen *at least r − 1* times from the chance that it will happen *at least r* times. He notices, but less distinctly than we might expect, the modern method which seems more simple and more direct, by which we begin with finding the chance that an event shall happen *just r* times and deduce the chance that it shall happen *at least r* times.[8]

De Moivre's *Doctrine of Chances; or, a Method of Calculating the Probabilities of Events in Play* was first published in 1718, a second edition appeared in 1738, and a third, two years after his death, in 1756. The third edition is available online through Google Books and happens to be the edition Todhunter was referring to. If we examine the work carefully, we see why he took an approach that might seem strange to us. In the Introduction, after discussing generalities he gets down to business, calculating the probabilities of various events involved in throwing an ordinary 6-sided die in various examples he labels "cases":

CASE Ist. To find the Probability of throwing an Ace in two throws of one Die.

CASE II^{d}. To find the Probability of throwing an Ace in three throws.

CASE III^{d}. To find the Probability of throwing an Ace in four throws.

CASE IVth. To find the Probability of throwing two Aces in two throws.

CASE Vth. To find the Probability of throwing two Aces in three throws.

CASE VIth. To find the Probability of throwing two Aces in four throws.

CASE VIIth. To find the Probability of throwing one Ace, and no more, in four throws.[9]

Cases I to VI are in accordance with gambling practice, whereby one wins if one obtains at least the desired number of successful throws. His method of determining the probabilities is recursive, like Pascal's original method of resolving the problem of points (which, incidentally, de Moivre takes up in Cases VIII to X). The key observation is that, to obtain at least m successes in

---

[8] Isaac Todhunter, *A History of the Mathematical Theory of Probability From the Time of Pascal to That of Laplace*, Cambridge University Press, Cambridge, 1865 (reprinted by Chelsea Publishing Company, New York, 1949, 1965), p. 142. "$r − 1$" is, of course, a typographical error and should read "$r + 1$".

[9] De Moivre, *Doctrine of Chances*, 3rd edition, 1756, pp. 9 - 16.

n trials, either one obtains one in the first trial and only needs $m-1$ successes in the remaining $n-1$ trials, or one does not succeed in the first trial and one must have m successes in the remaining $n-1$ trials. This translates to the recursion equation,

$$B(n,p,m) = p \cdot B(n-1,p,m-1) + (1-p) \cdot B(n-1,p,m),$$

for $m, n \geq 1$. Together with some initial values,

$$B(n,p,0) = 1, \text{ for } n > 0$$
$$B(n,p,m) = 0, \text{ for } m > n \geq 1,$$

one has a recursive procedure for computing $B(n,p,m)$ for integers $n \geq 1, m \geq 0$, and any value of p. Note that this procedure does not explicitly refer to A or to the representation (6).

1 Exercise. Use the recursion to compute the probabilities asked for in CASES I - VI.

Before proceeding to CASE VII, de Moivre discussed the general problem of finding the probability of obtaining at least m successes in n trials and derived a version of (6). Nonetheless, when considering CASE VII and the probability of exactly k successes in n trials, he does not simply refer to $_nC_k\, p^k(1-p)^{n-k}$ as the appropriate term of the sum, but subtracts the probability of coming up with at least $k+1$ successes from that of at least k successes—which would be the natural way to proceed given his exposition thus far.

In a modern, logically developed exposition of probability, A would be the fundamental notion presented first, B being determined by (6) and C, likewise, by its sum representation. However, B and C are equally fundamental, and are more common in applications. Thus, in a book of tables we ought to be more likely to see a table of values of B or C than one of A.[10]

If we look to the history, we don't initially see tables of any of these. There were tables produced during this period, and, indeed, several useful tables are included in the third edition (1756) of *Doctrine of Chances*: present value of annuities, mortality tables, and the sums of the base 10 logarithms of the first n numbers, i.e., the logarithms of $n!$, for $n = 10, 20, \ldots, 900$. And in his English translation of 1795 of the second part of *Ars conjectandi*, Maseres included several tables from other sources, including tables of square and cube roots from 1 to 180 taken from *The Calculator* (1747) of James Dodson (? - 1757), and a table of square roots and reciprocals of numbers 1 to 1000 taken from the *Miscellanea Mathematica* (1775) of Charles Hutton (1737 - 1823). But there are no tables of the binomial probabilities $A, B,$ or C.

De Moivre begins a section of *Doctrine of Chances* titled "A Method of approximating the Sum of the Terms of the Binomial $(a+b)^n$ expanded into a Series, from whence are deduced some practical Rules to estimate the Degree of Assent which is to be given to Experiments" with the following words:

---

[10] Note too that obtaining A from B or C generally requires a lot less calculation than obtaining B or C from A.

Altho' the Solution of Problems of Chance often requires that several Terms of the Binomial $(a+b)^n$ be added together, nevertheless in very high Powers the thing appears so laborious, and of so great a difficulty, that few people have undertaken that Task; for besides *James* and *Nicolas Bernoulli*, two great Mathematicians, I know of no body that has attempted it; in which, tho' they have shewn very great skill, and have the praise which is due to their Industry, yet some things were farther required; for what they have done is not so much an Approximation as the determining very wide limits, within which they demonstrated that the Sum of the Terms was contained.[11]

De Moivre bypassed the laboriousness of the calculation by means of an approximation that is very good for moderate values of p (not too close to 0 or 1) and moderate values of n (say, $n \geq 30$) and improves as n gets larger. In this he took the decisive step in a development that began in the fourth part of Jakob Bernoulli's *Ars conjectandi* and received a polished presentation at the hands of Pierre Simon de Laplace in the *Théorie analytique des probabilités* (1812). The final result is that binomial probabilities can be approximated by areas under the *Bell Curve*,

$$y = \frac{1}{\sqrt{\pi}} e^{-x^2}. \tag{7}$$

At this point in the present book there should be a couple of graphs, one of the function (7) to illustrate its bell shape and a plot of binomial probabilities to illustrate the same. Using a translation and some vertical and horizontal stretching, the two graphs can be made to approximate each other and one could then literally see for oneself how good or bad the fit is. I shall not present such illustrations here, but shall instead discuss how to do the graphing on the *TI-83 Plus* .

First, I must desert the chronology and bypass the construction of tables of binary probabilities and discuss how such probabilities are found on the calculator. Let us take as a simple example the case of 10 trials with probability $p = 1/2$ of success. We want to graph the points with coordinates,

$$\langle 0, A(10, .5, 0)\rangle, \langle 10, A(10, .5, 1)\rangle, \ldots, \langle 10, A(10, .5, 10)\rangle,$$

connecting successive points with the straight lines as we go. The straightforward and unsophisticated way is to calculate the individual probabilities $A(10, .5, 0), \ldots, A(10, .5, 10)$ via the calculator buttons and commands we already know and then graph the points and lines using the DRAW button. We can use the built-in LIST functions to do the calculations for us. For example,

seq(X,X,0,10)$\rightarrow$L$_1$

will create a list $\{0, 1, \ldots, 10\}$ of x-coordinates of the points wanted, and

---

[11] De Moivre, *op.cit.*, p. 243. The entire section has been reprinted in David Eugene Smith, *A Source Book in Mathematics*, Cambridge University Press, Cambridge, 1929 (reprinted by Dover Publishing Company, New York, 1959), pp. 566 - 575; and in David, *op. cit.*, pp. 254 - 267.

seq$((10 _n C_r X)*(.5^{\wedge}10),X,0,10) \rightarrow L_2$

will store the y-coordinates in L_2. We could then use the DRAW button to find the Line(command and successively enter

Line$(L_1(1),L_2(1),L_1(2),L_2(2))$
Line$(L_1(2),L_2(2),L_1(3),L_2(3))$

$\vdots$

Line$(L_1(10),L_2(10),L_1(11),L_2(11))$

making sure first to go into the equation editor to turn off any stored functions we don't want to simultaneously graph. Before doing all of this, of course, we should pick some suitable window setting. Choosing -1 for Xmin and 11 for Xmax will do nicely horizontally. Vertically, the largest term in L_2 is $L_2(6) \approx$ .24609. So choosing $-.2$ for Ymin and .3 for Ymax will work nicely. However, obtaining the graph this way (and for one who has not previously experimented with the DRAW commands, I recommend doing this) requires a lot of key pressing and one might prefer to write a little program to do all of this:

```
PROGRAM:BINGRAPH
:FnOff
:PlotsOff
:ClrDraw
:−1→Xmin
:11→Xmax
:−.2→Ymin
:.3→Ymax
:seq(X,X,0,10)→L₁
:seq((10 nCr X)*(.5^10),X,0,10)→L₂
:For(I,1,11)
:Pt-On(L₁(I),L₂(I),2)
:End
:For(I,1,10)
:Line(L₁(I),L₂(I),L₁(I+1),L₂(I+1))
:End
:DelVar I.
```

A few words of explanation: The first three commands guarantee that no functions, statistical plots, and draw figures other than those being programmed will appear on the graphics screen. FnOff can be found in the On/Off... submenu of the Y-VARS submenu of the variables menu accessed by the VARS button; PlotsOff is found by pressing the STAT PLOT button, and ClrDraw by pressing the DRAW button. Xmin, Xmax and so on are found in the Window... submenu accessed via the VARS button. The Pt-On(command, which is found in the POINTS submenu accessed by the DRAW button, puts a mark at the point whose coordinates are given. There are three types of points, number 2 being a small box. Note that I did not use this in our unprogrammed graph because there are 11 points to be turned on, hence 11 commands to enter by hand. In programming, however, this extra flourish required only three little lines of code.

As I said, this method of graphing the binomial probabilities is not the most sophisticated one. It certainly reveals the bell shape of the curve, so it serves its purpose and should not be dismissed out of hand. However, the *TI-83 Plus* is not just a graphing calculator, but was designed to do statistical work as well and has much of what we did already built into it. Before explaining how to access these features I should first explain some terminology. The whole array of probabilities associated with binomial probability is called the *binomial distribution* and the function $A(n, p, k)$ is termed the *binomial distribution function*. It has been pre-programmed into the calculator as binomialpdf(and can be accessed via the DISTR button.[12] The function binompdf(takes two or three arguments—the number n of trials, the probability p of success on a given trial, and, optionally, the number k of successes desired in n trials. The output on two arguments is a list of probabilities, and on all three arguments n, p, k is the probability of obtaining exactly k successes in n trials. Thus, for example,

binomialpdf(10,.5)

performs with fewer keystrokes the same task as

seq((10 $_nC_r$ X)*(.5^10),X,0,10).

And

binomialpdf(10,.5,5)

calculates $A(10, .5, 5)$, i.e., it does the same as

(10 $_nC_r$ 5)*(.5^10).

The built-in binomial distribution function not only simplifies slightly the task of finding probabilities, but it can also calculate probabilities where the straightforward representation fails. For example, suppose we want to know the probability of there being exactly 500 boys born out of a thousand births, assuming it equally likely that a boy or a girl is born on a given trial. Entering

(1000 $_nC_r$ 500)*(.5^1000)

results in an OVERFLOW error message. The first term, 1000 $_nC_r$ 500 is larger than any number the calculator can handle.[13] But entering

binompdf(1000,.5,500)

---

[12] Beware: There is also a binomialcdf(function programmed into the calculator, so be careful in choosing the correct function in the following. You want the one ending in "pdf" for "*probability distribution function*". The second function, ending in "cdf" for "*cumulative distribution function*", is essentially $C(n, p, m)$ described earlier.

[13] And the second term, .5^1000 is smaller in absolute value than any number the calculator can distinguish from 0.

results in .0252250182.

Getting back to the problem of graphing the binomial distribution, we can use binomialpdf(in place of the more cumbersome defining expression and enter successively

seq(X,X,0,10)→L_1
binompdf(10,.5)→L_2

to generate the lists L_1,L_2 of coordinates of the points we wish to graph. The actual graphing could now be done as before, but the calculator has a built-in statistical plotting feature to make this step easier as well. One hits the STAT PLOT button to enter the statistical plotting menu, then chooses Plot1. This brings one to a new screen with several entries. The first is the On/Off option. Positioning the cursor over On and hitting ENTER will turn the graph on. One can then scroll down to the Type option and choose the second pictogram for an xyLine plot and hit ENTER. The lines below the pictograms depend on the Type chosen. For the xyLine option, the choices are labelled Xlist, Ylist, and Mark. For Xlist and Ylist enter L_1 and L_2, respectively, and choose the little box for Mark as it will show up most clearly. One can now exit to the home screen and prepare to graph the distribution. One of the advantages of using the built-in statistical plotting is that one need not worry about choosing the appropriate window: one needs only hit the ZOOM button and scroll down to and select the ZoomStat item. An appropriate window is determined automatically and the plot is graphed. The bell shape of the curve stands out even more clearly than with the choice of window I made earlier.

To see the bell shape of the curve (7) is an easy matter. First one turns off the statistical plots and then enters

$Y_1=1/\sqrt(\pi)e^\wedge(-X^2)$.

As for the choice of window, ZDecimal in the ZOOM menu gives a serviceable, but rather flat and uninteresting graph. Going back to the ZOOM menu and choosing ZoomFit will stretch the graph vertically to give a more dramatic, if exaggerated, bell shape.

Graphing (7) and the binomial distribution together requires one to translate and stretch the former curve. The stretch and translation depend on two parameters called the *mean* (μ) and *standard deviation* (σ), or, equivalently, the mean (μ) and the *variance* (σ^2). And for this, I should save some space by considering at the outset the general binomial distribution given by n trials with probability p of success on any given trial. For such a distribution, the mean is just the weighted average of the number of successful trials—the expected number of successes in n trials:

$$\mu = 0 \cdot \binom{n}{0}p^0(1-p)^n + 1 \cdot \binom{n}{1}p^1(1-p)^{n-1} + \ldots + n \cdot \binom{n}{n}p^n.$$

It is a bit tricky, but it can be shown with a little algebra[14] that $\mu = np$. The variance is the average of the squares of the deviations of the numbers of

---

[14] See Appendix B for the details.

successes from the mean number of successes:

$$\sigma^2 = (0-\mu)^2 \binom{n}{0} p^0 (1-p)^n + (1-\mu)^2 \binom{n}{1} p^1 (1-p)^{n-1} + \ldots + (n-\mu)^2 \binom{n}{n} p^n.$$

With a bit of algebra, it can be shown[15] that $\sigma^2 = np(1-p)$. Storing appropriate values in N, P, we can graph the desired curve by first entering

N*P→M
M*(1−P)→S

and then defining

Y$_2$=1/$\sqrt{(2*S)}$Y$_1$((X−M)/$\sqrt{(2*S)}$).

If we start by storing 10 in N and .5 in P, turn off the graph of Y$_1$, turn on the statistical plot we did for the binomial distribution, and then graph using ZoomStat, we will see a remarkable fit between the binomial distribution and the transformed curve.

To get a feel for how well or how poorly the transformed curve fits the binomial distribution in general, one should examine numerous examples. To this end, I offer the following simple program:

```
PROGRAM:BELL1
:PlotsOff
:FnOff
:ClrDraw
:Disp "ENTER N:"
:Input N
:Disp "ENTER P:"
:Input P
:seq(X,X,0,N)→L₁
:binompdf(N,P)→L₂
:Plot1(xyLine,L₁,L₂,□)
:N*P→M
:M*(1−P)→S
:"1/√(2πS)e^(−(X−M)²/(2*S))"→Y₁
:ZoomStat.
```

The first thing requiring explanation here is the Plot1(command. The STAT PLOT button opens a different menu when accessed from the program editor from the one you see when the button is pressed elsewhere. The command Plot1(is in the PLOTS submenu visible on pressing STAT PLOT, the xyLine command is in the TYPE submenu, and the small box is in the MARK submenu. All three are also available in the CATALOG menu. The line Plot1(xyLine,L₁,L₂,□) does everything inside the program—turning on the graph, choosing the graph type, Xlist, Ylist, and mark—that we did previously by hand.

The command in the next to the last line of the program is of the form

---

[15] Ditto.

"*expression*" → Y_1

and attaches the expression trapped between the quotes to the function variable Y_1. Expressions can be attached to function variables and lists and are quite handy. I discuss them briefly in Appendix A. For now it suffices to know merely that it is the way of entering a function in the equation editor from within a program.

2 Exercise. Enter the program BELL1 into the *TI-83 Plus* and run it with various choices of ⟨N,P⟩, including the following
i.   ⟨10,.5⟩ , ⟨15,.5⟩
ii. ⟨10,.4⟩ , ⟨15,.4⟩
iii. ⟨10,.2⟩ , ⟨15,.2⟩ , ⟨20,.2⟩
iv. ⟨10,.05⟩ , ⟨15,.05⟩ , ⟨20,.05⟩
v.   ⟨10,.9⟩ , ⟨15,.9⟩ .
Describe your findings.

I am tempted to replace the line defining Y_1 by the lines

:"1/$\sqrt{(\pi)}$e^(−X²)" → Y_2
:"1/$\sqrt{(2*S)}$*Y_2((X−M)/$\sqrt{(2*S)}$)" → Y_1
:FnOff 2.

The first line emphasises the basic function (7), the second then stresses the transformations involved, and the third guarantees the auxiliary Y_2 will not be graphed.

There is an even better replacement. The function

$$y(x) = \frac{1}{\sqrt{2\pi}\sigma}\, e^{-\frac{1}{2}((x-\mu)/\sigma)^2} \tag{8}$$

is pre-programmed on the *TI-83 Plus* as normalpdf(, the *normal probability distribution function*. I will explain the name shortly. For now, I note that normalpdf(is the first item in the DISTR menu accessed by the DISTR button and that it can be taken as a function of a single variable x defined by fixing $\mu = 0, \sigma = 1$ in (8),

$$y(x) = \frac{1}{\sqrt{2\pi}}\, e^{-\frac{1}{2}x^2}, \tag{9}$$

or as a function of all three variables x, μ, σ. We can use this function and replace the line in question by

:"normalpdf(X,M,$\sqrt{(S)}$)" → Y_1,

or, computationally slightly more efficiently[16], by

---

[16] Recall that S initially defines the variance σ^2, not σ, and normalpdf(is a function of σ, not σ^2. With the two-line replacement the program will calculate the square root of the variance only once to obtain the standard deviation σ; with the single-line replacement, this calculation is repeated once for each of the 95 points plotted in graphing the curve.

:$\sqrt{(S)}{\rightarrow}S$
:"normalpdf(X,M,S)"$\rightarrow Y_1$.

A few words about probability distribution functions: Recall our paradigm of discrete probability spaces: a set, called the sample space, of *equally likely* outcomes. We did not assign probabilities to outcomes, but to sets of outcomes called events. Occasionally, one is tempted to describe a situation in a way in which the intended outcomes are not equally likely, as with Bertrand's Box Paradox, or declaring the outcomes of an experiment of tossing a coin 3 times to be $0H, 1H, 2H$ and $3H$. Up till now we declared this to be in error, that $0H, 1H, 2H$ and $3H$ were not outcomes, but events in the more correctly described sample space $\{TTT, HTT, \ldots, HHH\}$. The notion of a distribution function allows us to generalise our notion of probability and allow $\{0H, 1H, 2H, 3H\}$ to be a sample space of outcomes of various likelihoods[17]. A *distribution function* f on a sample space S is merely an assignment of numerical likelihoods $f(O)$ to the outcomes $O \in S$. It must satisfy two conditions:

  i. $f(O)$ is not negative
  ii. if $S = \{O_0, O_1, \ldots, O_{n-1}\}$, then $f(O_0) + f(O_1) + \ldots + f(O_{n-1}) = 1$.

Given a distribution function f on a sample space S, the probability of an event $E = \{O_{i_0}, \ldots, O_{i_{k-1}}\} \subseteq S$, is just the sum,

$$P(E) = f(O_{i_0}) + \ldots + f(O_{i_{k-1}}), \tag{10}$$

of the likelihoods of the elements of E.

This agrees with our older definition of a sample space S of equally likely outcomes, if for such a space we define

$$f(O) = P(\{O\}) = \frac{1}{|S|}.$$

Conditions i and ii are fairly obvious. And the agreement of $P(E)$ as defined by (10) with our earlier definition is a simple calculation:

$$P(E) = f(O_{i_0}) + \ldots + f(O_{i_{k-1}}) = \underbrace{\frac{1}{|S|} + \ldots + \frac{1}{|S|}}_{|E|} = \frac{|E|}{|S|}.$$

The notion of a distribution function generalises to the continuous case. A function f defined on the real numbers is a distribution function if it satisfies two conditions:

  i. $f(x) \geq 0$ for all real numbers x
  ii. the total area under the curve $y = f(x)$ is 1.

---

[17] The use of the word "likelihood" here is not a perfect choice, as it has been used since 1921 by statisticians in a different context with a different meaning. For now I simply want a word other than "probability" which is informally synonymous with "probability" and doesn't have the same formal meaning. The phrase "equally likely" in the present context suggests "likelihood".

Given a distribution function, the probability that a number chosen randomly according to the distribution lies in an interval $[\alpha, \beta]$ is just the area under the curve trapped between $x = \alpha$ and $x = \beta$. In the Calculus, this area is called the *definite integral* of f over the interval $[\alpha, \beta]$ and is denoted

$$\int_\alpha^\beta f(x)\, dx. \tag{11}$$

Each of the functions f defined by (7), (8), and (9) is a distribution function: For any real number x, e^{-x^2} is positive, whence i is satisfied. That condition ii,

$$\int_{-\infty}^\infty f(x)\, dx = 1,$$

is also satisfied is not trivial, as is testified to by the fact that attribution is given for the result:

> The formula $\int_0^\infty e^{-t^2}\, dt = \frac{1}{2}\sqrt{\pi}$ is ascribed by Gauss (*Theoria Motus Corporum Cœlestium*, p. 212) to Laplace. On making $e^{-t^2} = z$, the integral is transformed into $\frac{1}{2}\int dz(\log \frac{1}{z})^{-1/2}$ between the limits $z = 0$ and $z = 1$, the value of which $= \frac{1}{2}\sqrt{\pi}$, had been given by Euler, long before, in the *Petersburg Memoirs*. See Legendre, *Exercises du Calcul Intégral*, tom. i., p. 301.[18]

Much time is spent in the basic Calculus course learning how to evaluate expressions of the form (11). This is usually done by applying a panoply of rules to find a function F such that

$$\int_\alpha^\beta f(x)\, dx = F(\beta) - F(\alpha). \tag{12}$$

The method does not work for $f(x) = e^{-x^2}$: There is no expression $F(x)$ built up from $x, e^x, \ln x, \sin x, \cos x$, etc. using the usual algebraic operations for which (12) holds for all intervals—or even for all subintervals of some given interval of nonzero length. The evaluation of the integral over the particular intervals $(-\infty, \infty)$ and $(0, \infty)$ can be done using an inspired trick and the techniques studied in the latter portions of the Calculus course[19], but the more general instances of (11) for $f(x) = e^{-x^2}$ must be evaluated by numerical approximations. Thomas Galloway (1796 - 1851) describes how this is done. The reader who knows no Calculus will not be able to follow the steps, but he should appreciate a few points of the technique:

> The integral $\int e^{-t^2}\, dt$ is computed as follows. Developing the exponential e^{-t^2} in a series[20] of the ascending powers of t^2, and integrating the successive terms between the limits $t = 0$ and $t = \tau$, we find

---

[18] Thomas Galloway, *A Treatise on Probability; Forming the Article Under that Head in the Seventh Edition of the Encyclopædia Britannica*, Adam and Charles Black, North Bridge, Edinburgh, 1839, p. 142.

[19] *Cf.* Appendix B, below.

[20] One has

Celebrities Formula

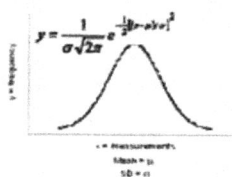

$$y = \frac{1}{\sigma\sqrt{2\pi}} e^{-\frac{1}{2}\left(\frac{x-\mu}{\sigma}\right)^2}$$

Abraham de Moivre
(1667 - 1754)

Destinatar

The only philatelic tribute to Abraham de Moivre in my collection is the above Romanian postal card from 2010. Curiously, it bears no monetary value in the imprinted stamp, no date on the front or back of the card, no country name, and no printer information. That could mean it is a *cinderella*, i.e., not issued by the postal authorities, or that it is for domestic use only.

The image of de Moivre in the cachet and the imprinted stamp is a commonly reproduced one. The other illustration is the bell curve and the formula for it. This part of the image is muddy in the original card, suggesting it was copied from an inferior computer printout or xerox.

There are several such cards issued along with this one, featuring such mathematicians of interest here as Pierre de Fermat, Jakob Bernoulli, Joseph Louis Lagrange, and Carl Friedrich Gauss. I would be tempted to show here the card featuring Bernoulli were not the portrait a very dark version of the portrait reproduced more clearly on pages 31 and 162. The cachet of that card has two formulæ of Bernoulli's, one of which, the infinite expansion of $x/(e^x - 1)$, we will use in Section 5 of Appendix B on page 455, below.

$$e^x = 1 + \frac{x}{1!} + \frac{x^2}{2!} + \frac{x^3}{3!} + \ldots,$$

whence

$$e^{-t^2} = 1 - \frac{t^2}{1!} + \frac{t^4}{2!} - \frac{t^6}{3!} + \ldots$$

By standard tools of the Calculus,

$$\int_0^\tau e^{-t^2}\, dt = \tau - \frac{\tau^3}{1!3} + \frac{\tau^5}{2!5} - \frac{\tau^7}{3!7} + \frac{\tau^9}{4!9} - \ldots$$

Note the typo in Galloway's book: the sign of τ^7 should be negative.

 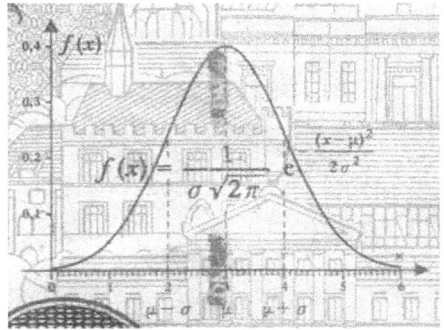

Carl Friedrich Gauss (1777 - 1855) was one of the greatest mathematicians of all time. His connexion with the Bell Curve was late but important, so much so that the normal distribution is often referred to as the *Gaussian distribution*. In 1955 West Germany commemorated the centennial of Gauss's death by issuing a stamp bearing his portrait and in 1977 both West and East Germany celebrated his 200th birthday by issuing postage stamps in his honour. The West German stamp pictured the "Gaussian plane", i.e., the geometric representation of the complex numbers, and the East German stamp featured his portrait and an artistic representation of the construction of the regular 17-gon by the young Carl Gauss. Of these I chose the East German stamp (above left) to picture here because of the postmark, which is the cancellation used on the first day of the issue of the stamp on 19 April 1977. (His actual birthday was 30 April—Walpurgisnacht!) The banknote, pictured below has a nice portrait and, in the background, the Bell Curve and the formula for the normal distribution function (inset above right) as well as some of the buildings of Göttingen.

AA5460267Y0

$$\int_0^\tau e^{-t^2}\, dt = \tau - \frac{\tau^3}{1\cdot 3} + \frac{\tau^5}{1\cdot 2\cdot 5} + \frac{\tau^7}{1\cdot 2\cdot 3\cdot 7} + \&c.;$$

a series which converges rapidly when τ is less than unity. In the contrary case, however, or when τ is greater than unity, the series is divergent[21], and it is necessary to proceed by a different method. Let the factor e^{-t^2} be multiplied and divided[22] by t; we have then

$$\int e^{-t^2}\, dt = \int \frac{1}{t} e^{-t^2} t\, dt,$$

and on integrating by parts[23]

$$\int \frac{1}{t} e^{-t^2} t\, dt = -\frac{e^{-t^2}}{2t} - \frac{1}{2} \int \frac{1}{t^2} e^{-t^2}\, dt.$$

Repeating the same process on the last integral, and so on with the last after each succeeding integration, the following series is obtained,

$$\int e^{-t^2}\, dt = -\frac{e^{-t^2}}{2t}\left\{ 1 - \frac{1}{2t^2} + \frac{3}{2^2 t^4} - \frac{3\cdot 5}{2^3 t^6} + \&c \right\}.$$

When $t =$ infinity the right hand side of this equation becomes 0; whence between the limits $t = \tau$ and $t =$ infinity, we have[24]

$$\int_\tau^\infty e^{-t^2}\, dt = -\frac{e^{-\tau^2}}{2\tau}\left\{ 1 - \frac{1}{2\tau^2} + \frac{3}{2^2 \tau^4} - \frac{3\cdot 5}{2^3 \tau^6} + \&c \right\},$$

a series which converges very rapidly when τ is greater than unity. Now the value of a definite integral between 0 and infinity is obviously equal to its two parts, of which the first is taken between 0 and τ, and the second between τ and infinity; that is to say[25],

$$\int_0^\tau e^{-t^2}\, dt = \int_0^\infty e^{-t^2}\, dt - \int_\tau^\infty e^{-t^2}\, dt.$$

But $\int_0^\infty e^{-t^2}\, dt$ is well known to have for its expression $\frac{1}{2}\sqrt{\pi}$, therefore

$$\int_0^\tau e^{-t^2}\, dt = \frac{1}{2}\sqrt{\pi} - \int_\tau^\infty e^{-t^2}\, dt,$$

---

[21] I am not sure exactly what is meant here. The series is convergent for all real values of τ. This convergence, however, is not exceptionally rapid for $\tau > 1$.

[22] A clever, but not uncommon, trick.

[23] A standard procedure in the Calculus.

[24] Two more typos: The leading minus sign on the right has incorrectly been carried down from the preceding formula and does not belong here; and "τ_2" should be "τ^2".

[25] The area under the curve between 0 and τ is obtained by subtracting the area between τ and ∞ from that between 0 and ∞.

so that the integral may be computed from either of the above series, according as τ is less or greater than 1.[26]

3 Project. Use your calculator to check Galloway's results numerically: Make a small table of values of

$$\int_0^\tau e^{-t^2}\, dt$$

by storing values of

fnInt(e^(−X²),X,0,x)

for $x = 0, .1, .2, \ldots, 2.0$ in L_1. (The function integration command, fnInt(, is in the MATH menu. It is discussed in greater detail in Appendix B.) Use the four-term expansions Galloway supplies. Use the first sum,

X−X^3/(1∗3)+X^5/(1∗2∗5)−X^7/(1∗2∗3∗7)

for X = 0, .1, .2, \ldots, .9; use the second sum

√(π)/2−e^(−X^2)/(2X)(1−1/(2∗X^2)+3/(2^2∗X^4)−(3∗5)/(2^3∗X^6))

for X = 1.1, 1.2, \ldots, 2.0; average the values obtained from these two expressions for X = 1; and store the results in proper order in L_2. Compare L_1 and L_2. If you know Calculus, determine the fifth terms of the power series and repeat the exercise.

Galloway finishes the section on tabulating the integral with the following remarks:

The integral $\int e^{-t^2}\, dt$ is of great importance in the higher mathematics. It occurs in the investigation of the path of a ray of light through the atmosphere, and of the law of the diffusion of heat in the interior of solid bodies, as well as in the determination of the degree of reliance that may be placed on the results of astronomical observations, and generally in most of the more difficult and important applications of the theory of probabilities. A table of its values from $t = 0$ to $t = 3$, for intervals each $= .01$, was given by Kramp, at the end of his *Analyse des Refractions Astronomiques*, Strasburg, 1799. In the *Berliner Astronomisches Jahrbuch* for 1834, there is also a table of its values from $t = 0$ to $t = 2$ (for the same intervals) multiplied by $2 \div \sqrt{\pi}$, with their first and second differences[27], for the purpose of facilitating interpolation. This last table, which appears to have been derived from that

---

[26] Galloway, *op.cit.*, pp. 141 - 142. This method of approximating the integral goes back to Laplace. I quote Galloway because his remarks are short, to the point, and already in English.

[27] If f is a tabulated function, with values calculated at intervals of size h, the difference function Δf is defined by

$$\Delta f(x) = f(x + h) - f(x),$$

and the second difference $\Delta^2 f$ is

$$\Delta^2 f(x) = \Delta f(x + h) - \Delta f(x).$$

of Kramp,...we have extended to $t = 3$, and given at the end of the present article. As the function which is thus tabulated will occur frequently in what follows, we shall in future, for convenience, in printing, denote it by Θ, that is to say, we shall assume

$$\Theta = \frac{2}{\sqrt{\pi}} \int_0^\tau e^{-t^2} \, dt = 1 - \frac{2}{\sqrt{\pi}} \int_\tau^\infty e^{-t^2} \, dt,$$

the two forms being equivalent in consequence of the above equation.[28]

The notation Θ did not catch on. The letter Φ proved to be more popular and is still in use. Nowadays, the function is called the *error function* and one sees

$$\mathrm{erf}(x) = \frac{2}{\sqrt{\pi}} \int_0^x e^{-t^2} \, dt$$

occasionally. The desire for a table of values of the error function, without the factor $2/\sqrt{\pi}$, was expressed by Laplace in a memoir published in 1786 and, indeed, as Galloway noted, Chrétien (or, Christian) Kramp (1760 - 1876) produced such a table in his *Analyse des réfractions astronomiques et terrestres*, published, according to the *Dictionary of Scientific Biography*, in Strasbourg and Leipzig in 1798. The function is sometimes referred to as *Kramp's transcendental* in his honour.[29] A quick glance at the thumbnails of the greater than 700 pages of the PDF file of the third (1820) edition of Laplace's monumental textbook did not reveal any tables of the error function. Other expositions, however, did include such. Todhunter's history cites two by Augustus de Morgan: in article 101 of *Theory of Probability* in the *Encyclopædia Metropolitana*, and in the first appendix to his *Essay on Probabilities* in the *Cabinet Cyclopædia*. Todhunter also cites Galloway's work quoted from above, and to this I might add two additional volumes that I was able to download from the Internet: Emanuel Czuber's (1851 - 1920) 1879 translation *Vorlesungen über Wahrscheinlichkeitsrechnung* [*Lectures on probability calculation*] of Anton Meyer's (1801 - 1857) *Théorie analytique des probabilités à posteriori*, and Joseph Louis François Bertrand's *Calcul des probabilités* of 1889.

I've not seen de Morgan's contribution to the *Encyclopædia Metropolitana*, but we can perhaps glean its significance from Galloway's appraisal of his predecessors:

> Since the time of Demoivre, the English treatises on the general theory of probability have neither been numerous, nor, with one or two exceptions, very important. Simpson's *Laws of Chance* (1740) contains

---

Differences, second differences, third differences, etc., are useful in interpolation—estimating the value of f at points between tabulated entries. Tables of functions often included the differences and, occasionally, the second differences as well.

[28] Galloway, *op. cit.*, pp. 142 - 143.

[29] To this, one might add that Kramp introduced the use of the exclamation point, $n!$, for the factorial, which ultimately replaced the typographically less convenient practice of placing n within a half open box: $\lfloor n$.

a considerable number of examples, in the solution of which the author displays his usual acuteness and originality, but as they belong entirely to that class in which the chances are known *a priori*, they give no idea of the most interesting applications of the theory. Dodson's *Mathematical Repository* contains a large selection of the same kind. The Essay in the *Library of Useful Knowledge*, by Mr. Lubbock[30], gives a more comprehensive and philosophical, though an elementary view of the subject; but by far the most valuable work in the language is the treatise in the *Encyclopedia Metropolitana*, by Professor De Morgan, 1837. In this very able production, Mr. De Morgan has treated the subject in its utmost generality, and embodied, within a moderate compass, the substance of the great work of Laplace.[31]

Modern opinion is much the same. Adrian Rice describes it as "a book-length article for the *Encyclopædia Metropolitana*" and adds that "while it contained no original results, [it] was nevertheless the first major exposition of the subject to be published in Britain, and as such, it constituted the first major work on modern probability theory to appear in the English language".[32] The key word here is "modern". Huyghens had been translated into English and de Moivre's book was originally published in English. In the 19th century, however, *the* most important work in probability was Laplace's *Théorie analytique* and de Morgan was the interpreter of this work in English.

I have seen de Morgan's other major work on probability, namely his contribution to the *Cabinet Cyclopædia*, also referred to as *Lardner's Cabinet Cyclopædia*, founded and edited by Dionysius Lardner (1793 - 1859). The *Cabinet Cyclopædia* ran for 133 volumes between the years 1829 and 1849. One of them was de Morgan's *An Essay on Probabilities, And on Their Applications to Life Contingencies and Insurance Offices*[33]. Whereas the first work was an advanced theoretical treatise and had heavy mathematical pre-requisites, this second work was practical and presupposed only a knowledge of basic arithmetic. It concerned the application of probability theory to insurance, a matter which he, as a consultant for the industry, had some expertise in. "His book on the subject, the first of its kind, remained highly regarded in insurance literature for well over a generation." [34] His table of values of the error function (Table I) appears at the end of the book following Appendix VI, but instructions on the use of the tables occupy Chapter IV (pp. 69 - 93). To discuss this, however, we should go back to the Introduction in which de Morgan discusses the table geometrically:

---

[30] Meant is *On Probability* (1830) by John Lubbock and John Drinkwater-Bethune, a short, but decent account of the theory.

[31] Galloway, *op. cit.*, pp. 14 - 15.

[32] Adrian Rice, "Augustus De Morgan", in: C.C. Heyde and E. Seneta, eds., *Statisticians of the Centuries*, Springer-Verlag, New York, 2001; here, p. 159.

[33] Longman, Orme, Brown, Green & Longmans, and John Taylor, London, 1838. This book is available in PDF form online.

[34] Rice, *op.cit.*, p. 160.

...we can get through much the greater part of our task...by means of a table of which it is not possible to explain either the principle, or the reason of its utility, to any but a mathematician. We can only explain its mere construction, as follows:— Let A,B be *one* (inch, for example); and take an indefinitely extended line AX, perpendicular to

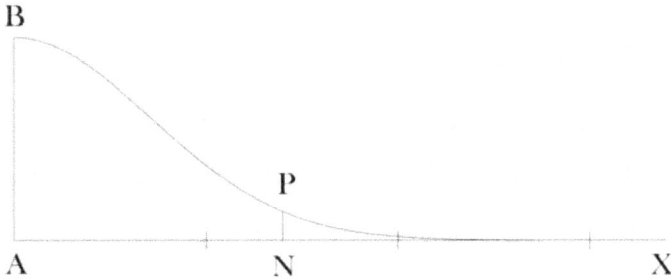

AB: from A toward X let a curve be conceived to be described, so that every *ordinate* NP shall be connected with its *abscissa* AN, by the following law. Measure AN in inches and parts of inches; and multiply the result by itself; and the product by .4342945. Find the number to which this product is the common logarithm, and divide 1.1284 by the result. The quotient is the fraction of an inch in NP [35], and in the table we find, not NP, but the area ANPB expressed as a fraction of a square inch. The curve itself is what is called an *asymptote* to AX, continually approaching, but never reaching, AX: and the whole area AX, continued forever, is one square inch. To this table I shall have continual occasion to refer: into it, in fact, is condensed almost the whole use I shall have to make of the higher mathematics.[36]

The table itself gives no name to the function, merely labelling the columns t, H, Δ, Δ^2, t ranging from .00 to 3.00 in increments of .01, and H, representing the values of the function, being given to 7 decimals, the first 5 separated from the last 2 by a brief space as de Morgan considered it rare that one would need greater than 5 decimal accuracy, and the columns Δ and Δ^2 of differences and second differences[37] of H, being given to 7 decimals, but with the leading 0's left undisplayed. The instructions on the use of first and second differences are sufficient to obtain accurate values of the function for arguments t given to 5 decimal places.

Galloway's tables are similar to de Morgan's, but for the labels. The tables themselves are headed

---

[35] *Exercise.* Show that, if x is the length AN, then the length NP is indeed $\frac{2}{\sqrt{\pi}}e^{-x^2}$.
[36] Pp. 16 - 17.
[37] *Cf.* Appendix B for more information on Δ and Δ^2.

TABLE

OF THE

VALUES OF THE INTEGRAL

$$\Theta = \frac{2}{\sqrt{\pi}} \int_0^\tau e^{-t^2}\, dt = 1 - \frac{2}{\sqrt{\pi}} \int_\tau^\infty e^{-t^2}\, dt$$

for intervals each $= .01$, from $\tau = 0$ to $\tau = 3$, with their first
and second differences.

The columns are headed $\tau, \Theta, \Delta, \Delta^2$ and the figures are given to 7 places, this time without the special separation after the fifth digit. Curiously, there are no entries for 2.98 and 2.99.

The Czuber/Meyer table is headed

$$\text{Tafel des Integrals } \ \Phi(\gamma) = \int_0^\gamma \frac{2}{\sqrt{\pi}} e^{-t^2}\, dt$$

and the columns are headed $\gamma, \Phi(\gamma), \Delta\Phi, \Delta^2\Phi$. γ ranges from 0.00 to 3.00 in increments of .01 and $\Phi(\gamma)$ is given to the usual 7 digits, presented more readably by separating the last 4 by a small space. Values for Δ^2 are not given for γ greater than 2.79, where de Morgan's values are primarily 1×10^{-7} and 2×10^{-7}.

Bertrand's table accidentally interchanges the numerator and denominator of the constant in the heading:

TABLE

DES

VALEURS DE L'INTEGRAL

$$\Theta(t) = \frac{\sqrt{\pi}}{2} \int_0^t e^{-t^2}\, dt\,.$$

Bertrand does not give the first and second differences of Θ, but his table is more extensive, giving values for $t = 0.00$ to 4.00 in increments of .01 and from 4.0 to 4.8 in increments of .1. The values of Θ are given to 7 decimals for $t \leq 3.45$ and to 11 decimals after that, as the first 6 digits are all 9's for $t > 3.45$.

The United States came into the act in the 1930's in response to the Great Depression with the Works Progress Administration sponsoring the Mathematical Tables Project, the purpose of which, like de Prony's project[38] for

---

[38] A famous tabulation project which set the standard for major computation efforts until the advent of electronic digital computers was that of Gaspard Clair François Marie Riche de Prony (1755 - 1839). De Prony led an assembly line production of tables of trigonometric and logarithmic tables, the bulk of the computation being performed by hairdressers rendered redundant by the revolution. *Cf.* Ivor Grattan-Guinness, "The computation factory: de Prony's project for making tables in the 1790s", in: Martin Campbell-Kelly, Mary Croarken, Raymond Flood, and Eleanor Robson, eds., *The History of Mathematical Tables: From Sumer to Spreadsheets*, Oxford University Press, Oxford, 2003.

The 19th century saw a great deal of activity in the theory of probability, and there were a number of texts elucidating it. At the beginning of the century there was Laplace's great, but difficult to read, synthesis, *Théorie analytique des probabilités*, Augustus de Morgan's book-length article for *Encyclopædia Metropolitana* (1837) was "the first major exposition of the subject to be published in Britain, and as such, it constituted the first major work on probability theory to appear in the English language" according to Adrian Rice's chapter on de Morgan in C.C. Heyde and E. Seneta, eds., *Statisticians of the Centuries* (Springer-Verlag, New York, 2001; here: p. 160). His *An Essay on Probabilities* (1838) was an instruction manual on applications to problems in insurance, and Rice informs us, "His book on the subject, the first of its kind, remained highly regarded in insurance literature for well over a generation".

Another English work of the period worthy of mention is *On Probability* (1830), a shorter work by John Lubbock and John Drinkwater-Bethune, which work was for years erroneously credited to de Morgan, who frequently acknowledged it was not his work. And, of course, there is Galloway's book of 1839 cited in the text.

LUXEMBOURG

Continental textbooks on probability include books by Simeon Denis Poisson (1837), Antoine Augustin Cournot (1843), Hermann Laurent (1873), Anton Meyer (1874), Joseph Bertrand (1888, 2nd edition 1907), and Henri Poincaré (1896). I represent them here with stamps depicting Meyer and Poincaré.

Born in Luxemburg, Meyer taught mathematics in various universities in neighbouring Belgium. In particular, he taught probability in Liège (Louk) from 1849 to 1857 and his lecture notes were published posthumously in 1874. The book caught on and in 1879 a translation from the French into German by Emanuel Czuber appeared.

Henri Poincaré (1854 - 1912) was one of the leading mathematicians of his day, best known for his work in mathematical physics and his popular philosophical expositions. His textbook on probability, *Calcul des probabilités* of 1896, though not as important as some of those noted, was successful enough to see an expanded second edition in 1912, which itself received a second printing in 1923.

The stamp bearing Meyer's portrait was issued in 1979 to celebrate 150 years of poetry in the Luxemburger language, Meyer being the father of Luxemburg poetry. The stamp bearing Poincaré's portrait is one of 6 semipostal charity stamps issued in 1952 to honour famous Frenchmen of the second half of the 19th century.

revolutionary France, was to provide employment for unskilled labour. Among the tables produced were the *Tables of Probability Functions*, the first volume (1941) of which bore the subtitle

$$H'(x) = \frac{2}{\sqrt{\pi}}\, e^{-x^2} \quad \text{and} \quad H(x) = \frac{2}{\sqrt{\pi}} \int_0^x e^{-\alpha^2}\, d\alpha. \tag{13}$$

Its tables gave values of H' and H to fifteen decimals in increments of .0001 between 0 and 1 and .001 between 1 and 5.6. There is also a table of values of H' and H to 8 decimals for x ranging from 4 to 10 in increments of .01. A second volume published in 1942 contained comparable tables for the functions[39]

---

[39] I haven't actually consulted these volumes, but am relying on reviews by Virgil Snyder in *The American Mathematical Monthly* vol. 48 (1942), p. 201, and vol. 49 (1943), pp. 31 - 32. General information about the Mathematical Table Project of

$$I(x) = \frac{1}{\sqrt{2\pi}} e^{-\frac{1}{2}x^2} \quad \text{and} \quad H(x) = \frac{1}{\sqrt{2\pi}} \int_{-x}^{x} e^{-\frac{1}{2}y^2}\, dy. \tag{14}$$

This might be a good point to pause and reflect on the fact that the title of this chapter refers to binomial probability, a topic from which I appear to have strayed in some sort of stream of consciousness narration. I haven't really done so, as the error function is used to approximate binomial probabilities. It is high time that I illustrated this.

We have already seen that $A(n,p,k)$ is approximated by

$$y(x) = \frac{1}{\sqrt{2\pi}\sigma} e^{-(x-\mu)^2/(2\sigma^2)} = \frac{1}{\sqrt{2\pi}\sigma} e^{-\frac{1}{2}((x-\mu)/\sigma)^2}, \tag{15}$$

for $\mu = np, \sigma^2 = np(1-p)$. The cumulative probabilities B, C can be approximated by areas under the curve

$$y = \frac{1}{\sqrt{2\pi}\sigma} e^{-\frac{1}{2}((x-\mu)/\sigma)^2}.$$

In fact, if $A(n,p,k)$ approximates

$$\frac{1}{\sqrt{2\pi}\sigma} e^{-\frac{1}{2}((k-\mu)/\sigma)^2},$$

then $A(n,p,k)$, being the area of a rectangle of height $A(n,p,k)$ and width 1, approximates the area under the curve $y = y(x)$ defined by (15) on the interval $[k-1/2, k+1/2]$, i.e.,

$$A(n,p,k) \approx \int_{k-1/2}^{k+1/2} y(x)\, dx.$$

I should probably draw a picture to illustrate this, but I think again an appeal to the graphing calculator is called for. For small values of n, the following program will illustrate the approximation to the area under the bell curve by rectangles in a histogram representing a binomial distribution:

```
PROGRAM:BELL2
:PlotsOff
:FnOff
:ClrDraw
:Disp "ENTER N:"
:Input N
:Disp "ENTER P:"
:Input P
:seq(X,X,0,N)→L₁
:binompdf(N,P)→L₂
:−.5→Xmin
```

the Works Progress Administration is given in David Alan Grier, "Table making for the relief of labour", in: Martin Campbell-Kelly, *op. cit.*

```
:N+.5→Xmax
:1→Xscl
:max(L₂)→ K
:−.1*K→Ymin
:1.1*K→Ymax
:Plot2(Histogram,L₁,L₂)
:N*P→ M
:M*(1−P)→S
:"1/√(2πS)e^(−(X−M)²/(2*S))"→Y₁
:DispGraph .
```

4 Exercise. Repeat Exercise 2 using the new program.

Now, repeating the observation cited in an instance by Galloway that the total area under the curve over two adjacent intervals is the area under the curve over the whole interval,

$$\int_\alpha^\beta f(x)\,dx + \int_\beta^\gamma f(x)\,dx = \int_\alpha^\gamma f(x)\,dx,$$

we see that

$$C(n,p,m) = A(n,p,0) + A(n,p,1) + \ldots + A(n,p,m)$$
$$\approx \int_{-\frac{1}{2}}^{\frac{1}{2}} y(x)\,dx + \int_{\frac{1}{2}}^{\frac{3}{2}} y(x)\,dx + \ldots + \int_{m-\frac{1}{2}}^{m+\frac{1}{2}} y(x)\,dx$$
$$\approx \int_{-\frac{1}{2}}^{m+\frac{1}{2}} y(x)\,dx,$$

and similarly

$$B(n,p,m) \approx \int_{m-\frac{1}{2}}^{n+\frac{1}{2}} y(x)\,dx.$$

Thus one can approximate any binomial probability by looking up the value of the integral (11) for f defined by (15). This is still an infinity of tables, one for each pair μ, σ, and the tables would have to tabulate the integrals for each pair of endpoints α, β. It turns out one can fix one of the endpoints, say α: for

$$\int_\alpha^\beta f(x)\,dx = \int_{\alpha_0}^\beta f(x)\,dx - \int_{\alpha_0}^\alpha f(x)\,dx.$$

The modern preference is to choose $\alpha_0 = -\infty$, while Kramp and his early successors chose $\alpha_0 = 0$. Furthermore, by *normalising*, or *standardising*, the data, replacing x, t, or whatever variable one is using by its *standard score*, or *z-score*,

$$z_x = \frac{x - \mu}{\sigma},$$

the integral of (15) becomes an integral of (9). This is a routine transformation of the Calculus, wherein one derives

$$\frac{1}{\sqrt{2\pi}\sigma}\int_{\alpha}^{\beta}e^{-\frac{1}{2}((x-\mu)/\sigma)^2}\,dx = \frac{1}{\sqrt{2\pi}}\int_{(\alpha-\mu)/\sigma}^{(\beta-\mu)/\sigma}e^{-\frac{1}{2}z^2}\,dz.$$

Another substitution reduces this to the error function: setting $u = z/\sqrt{2}$ yields

$$\frac{1}{\sqrt{2\pi}}\int_{(\alpha-\mu)/\sigma}^{(\beta-\mu)/\sigma}e^{-\frac{1}{2}z^2}\,dz = \frac{1}{\sqrt{\pi}}\int_{(\alpha-\mu)/(\sqrt{2}\sigma)}^{(\beta-\mu)/(\sqrt{2}\sigma)}e^{-u^2}\,du,$$

whence

$$\frac{1}{\sqrt{2\pi}\sigma}\int_{\alpha}^{\beta}e^{-\frac{1}{2}((x-\mu)/\sigma)^2}\,dx = \frac{1}{2}\,\mathrm{erf}\left(\frac{\beta-\mu}{\sqrt{2}\sigma}\right) - \frac{1}{2}\,\mathrm{erf}\left(\frac{\alpha-\mu}{\sqrt{2}\sigma}\right).$$

5 Example. A thousand babies are born. What is the probability that between 500 and 520 are boys? In solving this we make the not quite accurate simplifying assumption that the probability of having a boy is $\frac{1}{2}$. The sought for probability as given by binomcdf(is obtained by entering

binomcdf(1000,.5,520)−binomcdf(1000,.5,499)

in the calculator. The result is .4152293824. For the normal approximation, we calculate $\mu = np = 1000 \cdot \frac{1}{2} = 500, \sigma^2 = 1000 \cdot \frac{1}{2} \cdot \frac{1}{2} = 250$, whence $\sigma = \sqrt{250}$. Taking $\alpha = 499.5, \beta = 520.5$, and calculating z-scores we have

$$z_\alpha = \frac{499.5 - 500}{\sqrt{250}} = -.0316227766$$

$$z_\beta = \frac{520.5 - 500}{\sqrt{250}} = 1.296533841,$$

and after dividing by $\sqrt{2}$ we have the arguments for erf:

$$z_\alpha/\sqrt{2} = -.02236$$
$$z_\beta/\sqrt{2} = .91679$$

to 5 decimals, the accuracy promised by de Morgan. Being too lazy to perform the interpolation, I just rounded these figures to $-.02$ and $.92$, respectively, and looked them up in Galloway's table. The area under the curve between $-.02$ and $.92$ is the sum of the areas under the curve between $-.02$ and 0 and between 0 and $.92$. The latter is $\frac{1}{2}\,\mathrm{erf}(.92)$. The former, by the symmetry of the bell curve with respect to the y-axis, is the same as that lying under the curve between 0 and $.02$, i.e., $\frac{1}{2}\,\mathrm{erf}(.02)$. Using Galloway's table the area in question is found to be

$$\frac{1}{2}\,\mathrm{erf}(.92) + \frac{1}{2}\,\mathrm{erf}(.02) = \frac{1}{2}(.8067677) + \frac{1}{2}(.0225644) = .41466605,$$

which is not a bad approximation to .4152293824. Interpolating instead of rounding as I did will improve the estimate.

My example and the use of the error function may make this seem more complicated than it is. A more modern approach offers two simplifications. First, by using

$$\frac{1}{\sqrt{2\pi}} \int e^{-\frac{1}{2}z^2} \, dz,$$

one need not divide the z-scores by $\sqrt{2}$. Second, by choosing $-\infty$ as the lower limit of integration one doesn't have to worry about what to do with negative z-scores.

Of course, today the integral

$$\frac{1}{\sqrt{2\pi}} \int_\alpha^\beta e^{-\frac{1}{2}((x-\mu)/\sigma)^2} \, dx$$

can be calculated directly as a function of $\alpha, \beta, \mu, \sigma$ on our calculators. In the example just given, we have but to enter

normalcdf$(499.5,520.5,500,\sqrt(250))$

to get .4152177364, agreeing to 4 digits with the actual probability.

The normal distribution is a good approximation to the binomial distribution for large values of n and values of p close to $\frac{1}{2}$. Moreover, one only needs a single table, whether a table of values of the function H of (13), the H of (14), or

$$H(x) = \frac{1}{\sqrt{2\pi}} \int_{-\infty}^x e^{-\frac{1}{2}y^2} \, dy.$$

However, as p moves closer to 0 or 1, one needs ever larger values of n to get a good approximation.

6 Example. Let $n = 25, p = .001$, and $k = 1$. Then $A(25, .001, 1)$, the probability of exactly one success in 25 trials, is found on the calculator to be

binompdf$(25,.001,1) = .0244068497$.

Using the normal distribution function, we would approximate $A(25, .001, 1)$ by either

$$\frac{1}{\sqrt{2\pi}} e^{-((1-25*.001)/\sqrt{25*.001*.999})^2}$$

or

$$\frac{1}{\sqrt{2\pi}} \int_{.5}^{1.5} e^{-((1-25*.001)/\sqrt{25*.001*.999})^2} \, dx.$$

Using the calculator, we obtain the first estimate as

normalpdf$(1,25*.001,\sqrt(25*.001*.999)) = 1.370465765E-8$,

i.e., .00000001370465765, and the second as

normalcdf$(.5,1.5,25*.001,\sqrt(25*.001*.999)) = .0013250644$.

Neither of these is even of the same order of magnitude as the correct answer, nor of the same order of magnitude as each other. The normal distribution is useless in this case.

There is another distribution, somewhat easier to calculate by hand than the binomial one, that approximates the binomial distribution well when p is small, μ moderate, and n large. This is the *Poisson distribution*, the beginnings of which can be found in de Moivre's *Doctrine of Chances*, but which was developed by Simeon Denis Poisson (1781 - 1840). For fixed p and mean μ, the probability of exactly k successes in n trials, where $\mu = np$, is given by

$$p(\mu, k) = e^{-\mu}\frac{\mu^k}{k!},$$

and is represented on the *TI-83 Plus* by the built-in function poissonpdf(which is found in the DISTR menu. For the example just given, with $p = .001, n = 25$, we have $\mu = np = .025$, and we can enter

poissonpdf(.025,1)

in the calculator to get .0243827478, which is acceptably close to .02441.

Between them, the normal and Poisson distributions offer generally adequate approximations to the binomial distribution, which is less readily calculated. Nonetheless, tables of the binomial distribution were eventually needed and published.

John Fertig reviewed a 387 page volume of *Tables of the Binomial Probability Distribution*, published by the National Bureau of Standards (Washington), which had taken over from the Works Progress Administration:

> Public health statisticians often need to expand the binomial distribution for probability purposes when working with small samples, such as comparing a sample rate with a well established rate, or to determine from a sample rate suitable limits for the appropriate universe value. Unfortunately because of the labor involved in the expansion, the normal curve approximation is often employed, even with small samples ... They are the most extensive tables thus far available and should prove very useful to public health workers.[40]

The first table of the book tabulated values of $A(n, p, k)$ for $n = 2, 3, \ldots, 49$ and p ranging from .01 to .50 in increments of .01. The second table tabulates the cumulative distribution function.

In 1953 the Ordnance Corps of the United States Army published a 577 page volume of *Tables of the Cumulative Binomial Probabilities* and in 1955 the staff of the Computation Laboratory at Harvard University published a volume of *Tables of the Cumulative Binomial Probability Distribution* containing a 61 page introduction and 503 pages of tables. The "Harvard Binomial Tables" have been described as the best-known set of cumulative binomial tables[41]. They calculate values of $B(n, p, m)$ to 6 decimal places for various values of n up to 1000 and 60 values of p.

---

[40] *American Journal of Public Health* 41 (1951), pp. 1011 - 1012.

[41] Gary Tietjen, "Recursive schemes for calculating cumulative binomial and Poisson probabilities", *The American Statistician* 48 no. 2 (1994), pp. 136 - 137. *Cf.* page 145, below, for the actual quotation.

In 1963 new tables were published by Sol Weintraub. Under the heading *"NEED FOR NEW TABLES"*, he explains

> The author first felt the need for more accurate tables of the cumulative binomial probability distribution while working on radar detection probabilities. The existing tables were found inadequate for the following reasons:
>
> 1. No table gives values to more than seven decimal places. In work with very low or very high probabilities, significant figures often do not appear until the sixth or seventh place, so that hand computation or interpolation becomes necessary.
> 2. A problem arises in obtaining individual binomial distribution values from the cumulative distribution: if the value in the cumulative distribution for a given r is .9999 and the value for $r+1$ is .9999, subtracting the second from the first to obtain the value desired gives zero, which is certainly not the result which should be obtained. On the other hand, if the values are .9999998310 and .9999962501, the computation can be carried out with significant results.
> 3. The argument p is usually entered only to two decimal places, and this calls for extensive interpolative algorithms to obtain values for the desired p.
>
> The tables presented herein tabulate values of the cumulative binomial distribution to ten decimal places: p is given to four places from .0001 to .0009, and to three places from .001 to .100. If the need should arise for accurate tables with higher values of p, the computer program will be used to generate these new tables, and they will be published in a separate volume.[42]

Following the introduction, the volume contains 818 pages of tables of cumulative probabilities given to 10 decimal accuracy. The introduction offers a few problems that can be solved using the tables, some mathematical background, a description of the method of computing the values of the tables, and an error analysis.

As this is supposed to be a chapter on probability and not a chapter on the history of tables, I quote the problems he cites:

> To further illustrate the use of the tables, we will consider the following simple examples:
>
> 1. A manufacturer of electronic tubes guarantees with 99 per cent certainty that each of his tubes will last at least a year. A builder of an intricate missile part uses 100 of these tubes in his design. What are the chances that at least 80 of the 100 tubes will last out the year?

---

[42] Sol Weintraub, *Tables of the Cumulative Binomial Probability Distribution for Small Values of p*, Macmillan Company, New York, 1963, p. ix.

---

2. A player plays "red" at the roulette wheel 40 times. What is the probability that "red" will come up at least 25 times in 40 spins?

3. When there are dense clouds, there is an 80 per cent probability that an airliner will appear on a radar screen. What are the chances that the plane will be detected at least 5 times in 10 radar scans?[43]

The only problem he actually considers is the second:

> Thus, in the example of the gambler's problem, if there are as many "reds" as "blacks" on the roulette wheel, his "single-try probability of success" is $\frac{1}{2}$ or 50 per cent. This is the first ingredient; its probability will be called "p."
>
> He wants to be successful at least 25 times. This is the second ingredient; call it "r."
>
> He will play 40 times, and we denote this third parameter by "n."
>
> He can then find his answer easily in the tables by looking up
>
> $$n = 40, r = 25, p = \frac{1}{2};$$
>
> he will find that his chance of obtaining his goal is .07693, or 7.693%.

I am not sure that this is a good advertisement for his tables as his probabilities only go up to .1, not to .5. Presumably he used the Harvard tables.

We, of course do not need tables because we have calculators. Here, $n = 40, p = .5, m = 25$ and we want $B(40, .5, 25) = 1 - C(40, .5, 24)$, so we enter

$$1-\text{binomialcdf}(40,.5,24)$$

in the calculator and obtain a probability of .0769299726. The normal approximation can be calculated by entering

$$\text{normalcdf}(24.5,40.5,40*.5,\sqrt{(40*.5*.5)}).$$

The result, .0773645088, is not too shabby an estimate. (It looks bad because so many digits disagree, but both give .077 when rounded to 3 decimal places and the actual error is less than .0005.)

7 Exercise. Solve Weintraub's remaining problems 1 and 3 by using the binomialcdf(function on your calculator. Then approximate the solution using the normalcdf(function. Compare the results.

The description of the program is of some interest. Here, Weintraub uses $e(n, r, p)$ for the individual binomial probability $A(n, p, r)$ of exactly r successes in n trials and $E(n, r, p)$ for the probability $B(n, p, r)$ of at least r successes:

> The *exact* binomial probability $e(n, r, p)$ is given by the formula

$$e(n, r, p) = \frac{n!}{r!(n-r)!} q^{n-r} p^r \tag{16}$$

---

[43] *Ibid.*, p. x.

and the *cumulative* binomial by the formula

$$E(n,r,p) = \sum_{x=r}^{n} e(n,x,p). \tag{17}$$

Now, since the sum of all probabilities must be 1, we have

$$\sum_{r=0}^{n} e(n,r,p) = 1. \tag{18}$$

Equation (18) may be written

$$E(n,0,p) = 1. \tag{19}$$

From the above equations it follows that

$$E(n,r,p) = 1 - e(n,0,p) - e(n,1,p)\ldots - e(n,r-1,p). \tag{20}$$

We must now find a recursive formula for the values of $e(n,r,p)$. Using the defining equation (16), we form

$$\frac{e(n,r+1,p)}{e(n,r,p)} = \frac{C_{r+1}^n q^{n-r-1} p^{r+1}}{C_r^n q^{n-r} p^r}. \tag{21}$$

The right side of equation (21) yields (after some manipulation):

$$e(n,r+1,p) = e(n,r,p)\frac{(n-r)}{(r+1)}\frac{(p)}{(q)}. \tag{22}$$

Equation (22) was used in the program for successive values of $e(n,r,p)$ using the starting value

$$e(n,0,p) = q^n. \tag{23}$$

In equation (22), p/q was computed first. Hence each new value of r could be found with only 2 multiplications and 1 division.
It was deemed more advisable to use the foregoing method rather than the one that computes across n, namely:

$$E(n+1,r,p) = pE(n,r-1,p) + (1-p)E(n,r,p) \tag{24}$$

because the editing was synchronous with n, and use of (24) would have required too much storage and/or tape movement.[44]

The recursion (24) which Weintraub finds less advisable than the method used is the recursion (4) used by de Moivre in the 18th century. Tietjen expresses himself a little more directly on the disadvantages of this recursion in describing the Harvard tables:

---

[44] *Ibid.*, pp. xv - xvi. I have taken the liberty of renumbering the equations to match those of the current chapter.

The best known set of cumulative binomial tables, "The Harvard Binomial Tables" (Computational Laboratory 1955), uses the three-term recursion:

$$E(n+1, r, p) = pE(n, r-1, p) + qE(n, r, p).$$

There is more bookkeeping and calculation in this formula than necessary. The computational difficulties (factorials and vanishingly small exponents) can be avoided entirely by using the ratio $a_j = p(n-r+1)/qr$ of each term of the cumulative binomial to the preceding term.[45]

Continuing from here, Tietjen gives an 11-line FORTRAN program for calculating $B(n, p, m)$ for given n, p, m, which proceeds by first calculating $C(n, p, m-1)$ and then subtracting this from 1, just as Weintraub does.

For moderate values of n, the calculation seems to be fairly quick. On the *TI-83 Plus* we can test this fairly simply by putting the calculator in sequence mode and going into the equation editor to enter

nMin=0
u(n)=u($n-1$)*(N$-n$+1)/n*P/Q
u(nMin)=Q^N,

and then storing appropriate values[46] of $p, n, 1-p$ into P, N, Q, respectively. Try $p = .5, n = 40, 1-p = .5$ and then enter

u(0,40)→L$_1$.

In short order all 41 values are calculated. Entering

binomialpdf(40,.5)→L$_2$

is not noticeably faster. The two lists appear to be the same and L$_1$−L$_2$ will reveal differences only some number of decimals beyond what is displayed. Repeating the experiment for $n = 100$ (making sure to go back into the equation editor and re-entering Q^N for u(nMin)) again gave remarkable agreement. For large values of n, however, the exponential Q^N will underflow and be identified with 0. But, for large values of n, the normal distribution would give adequate results for cumulative probabilities.

Digressing on the issue of the history of tables and table-making, it is interesting to see a reversal of the trend. With de Prony, Babbage, the military calculations of the two world wars, and the massive program of the Works Progress Administration, the move was away from multiplication and, insofar as it was possible, to addition only. Now suddenly iterative steps involving only multiplication are advocated. (There was, of course, one large summation near the end to obtain the cumulative result.)

---

[45] Tietjen, *op. cit.*, p. 136.
[46] I hope the two uses of the letter n is not confusing. Outside the calculator it is the total number of trials. The calculator reserves n for the variable of the recursion in sequence mode. In the present case this is the r in $e(n, r, p)$ or k in $A(n, p, k)$.

The error analysis in Weintraub's book, performed by "Mr. Lee J. Cohen, senior mathematician of Applied Data Research, Inc.", points to an important difference between tables and pocket calculators. Novel tables are often accompanied by explanations. Ptolemy explained the method of calculation and provided proofs of the geometric theorems on which his calculations were based. Napier wrote his *Constructio* to explain the construction of his logarithm tables. De Prony attached an entire volume of explanation to his collection of tables. Etc. And Weintraub continued this tradition by carefully explaining the workings of the algorithm used and adding Cohen's analysis of the accuracy of the final results. The calculator comes with no such information and one must take it on faith that the results that appear onscreen for pre-programmed functions are accurate in all the displayed digits, something that was not the case with the early calculators that were marketed in the 1960's, when calculator displays showed all the digits used in the calculation and one was advised to round the results to one or two fewer significant digits. I recall, for example, an engineering student proudly showing off his new calculator that performed exponentiation as well as the customary addition, subtraction, multiplication, and division. On his entering 2^2 the screen read 3.9999999, which is quite accurate and on rounding becomes the more acceptable (and correct!) 4.000000. This was in the mid-1960's and I am certain I do not remember the number of significant digits correctly: I doubt the calculator of this incident displayed as many digits as I've indicated. The point is, however, that early calculators displayed more digits than could be guaranteed and one had to ignore the last ones by rounding. In short order the calculators did this for you[47], displaying fewer digits than were stored. Try this: enter

binompdf(40,.5,25)

on the *TI-83 Plus*. You should get

.0365847382.

Now enter

Ans−.0365847382.

The result is not 0, but

5.492E−12,

meaning that the stored result is

.036584738205492

and the calculator has rounded the result to fewer significant digits for display.

Returning to probability itself, I finish this chapter with some exercises on binomial probability:

---

[47] But not always! *Cf.* the discussion of integration in section 2 of Appendix B, below for an example where the *TI-83 Plus* does not do this.

8 Exercise. Weintraub's analysis of the roulette problem is an oversimplifica-
tion that ignores green. Roulette wheels have 18 red, 18 black, and 1 (Europe)
or 2 (United States) green slots. In either case, the probability of coming up
"red" on an individual spin is not $\frac{1}{2}$. Determine, either for a European or an
American roulette wheel, the probability of a player winning at least 25 spins
out of 40 when betting on "red".

Here are a few problems from de Morgan[48]. Insofar as possible solve them
by means of formulæ for $A, B,$ or C, then solve them using the binomial distribution functions on your calculator, and finally approximate the results using
the normal distribution:

9 Exercise. In 6000 throws with a die, what is the chance that the number of
aces shall not differ from 1000 by more than 50; that is, shall lie betweeen 950
and 1050, inclusive.

10 Exercise. In 200 tossses [of a coin], what is the chance that the number of
heads shall lie between 97 and 103, both inclusive?

11 Exercise. In 12 tosses [of a coin], what is the chance of the heads being
either 5, 6, or 7 in number?

The problems with which we opened the chapter were more difficult. Galloway discusses the situation as follows:

> ... let it be proposed to determine how often a common die must be
> thrown in order to give the probability of ace turning up *once at least*,
> equal to a given fraction u. Here the probability of throwing ace on any
> throw being $\frac{1}{6}$, we have $p = \frac{1}{6}$, and $q = \frac{5}{6}$. Now, as every term of the
> development of $(p+q)^h$, excepting the last, gives a combination in which
> ace occurs once or oftener, the question requires a value to be found
> for h, such, that the sum of the first h terms of that development, shall
> be equal to u. This may be done, in general, by the common methods
> of trial and error; but in the present case, the last term being the only
> one not included among those which contain a chance of throwing ace,
> it is evident that it is only necessary to find the last term alone in order
> to have the probability of *not* throwing ace in h trials, which by the
> question is $1 - u$. The last term of the development is q^h; therefore we
> must have the equation $q^h = 1 - u$; whence $h \log q = \log(1 - u,)$ and
> $h = \log(1 - u) \div \log q$. Let $1 - u = \frac{\beta}{\gamma}$ and $q = \frac{b}{c}$, we shall then have
> $\log(1 - u) = \log \beta - \log \gamma$, and $\log q = \log b - \log c$; whence
>
> $$h = \frac{\log \beta - \log \gamma}{\log b - \log c}.$$
>
> Substituting in this general formula the particular numbers given in the
> question, namely $b = 5, c = 6$; and supposing $u = \frac{1}{2}$, and consequently
> $\beta = 1, \gamma = 2$, we have

---

[48] *An Essay...*, *op. cit.*, p. 77.

$$h = \frac{\log 2}{\log 6 - \log 5};$$

whence, by computing from the logarithmic tables, $h = 3.8$. From this it follows, that in four trials the probability of throwing ace once at least, is greater than the probability of not throwing it at all.

If the question had been to determine in how many throws with two dice one may undertake, on an equality of chance, to throw aces at least once, we should have had $p = \frac{1}{36}, q = \frac{35}{36}$, and consequently $b = 35$, and $c = 36$. Substituting these numbers in the general formula, and observing, that in this case also $\beta = 1, \gamma = 2$, we get

$$h = \frac{\log 2}{\log 36 - \log 35} = 24.6.$$

The probability of not throwing aces once is therefore greater than the opposite probability or that of throwing aces once or oftener, when the number of throws is 24, but less when the number is 25.

These two questions are celebrated in the early history of the theory of Probability, from the circumstance that the Chevalier de Méré, by whom they were proposed to Pascal, declared the two results above stated to be inconsistent with each other, and thence took occasion to question the accuracy of the theory of combinations by means of which they had been obtained. He reasoned thus: Since the probability of throwing ace with one die is $\frac{1}{6}$, and that of throwing aces with two dice $\frac{1}{6}$ of $\frac{1}{6} = \frac{1}{36}$; therefore, if there be a given probability in favour of throwing ace in four throws with one die, there must likewise be the same probability of throwing aces with two dice in $6 \times 4 = 24$ throws; in other words, the chances in favour of an event E in a single trial, being six times more numerous than those in favour of F, there will be as many chances in favour of F in six trials as there are in favour of E in one. The error consists in supposing that the number of trials must increase or diminish exactly in the inverse ratio of the probability of obtaining the proposed point.[49]

Note that Galloway's problem of throwing aces with two dice is the same as the problem of Proposition XI of Huyghens (two sixes). The point I wish to make is that the "common method of trial and error" eschewed by Galloway in favour of his clever algebraic argument is now quite simply implemented on the calculator. First, choose a number n that you think will be an upper bound on the number of needed trials. To guess this, simply ask what the expected number of throws needed for success is. Following the discussion on pages 37 - 38, if p is the probability of success on a given throw and $q = 1 - p$, this is:

$$n = 1 \cdot p + 2 \cdot p \cdot q + 3 \cdot p \cdot q^2 + 4 \cdot p \cdot q^3 + \dots$$

---

[49] Galloway, *op. cit.*, pp. 47 - 49. The latter two displayed formulæ were inline in the original. I have displayed them for the sake of readability. I have also left unitalicised a couple of his semi-colons.

$$= p(1 + 2q + 3q^2 + 4q^3 + \ldots)$$
$$= p \cdot \frac{1}{(1-q)^2} = \frac{p}{p^2} = \frac{1}{p}.$$

For $p = 1/36$, this yields $n = 36$.

Now create a list L_1 of the numbers of trials[50] from 1 to 36:

seq(X,X,1,36)$\rightarrow L_1$.

The probability of throwing a pair of aces (or sixes) in n trials is 1 minus the probability of throwing no such pairs. So make a list of these probabilities:

seq(1$-$binompdf(X,1/36,0),X,1,36)$\rightarrow L_2$.

Now press the STAT button and choose the first option to enter the List Editor and scroll down until a probability, in this case .50553, greater than or equal to 1/2 appears in L_2. The corresponding entry in L_1 is 25.

12 Exercise. (Galloway) Find how many times one should wager to throw 10 heads with one coin.

And here is a problem to think about:

13 Exercise. Recall Todhunter's typographical error cited on page 48, above: Buffon's child played the Petersburg game 2048 times with the coin coming up heads on the first throw 1061 times. Todhunter transposed two of the digits and declared the child to have played the game 2084 times. What is the probability of coming up heads at least 1060 times in 2048 throws? In 2084 throws? Which hypothesis, 2048 playings of the game or 2084 playings, better fits the data, i.e., 1061 heads on the first toss? Put differently, without the additional information that Todhunter cites de Morgan as his source and de Morgan gives 2048 plays of the game, which figure 2048 or 2084 is more likely to yield 1061 heads? Is it extremely unlikely that either number of trials could yield a result as far away from the expected value as 1061 is?

---

[50] Actually, this step can be omitted as the number is displayed on the bottom of the screen during the scrolling in the List Editor. I just find it convenient to have L_1.

Part II

Law of Large Numbers

5

Bernoulli's Theorem

1 Bernoulli's Description of His Theorem

Jakob Bernoulli's reputation as one of the giants of probability theory is already secured by the first three parts of the *Ars conjectandi*. But it is the fourth part, chapters IV and V (which justify the title of the book— the "art of conjecture"), which is most important. It is best to let Bernoulli explain in his own words[1]. Here, in full, is his Chapter IV, explaining what is to follow:

---

[1] Well, in a translation of a translation of his own words. My German at least existing I translate the following from the German of Haussner's translation, *Wahrscheinlichkeitsrechnung (Ars conjectandi)* (Leipzig, 1899), rather than from the Latin original. Portions of this selection (part 4, chapter IV) from *Ars conjectandi* were translated anew by Ivo Schneider as text 2.3 in his source book, *Die Entwicklung der Wahrscheinlichkeitstheorie von den Anfängen bis 1933; Einführungen und Texte*, Wissenschaftliche Buchgesellschaft, Darmstadt, 1988. After completing my own translation from the German, I was reminded that a beautiful, literary translation with commentary on portions of this chapter is given in James R. Newman, ed., *The World of Mathematics, vol. III*, Simon and Schuster, New York, 1956. Additionally, I discovered an English translation of the full fourth part of the *Ars conjectandi* by Oscar B. Sheynin from 2005 online, and an English translation of the entire work at Amazon.com: Edith Dudley Sylla, *The Art of Conjecturing, together with a Letter to a Friend on Sets in Court Tennis*, Johns Hopkins University Press, Baltimore, 2006. Sheynin translated from translations, beginning with J.V. Uspensky's 1913 translation into Russian, and consulting other translations and even the Latin original when necessary. I have subsequently used these as a control on my own translation, making such corrections as needed. That translation is not an exact science I illustrate at one or two occasional points through footnoted comments. Such remarks should give the reader some indication, not only of the level of confidence to place specifically on my translation, but, perhaps, some idea of the level of confidence to place on translations in general.

Chapter IV

On the two ways to determine the number of cases. What to think about the method, to determine it through observations. The main problem in this connexion and others.

In the preceding chapter it was shown, from the numbers of cases in which grounds for something exist or do not exist, how they can indicate or not indicate or even indicate its opposite, how one can determine and appraise their evidentiary value and their proportional probabilities. It is thus established that, for a correct formation of conjectures on any matter, nothing more is necessary for us to do than that we first establish exactly the number of cases and then determine how much more easily one of the cases can occur than the other. And here it seems to us is exactly where the difficulty lies, that only for the rarest phenomena and almost nowhere other than in games of chance is this possible: games of chance, however, were set up by their inventors so that the participants should have equal prospects of winning, that the numbers of the cases in which profit or loss must be given were determined and known at the outset, and that all cases can occur with equal likelihood. However, with most other phenomena which depend on the forces of nature or the capriciousness of men this is in no way the case. So, for example, the numbers of the cases with dice are known, because there are for each individual die exactly as many cases as it has surfaces; all these cases are also equally likely; because of the identical shapes of all the surfaces and because of the uniform distribution of mass of the die there is no reason to assume that the die should more easily fall on one of its surfaces than on another, which would be the case if the die's surfaces possessed different shapes and one portion of the die was made from heavier materials than the other part. So too the numbers of the cases for the drawing of a white or black pebble[2] from an urn are known and all pebbles can be drawn equally readily, because it is known how many pebbles of each sort are in the urn, and because no grounds can be given why this or that pebble should be more likely to be drawn than any other one. What mortal, however, could ever determine and report the number of diseases (i.e., however many cases) which befall the human body in all its parts and in each age, and can bring about death, about how much more likely this or that disease, the plague than dropsy[3], dropsy than fever, will ruin a man, in order therefrom to derive a conjecture on the ratio of living and dying of future generations? Or who could hope to enumerate the innumerable cases of changes which the air is subject to daily in order therefrom to predict

---

[2] Sheynin says "tickets" at this point; Newman says "balls"; and Sylla says "slips of paper".

[3] Sheynin: "rabies"; Newman: "dropsy"; Sylla: "dropsy". Haussner's German clearly translates to "dropsy", while Schneider omits this sentence.

today already what nature it will have after a month or indeed after a year? Or, further, who might have studied the nature of the human spirit or the admirable construction of our bodies far enough in order to hope to determine the cases in which this player or that player can win or lose in games which depend in whole or in part on the sharpness of mind or on the physical skilfulness of the players? As these and similar things depend on entirely hidden causes, which moreover constantly cheat our insight through the infinite multiplicity of their interaction, so it would be completely senseless to want to investigate something in this way.

But another path lies open to us for finding what is sought and that is to determine *a posteriori* from the success which would be observed in numerous cases of similar examples that which we cannot determine *a priori*. For this it must be assumed that each individual event can occur or not in however many cases as would have previously been observed under identical circumstances to occur or not to occur. Then, for example, if one had observed that of 300 men of the age and constitution of Titius 200 had died before the passing of 10 years, the rest however having lived longer, one can conclude with sufficient certainty that there are twice as many cases in which Titius too must serve nature its due within the next decade as cases in which he can survive this period of time. Likewise, if someone has observed the weather for many years and noted how often it was clear or rainy, or if someone very often watched two players and had seen how often this or that one won, so he could in a plausible way determine exactly through that the ratio which the numbers of cases in which the same event can or cannot occur under the preceding identical circumstances have to each other. This empirical method, to determine the number of cases through observation, is neither new nor unusual; for the famous author of the work "L'art de penser" [4], an astute and talented man, has already in Chapter 12 and following, in the last part of his work, described a quite similar procedure, and all men observe the same procedure in their daily lives. Also it is clear to every man, t h a t i t d o e s n o t s u f f i c e t o m a k e o n l y o n e o r a n o t h e r o b s e r -

---

[4] *La Logique, ou l'art de penser*, commonly referred to as the *Port-Royal Logic*, was anonymously written by Antoine Arnauld (1612 - 1694) and Pierre Nicole (1625 - 1695) and published in 1662. It became a popular textbook in logic and was used as such into the 20th century. English translations go back to at least the 19th century, that by Thomas Spencer Baynes is available online in its second (1861) edition under the title *The Port-Royal Logic*. Arnauld and Nicole were members of a religious group of *Jansenists* centered at the convent Port-Royal Des Champs, from which the book and its companion, *The Port Royal Grammar*, got its popular name. The most famous member of the group was Pascal, who is believed to have contributed to the volume as well. The "famous author" referred to is Arnauld, whose name is more prominently associated with the book than that of his co-author. A more recent translation of the *Port-Royal Logic* was published in 1964 by Bobbs-Merrill.

vation in order in this way to make a judgement on any occurrence, but rather that a large number of observations is necessary. Occasionally too a truly simple man, in consequence of some natural instinct, has found with no prior instruction (which is truly miraculous), that one runs less danger of straying from the truth the more relevant observations he's made.[5] Although this will be seen by everyone as inherent in the nature of things, a proof based on scientific principles is not obvious and it suits me on that account to furnish such here. I would think to accomplish too little if I wished only to proceed with the proof of this point which everyone knows. One must take into account much more that perhaps nobody has previously even thought about. It remains namely yet to investigate whether, through increasing the number of observations, the probability also steadily increases that the ratio of the number of favourable to the number of unfavourable observations reaches the true ratio, and indeed to the extent that this probability finally exceeds any arbitrary degree of certainty, or whether the problem just, so to speak, has an asymptote, i.e., whether there is a definite degree of certainty of having found the true ratio of cases, which can never be exceeded even by an arbitrary increase in the number of observations, e.g., that we can never be more than $\frac{1}{2}$, $\frac{2}{3}$ or $\frac{3}{4}$ certain that we have ascertained the true ratio of the cases. To clarify what I mean through an example, I assume without your foreknowl-

---

[5] One might compare this with the translation of this line from the Latin given in Stephen M. Stigler, *The History of Statistics; The Measurement of Uncertainty before 1900*, Harvard University Press, Cambridge (Mass.), 1986, p. 65: "For even the most stupid of men, by some instinct of nature, by himself and without any instruction (which is a remarkable thing), is convinced that the more observations have been made, the less danger there is of wandering from one's goal." Sheynin reads, "Because even the most stupid person, all by himself and without any preliminary instruction, being guided by some natural instinct (which is extremely miraculous), feels sure that the more such observations are taken into account, the less is the danger of straying from one's goal". Sylla reads "the most foolish person". The phrase in Haussner's German translation is "ein recht einfältiger Mensch", which could be rendered "a truly stupid man" or "a truly simple man", both without a superlative. Haussner acknowledged not offering a literal translation, but in part 4 only in a few proofs in Chapter V for the sake of intelligibility. Schneider, in his partial translation of this chapter, skipped this particular sentence. Newman's English translation, by the way, politely phrases it, "a person with no education".

edge there to be 3000 white and 2000 black pebbles[6] in an urn and you want to determine the ratio of these through trials in which you draw one pebble after another (in such a manner, however, that you return each withdrawn pebble before drawing another, thereby the number of pebbles in the urn will not be decreased), and you observe how often a white and how often a black pebble is drawn. The question arises, whether this could be done so often that it thereby becomes ten, a hundred, a thousand times, etc., more probable (i.e., finally morally certain), that the number of draws with which you drew a white pebble has to the number of those with which you drew a black one the same ratio $1\frac{1}{2}$ which the numbers of pebbles (or of cases) themselves have to each other, than that these numbers form some other, different ratio. If this is not the case, I confess that our attempt to determine the number of cases through observation is in a bad way. If, however, it is the case and one finally attains moral certainty in this way (that this really is so I will prove in the following chapter), then we can find the number of cases *a posteriori* almost as exactly as if it were *a priori* known to us. Now, for daily life, where moral certainty is taken to be absolutely certain, by Axiom 9 of Chapter II, this is more than sufficient to support our conjecture in any arbitrary chance-domain no less scientifically than in the case of gambling. For, if instead of the urns we think ourselves presented with the air or the human body, which host a vast number of diverse changes and diseases exactly as the urn has pebbles, so we can also in like fashion determine in these areas through observation approximately how much more easily this or that event occurs.

However, so that this may not be misunderstood, it is to be noted, **that we cannot absolutely determine the ratio between the numbers of the cases, which we undertake to determine** (because quite the opposite would happen and it would be all the more improbable that the correct ratio be found the more observations that are made[7]), **rather only we obtain a definite approximation, i.e., we would confine it between two bounds which can be taken arbitrarily close to one another.** If we assume, in the example of the urn with pebbles introduced above, two ratios, e.g., $\frac{301}{200}$ and $\frac{299}{200}$ or $\frac{3001}{2000}$ and $\frac{2999}{2000}$, or etc., of which one is a little smaller, the other a little larger, than $1\frac{1}{2}$, then it turns out that with that probability it will be more probable that the ratio found through frequently repeated observations lies within these limits of the

---

[6] Sheynin notes that Bernoulli wrote "stones" in Latin and Haussner wrote "Steinchen", i.e., "small stones" or "pebbles". Sheynin uses "pebbles". Sylla uses "tokens".

[7] Sheynin points out that the probability of there being exactly $nr/(r+s)$ successes in n trials is approximately $1/\sqrt{2\pi nrs}$, which goes to 0 as n approaches infinity.

ratio $1\frac{1}{2}$ than outside these same.

This is the problem which I have resolved to publish here, after I have already carried it around with me for 20 years; its novelty, as well too as its extraordinarily great uses in connexion with its likewise so great a difficulty grants all the remaining chapters of this doctrine a gain in importance and significance.[8] However, before I go into its solution[9] I wish quickly to set aside the objections which some scholars[10] have raised against it.

1. First, they make the objection that the ratio between the pebbles is of a nature different from that between the diseases and the changes in the air; the number of the former is definite, the number of these, however, is indefinite and uncertain.

To this I answer: With regard to our knowledge, both are equally uncertain and indefinite. That, however, some thing can in itself, and by its nature be uncertain and vague, can be just as little understood by us as we can understand that God has at the same time created and not created something; for everything which God has made, he has, exactly because he has made it, also determined.

2. Second, they object that the number of pebbles is finite, while that of diseases is infinite.

I reply hereupon: The latter number is, rather, amazingly large, but not infinite; however, accepting that it be infinitely large, it is known that even between two infinitely large numbers a definite ratio can exist, which may be expressed through finite numbers either exactly or at least as exactly as may be desired. Thus, the circumference of a circle al-

---

[8] This sentence offers a bit of a challenge. Both Haussner's (*op. cit.*, p. 92) and Schneider's (*op. cit.*, p. 67) translations seem to me to say that this theorem elevates the importance and significance of the remaining chapters of the book. As luck would have it, F.N. David (*Games, Gods and Gambling; A History of Probability and Statistical Inference*, Charles Griffin, Ltd., London, 1962 (reprinted by Dover Publications, Mineola (New York), 1998, p. 136)) has translated this sentence into English: "This is therefore the problem that I now want to publish here, having considered it closely for a period of twenty years, and it is a problem of which the novelty, as well as the high utility, together with its grave difficulty, exceed in weight and value all the remaining chapters of my doctrine." This is a more forceful statement and makes good sense. Todhunter (*A History of the Mathematical Theory of Probability From the Time of Pascal to That of Laplace*, Cambridge University Press, Cambridge, 1865 (reprinted by Chelsea Publishing Company, New York, 1949, 1965), p. 71) quotes this line in the original Latin and my software suggests that Bernoulli's Theorem "can add weight and value to the remaining chapters of the doctrine". My preferred reading is David's, though it may be a bit more direct than Bernoulli intended.

[9] Karl Pearson (*Biometrika* 17 (1925), p. 206): "Before I treat of this 'Golden Theorem'..."; Sylla (*op. cit.*, p. 329): "But before I convey its solution, ..."

[10] Leibniz had raised these objections in correspondence with Bernoulli that may be found in Leibniz's collected works. Excerpts in German appear as Text 2.2.2 in Schneider, *op. cit.*

GOTTFRIED WILHELM LEIBNIZ 1646-1716 DEUTSCHLAND

Gottfried Wilhelm von Leibniz, or Leibnitz, independent discoverer after Newton of the collection of algorithms lying at the heart of the Calculus, was at most a minor player in the development of probability. In his dissertation *De arte combinatoria* (1666) he did touch on combinatorial problems of the kind more fully treated already by Pascal in the latter's *Traité du triangle arithmétique* (1654). And he did correspond with Jakob Bernoulli on the subject, providing encouragement and criticism.

Perhaps most interesting mathematically was his attempt to apply simple probabilistic reasoning to the summation of the divergent series

$$1 - 1 + 1 - 1 + 1 - 1 + \ldots$$

The partial sums are

$$1, \quad 1 - 1 = 0, \quad 1 - 1 + 1 = 1, \quad 1 - 1 + 1 - 1 = 0, \quad \ldots,$$

every other one, i.e., half of the sums being 1 and half being 0. The expected value of the summation should thus be

$$\frac{1}{2} \cdot 1 + \frac{1}{2} \cdot 0 = \frac{1}{2},$$

a value assigned by other reasonings.

The various Germanies have depicted Leibniz on postage stamps on various occasions. The first such stamp was issued by the German Reich in 1926 in a set of 9 definitives bearing portraits of famous Germans. The second, issued by the German Democratic Republic in 1950 celebrating the 250th anniversary of the Berlin Academy of Sciences is pictured above left. To its right is a stamp issued by the Federal Republic in 1996 celebrating his 350th birthday and offers direct evidence of his mathematical interests.

ways has a definite ratio to its diameter, which to be sure will be given exactly only through infinitely many decimals of Ludolph's number, but was all the same confined by Archimedes, Metius and Ludolph[11], between bounds which are fully sufficient for application. T h u s n o t h i n g   h i n d e r s   u s   f r o m   d e t e r m i n i n g   t h e   r a t i o b e t w e e n   t w o   i n f i n i t e l y   l a r g e   n u m b e r s   w h i c h   c a n b e   r e p r e s e n t e d   v e r y   n e a r l y   e x a c t l y   t h r o u g h f i n i t e   n u m b e r s   v i a   a   f i n i t e   n u m b e r   o f   o b s e r v a - t i o n s .

3. Third, they make the objection that the number of diseases is not constantly the same but rather that new ones emerge daily.

To this I reply: That in the course of time the diseases can increase in number, I don't deny it, and certainly anyone who wanted to draw conclusions on antediluvian time from modern observations would powerfully stray from the truth. But from this nothing more follows than t h a t   f r o m   t i m e   t o   t i m e   n e w   o b s e r v a t i o n s m u s t   b e   m a d e ;   also with the pebbles new observations would be necessary if one must assume that their number in the urn had changed.[12]

2 Bernoulli's Example Illustrating His Theorem

Bernoulli kept a scientific diary, *Meditationes*, which was first published in his collected works in 1975. Around 1689, in Article 133 of this diary, he entered a simple example illustrating what he would later prove in Article 151 and again in Chapter V of part 4 of *Ars conjectandi*. Suppose, he said, he played a game against an opponent in which they had equal chances of winning. If he played 3 games, he could win

3, 2, 1, 0 games in 1, 3, 3, 1 cases, respectively,

for a total of 8 cases; with 6 games he could win

6, 5, 4, 3, 2, 1, 0 games in 1, 6, 15, 20, 15, 6, 1 cases, respectively,

for a total of 64 cases; and he repeated the enumeration for 9 and 12 games, respectively. He would expect to win around half the games, thus between, say,

---

[11] Archimedes (*c.* 287 B.C. - 212 B.C.) trapped π between $3\frac{1}{7}$ and $3\frac{10}{71}$, the greater bound being the familiar $\frac{22}{7}$; Adriaen Anthonisz, called Metius (*c.* 1543 - 1620), gave the bounds $3\frac{15}{106}$ and $3\frac{17}{120}$ from which he obtained the next most famous approximation,

$$3 + \frac{15+17}{106+120} = 3\frac{16}{113} = \frac{355}{113},$$

an approximation known centuries earlier to the Chinese mathematician Zŭ Chōngzhī (*c.* 429 - *c.* 500); and Ludolph van Ceulen (1540 - 1610) calculated π to 35 decimals, a feat which led mathematicians to refer to π in his honour as *Ludolph's number*.

[12] Haussner, *op. cit.*, part IV, pp. 88 - 93.

1/3 and 2/3 of the games. Counting cases, he noted that the probabilities of winning more than 2/3 of the time or less than 1/3 of the time decreased when more games were played: The probabilities of, say, winning more than 2/3 of the games are

1/8 for 3 games
7/64 for 6 games
46/512 for 9 games
299/4096 for 12 games,

and $1/8 > 7/64 > 46/512 > 299/4096$. He concluded

> that one can arrive through an increase in the number of games at a probability which is smaller than any arbitrarily given one, namely because the ratio of the individual [probabilities] to the succeeding ones is greater than 8 to 7. Therefore I can observe so often that I unlock the ratio of chances with a probability bordering on certainty, as if I had this *a priori*. Question: How many observations must be made so that it is a hundred thousand times more probable that the ratio of the games won by me to those lost approaches equality better than 101 to 99?[13]

A few words of explanation are in order. Let me begin by stating that all of this is leading up to *Bernoulli's Theorem*, proven in Chapter V of the fourth part of *Ars conjectandi*:

> **Theorem One wants the ratio of the number of favourable cases to the number of unfavourable ones to be exactly or approximately $\frac{r}{s}$, thus in ratio to the number of all cases as $\frac{r}{r+s} = \frac{r}{t}$ — if $r + s$ is set equal to t, which last ratio is caught between the bounds $\frac{r+1}{t}$ and $\frac{r-1}{t}$. Now, as is to be proven, so many observations can be made that it is arbitrarily many times (e.g., c times) more probable that the ratio of favourable to all observations taken lies within these bounds than outside the same, thus is neither larger than $\frac{r+1}{t}$ nor smaller than $\frac{r-1}{t}$.[14]**

Here, r denotes the number of favourable cases, s the number of unfavourable ones, and $t = r + s$ the total number of possible cases. These are the theoretical constants, not the observed values. In his example, $r = 1, s = 1$, and $t = 2$. Or, $r = 2, s = 2$, and $t = 4$. To fit the bounds $1/3, 2/3$ into his Theorem, we could take $r = 3, s = 3, t = 6$:

$$\frac{r+1}{t} = \frac{3+1}{6} = \frac{2}{3}, \quad \frac{r-1}{t} = \frac{3-1}{6} = \frac{1}{3}.$$

---

[13] Schneider, *op. cit.*, pp. 122 - 123. The Latin original of the full article, as well as that of Article 151 presenting the proof, can be found in the third volume of Bernoulli's collected works. I quote Schneider's German translation because I know no Latin, but some German.

[14] Haussner, *op. cit.*, p. 104.

ТАШКЕНТ · 1986

I ВСЕМИРНЫЙ КОНГРЕСС
ОБЩЕСТВА МАТЕМАТИЧЕСКОЙ СТАТИСТИКИ
И ТЕОРИИ ВЕРОЯТНОСТЕЙ ИМЕНИ БЕРНУЛЛИ

Professional organisations for probability and statistics include the International Statistical Institute (ISI) and the Bernoulli Society for Mathematical Statistics and Probability. The Bernoulli Society, founded in 1975 as a section of the ISI, was named in honour of the four Bernoullis who contributed to the then newly emerging field of probability—Jakob, Johann, Nikolaus, and Daniel. The Society publishes *Bernoulli*, a quarterly journal devoted to probability and statistics, and organises international conferences on the subject. *The First World Congress of the Bernoulli Society for Mathematical Statistics and Probability* was held in Tashkent, USSR, in September of 1986. The Soviet post office issued a special envelope on the occasion with the above cachet featuring Jakob Bernoulli's portrait; the triple $(\Omega, \mathcal{F}, \mathrm{P})$ representing a probability space consisting of a sample space Ω, a set $\mathcal{F}$ of events, and an assignment P of probabilities to events; and the formula

$$\mathrm{P}\left(\left|\frac{m}{n} - \mathfrak{p}\right| < \varepsilon\right) > 1 - \delta,$$

a symbolic expression of a modern formulation of Bernoulli's Theorem.

In his example, Bernoulli cheats a little. We might expect this illustration of his theorem to proceed by calculating the probabilities p_n that the number of successes in n trials lies between $n/3$ and $2n/3$ and verifying that the sequence $p_1, p_2, p_3, \ldots$ tends to 1, whence the odds $p_n/(1-p_n)$ eventually exceed any given constant c. Bernoulli more or less does this, almost calculating $q_n = 1-p_n$ and showing q_n to approach 0 as n gets larger and larger. I say "almost" because he departs from this strategy in two ways. First, there is the trivial matter of calculating separately the probabilities that the number of successes in n trials is less than $n/3$ or greater than $2n/3$, respectively. Since $r = s$ in his example, the two probabilities are equal and thus he needs only to calculate one of these two to obtain $q_n/2$. Second, and more seriously, he calculates, not $q_1/2, q_2/2, q_3/2, \ldots$, but $q_3/2, q_6/2, q_9/2, \ldots$. In short, he considers the sequence $\{q_{3n}\}$ instead of $\{q_n\}$. To see why he does this, calculate a few initial values of the sequence $q_1/2, q_2/2, q_3/2, \ldots$. This is easily done on the calculator.

First, go into the equation editor and enter

$Y_1 = \text{int}(X/3) - (X/3 = \text{int}(X/3))$.

The function int(, which is found in the NUM menu accessed by the MATH button on the *TI-83 Plus*, calculates the *greatest integer function* $[x]$, i.e., applied to an argument, int(yields the greatest integer less than or equal to the argument. The desired probability,

$$\sum_{k<n/3} \binom{n}{k} p^k (1-p)^{n-k},$$

equals

$$\sum_{k=0}^{[n/3]} \binom{n}{k} p^k (1-p)^{n-k} \tag{1}$$

if $n/3 \neq [n/3]$; and it equals

$$\sum_{k=0}^{[n/3]-1} \binom{n}{k} p^k (1-p)^{n-k} \tag{2}$$

if $n/3 = [n/3]$, i.e., if 3 divides n.[15] For input n, Y_1 first calculates $[n/3]$. Then it performs a test to see if $[n/3] = n/3$. The result of the test is 1 if equality holds and 0 otherwise. Y_1 then subtracts this result from $[n/3]$, thus yielding the upper bounds on the summations in (1) and (2).

Now calculate these sums using binomialcdf(for a number of values of n (using $p = .5$ for this example):

seq(binomialcdf(X,.5,Y_1(X)),X,1,12)$\rightarrow L_1$.

If you examine the values listed, you will see that they oscillate. You will get a clearer picture of this oscillation if you graph a scatter plot using L_1 as the list of y-values. Store the corresponding x-values in L_2:

---

[15] Cf. Appendix B for an explanation of the Σ-notation.

seq(X,X,1,12)$\to L_2$.

Enter the STAT PLOT editor, choose one of the plots, turn it on, scroll down to Type, choose the first pictogram and hit ENTER. Then enter L_2 for Xlist and L_1 for Ylist. Hitting the ZOOM button and choosing ZoomStat will graph the pairs $\langle 1, q_1/2\rangle, \langle 2, q_2/2\rangle, \ldots, \langle 12, q_{12}/2\rangle$ and you will *see* why Bernoulli chose every third element of the sequence: The sequence $q_1, q_2, q_3, \ldots$ oscillates, but the subsequences $q_1, q_4, q_7, \ldots$; $q_2, q_5, q_8, \ldots$; and $q_3, q_6, q_9, \ldots$ all tend monotonically to 0. They, in fact, tend exponentially to 0. For example, as Bernoulli remarked in the quote above,

$$\frac{q_{3(n+1)}}{2} \leq \frac{7}{8} \frac{q_{3n}}{2},$$

whence

$$\frac{q_{3(n+1)}}{2} \leq \left(\frac{7}{8}\right)^n \frac{q_3}{2} = \frac{1}{8}\left(\frac{7}{8}\right)^n.$$

1 Project. Explore the situation further by writing a program BERNULLI for the *TI-83 Plus* generalising what we have just done for Bernoulli's example. Given positive $r > 1$ [16], s, $t = r + s$, and a number n of trials, redefine

$$q_n = \sum_{k < n(r-1)/t} \binom{n}{k} p^k (1-p)^{n-k}.$$

BERNULLI will first ask for values R, S, and N. Then it should store the expression

int((R−1)X/(R+S))−((R−1)X/(R+S)=int((R−1)X/(R+S)))

in Y_1, use this to generate the probabilities of obtaining at most $Y_1(n)$ successes in n trials for $n = 1, 2, \ldots, N$, and store the list in L_1. Finally the program should store the list $\{1, 2, \ldots, N\}$ in L_2 and graph a scatter plot of the data using L_2 as the list of x-values and L_1 as the list of corresponding y-values. Run the program using the values:
a. R = 3, S = 3, N = 12
b. R = 2, S = 3, N = 20
c. R = 3, S = 5, N = 16
d. R = 3, S = 4, N = 28
e. R = 5, S = 4, N = 27.
What do you observe? (For the last example you might prefer to enter the STAT PLOT editor and change the Type to xyLine.)

3 Proof of Bernoulli's Theorem

[This section discusses Bernoulli's proof of his celebrated Theorem and is a bit more technical than anything that has gone before. There is something to

---

[16] The Theorem is valid for $r = 0, 1$, but in these cases $(r - 1)/t \leq 0$ and there are no k's less than $n(r - 1)/t$: the sums $q_0, q_1, \ldots$ are all empty, thus equal to 0.

be gained by its careful study, but the ratio of gain to effort may be deemed unacceptably low for nonmathematicians. I leave it to the reader to decide whether to read this section or to skip ahead to the next section—or, better, to skip to the concluding quotations starting on page 174, below.]

The proof Bernoulli gave of his Theorem in *Ars conjectandi* proceeds along lines different from those suggested by Bernoulli's example. I have tried to reproduce a proof of Bernoulli's Theorem following these lines and the pattern evident from running the program BERNULLI a number of times, but small complications kept arising, convincing me that the proof in *Ars conjectandi* is simpler and better suited for an exposition.

There are, however, a couple of weak points to Bernoulli's proof which limit the generality of the result established. He was strongly wedded to the determination of probability through the enumeration of equally likely outcomes or cases. Thus, instead of considering the general problem of repeatable trials with arbitrary probability p of success on a given trial, he considers only probabilities p arising from a situation where there are $t = r + s$ equally likely outcomes, of which r are favourable and s unfavourable. Hence he only proves his Theorem for rational values of p. Second, in showing that, for a large number N of trials, the probability that the number of successes lies between

$$\frac{r-1}{t}N \quad \text{and} \quad \frac{r+1}{t}N$$

is much greater than that the number lies outside this range, he doesn't prove the result for all N, but only for values of N that are multiples of t. Modern expositions based on his ideas avoid this limitation, as indeed they must if the result is to hold in the case of an irrational value of p for which there is no corresponding value of t.[17]

In essence, Bernoulli's proof[18] proceeds by proving some lemmas concerning the relative sizes of the terms in the binomial distribution,

$$\binom{N}{0}p^0 q^N, \ \binom{N}{1}p^1 q^{N-1}, \ \ldots, \ \binom{N}{N}p^N q^0,$$

which are the terms, in reverse order, of the binomial expansion of $(p+q)^N$. As we saw in running the programs BINGRAPH, BELL, and BELL2 back in Chapter 4, the terms of the distribution increase with increasing values of k, the number of successes in N trials, until a maximum is reached, and then decrease thereafter. Bernoulli's lemmas codify this quantitatively for N of the form nt and for terms of the expansion $(r+s)^N = (p+q)^N t^N$.

Bernoulli's Theorem is trivially true if either r or s is 0. Thus we assume as given two positive integers r, s, their sum t, and a positive integer n.

---

[17] *Cf.*, e.g., William Feller, *An Introduction to Probability Theory and Its Applications*, vol. I, 2nd edition, John Wiley & Sons, New York, 1957, pp. 139 - 142.

[18] In presenting Bernoulli's proof in what follows, I stray a little from Haussner's translation, which itself strayed a little from Bernoulli. The proof presented here is faithful to Bernoulli in outline if not in detail. Those wanting to see the original details are referred to Sheynin, *op. cit.*, or Sylla, *op. cit.*

Bernoulli proves five lemmas in all, the first two of which are trivial and combine to state that

$$(r+s)^{nt} = L_{ns} + L_{ns-1} + \ldots + L_1 + M + R_1 + \ldots + R_{nr-1} + R_{nr},$$

where

$$M = \binom{nt}{ns} r^{nr} s^{ns} = \binom{nt}{nr} r^{nr} s^{ns}$$

$$L_i = \binom{nt}{ns - i} r^{nr+i} s^{ns-i}, \quad \text{for } i = 1, 2, \ldots, ns$$

$$R_i = \binom{nt}{ns + i} r^{nr-i} s^{ns+i}, \quad \text{for } i = 1, 2, \ldots, nr.$$

In words, his first two lemmas assert there to be $nt + 1$ terms in the expansion of $(r + s)^{nt}$, ns lying to the left of M and nr to the right when the terms are listed in decreasing order of the exponent on r:

$$\underbrace{\binom{nt}{0} r^{nt} s^0 + \ldots + \binom{nt}{ns - 1} r^{nr+1} s^{ns-1}}_{ns \; L's}$$

$$+ \underbrace{\binom{nt}{ns} r^{nr} s^{ns}}_{M} + \underbrace{\binom{nt}{ns + 1} r^{nr-1} s^{ns+1} + \ldots + \binom{nt}{nt} r^0 s^{nt}}_{nr \; R's}.$$

A couple of quick comments before continuing: In the last chapter we listed the probabilities of the binomial distribution in increasing order of the number of successes: $k = 0$, then 1, then 2, etc. Bernoulli uses the natural algebraic order of terms in the expansion $(r + s)^N$, which reverses this: $k = N$, then $N - 1$, then $N - 2$, etc. Consistency in exposition is not always compatible with faithfulness to the historical development. In any event, I hope the reader will not find this sudden notational reversal confusing.

The letter "M" can be taken to stand for "mean" as it is the probability of scoring a mean number of successes,

$$M = \binom{nt}{\mu} r^\mu s^{nt-\mu},$$

with $\mu = (nt)p = (nt)\frac{r}{t} = nr$. Choosing N to be a multiple of t guarantees μ to be an integer and the term M to exist. The existence of such a term makes for the simplest presentation of the argument and may well be the reason why Bernoulli chose N to be a multiple of t. It is not really necessary in carrying out the proof, but it does simplify matters.

Note that, if we set $i = 0$ in the expressions for L_i and R_i, we obtain M. We may thus write $M = L_0 = R_0$.

Bernoulli's third lemma has real content. It is the formal codification of the fact remarked on above that the terms of the binomial expansion increase to a maximum and decrease thereafter.

2 Lemma. *M is the largest term in the expansion,*

$$L_{ns} + L_{ns-1} + \ldots + L_1 + M + R_1 + \ldots + R_{nr-1} + R_{nr},$$

in fact,

$$L_{ns} < L_{ns-1} < \ldots < L_1 < L_0 \quad and \quad R_0 > R_1 > \ldots > R_{nr-1} > R_{nr}. \quad (3)$$

Moreover,

$$\frac{L_0}{L_1} < \frac{L_1}{L_2} < \ldots < \frac{L_{ns-1}}{L_{ns}} \quad and \quad \frac{R_0}{R_1} < \frac{R_1}{R_2} < \ldots < \frac{R_{nr-1}}{R_{nr}}. \quad (4)$$

Proof. Consider the ratio of two successive terms in the binomial expansion of $(r+s)^{nt}$:

$$\frac{\binom{nt}{k} r^{nt-k} s^k}{\binom{nt}{k+1} r^{nt-k-1} s^{k+1}} = \frac{\dfrac{(nt)!}{(nt-k)!k!}}{\dfrac{(nt)!}{(nt-k-1)!(k+1)!}} \cdot \frac{r}{s}$$

$$= \frac{(nt-k-1)!(k+1)!}{(nt-k)!k!} \cdot \frac{r}{s}$$

$$= \frac{k+1}{nt-k} \cdot \frac{r}{s}.$$

Now, the function

$$f(x) = \frac{x+1}{nt-x} \cdot \frac{r}{s}$$

is strictly increasing for positive inputs: if $nt > y > x$, then

$$f(y) = \frac{y+1}{nt-y} \cdot \frac{r}{s} > \frac{x+1}{nt-y} \cdot \frac{r}{s} > \frac{x+1}{nt-x} \cdot \frac{r}{s} = f(x).$$

The term R_i corresponds to $k = ns + i$, whence

$$\frac{R_i}{R_{i+1}} = f(ns+i) < f(ns+i+1) = \frac{R_{i+1}}{R_{i+2}},$$

and the second string of inequalities of (4) is established. For the first string, note that L_i corresponds to $k = ns - i$, whence

$$\frac{L_{i+2}}{L_{i+1}} = f(ns-i-2) < f(ns-i-1) = \frac{L_{i+1}}{L_i},$$

and, inverting the fractions,

$$\frac{L_i}{L_{i+1}} < \frac{L_{i+1}}{L_{i+2}}.$$

Thus we have (4).

To prove (3), solve the equation $f(x) = 1$:

$$\frac{x+1}{nt-x} \cdot \frac{r}{s} = 1 \Rightarrow \frac{rx+r}{nts-xs} = 1$$

$$\Rightarrow rx + r = nts - xs$$

$$\Rightarrow (r+s)x = nts - r$$

$$\Rightarrow x = \frac{nts-r}{r+s} = \frac{nts-r}{t} = ns - \frac{r}{t},$$

whence $ns - 1 < x < ns$. Thus, for $k \geq ns$, $f(k) > 1$:

$$\frac{R_i}{R_{i+1}} > 1, \quad \text{i.e., } R_i > R_{i+1};$$

and, for $k \leq ns - 1$, $f(k) < 1$:

$$\frac{L_{i+1}}{L_i} < 1, \quad \text{i.e., } L_{i+1} < L_i,$$

and we have proven (3)   □

The fourth Lemma of Bernoulli concerns the relative sizes of M and terms distant from it in the expansion.

3 Lemma. *The ratios M/L_n and M/R_n can be made as large as we please by choosing n sufficiently large. In modern notation,*

$$\lim_{n \to \infty} \frac{M}{L_n} = \lim_{n \to \infty} \frac{M}{R_n} = \infty.$$

False proof. Note that

$$\frac{M}{L_n} = \frac{L_0}{L_n} = \frac{L_0}{L_1} \cdot \frac{L_1}{L_2} \cdots \frac{L_{n-1}}{L_n}$$

$$> \frac{L_0}{L_1} \cdot \frac{L_0}{L_1} \cdots \frac{L_0}{L_1}, \quad \text{by Lemma 2}$$

$$> \left(\frac{L_0}{L_1}\right)^n.$$

But, by Lemma 2, $L_0 > L_1$, whence

$$\frac{M}{L_n} > a^n,$$

for some $a > 1$ and a^n can be made as large as we choose by making n large enough.   □

Do you see the error? Certainly, for any constant $a > 1$, a^n can be made as large as we please by choosing n large enough. The problem here is that L_0/L_1 is not a constant, but a function of n. We have only shown that there is a function $f(n)$ such that

$$\frac{M}{L_n} > f(n)^n,$$

and, without further information on f, there is no guarantee that $f(n)^n$ gets large without bound. Indeed, the archetypical counterexample is the limit, incidentally determined by Bernoulli himself,

$$\lim_{n \to \infty} \left(1 + \frac{1}{n}\right)^n = e. \tag{5}$$

This is one of two difficult limit computations of the beginning Calculus course (the other being $\lim_{x \to 0}(\sin x)/x = 1$) and should only be familiar to one who has taken a course in the Calculus already. If you haven't had such a course, enter the following into your calculator

```
seq((1+1/X)^X,X,1,12)→L₁
seq((1+1/X)^(X+1),X,1,12)→L₂
```

and enter the List Editor to compare the two lists. While that might not convince you that the limit is e, it should convince you that the limit exists and is close to e.

So the proof of Bernoulli's fourth lemma turns out to be a subtle matter. He, in fact, offers two proofs. One, which has only become acceptable by modern standards since the invention of Nonstandard Analysis in the 1960's, assumes n to be an infinite integer. As the use of such infinite numbers was new, Bernoulli followed up with a more detailed analysis for "those who are not acquainted with inquiries involving infinities" [19].

Proof of Lemma 3. We have

$$\frac{M}{L_n} = \frac{\binom{nt}{ns} r^{nr} s^{ns}}{\binom{nt}{ns-n} r^{nr+n} s^{ns-n}}$$

$$= \frac{(nt-ns+n)(nt-ns+n-1)\cdots(nt-ns+1)}{ns(ns-1)\cdots(ns-n+1)} \cdot \frac{s^n}{r^n}$$

$$= \frac{(nr+n)(nr+n-1)\cdots(nr+1)}{ns(ns-1)\cdots(ns-n+1)} \cdot \frac{s^n}{r^n}$$

$$= \frac{nr+n}{ns-n+n} \cdot \frac{nr+n-1}{ns-n+(n-1)} \cdots \frac{nr+1}{ns-n+1} \cdot \frac{s^n}{r^n}.$$

Now consider the function

$$f(x) = \frac{nr+x}{ns-n+x} = \frac{a+x}{b+x},$$

---

[19] I quote Sheynin's translation. The online version of Sheynin, *op. cit.*, does not number the pages, so I will simply state that I take these words from the first line of the Explanatory Comment following the proof of Lemma 5.

with $a = nr$, $b = ns - n$. Now the graph of $y = f(x)$ is an hyperbola unless $a = b$, in which case the graph is the straight line $y = 1$. For $a \neq b$, the portion of the curve for $x > 0$ lies above the line $y = 1$ if $a > b$ and below this line if $a < b$. In either case the line $y = 1$ is an asymptote and convergence of the curve to the line is monotone: f is strictly decreasing if $a > b$ and increasing if $a < b$. Now

$$a \overset{<}{\underset{>}{=}} b \quad \text{iff} \quad nr \overset{<}{\underset{>}{=}} ns - n \quad \text{iff} \quad r \overset{<}{\underset{>}{=}} s - 1$$

and we have three cases to consider, according as r is less than, equal to, or greater than $s - 1$.

The reader whose analytic geometry is a bit rusty, but who knows the Calculus, can replace the appeal to our knowledge of hyperbolas by a simple differentiation:

$$f'(x) = \frac{1 \cdot (b + x) - (a + x) \cdot 1}{(b + x)^2} = \frac{b - a}{(b + x)^2},$$

whence

$$f'(x) \overset{<}{\underset{>}{=}} 0 \quad \text{iff} \quad b \overset{<}{\underset{>}{=}} a.$$

By whichever argument one chooses, we have three cases to consider. I will start with the easiest.

Case 1. $(r = s - 1)$. For each k,

$$f(k) = \frac{nr + k}{ns - n + k} = \frac{n(s - 1) + k}{n(s - 1) + k} = 1.$$

Thus

$$\frac{M}{L_n} = 1 \cdot 1 \cdots 1 \cdot \frac{s^n}{r^n} = \left(\frac{s}{s - 1}\right)^n,$$

which can be made as large as we please by choosing n sufficiently large.

Case 2. $(r < s - 1)$. Then f is increasing and the smallest of the factors $f(1), f(2), \ldots, f(n)$ is given by $k = 1$:

$$f(1) = \frac{nr + 1}{ns - n + 1} > \frac{nr}{ns - n + 1} = \frac{r}{s - 1 + \frac{1}{n}},$$

and

$$f(1) \cdot \frac{s}{r} > \frac{s}{s - 1 + \frac{1}{n}} > \frac{s}{s - 1 + \frac{1}{2}} = \frac{s}{s - \frac{1}{2}} > 1. \qquad (6)$$

Thus

$$\frac{M}{L_n} > \left(\frac{s}{s - \frac{1}{2}}\right)^n,$$

which again can be made as large as we please by choosing n sufficiently large.

Case 3. ($r > s - 1$). Then f is strictly decreasing and the smallest value of f among the factors occurs at $k = n$:

$$f(n) = \frac{nr + n}{ns - n + n} = \frac{r + 1}{s},$$

whence

$$f(n) \cdot \frac{s}{r} = \frac{r + 1}{s} \cdot \frac{s}{r} = \frac{r + 1}{r} > 1.$$

Thus

$$\frac{M}{L_n} > \left(\frac{r + 1}{r}\right)^n,$$

which can once again be made as large as we please by choosing n sufficiently large.

The ratio M/R_n can be handled similarly or reduced to the result for the terms on the left by interchanging the rôles of r and s. ☐

4 Exercise. i. Let λ be given. Show that

$$\left(\frac{s}{s - 1}\right)^n > \lambda, \quad \left(\frac{s}{s - \frac{1}{2}}\right)^n > \lambda, \quad \left(\frac{r + 1}{r}\right)^n > \lambda$$

provided that

$$n > \frac{\ln \lambda}{\ln s - \ln(s - 1)}, \quad n > \frac{\ln \lambda}{\ln s - \ln(s - \frac{1}{2})}, \quad n > \frac{\ln \lambda}{\ln(r + 1) - \ln r},$$

respectively.

ii. Store $3, 2, 100$ in the variables R, S, L, respectively, in your calculator and find

$$\max\left\{\frac{\ln \lambda}{\ln s - \ln(s - 1)}, \frac{\ln \lambda}{\ln s - \ln(s - \frac{1}{2})}, \frac{\ln \lambda}{\ln(r + 1) - \ln r}\right\}.$$

You should get 16.00784556, meaning that M/L_n will be greater than 100 for $n \geq 17$, i.e., for $nt \geq 17 \cdot 5 = 85$.

iii. Enter the following program in your calculator:

```
PROGRAM:BTHEOREM
:1→B
:1→N
:R+S→T
:R/T→P
:Repeat B>100
:N+1→N
:binomialpdf(NT,P,NR)/binompdf(NT,P,NR−N)→B
:End
:Disp N .
```

This will find the smallest value n_0 of n such that $n_0 t = 5n_0$ trials will yield a ratio $M/L_{n_0} > 100$. How does the value obtained compare with that found in part ii? Calculate the ratios M/L_n for $n = n_0$ and $n = n_0 - 1$.

Bernoulli's fifth lemma is a simple corollary to the fourth:

5 Lemma. *For any positive constant c, n can be chosen so large that*

$$\sum_{k=1}^{n} L_k + M + \sum_{k=1}^{n} R_k > c \left(\sum_{k>n+1} L_k + \sum_{k>n+1} R_k \right). \tag{7}$$

In words, given c, no matter how large, n can be chosen so that the odds that the number k of successes satisfies

$$nr - n \le k \le nr + n$$

are greater than c to 1. Or, dividing by nt, the odds that $\frac{k}{nt}$ lies between $\frac{r-1}{t}$ and $\frac{r+1}{t}$ are greater than c to 1. Stated in terms of probability, the probability that $\frac{k}{nt}$ lies between $\frac{r-1}{t}$ and $\frac{r+1}{t}$ can be made greater than $\frac{c}{c+1}$ by choosing n large enough. That is, with the proof of Lemma 5, Bernoulli's Theorem will have been proved.

Proof of Lemma 5. Let c be given. As already remarked, there is nothing to prove if either r or s is 0. Let us also assume, for the time being, that neither r nor s is 1.

We first want to compare

$$L_1 + L_2 + \ldots + L_n \tag{8}$$

with

$$\sum_{i=n+1}^{ns} L_i = L_{n+1} + \ldots + L_{2n} + L_{2n+1} + \ldots + L_{ns}$$

$$= \sum_{k=1}^{s-1} \left(L_{kn+1} + \ldots + L_{kn+n} \right). \tag{9}$$

We shall do this by comparing (8) with each summand $L_{kn+1} + \ldots + L_{kn+n}$ of (9). To this end, suppose, for some λ yet to be determined, we have found n so large that

$$\frac{M}{L_n} > \lambda.$$

By Lemma 2, for $i < j$,

$$\frac{L_i}{L_{i+1}} < \frac{L_j}{L_{j+1}},$$

whence

$$\frac{L_i}{L_j} < \frac{L_{i+1}}{L_{j+1}}.$$

In particular,

$$\lambda < \frac{M}{L_n} = \frac{L_0}{L_n} < \frac{L_1}{L_{n+1}} < \frac{L_2}{L_{n+2}} < \ldots,$$

and for each m, $L_m > \lambda L_{m+n}$. Thus, for $k < s$,

$$L_{kn+i} > \lambda L_{(k+1)n+i}$$

and

$$L_{kn+1} + L_{kn+2} + \ldots + L_{kn+n} > \lambda L_{(k+1)n+1} + \lambda L_{(k+1)n+2} + \ldots + \lambda L_{(k+1)n+n}.$$

Thus,

$$\left. \begin{aligned} L_1 + L_2 + \ldots + L_n &> \lambda \left(L_{n+1} + L_{n+2} + \ldots + L_{2n} \right) \\ &> \lambda^2 \left(L_{2n+1} + L_{2n+2} + \ldots + L_{3n} \right) \\ &\vdots \\ &> \lambda^{s-1} \left(L_{(s-1)n+1} + L_{(s-1)n+2} + \ldots + L_{sn} \right) \end{aligned} \right\}. \tag{10}$$

If λ is chosen greater than or equal to 1, then, for each k, $\lambda^k \geq \lambda$ and

$$L_1 + L_2 + \ldots + L_n > \lambda \left(L_{kn+1} + L_{kn+2} + \ldots + L_{kn+n} \right).$$

Hence

$$(s-1)\left(L_1 + L_2 + \ldots + L_n \right) > \sum_{k=1}^{s-1} \lambda \left(L_{kn+1} + L_{kn+2} + \ldots + L_{kn+n} \right)$$

$$> \lambda \sum_{k=n+1}^{ns} L_k, \text{ by (9)}$$

and

$$\sum_{k=1}^{n} L_i > \frac{\lambda}{s-1} \sum_{k=n+1}^{ns} L_i.$$

Similarly, one can show

$$\sum_{k=1}^{n} R_i > \frac{\lambda}{r-1} \sum_{k=n+1}^{nr} R_i, \tag{11}$$

whence

$$\sum_{k=1}^{n} L_i + M + \sum_{k=1}^{n} R_i > \frac{\lambda}{s-1} \sum_{k=n+1}^{ns} L_i + \frac{\lambda}{r-1} \sum_{k=n+1}^{nr} R_i$$

$$> \frac{\lambda}{\max\{r-1, s-1\}} \left(\sum_{k=n+1}^{ns} L_i + \sum_{k=n+1}^{nr} R_i \right),$$

and, for $r, s \neq 1$ we need only choose $\lambda > c \cdot \max\{r-1, s-1\}$, e.g., $\lambda > ct$, to conclude the Lemma in the case where neither r nor s is 1.

Should one of the numbers r, s be 1, the other is not and only one of the sums appears on the right in (7). For example, if $s = 1$, then we apply (11) to get

$$\sum_{k=1}^{n} L_i + M + \sum_{k=1}^{n} R_i > \frac{\lambda}{r-1}\left(0 + \sum_{k=n+1}^{nr} R_i\right),$$

and choosing $\lambda > c(r-1)$ suffices to conclude the Lemma.   □

As with Lemma 3, Bernoulli first gave a quick proof of Lemma 5 using an infinite value of n. He followed this proof with more direct proofs of his last two lemmas yielding explicit finite estimates for n. As an example he supposes repeating an experiment with probability 3/5 of success on a given trial:

> In a particular application of this theorem to numbers, one easily recognises that the larger the numbers chosen for r, s, t (where the ratio $\frac{r}{s}$ must not change), the tighter together the limits $\frac{r+1}{t}$ and $\frac{r-1}{t}$ move. If thus the ratio $\frac{r}{s}$ equals $\frac{3}{2}$ for example, I do not set $r = 3$ and $s = 2$, but rather $r = 30$ and $s = 20$, thus $t = r + s = 50$, or $r = 300$ and $s = 200$, thus $t = 500$. In the first case the limits are
>
> $$\frac{r+1}{t} = \frac{31}{50} \quad \text{and} \quad \frac{r-1}{t} = \frac{29}{50}.$$
>
> If I then take $c = 1000$, then...
>
> $$nt = \ldots = 25550.$$
>
> Thus it is, by the above proven theorem, more than 1000 times more probable that in 25550 observations the ratio of favourable to all observations lies within the limits $\frac{31}{50}$ and $\frac{29}{50}$ (inclusive) than outside the same... [20]

Bernoulli adds that 31258 observations yield odds 10000 to 1 and 36966 observations yield odds of 100000 to 1 that the ratio of favourable to all observations will lie within the limits $\frac{31}{50}$ and $\frac{29}{50}$ (inclusive).

Because I have followed Haussner's slight deviation from Bernoulli's proof, I cannot reproduce these values. In fact, the proof given here yields larger bounds. An extra trick, due to Nikolaus Bernoulli will allow us to reduce these slightly. Here, we have $c = 1000, 10000$, or $100000, r = 30, s = 20, t = 50$. For Lemma 5 we chose $\lambda > c \cdot \max\{r - 1, s - 1\}$, which for $c = 1000$ means $\lambda > 1000 \max\{29, 19\} > 29000$, say $\lambda = 29001$. The proof of Lemma 3 yielded $M/L_n > \lambda$ provided

$$n > \frac{\ln\lambda}{\ln s - \ln(s-1)}, \frac{\ln\lambda}{\ln s - \ln(s-1/2)}, \frac{\ln\lambda}{\ln(r+1) - \ln r},$$

and $M/R_n > \lambda$ provided

$$n > \frac{\ln\lambda}{\ln r - \ln(r-1)}, \frac{\ln\lambda}{\ln r - \ln(r-1/2)}, \frac{\ln\lambda}{\ln(s+1) - \ln s}.$$

Now the minimum of these denominators is

---

[20] Haussner, *op. cit.*, pp. 106 - 107.

$$\ln r - \ln(r - \frac{1}{2}) = \ln 30 - \ln 29.5,$$

whence the maximum of these ratios is

$$\frac{\ln \lambda}{\ln r - \ln(r - \frac{1}{2})} = \frac{\ln 29001}{\ln 30 - \ln 29.5} = 611.3532015. \tag{12}$$

This gives us $n \geq 612$, i.e. $N = nt = 50n \geq 30600$. For $c = 10000$, we get $N = 50n \geq 37450$, and for $c = 100000$, $N = 50n \geq 44300$.

We can improve this by returning to the proof of Lemma 5 and rewriting formulæ (10) as

$$\left(\frac{1}{\lambda}\right)^k (L_1 + L_2 + \ldots + L_n) > \left(L_{kn+1} + L_{kn+2} + \ldots + L_{(k+1)n}\right)$$

and successively concluding

$$\left(\sum_{k=1}^{s-1} \frac{1}{\lambda^k}\right)(L_1 + L_2 + \ldots + L_n) > \sum_{k=n+1}^{ns} L_k$$

$$\left(\sum_{k=1}^{\infty} \frac{1}{\lambda^k}\right)(L_1 + L_2 + \ldots + L_n) > \sum_{k=n+1}^{ns} L_k, \quad \text{provided } \lambda > 1$$

$$\frac{1}{\lambda - 1}(L_1 + L_2 + \ldots + L_n) > \sum_{k=n+1}^{ns} L_k$$

$$L_1 + L_2 + \ldots + L_n > (\lambda - 1)\sum_{k=n+1}^{ns} L_k.$$

Thus we will have

$$L_1 + L_2 + \ldots + L_n > c\sum_{k=n+1}^{ns} L_k$$

provided $\lambda - 1 \geq c$, i.e., $\lambda \geq c+1$. For $c = 1000$, this means we can take $\lambda = 1001$ in (12) and conclude we need only take $n > 411.0612331$, i.e., $N = 50n \geq 20600$ and we have reduced Jakob Bernoulli's estimate from 25550 to 20600, still not as good as de Moivre would later yield[21], but a definite improvement. And, for $c = 10000$ and $c = 100000$, we get $N \geq 27450$ and $N \geq 34300$, respectively.

Another inefficiency of the proof as presented here occurs in Case 2 of the proof of Lemma 3, where I used inequaltiy (6) to replace

$$\frac{s}{s - 1 + \frac{1}{n}}$$

by the smaller

$$\frac{s}{s - \frac{1}{2}} > 1,$$

---

[21] *Cf.* the opening remarks of the next section.

to get an exponential lower bound

$$\frac{M}{L_n} > \left(\frac{s}{s - \frac{1}{2}}\right)^n.$$

This is fine for quickly verifying that M/L_n can be as large as we wish by choosing n large, but for the practical matter of determining a bound on n making

$$\frac{M}{L_n} > \lambda$$

for a given λ, it is inefficient. An improved estimate is obtained by considering

$$\frac{M}{L_n} > \left(\frac{s}{s - 1 + \frac{1}{n}}\right)^n > \lambda$$

and observing that this inequality holds provided

$$n > \frac{\ln \lambda}{\ln s - \ln(s - 1 + \frac{1}{n})}.$$

The problem of estimating n when it occurs on both sides of the inequality is algebraically a bit daunting. With our modern calculators, however, it is hardly a computationally formidable task:

6 Project. To determine n, we would start with an initial estimate n_0 and then compare it to

$$n > \frac{\ln \lambda}{\min(\{\ln r - \ln(r - 1 + \frac{1}{n_0}), \ln s - \ln(s - 1 + \frac{1}{n_0})\}}.$$

To this end, store 30 in R, 20 in S, and either 29001 or 1001 in L. Go into the equation editor and enter

Y$_1$=ln(L)/min({ln(R)−ln(R−1+1/X),ln(S)−ln(S−1+1/X)}).

X will assume various values of n_0 and Y$_1$(X) will give us estimates on lower bounds for n. Y$_1$(2) is the crude approximation given by (6) and offers an upper bound on n_0. Now for $x \geq 2$, Y$_1$(x) is strictly decreasing, so as one chooses ever larger values of n_0, one will get ever smaller values of n. As there is no point in choosing n_0 greater than n, we look to see where $x = $ Y$_1$(x) for an approximation to n_0 and n. To this end, enter

Y$_2$=X

in the equation editor and find the point of intersection of Y$_1$ and Y$_2$ in the usual fashion[22]. The one point I might make here is the choice of the window. Obviously one can choose 2 or any number slightly to the left of it as Xmin. Y$_1$(2) ($= 611.35\ldots$ for $\lambda = 29001$, and $= 411.06\ldots$ for $\lambda = 1001$) offers an upper

---

[22] I assume familiarity with the use of the graphing calculator for such purposes. Those unfamiliar with such are referred to the calculator manual.

bound on n_0, so one can choose, say, 650 as Xmax. Choosing 50 for Xscl will not make too many tick marks appear at the bottom of the screen. You can let the calculator choose the Y parameters by pressing the ZOOM button and choosing ZoomFit. Show in this way that if we take $\lambda = 29001$ we get the bound $N \geq 15250$; and if we take $\lambda = 1001$ we get the bound $N \geq 10250$.

4 Criticism of Bernoulli's Proof

The *Ars conjectandi* stops fairly abruptly after Bernoulli's proof. The book was never completed. Explanations offered include his failing health, responsibilities at the university in Basel, and critiques like the following:

> Bernoulli's upper bound was a start but it must have been a disappointing one, both to Bernoulli and to his contemporaries. To find that 25,550 experiments are needed to learn the proportion of fertile[23] to sterile cases within one part in fifty is to find that nothing reliable can be learned in a reasonable number of experiments. The entire population of Basel was then smaller than 25,550; Flamsteed's 1725 star catalogue listed only 3,000 stars. The number 25,550 was more than astronomical; for all practical purposes it was infinite. I suspect that Jacob Bernoulli's reluctance to publish his deliberations of twenty years on the subject was due more to the magnitude of the number yielded by his first and only calculation than to any philosophical reservations he may have had about making inferences based upon experimental data. ... From this viewpoint Bernoulli's noble start *was* a failure, but one that held out hope for De Moivre.[24]

One of the great statisticians of the early 20th century, Karl Pearson, offered the same criticism, albeit somewhat more devastatingly:

> James Bernoulli... adopted a very crude method of inequalities which I will shortly reproduce. Let us suppose that N is the true number of observations which must be made to give the required value of the probability, then clearly if m be > 1, mN will give a probability falling within the required limits. But the practical value of the solution must depend on m being nearly unity. This is far from the case with Bernoulli's solution. He gets most exaggerated values for the needful number of observations, and for this reason his solution must be said to be from the practical standpoint a failure; it would ruin either an insurance society or its clients, if it were adopted. All Bernoulli achieved was to show

---

[23] Haussner uses "günstig", or "favourable", and I have translated him. Schneider uses "fruchtbar", or "fertile" in another passage, and Sylla uses "fertile". Since "favourable" is the term currently used in this context, I prefer its use as it quickly conveys the meaning of the passage.

[24] Stigler, *op. cit.*, p. 77. Stigler continues, noting that Bernoulli's insistence on "moral certainty" was too high a standard, the common modern standard taking $c = 20$.

that by increasing the number of observations the results would un-doubtedly fall within certain limits, but he failed entirely to determine what the *adequate* number of observations were for such limits. This was entirely De Moivre's discovery.[25]

Pearson, essentially assuming the rôle of advocate for the cause of Abraham de Moivre whose contributions to the problem were, at the time, commonly, directly or indirectly, being inaccurately attributed to Bernoulli, drew up some tables determining the estimates n of the number N of trials needed to assure the odds $c : 1$ that the observed relative frequencies in Bernoulli's example $r = 30, s = 20, t = 50$ were within $1/50$ of $3/5$. I reproduce the first of these tables[26] as *Table 1*, below. Pearson says of this table:

Odds $c : 1$	Bernoulli's Result	De Moivre's Result
1000:1	25,550 trials	6,498 trials
10,000:1	31,258 "	9,082 "
100,000:1	36,966 "	11,704 "

Table 1. Bernoulli/De Moivre Comparison

These numbers, I think, show that while Bernoulli had a vision of "Bernoulli's Problem," he failed to solve it in anything like a reasonable manner. His discussion is not only lengthy and somewhat obscure, but is too loose to give any valid solution. He appreciated, what most exper-imenters knew already, that one gets nearer to the truth by increasing the size of the sample, but he really failed to obtain any adequate mea-sure of the approach. Yet Bernoulli himself thought his results of "high utility." [27]

He adds

Bernoulli admits that as a matter of experience even in his day, the increase in accuracy by the increase of trials was recognised. But he was probably the first to more or less clearly state the problem. It must be admitted, I think, that his twenty years consideration of the matter did not lead him to a solution of much value.

(3) It may not be without interest to deal a little fully with what seem

[25] Karl Pearson, "James Bernoulli's Theorem", *Biometrika* 17, Nos. 3 - 4 (1925), pp. 201 - 210; here, p. 202.

[26] *Ibid.*, p. 206.

[27] *Ibid.*

to be the weak points in Bernoulli's method of inequalties. They are essentially two-fold...[28]

There follow specific points at which Bernoulli's estimate could be improved, Pearson saying of one, "His approximation... is so appallingly crude that he must get bad results".[29]

Jakob Bernoulli's estimates are indeed bad, but as we have seen in Project 6, above, a little fiddling brings them down quite a bit. They are still crude and larger than de Moivre's, disappointingly so, but are they still appallingly so?

I am inclined to think Pearson overstates his criticism of Bernoulli. Certainly, where applications are concerned Bernoulli's estimates are not much more useful than the intuitive realisation that relative frequency aproaches probability as the number of repetitions grows large—a realisation that Bernoulli explicitly acknowledges was already widespread. But he was the first to realise that quantitative estimates could be made, and, in making such, pointed the way for his successors, including de Moivre. I suspect Pearson overstates things because such stress was needed to "correct the record". From 1921 until his retirement in 1933 Pearson lectured on the history of statistics.[30] In the opening chapter of his lecture notes he declares,

> There is a fundamental theorem in statistics called Bernoulli's Theorem. For 150 years nobody appears to have looked at what James Bernoulli actually wrote, but because Laplace in discussing a similar theorem cited Bernoulli, a most fundamental principle of statistics has been attributed to Bernoulli instead of to its real discoverer De Moivre.[31]

When he reaches de Moivre, Pearson says

> In other words, De Moivre in 1733 reached the normal curve as a limit to the skew binomial[32] and gave the correct measure of dispersion, $\sqrt{(npq)}$, which we now term the standard deviation. I must confess that this was a great revelation to me, when I discovered it in De Moivre. Many years ago a study of Laplace convinced me that to attribute the so-called curve of errors to Gauss was a grave mistake, since Laplace antedated Gauss. I proposed to call it the Laplace-Gaussian or better 'Normal Curve'. The latter name may stand, but its originator was not even Laplace, it was De Moivre, who antedates Laplace by upwards of 40

---

[28] *Ibid.*, p. 207.

[29] *Ibid.*

[30] Eventually his son Egon (1885 - 1980) edited and published these lectures: Karl Pearson, *The History of Statistics in the 17th & 18th Centuries against the changing background of intellectual, scientific and religious thought*, Charles Griffin & Co. Ltd., London, 1978.

[31] *Ibid.*, p. 1. "Bernoulli's Theorem" had come to refer not just to Bernoulli's Theorem as presented here, but more generally to de Moivre's refinement.

[32] I.e., the binomial distribution for all values of p, not just $p = 1/2$, as de Moivre is often represented as having limited his proof to.

years. This is a tremendous feather in De Moivre's cap and we have got
to revolutionise our historical ideas of the normal curve.[33]

And when he reaches the Bernoulli brothers and their disputes, he writes

> Mathematicians have always been rather of a jealous nature, and un-
> doubtedly jealousy was a family characteristic of the Bernoullis. There
> is some excuse for the mathematicians, for their reputations stand for
> posterity largely not on what they did, but on what their contempo-
> raries attributed to them, and I have only to cite the case of the normal
> curve being attributed to Gauss and the coefficient of correlation to
> Bravais to show how easy it is to overrate one man at the expense of
> another, and how vain the effort of a later historian may be to rectify
> matters.[34]

The mention of the physicist Auguste Bravais may help explain the sense of
urgency Pearson felt about setting the record straight. On page 1, directly after
our first quote from these lectures he admits his own guilt in falsifying history:

> In quite recent times, by an error for which I personally am to some
> extent responsible, the coefficient of correlation has been attributed
> to Bravais. This is cited in many textbooks and 'Bravais' formula' is
> spoken of. Usually the memoir in which he published it is cited. That
> memoir actually gives no more than Gauss had given years previously,
> and there is no coefficient of correlation in it! Thus Bravais has got
> the credit of what was really due to Galton. I expect that error will
> go down to the last trump and you will find it everywhere—except in
> Professor Kelley's recent textbook of statistics for psychologists. It will
> be interesting to see if that will have any success in stemming the flood
> of error.[35]

In editing his father's lectures for publication about a half century later, E.S.
Pearson inserted a parenthetical note: "It has certainly been stemmed by now!"
In like manner, whatever historical slight has been given de Moivre, it has been
stemmed by now, and the forcefulness of Pearson's prose is no longer necessary.

The upshot of all this is that Bernoulli's Theorem needed sharpening and
later mathematicians, first his nephew Nikolaus, a little later de Moivre, and
thereafter others provided such.

There are two additional weak points to the proof that can be cited here.
They have both been mentioned earlier. First, Bernoulli does not prove his Theorem for all sufficiently large numbers N of trials, but only for those numbers
N that are multiples of t. Second, his proof does not cover the case of irrational
probabilities. These weaknesses are not subject to the strongly stated criticism
we have seen for the inadequacy of Bernoulli's bounds.

---

[33] Pearson, *History...*, *op. cit.*, p. 156.

[34] *Ibid.*, p. 226.

[35] *Ibid.*, p. 1. Incidentally, the "last trump" is not a reference to playing cards, as our
current preoccupation with probability might suggest, but to the "last trumpet",
i.e., the "trump of doom".

The difference between rational and irrational in mathematics can be the difference between pre-calculus and the Calculus, between algebraic manipulation and the careful analysis of estimates and limits. Perhaps one doesn't make much of Bernoulli's failure in this respect because doing so would be ahistorical: irrational probabilities would not have arisen in a day in which probabilities were given as ratios of the numbers of favourable and all cases in situations which reduced to the consideration of finite numbers of equally likely outcomes. Another reason might be the fact that modern proofs require no special complications to cover the irrational case along with the rational one.

As for the criticism that Bernoulli only proves his Theorem for N of the form nt, this is not a big point. At the cost of some notational complication, Bernoulli's argument can be reformulated to cover the case of $nt + k$ trials for $0 < k < t$. Also, the general result reduces to the special case by a standard trick from the Calculus. I refer the curious reader to Appendix B for the details.

The only other criticism of Bernoulli's proof that comes to mind is Pearson's earlier quoted remark that the proof is "not only lengthy and somewhat obscure but is too loose to give any valid solution". This, however, is not criticism of the proof, but of the presentation thereof. The basic outline of the proof is clear enough: given the list of lemmas to prove, any modern mathematician should have little difficulty filling in the details. In fact, as Stigler reports, J.V. Uspensky, in his textbook, "presented Bernoulli's original proof, despite its length, as more natural than later, simpler proofs".[36]

Bernoulli's Theorem itself also has its shortcomings and for over two centuries researchers would be returning to it again and again, strengthening and generalising it. Additionally, beginning with Leibniz, even before publication, there was criticism of the application of the result.

5 An Application of Bernoulli's Theorem: Birth Rates

Pearson's remarks to the contrary, it is hard to overstate the importance of Bernoulli's Theorem, whatever its imperfections, in the history of probability. The following bit of hyperbolé by John Maynard Keynes is not that far from the truth:

> BERNOULLI'S Theorem is generally regarded as the central theorem of statistical probability. It embodies the first attempt to deduce the measures of statistical frequencies from the measures of individual probabilities and it is a sufficient fruit of the twenty years which Bernoulli alleges that he spent in reaching his result, if out of it the conception first arose of general laws amongst masses of phenomena, in spite of the uncertainty of each particular case.[37]

---

[36] Stigler, *op. cit.*, p. 66. Uspensky's textbook is *Introduction to Mathematical Probability*, McGraw-Hill, New York, 1937.

[37] John Maynard Keynes, *A Treatise on Probability*, Macmillan and Co., London, 1921; reprinted by Dover Publications, Inc., Mineola, NY, 2004; p. 337.

A decade later Harald Westergaard stated of Bernoulli's Theorem:

> ... at all events this achievement was a masterpiece. If statisticians had followed the track, they might at a very early epoch of the evolution of political arithmetic[38] have reached the means of controlling and criticising many results which they were now only able to treat in a very vague manner.[39]

Municipalities and insurance companies had been compiling statistics for some time before Bernoulli proved his "golden theorem", and there had been crude applications of probabilistic reasoning before him, but his Theorem marks the first solid step toward a mathematical theory of statistics. One might take exception to Keynes's reference to Bernoulli as "the real founder of mathematical probability"[40], but no one besides Pearson would disagree if "probability" were to be replaced by "statistics". Bernoulli's Theorem marks the birth of Mathematical Statistics.

One of the first fruits of the birth of this Mathematical Statistics was a discussion of the statistics of births. While the *Ars conjectandi* was lying unpublished, Dr. John Arbuthnot, Physician in Ordinary to Her Majesty, Fellow of the College of Physicians, and Fellow of the Royal Society[41], turned his attention to the statistics of birth:

> A Mong innumerable Footsteps of Divine Providence to be found in the Works of Nature, there is a very remarkable one to be observed in the exact Ballance that is maintained, between the Numbers of Men and Women; for by this means it is provided, that the Species may never fail, nor perish, since every Male may have its Female, and of a proportionable Age. This Equality of Males and Females is not the Effect of Chance but Divine Providence, working for a good End, which I thus demonstrate:
>
> Let there be a Die of Two sides, M and F, (which denote Cross and Pile[42]), now to find all the Chances of any determinate Number of such Dice, let the Binome M+F be raised to the Power, whose Exponent is the Number of Dice given; the Coefficients of the Terms will shew all the Chances sought. For Example, in Two Dice of Two sides M+F the Chances are $M^2+2MF+F^2$, that is, One Chance for M double, One for F double, and Two for M single and F single; in Four such Dice there are Chances $M^4+4M^3F+6M^2F^2+4MF^3+F^4$, that is One Chance for M quadruple, One for F quadruple, Four for triple M and single F,

---

[38] "Political arithmetic" was an early name given to the field of statistics. Indeed, the name "statistics" deriving from the word "state" reveals its political origins.

[39] Harald Westergaard, *Contributions to the History of Statistics*, P.S. King & Son Ltd., London, 1932; reprinted by Augustus M. Kelley · Publishers, New York, 1969; pp. 103 - 104 in the reprint.

[40] Keynes, *op. cit.*,p. 41.

[41] Outside science he is known for his satire and the creation of the character John Bull.

[42] *Cf.* footnote 22 on page 45, above.

Four for single M and triple F, and Six for M double and F double; and universally, if the Number of Dice be n, all their Chances will be expressed in this Series $M^n + \frac{n}{1} \times M^{n-1}F + \frac{n}{1} \times \frac{n-1}{2} \times M^{n-2}F^2 + \frac{n}{1} \times \frac{n-1}{2} \times \frac{n-2}{3} \times M^{n-3}F^3 +$, &c.

It appears plainly, that when the Number of Dice is even there are as many M's as F's in the middle Term of this Series, and in all the other Terms there are most M's or most F's.

If therefore a Man undertake with an even Number of Dice to throw as many M's as F's, he has all the Terms but the middle Term against him; and his Lot is to the Sum of all the Chances, as the coefficient of the middle Term is to the power of 2 raised to an exponent equal to the Number of Dice: so in Two Dice his Lot is $\frac{2}{4}$ or $\frac{1}{2}$, in Three Dice $\frac{6}{16}$ or $\frac{3}{8}$, in Six Dice $\frac{10}{64}$ or $\frac{5}{16}$, in Eight $\frac{70}{256}$ or $\frac{35}{128}$ &c.

To find this middle term in any given Power or Number of Dice, continue the Series $\frac{n}{1} \times \frac{n-1}{2} \times \frac{n-2}{3}$, &c. till the number of terms are equal to $\frac{1}{2}n$. For Example, the coefficient of the middle Term of the tenth Power is $\frac{10}{1} \times \frac{9}{2} \times \frac{8}{3} \times \frac{7}{4} \times \frac{6}{5} = 252$, the tenth Power of 2 is 1024, if therefore A undertakes to throw with Ten Dice in one throw an equal Number of M's and F's, he has 252 Chances out of 1024 for him, that is his Lot is $\frac{252}{1024}$ or $\frac{63}{256}$, which is less than $\frac{1}{4}$.

It will be easy by the help of Logarithms, to extend this Calculation to a very great Number, but that is not my present Design. It is visible from what has been said, that with a very great Number of Dice, A's Lot would become very small; and consequently (supposing M to denote Male and F Female) that in the vast Number of Mortals, there would be but a small part of all the possible Chances, for its happening at any assignable time, that an equal Number of Males and Females should be born.

It is indeed to be confessed that this Equality of Males and Females is not Mathematical but Physical, which alters much the foregoing Calculation; for in this Case the middle Term will not exactly give A's Chances, but the Chances will take in some of the Terms next the middle one, and will lean to one side or the other. But it is very improbable (if mere Chance govern'd) that they would never[43] reach as far as the Extremities: But this Event is wisely prevented by the wise Oeconomy of Nature; and to judge of the wisdom of the Contrivance, we must observe that the external Accidents to which are Males subject (who must seek their Food with danger) do make a great havock of them, and that this loss exceeds far that of the other Sex, occasioned by Diseases incident to it, as Experience convinces us. To repair that Loss, provident Nature, by the Disposal of its wise Creator, brings forth more Males than Females; and that in almost a constant proportion. This appears from the annexed Tables, which contain Observations for 82 Years of the Births in *London*. Now, to reduce the Whole to a Calculation, I propose this:

---

[43] See the discussion beginning on page 193, below.

Problem. A lays against B, that every Year there shall be born more Males than Females: To find A's Lot, or the Value of his Expectation.

It is evident from what has been said, that A's Lot for each Year is less than $\frac{1}{2}$; (but that the Argument may be stronger) let his Lot be equal to $\frac{1}{2}$ for one Year. If he undertakes to do the same thing 82 times running, his Lot will be $(\frac{1}{2})^{82}$, which will be found easily by the Table of Logarithms to be $\frac{1}{4836\,0000\,0000\,0000\,0000\,0000}$ [44]. But if A wager with B, not only that the Number of Males shall exceed that of Females, every Year, but that this Excess shall happen in a constant Proportion, and the Difference lye within fix'd limits; and this not only for 82 Years, but for Ages of Ages, and not only at *London*, but all over the World; (which 'tis highly probable is Fact, and designed that every Male may have a Female of the same Country and suitable Age) then A's Chance will be near an infinitely small Quantity, at least less than any assignable Fraction. From whence it follows, that it is Art, not Chance that governs.

There seems no more probable Cause to be assigned in Physicks[45] for this Equality of the Births, than that in our first Parents Seed there were at first formed an equal Number of both Sexes.

Scholium. From hence it follows, that Polygamy is contrary to the Law of Nature and Justice, and to the Propagation of Human Race; for where Males and Females are in equal number, if one Man takes Twenty Wives, Nineteen Men must live in Celibacy, which is repugnant to the Design of Nature; nor is it probable that Twenty Women will be so well impregnated by one Man as by Twenty.[46]

Arbuthnot's notice in the *Philosophical Transactions* caught the eye of Nikolaus Bernoulli, whose subsequent letter to Pierre Rémond de Montmort of 11 October 1712 on the subject was printed in the second edition of Montmort's *Essai d'Analyse*:

I'm going to share with you one [finding] I made recently occasioned by an argument for Divine Providence they have inserted into the Philosophical Transactions. Someone had already spoken to me about this argument in Holland without saying that it had been printed somewhere. It is an argument based on the regularity observed among the children of one & of the other sex who are born every year in London. They say that if chance governed the world, it would be impossible that the numbers of males & females would approach as closely for several years running, as they did for 80 years, & one gives for a reason that

---

[44] There should be one more 0 in the denominator. Stigler, *op. cit.*, p. 226, silently corrects this.

[45] I.e., medicine or physiology—not physics.

[46] John Arbuthnot, "An Argument for Divine Providence, taken from the constant Regularity observ'd in the Births of both Sexes", *Philosophical Transactions of the Royal Society* 27 (1710 - 1712), pp. 186 - 190; here, pp. 186 - 189. I have had to make a few minor typographic changes in reproducing the text.

throwing a large number of counters, e.g., 10000 at random, it is highly unlikely that half falls cross & half pile, & even less likely that this happen many times. As I reiterated the same thing here, & they asked me my opinion on it, I was obliged to refute this argument, & to prove that there is a high probability that the number of males & females arrive each year between limits even smaller than those we have observed for 80 consecutive years. You feel well, Sir, that it would be a ridiculous thing, if we wanted to prove that it is likely that the number of boys will be precisely equal to the number of girls, but that the ratio between the numbers of the one & the other approaches very nearly the ratio of equality, is that of which I believe you will be persuaded. I found by examining the Catalogue of children born in London from 1629 until 1710 inclusive, there are more males than females; & on taking an average, the ratio of the males to females is very close to the ratio of 18 to 17, slightly larger; hence I conclude that the probability to be born a boy, is to the probability to be born a girl as 18 is to 17, & that among 14000 children, which is roughly the number of children born per year in London, there will be approximately 7200 males & 6800 females. Now the year in which was born the greatest number of males, compared to females, was the year of 1661, in which there were born 4748 males & 4100 females; & the year when there was the smallest number of males, compared to females, is the year 1703, where there were born 7765 males & 7683 females. I say that these limits are so large, you can bet at more than 300 to 1 that among 14,000 children the numbers of males & females fall within these limits rather than without.[47]

Nikolaus Bernoulli expanded his remarks in another letter to Montmort a few months later on 23 January 1713. This letter is notable for providing the first improvement in the proof of his uncle's celebrated Theorem:

I send you the Catalogue of the Children of both sexes born in London from 1629 until 1710, with my demonstrations that I wrote to you concerning the argument by which one wants to prove that it is a miracle that the numbers of children of each gender born in London did not differ more from each other for 82 consecutive years, & that by chance it would be impossible that for so long they were always confined within the limits as small as those we have observed in the Catalogue of 82 years. I maintain that there is no cause for surprise, & there is a high probability that the numbers of males & females fall between limits even smaller than those we have observed. To prove this, I suppose the number of all children born each year in London is 14000, among whom there ought to be born 7200 males & 6800 females, if the numbers of children of each sex followed exactly the ratio of 18 to 17, which expresses the proportion between the likelihoods of the birth of a boy &

---

[47] Pierre Rémond de Montmort, *Essai d'Analyse sur les Jeux de Hazard*, second edition, 1713, pp. 373 - 374.

that of the birth of a daughter; yet as the number of boys is sometimes greater, sometimes smaller than 7200, taking a limit: For example, the year 1703, in which the number of girls was the closest to that of boys, there were born in that year 7765 males & 7683 females, reducing the sum to 14000, this is 7037 males & 6963 females; the number of females has surpassed the number 6800 by 163, & the number of males was even lower than 7200. But I will prove that there is much to bet that among 14000 children, the number of males born will be neither more nor less than 7200 by [as much as] 163; i.e., that the ratio of males to females is not greater than 7363 to 6637, or smaller than 7037 to 6963. To this end imagine 14000 dice of 35 faces each, 18 of which are white & 17 black. You know that the terms of the binomial $18 + 17$, raised to 14000, give us all possible cases for obtaining any number of white faces wanted with the 14000 dice; namely, the first term that of all cases for obtaining all white faces; the second, for obtaining one black face & 13999 white; the 3rd, for obtaining 2 black faces & 13998 white, &c. So that the 6801th term expresses all cases for obtaining 6800 black faces & 7200 white; the 6638th term the cases for obtaining 6637 black faces & 7363 white; & the 6964th term the cases for obtaining 6963 black faces & 7037 white. The task therefore is to find what ratio exists between the sum of all terms from the 6638th until the 6964th term inclusive, & the sum of all of the terms that are taken below the 6638th, then beyond the 6964th. But as these terms are prodigiously large, it takes a singular device for finding this ratio: here is how I handle it.[48] In general, in place of 14000, let the number of children $= n$, the odds that the birth of a male & that of a female are as m to f, in place of the ratio 18 to 17; & in place of the limit 163, let there be taken any limit l; moreover let $p = \frac{n}{m+f}$, or $n = mp + fp$;[49] in our example mp is $= 7200$ & $fp = 6800$. I look first at a very close approximation to the ratio of the term whose number is $fp + 1$ to the term whose number is $fp - l + 1$. By the law of progression of these terms, the term $fp + 1$ is $=$

$$\frac{n}{1} \times \frac{n-1}{2} \times \frac{n-2}{3} \times \ldots \frac{n-fp+1}{fp} \times m^{n-fp} f^{fp},$$

& the term $fp - l + 1$ is $=$

$$\frac{n}{1} \times \frac{n-1}{2} \times \frac{n-2}{3} \times \ldots \frac{n-fp+l+1}{fp-l} \times m^{n-fp+l} f^{fp-l};$$

---

[48] The algebra gets a little hairy at this point and the reader, particularly one who skipped the preceding section, may wish to skip ahead to page 193.

[49] p does not denote here the probability of a male birth as one might expect. The notation is probably best clarified by matching variables with Jakob's Bernoulli's usage:

Jakob	r	s	t	n	nt
Nikolaus	m	f	$m+f$	p	n

thus the ratio of the former to the latter is as

$$\frac{n-fp+l}{fp-l+1} \times \frac{n-fp+l-1}{fp-l+2} \times \frac{n-fp+l-2}{fp-l+3} \times \cdots \frac{n-fp+1}{fp} \times \left(\frac{f}{m}\right)^l$$

is to 1, or, putting mp in place of $n-fp$ this ratio is as

$$\frac{mp+l}{fp-l+1} \times \frac{mp+l-1}{fp-l+2} \times \frac{mp+l-2}{fp-l+3} \times \cdots \frac{mp+1}{fp} \times \left(\frac{f}{m}\right)^l$$

is to 1; I assume that the factors of the first term of this ratio except the last $(f/m)^l$ were in geometric progression and their logarithms in arithmetic progression; this assumption is very close to the truth, especially when n is a large number; the sum therefore of all their logarithms will be

$$\frac{1}{2}l \times \left(\log \frac{mp+l}{fp-l+1} + \log \frac{mp+1}{fp}\right),$$

i.e., the sum of the logarithms of the first & last factors, multiplied by half the number of all the terms, with which if one adds the logarithm of $(f/m)^l$, i.e., $l \times \log \frac{f}{m}$, one has

$$\frac{1}{2}l \times \left(\log \frac{mp+l}{fp-l+1} + \log \frac{mp+1}{fp}\right) + l \times \log \frac{f}{m}, \text{ or}$$

$$\frac{1}{2}l \times \left(\log \frac{mp+l}{fp-l+1} + \log \frac{mp+1}{mp} + \log \frac{fp}{mp}\right)$$

for the logarithm of the ratio sought. And consequently the ratio of them is as

$$\left(\frac{mp+l}{fp-l+1} \times \frac{mp+1}{mp} \times \frac{fp}{mp}\right)^{\frac{1}{2}l}$$

is to 1.[50]

Now that we have finally reached the end of a paragraph, we might break off for a moment to see what he is saying. The term

$$\binom{n}{k}m^{n-k}f^k$$

in the expansion $(m+f)^n$, for $k = fp$ is

$$\frac{n(n-1)\cdots(n-fp+1)}{1 \cdot 2 \cdots (fp)} \cdot m^{n-fp}f^{fp}, \tag{13}$$

while that for $k = fp - l$ (the $(fp-l+1)$-th term) is

---

[50] Montmort, *op. cit.*, pp. 388 - 390. I have taken the liberty of correcting a few typos, displaying some inline formulæ for the sake of readability, and updating some of the mathematical notation, but generally I have tried to preserve the flavour of Bernoulli's comments.

$$\frac{n(n-1)\cdots(n-fp+l+1)}{1\cdot 2\cdots(fp-l)}\cdot m^{n-fp+l}f^{fp-l}. \tag{14}$$

The ratio of these terms, (13) to (14), is thus

$$\frac{(n-fp+l)(n-fp+l-1)\cdots(n-fp+1)}{(fp-l+1)(fp-l+2)\cdots(fp)}\cdot\left(\frac{f}{m}\right)^{l}, \tag{15}$$

the numerator of the fraction on the left consisting of the l terms of the numerator of (13) not occurring in the numerator of (14), and the denominator of that fraction consisting of the l terms of the denominator of (14) not occurring in the denominator of (13). He now replaces $n-fp$ in the numerator by mp to express the ratio as

$$\frac{mp+l}{fp-l+1}\cdot\frac{mp+l-1}{fp-l+2}\cdots\frac{mp+1}{fp}\left(\frac{f}{m}\right)^{l}. \tag{16}$$

Bernoulli now claims that all but the last factor of (16), namely the ratios

$$\frac{mp+1}{fp},\frac{mp+2}{fp-1},\ldots,\frac{mp+l-1}{fp-l+2},\frac{mp+l}{fp-l+1}, \tag{17}$$

are almost in geometric progression. This is not obvious by looking at the terms themselves, but we can examine the claim on our calculators. To this end, I have written a small program that will test this. It assumes one has stored values for m, f, n, l in the variables M, F, N, and L:

```
PROGRAM:GPTEST
:N/(M+F)→P
:seq((MP+X)/(FP−X+1),X,1,L)→L₁
:ln(L₁)→L₂
:ΔList(L₂)→L₃
:ΔList(L₃)→L₄
:Disp max(L₄).
```

The first line of the program stores p in the variable P. The second line creates the list of ratios (17) and stores it in L_1. One can take the ratios of successive terms of L_1 to check that this new sequence is nearly constant and that the terms of L_1 are thus nearly in geometric progression. It is simpler, however, to take logarithms (whence the list L_2) and their differences (L_3). This is the purpose of the next two lines. The function ΔList(is found in the OPS menu accessed via the LIST button and will take a list $\{a_0, a_1, \ldots, a_d\}$ of dimension $d+1$ and produce the d-dimensional list $\{a_1 - a_0, a_2 - a_1, \ldots, a_d - a_{d-1}\}$ of differences of elements of the given list. If L_1 is nearly in geometric progression, L_2 will almost be in arithmetic progression, and L_3 nearly constant. Indeed, for the values at hand, $n = 14000, m = 18, f = 17, l = 50$, one finds the elements of L_3 all to be approximately .000124. For longish lists, checking the constancy requires a lot of scrolling. Thus I added the last two lines: If L_3 is nearly constant, L_4 will be close to 0 and the program displays the maximum element of L_4. The result, .000000002904698, is indeed quite small. Thus, in this case, the logarithms are indeed almost in arithmetic progression, whence the ratios themselves almost form a geometric progression, and their product should approximate the product of the geometric progression of l terms beginning and ending with

$$\frac{mp+1}{fp} \quad \text{and} \quad \frac{mp+l}{fp-l+1},$$

respectively, which is

$$\left(\frac{mp+l}{fp-l+1} \cdot \frac{mp+1}{fp} \right)^{l/2}. \tag{18}$$

Multiplying by the last term of the product (16) yields

$$\left(\frac{mp+l}{fp-l+1} \cdot \frac{mp+1}{fp} \right)^{l/2} \cdot \left(\frac{f}{m} \right)^l = \left(\frac{mp+l}{fp-l+1} \cdot \frac{mp+1}{fp} \cdot \frac{f^2}{m^2} \right)^{l/2}$$

$$= \left(\frac{mp+l}{fp-l+1} \cdot \frac{mp+1}{mp} \cdot \frac{f}{m} \right)^{l/2}$$

$$= \left(\frac{mp+l}{fp-l+1} \cdot \frac{mp+1}{mp} \cdot \frac{fp}{mp} \right)^{l/2},$$

as concluded by Nikolaus Bernoulli in the last line of the paragraph.

Before proceeding any further, you might want to run a quick check by entering

$$\mathsf{prod(L_1)} \tag{19}$$

and

$$\mathsf{((MP+L)/(FP-L+1)*(MP+1)/(FP))^{\wedge}(L/2)}, \tag{20}$$

and comparing the two results.

On comparing the results of these two calculations, you might notice that (19) and (20) are very close, but (20) is slightly larger than (19), i.e., (18) is a little larger than the product of the ratios in (7):

$$\left(\frac{mp+l}{fp-l+1} \cdot \frac{mp+1}{mp} \cdot \frac{fp}{mp} \right)^{\frac{l}{2}} > \frac{mp+l}{fp-l+1} \cdot \frac{mp+l-1}{fp-l+2} \cdots \frac{mp+1}{fp} \left(\frac{f}{m} \right)^l.$$

This inequality is actually more important for what follows. Yet Bernoulli doesn't give it much emphasis when he mentions it without proof in the next paragraph:

If we want to approach more closely to the true value, we can divide the sequence of factors

$$\frac{mp+l}{fp-l+1} \times \frac{mp+l-1}{fp-l+2} \times \frac{mp+l-2}{fp-l+3} \times \&\mathrm{c}.$$

into several parts, & assume that the factors of each part are in ge-ometric progression; but we need not do this, because all the values which we will find by these various assumptions will differ very little, the one from the other; & even though by this first assumption I make this ratio a little larger than it is, this excess will be inconsiderable in comparison with what I will neglect in the sequel.[51]

The approximation and inequality form the first of two innovations Nikolaus Bernoulli made in his uncle's proof. The other, coming up, introduces another geometric progression:

If we now take up the terms which immediately precede the terms numbered $fp + 1$ & $fp - l + 1$; to wit those whose numbers are fp & $fp - l$, the ratio of the former to the latter will be as

$$\frac{mp + l + 1}{fp - l} \times \frac{mp + l}{fp - l + 1} \times \frac{mp + l - 1}{fp - l + 2} \times \cdots \frac{mp + 2}{fp - 1} \times \left(\frac{f}{m}\right)^l$$

to 1, & consequently larger than

$$\frac{mp + l}{fp - l + 1} \times \frac{mp + l - 1}{fp - l + 2} \times \frac{mp + l - 2}{fp - l + 3} \times \cdots \frac{mp + 1}{fp} \times \left(\frac{f}{m}\right)^l$$

or

$$\left(\frac{mp + l}{fp - l + 1} \times \frac{mp + 1}{mp} \times \frac{fp}{mp}\right)^{\frac{1}{2}l}$$

to 1, since each factor of the first series is larger than the corresponding factor of the second. For the same reason the term numbered $fp - 1$ has to the term, whose number is $fp - l - 1$, a larger ratio than the term fp has to the term $fp - l$; & the term $fp - 2$ has to term $fp - l - 2$ a larger ratio than the term $fp - 1$ has to the term $fp - l - 1$, & so forth retreating all the way to the first. Therefore if we divide all the terms preceding the $fp + 1$ term into classes, each of which contains terms equalling l in number, starting with the term whose number is fp; the first term of the first class will have to the first term of the second class a ratio larger than the term $fp + 1$ has to term $fp - l + 1$; & the second term of the first class has to the second term of the second class an even larger ratio; & the third term of the first class to the third term of the second class an even larger ratio, & so forth; thus too all the terms of the first class taken together have to all the terms of the second class a ratio greater than that of the term $fp + 1$ to the term $fp - l + 1$. By the same reasoning the terms of the second class have to the terms of the third class, likewise, all the terms of of the third class to those of the fourth, &c. a larger ratio than the term $fp + 1$ to the term $fp - l + 1$; i.e., than

$$\left(\frac{mp + l}{fp - l + 1} \times \frac{mp + 1}{mp} \times \frac{fp}{mp}\right)^{\frac{1}{2}l}$$

to 1. Thus if we name

---

[51] *Ibid.*, p. 390.

$$\left(\frac{mp+l}{fp-l+1} \times \frac{mp+1}{mp} \times \frac{fp}{mp}\right)^{\frac{1}{2}l} = q,$$

& the sum of the terms of the first class $= s$, the sum of the terms of the second class will be smaller than $\frac{s}{q}$; & the sum of the terms of the third class smaller than $\frac{s}{qq}$, & that of the terms of the fourth class smaller than $\frac{s}{q^3}$, &c. Thus the sum of all the classes, except the first, even should the number of classes be infinite, will be smaller than

$$\frac{s}{q} + \frac{s}{qq} + \frac{s}{q^3} + \frac{s}{q^4},$$

continued to infinity, i.e., smaller than $\frac{s}{q-1}$; hence it follows that the sum of the first class, i.e., of all the terms which are between the term $fp+1$ & the term $fp-l+1$, herein also including the term $fp-l+1$, have to the sum of all preceding [terms] a ratio greater than $q-1$ to 1, or putting in for q its value, than

$$\left(\frac{mp+l}{fp-l+1} \times \frac{mp+1}{mp} \times \frac{fp}{mp}\right)^{\frac{1}{2}l} - 1$$

to 1; accordingly by putting m in for f and f for m; the sum of all the terms which are between the $fp+1$ term & the $fp+l+1$ term, herein including the term $fp+l+1$, has to the sum of all the other terms to the last a larger ratio than

$$\left(\frac{fp+l}{mp-l+1} \times \frac{fp+1}{fp} \times \frac{mp}{fp}\right)^{\frac{1}{2}l} - 1$$

to 1. Thus finally the sum of all the terms from the term $fp-l+1$ to the term $fp+l+1$, inclusive, without counting the term $fp+1$ which is in the middle, has to the sum of all the other terms at least a larger ratio as the smaller between the two quantities

$$\left(\frac{mp+l}{fp-l+1} \times \frac{mp+1}{mp} \times \frac{fp}{mp}\right)^{\frac{1}{2}l} \quad \& \quad \left(\frac{fp+l}{mp-l+1} \times \frac{fp+1}{fp} \times \frac{mp}{fp}\right)^{\frac{1}{2}l}$$

minus 1 to 1 [52]; *which it is necessary to find.*[53]

The second geometric progression,

$$\frac{s}{q}, \frac{s}{q^2}, \frac{s}{q^3}, \dots,$$

is the second important innovation Nikolaus Bernoulli brought to his uncle's proof, as we saw earlier on page 174.

Nikolaus Bernoulli did not apply his estimates to his uncle's example, but to the birth rates:

---

[52] I.e., subtract 1 from each term and then take the respective ratios to 1.
[53] Montmort, *op. cit.*, pp. 390 - 392.

We now apply this to our example, where $n = 14000, mp = 7200, fp = 6800, l = 163$, & we find

$$\frac{1}{2}l \times \left(\log \frac{mp+l}{fp-l+1} + \log \frac{mp+1}{mp} + \log \frac{fp}{mp} \right)$$

$$= \frac{163}{2} \times \left(\log \frac{7363}{6638} + \log \frac{7201}{7200} + \log \frac{6800}{7200} \right)$$

$$= \frac{163}{2} \times (0.0450176 + 0.0000603 - 0.0248236)$$

$$= 1.6507254;$$

the number of which this is the logarithm is $44\frac{74}{100}$. Putting fp in place of mp, & mp in place of fp, we find

$$\frac{1}{2}l \times \left(\log \frac{fp+l}{mp-l+1} + \log \frac{fp+1}{fp} + \log \frac{mp}{fp} \right)$$

$$= \frac{163}{2} \times \left(\log \frac{6963}{7038} + \log \frac{6801}{6800} + \log \frac{7200}{76800} \right)$$

$$= \frac{163}{2} \times (-0.0046529 + 0.0000639 + 0.0248236)$$

$$= 1.6491199;$$

the number of which this is the logarithm is $44\frac{58}{100}$; from which we conclude that the probability that among 14000 children the number of males not being larger than 7363, nor smaller than 7037, is to the probability that the number of males falls outside these limits in a ratio at least as large as $43\frac{58}{100}$ to 1. Thus we can already wager with advantage that in 82 trials the number of males will not fall three times outside these limits.[54] Now in examining the Catalogue of children born during 82 years in London, we find that the number of males has been 11 times greater than 7363; to wit in 1629, 39, 42, 46, 49, 51, 59, 60, 61, 69, 76; we also easily find that we can wager more than 226 to 1 that the number of males will not fall in 82 years 11 times outside these limits. We must remark also that if I had taken an other limit larger than 163, but still smaller than the largest which are found in the Catalogue, I would have found a probability much larger than 43 to 1, that the number of children of each sex falls each year within these limits rather than without. Thus concerning the subject, it is not at all to be astonished that the numbers of children of each sex are not themselves more distant the one from the other, which I have wanted to demonstrate. I remember that my late Uncle has demonstrated a similar thing in his treatise *De Arte conjectandi*, which is at present being printed at Basel, to wit, that if one wants to discover by the frequent repetition

---

[54] If the odds of falling within the limits are 43.58 to 1, the probability of falling outside is $1/44.58$, and the expected number of such occurrences is $1/44.58 \times 82 = 1.834\ldots$, i.e., fewer than 3 such.

of experiments the number of which a certain event can happen or not, one can augment the observations in such a manner that finally the probability that we have discovered the true ratio which exists between the numbers of cases, is greater than any given probability. When this Book appears we will see if in this sort of matter I have found an approximation as accurate as him. I have the honour of being with great esteem,

Sir,

Your very humble & very
obedient Servant
N. Bernoulli

Todhunter informs us that Nikolaus Bernoulli also wrote to Leibniz on the topic and again to 'sGravesande when the latter concluded via a laborious calculation that, if the chances of male and female births were equal, the probability of the number of male births out of 11429 lying between 5745 and 6128 for 82 years in successession would be $(1/4)^{82}$. Therefore, he concluded the chances of male and female births not to be equal. Bernoulli subsequently wrote to 'sGravesande giving "a proof of the famous theorem of James Bernoulli; the proof is substantially the same as that given by Nicolas Bernoulli to Montmort and published by the latter". 'sGravesande replied and Bernoulli answered, offering the following summary:

> Mr. *Arbuthnot* has composed his argument in two matters; 1st. in that which, supposing an equality of births among girls and boys, there is a small probability that the numbers of boys and girls be found within limits very close to equality: 2nd. that there is a small probability that the number of boys surpasses a large number of times in succession the number of girls. It is the first part that I refute, and not the second.

Todhunter adds

> But this does not fairly represent Arbuthnot's argument. Nicolas Bernoulli seems to have imagined, without any adequate reason, that the theorem known by his uncle's name was in some way contradicted by Arbuthnot.[55]

Armed with knowledge of Jakob Bernoulli's Theorem, it is easy to dismiss Arbuthnot out of hand. Recall his words quoted on page 183:

> It is indeed to be confessed that this Equality of Males and Females is not Mathematical but Physical, which alters much the foregoing Calculation; for in this Case the middle Term will not exactly give A's Chances, but the Chances will take in some of the Terms next the middle one, and will lean to one side or the other. But it is very improbable (if mere Chance govern'd) that they would never reach as far as the Extremities.

---

[55] Todhunter, *op. cit.*, pp. 198 - 199. I also recommend Pearson's *History...*, *op. cit.*, pp. 301 - 303, for a fuller and quite entertaining discussion of 'sGravesande's work.

Having already read Jakob Bernoulli's proof that it is highly improbable that the extremities will be reached, the reader may, as I did on first reading, have wondered about this passage. Arbuthnot wrote in English, so I have not mistranslated. Perhaps it was a printer's error, some words jumbled? A moment's hesitation, a shrug of the shoulders, and one reads on, eventually achieving enlightenment or forgetting the line entirely. But the passage is not corrupt: he means "never", and he is correct. The problem is the vagueness of the language: "some of the Terms" and "the Extremities" mean different things in the works of Jakob Bernoulli and John Arbuthnot.

Using Nikolaus Bernoulli's notation, the probability of obtaining exactly k successes in $n = mp + fp$ trials for $mp - l \leq k \leq mp + l$, is

$$\sum_{k=mp-l}^{mp+l} \binom{n}{k} \left(\frac{m}{m+f}\right)^k \left(\frac{f}{m+f}\right)^{n-k}. \tag{21}$$

Arbuthnot (and 'sGravesande) is correct if we fix l and let n get larger and larger, for then (21) will tend to 0. However, if we let l vary by being a constant times n, as Jakob Bernoulli did, the probability (21) will tend to 1.

Later De Moivre, applying the normal distribution, would take l to be a multiple of $\sqrt{n}$, in which case the probability (21) can, depending on the multiplier, approach whatever probability one wishes as n tends to infinity. At least in his letter to Montmort, Nikolaus Bernoulli does not address the issue of the relative sizes of n and l as n tends to infinity, but merely sticks to the values at hand.

As for the relative values of their computations, I think Todhunter was a little too quick in rejecting Nikolaus Bernoulli's refutation of Arbuthnot's first argument. It is indeed numerically correct, but it establishes nothing. And, as to the unrefuted second point, it establishes with high probability that the probability of a male or a female birth is not 1/2, but greater for the male and lesser for the female. Yet, unlike 'sGravesande, he fails to draw the conclusion or to realise that Chance can account for the regularities observed. Instead, he concluded that, given that the probabilities were nearly 1/2, other, divine, forces were at work. God, Newton's Clockmaker who occasionally would adjust the celestial mechanism, would also step in and regulate birth rates. Nikolaus Bernoulli, and modern scientists would leave God out of the picture. His intervention was not required and thus should not be assumed.

De Moivre, the next great probabilist to consider the issue was a most religious man. A Protestant, he left his native France for the shores of England to escape persecution by the Catholics who had imprisoned him in the Priory of St. Martin for 2 1/2 years. He could not help but let his religious beliefs tinge his discussion:

> ...And thus in all Cases it will be found, that *altho' Chance produces Irregularities, still the Odds will be infinitely great, that in process of Time, those Irregularities will bear no proportion to the recurrency of that Order which naturally results from* ORIGINAL DESIGN.

REMARK II.

As, upon the Supposition of a certain determinate Law according to which any Event is to happen, we demonstrate that the Ratio of Happenings will continually approach to that Law, as the Experiments or Observations are multiplied: so, *conversely*, if from numberless Observations we find the Ratio of the Events to converge to a determinate quantity, as to the Ratio of P to Q; then we conclude that this Ratio expresses the determinate Law according to which the Event is to happen.

For let that Law be expressed not by the Ratio P : Q, but by some other, as R : S; then would the Ratio of the Events converge to this last, not to the former: which contradicts our *Hypothesis*. And the like, or greater, Absurdity follows, if we should suppose the Event not to happen according to any Law, but in a manner altogether desultory and uncertain; for then the Events would converge to no fixt Ratio at all.

Again, as it is thus demonstrable that there are, in the constitution of things, certain Laws according to which Events happen, it is no less evident from Observation, that those Laws serve to wise, useful and beneficent purposes; to preserve the stedfast Order of the Universe, to propagate the several Species of Beings, and furnish to the sentient Kind such degrees of happiness as are suited to their State.

But such Laws, as well as the original Design and Purpose of their Establishment, must all be *from without*; the *Inertia* of matter, and the nature of all created Beings, rendering it impossible that any thing should modify its own essence, or give to itself, or to any thing else, an original determination or propensity. And hence, if we blind not ourselves with metaphysical dust, we shall be led, by a short and obvious way, to the acknowledgment of the great MAKER and GOVERNOUR of all; *Himself all-wise, all-powerful* and *good.*

Mr. *Nicolas Bernoulli*, a very learned and good Man, by not connecting the latter part of our reasoning with the first, was led to discard and even to vilify this Argument from *final Causes*, so much insisted on by our best Writers; particularly in the Instance of the nearly equal numbers of *male* and *female* Births, adduced by that excellent Person the late Dr. *Arbuthnot*, in *Phil. Trans.* N°. 328.

Mr. *Bernoulli* collects from Tables of Observations continued for 82 years, that is from *A.D.* 1629 to 1711, that the number of Births in *London* was, at a *medium*, about 14000 yearly: and likewise, that the number of *Males* to that of *Females*, or the facility of their production, is nearly as 18 to 17. But he thinks it the greatest weakness to draw any Argument from this against the Influence of *Chance* in the production of the two Sexes. For, says he,

"Let 14000 Dice, each having 35 faces, 18 white and 17 black, be thrown up, and it is great Odds that the numbers of white and black faces shall come as near, or nearer, to each other, as the numbers of Boys and Girls do in the Tables."

To which the short answer is this: Dr. *Arbuthnot* never said, "that supposing the facility of the production of a Male to that of the production of a female to be already *fixt* to nearly the Ratio of equality, or to that of 18 to 17; he was *amazed* that the Ratio of the numbers of Males and Females born should, for many years, keep within such narrow bounds:" the only Proposition against which Mr. *Bernoulli*'s reasoning has any force.

But he might have said, and we do still insist, that "as, from the Observations, we can with Mr. *Bernoulli*, infer the facilities of production of the two Sexes to be nearly in a Ratio of equality; so from this Ratio once discovered, and *manifestly serving to a wise purpose*, we conclude the Ratio itself, or if you will the *Form of the Die*, to be an Effect of *Intelligence* and *Design*." As if we were shewn a number of Dice, each with 18 white and 17 black faces, which is Mr. *Bernoulli*'s supposition, we should not doubt but that those Dice had been made by some Artist; and that their form was not owing to *Chance*, but was adapted to the particular purpose he had in View.

Thus much was necessary to take off any impression that the authority of so great a name might make to the prejudice of our argument. Which, after all, being level to the lowest understanding, and falling in with the common sense of mankind, needed no formal Demonstration, but for the scholastic subtleties with which it may be perplexed; and for the abuse of certain words and phrases; which sometimes are imagined to have a meaning merely because they are often uttered.

Chance, as we understand it, supposes the *Existence* of things, and their general known *Properties:* that a number of Dice, for instance, being thrown, each of them shall settle upon one or other of its Bases. After which, the *Probability* of an assigned Chance, that is of some particular disposition of the Dice, becomes as proper a subject of Investigation as any other quantity or Ratio can be.

But *Chance*, in atheistical writings or discourse, is a sound utterly insignificant: It imports no determination to any *mode of Existence*; nor indeed to *Existence* itself, more than to *non-existence*; it can neither be defined nor understood: nor can any Proposition concerning it be either affirmed or denied, excepting this one, "That it is a mere word." [56]

The like may be said of some other words in frequent use; as *fate, necessity, nature,* a *course of nature* in contradistinction to the *Divine energy:* all which, as used on certain occasions, are mere sounds: and yet, by artful management, they serve to found specious conclusions: which however, as soon as the latent fallacy of the *Term* is detected,

---

[56] The question arises: Would de Moivre have rejected Pascal's Wager discussed in Chapter 2, section 3, above? If chance supposes the existence of things, God's existence will not be subject to chance and probabilistic reasoning. On the other hand, his strongly held religious beliefs, which blinded him to the point Nikolaus Bernoulli was making, could easily have led him to regard Pascal's argument none too critically.

appear to be no less absurd in themselves, than they commonly are hurtful to society.

I shall only add, That this method of Reasoning may be usefully applied in some other very interesting Enquiries; if not to force the Assent of others by a strict Demonstration, at least to the Satisfaction of the Enquirer himself: and shall conclude this Remark with a passage from the *Ars Conjectandi* of Mr. *James Bernoulli*, *Part* IV. *Cap.* 4. where that acute and judicious Writer thus introduceth his Solution of the Problem for *Assigning the Limits within which, by the repetition of Experiments, the Probability of an Event may approach indefinitely to a Probability given*, "Hoc igitur est illud Problema &c." *This, says he, is the Problem which I am now to impart to the Publick, after having kept it by me for twenty years: new it is, and difficult; but of such excellent use, that it gives a high value and dignity to every other Branch of this Doctrine.* Yet there are Writers, of a Class indeed very different from that of *James Bernoulli*, who insinuate as if the *Doctrine of Probabilities* could have no place in any serious Enquiry; and that Studies of this kind, trivial and easy as they be, rather disqualify a man for reasoning on every other subject. Let the Reader chuse.[57]

6 Discussion of de Moivre's Remarks

There is so much in these remarks to discuss I scarcely know where to begin. The three main things I wish to discuss are hypothesis testing, intelligence and design, and the low repute of probability.

Type I and Type II Errors

The modern description of the given calculation is stated in terms of *hypothesis testing*. One formulates an hypothesis, in Arbuthnot's and 'sGravesandes's case that the probability of a child being born male is $1/2$, in Nikolaus Bernoulli's case that this probability is $18/35$. Then one performs some number of experiments. If the result is outside the pale of probability as erected by the hypothesis, one rejects the hypothesis; otherwise one fails to reject it. The modern statistician never accepts an hypothesis: he only rejects or fails to reject.[58] Thus, Arbuthnot and 'sGravesande rejected the probability $1/2$ hypothesis, and Nikolaus Bernoulli failed to reject the $18/35$ hypothesis. The modern statistician would not accept this hypothesis because there may be other non-rejectable ones: $18/35 \approx .5142857183$; accepting it implies a rejection of $2057/4000 = .51425$ (or even $13/25 = .52$), which might also fail to be rejected on its own. Yet we cannot simultaneously accept both.

---

[57] De Moivre, *op. cit.*, pp. 251 - 254.

[58] If one does not like the passive phrase "fail to reject", and prefers the more active verb "accept", I suppose it might be permissible, provided the adverb "tentatively" is attached to the preferred active verb.

7 Exercise. Consider Exercise 13 of Chapter 4 concerning Buffon's coin tossing experiment. We have two hypotheses: he had the child toss the coin 2048 times and he had the child toss the coin 2084 times. He came up with heads on the first toss 1061 times. Should either hypothesis be rejected?

Modern statisticians have resigned themselves to the fact that errors will be made. The moral certainty aimed at by Jakob Bernoulli is recognised as no more than a myth. Instead of eliminating error, the goal is to minimise it and the first step toward this is the classification of errors. The second step is the intelligent design of tests that will minimise these errors. The work on this was largely due to Egon Pearson and the Polish-Russian-English (and later: -American) Jerzy Neyman, who developed much of the theory in a series of papers in the decade 1928 - 1938, beginning with a classification of two basic types of errors and culminating in Neyman's introduction of *confidence intervals*, a staple of service courses in statistics that we will not attempt to discuss here. Indeed, I shall limit our discussion to the Pearson-Neyman classification of errors, which recognises two types of errors: If one rejects a true hypothesis, one has committed a *type I error*; while if one fails to reject a false hypothesis, one has committed a *type II error*. Arbuthnot correctly rejected the probability 1/2 hypothesis, but did so not because he concluded the probability was larger, but because he concluded the issue was not a matter of probability. He made the right choice for the wrong reason, something sometimes called a *type III error*. And he was wrong about chance not being involved, thus rejecting a true hypothesis that it was a matter of chance, and thereby committing a type I error.[59]

Arbuthnot's birth rates afford us a simple example with which to illustrate the procedure involved in testing an hypothesis. Here, we have the basic hypothesis

$$H_0: \quad p = \tfrac{1}{2},$$

where p denotes the probability that a child will be born a boy. H_0 is called the *null hypothesis*[60]. Against this, we have the hypothesis,

$$H_1: \quad p \neq \tfrac{1}{2}.$$

H_1 is called the *alternative hypothesis*. We test H_0 by assuming it and running an experiment, in this case, checking one or more entries in the register of births in London from 1629 until 1710. This table is reproduced in Appendix D, below, from Arbuthnot's paper. Our test will consist of choosing a single year from the table and determining if the result obtained is in accordance with the null hypothesis.

To be absolutely fair, we should choose the year at random by entering

---

[59] Or, did he fail to reject the false hypothesis that chance was not involved, whence the error was type II? In actual practice, there is a ritual to follow that removes ambiguity.

[60] The term "null hypothesis" originated with Ronald Aylmer Fisher (1890 - 1962) in 1935. *Cf.* H.A. David and A.W.F. Edwards, *Annotated Readings in the History of Statistics*, Springer-Verlag New York, Inc., New York, 2001, p. 222.

1629+randInt(0,81).

The function randInt(m,n) randomly generates an integer between m and n, inclusive, for integer inputs m,n. Adding 1629 to randInt(0,81) will thus randomly choose a year from the list 1629, 1630, ..., 1710. My calculator chose the year 1651, in which 3231 males and 2840 females were born—6071 births in all. Assuming $p = \frac{1}{2}$, the expected number of males in 6071 births would be 3035.5 and the probability that the number of males born lies between $3230 = 3035 + 195$ and $2840 = 3035 - 195$ is

binomcdf(6071,1/2,3230)−binomcdf(6071,1/2,2839),

which the calculator tells us is .9999994818, whence the probability of lying outside these limits is roughly .0000005. The result of our experiment is so highly improbable under the stated hypothesis H_0 that we must reject it.

Let us now see what the same experiment would have advised us to do had we tested Bernoulli's null hypothesis,

$$H_0: \quad p = \tfrac{18}{35},$$

against its alternative

$$H_1: \quad p \neq \tfrac{18}{35}.$$

H_0 yields $6071 \cdot 18/35 \approx 3122$ as the expected number of male births. $3231 - 3122 = 109$ and we check the probability that the number of male births lies between $3122 - 108 = 3014$ and $3122 + 108 = 3230$:

binomcdf(6071,18/35,3230)−binomcdf(6071,18/35,3014),

which is .9944543583. Thus the probability of differing by as much as 109 from the expected value is $1 - .9944543513 = .0055\ldots$ This, too, is quite small and we would probably reject the null hypothesis on the basis of this experiment. If we accept, or carry out, Bernoulli's calculations for the years 1629 - 1710 yielding an average proportion of 18 to 17 of the numbers of male and female births, this means we would be making a type I error.

Had we chosen 1653 instead of 1651, we would have found 3196 males and 2959 females, 6155 births in all, with an expected 3165 males born according to Bernoulli's hypothesis. The improbability of the difference from the expected value being as large as $3196 - 3165 = 31$ is

binomcdf(6155,18/35,3195)−binomcdf(6155,18/35,3135),

which calculates to .5557810996, and the probability of being so far from the expected value is .4442189004. We should expect such a difference more than 30 times in 82 years if Bernoulli's hypothesis is true and the experiment gives us no cause to reject the null hypothesis.

In all three experiments, the probabilities of the experimental results, assuming the null hypotheses in question, were clearly so small as to mandate rejection or so large as to rule out rejection. None offered borderline cases, requiring additional deliberation. In practice, one avoids this by specifying in advance a cut-off probability—usually .01 or .05—below which one rejects the null hypothesis and above which one fails to reject it.

8 Project. Today's calculators, with their list-handling capabilities and programmability, make easy some extended computations we would have found infeasible, or, at least, unreasonably burdensome, back when I was a student. Write a program HYPTESTS which will perform all 82 experiments as follows. Prior to running it, copy the data from Arbuthnot's table into two lists ∟MALE and ∟FMALE, store 18/35 in P, and store .01 or .05 in the variable A. HYPTESTS will produce several lists:

> ∟TOTAL: ∟MALE+∟FMALE
> ∟EXPVL: int(P∗∟TOTAL)
> ∟DIFF: abs(∟MALE−∟EXPVL)
> ∟LOWB: ∟EXPVL−∟DIFF+1
> ∟UPBD: ∟EXPVL+∟DIFF−1
> ∟PROBS: binomcdf(∟TOTAL,P,∟UPBD)−binomcdf(∟TOTAL,P,∟LOWBD))
> ∟IMPRB: 1−∟PROBS, and
> ∟RSULT: ∟IMPRB<A .

∟RSULT stores the results (1 for rejection, 0 for failure of rejection) of all 82 experiments. Run HYPTESTS assuming the true value is 18/35. How many type I errors can be found? How many type II errors. Rerun HYPTESTS after storing 1/2 in P. Ditto with $(1/2 + 18/35)/2 = 71/140$.

9 Exercise. When Laplace later analysed birth data for Paris, he obtained a probability slightly lower than 18/35, but still greater than 1/2. Go to the reference section of your library and test Bernoulli's null hypothesis against more recent data. You may either pick a single year at random or run HYPTESTS for a series of years if you have several years of data. [WARNING: binomcdf(can take a list as its third argument, but not as its first. Thus, the expression given for ∟PROBS will result in an error message if you try to use it directly. You must either normalise the lists ∟MALE and ∟FMALE, *á la* Bernoulli, to obtain a constant sum T and replace ∟TOTAL by T, or construct ∟PROB with I-th entry of

> binomcdf(∟TOTAL(I),P,∟UPBD(I))−binomcdf(∟TOTAL(I),P,∟LOWBD(I)) ,

by using the function seq(.]

The distinction between type I and type II errors may seem rather obvious and, thus, pointless. But it does offer a useful framework in discussions of decision making and, as such, has spread well beyond the confines of statistics. The economist Walter E. Williams has several times used the distinction in criticising the activities of the U.S. Food and Drug Administration:

> Another example of type I and type II errors hits closer to home. Food and Drug Administration (FDA) officials, in their drug approval process, can essentially make two errors. They can approve a drug that has unanticipated dangerous side effects (type II). Or, they can disapprove, or hold up approval of, a drug that's perfectly safe and effective (type

I). In other words, they can err on the side of under-caution or on the side of over-caution.[61]

Williams is not alone in addressing this issue and one can find discussion of these types of errors involved in drug approval online or in the literature. The most famous type II error in this respect was the European use in the 1960's of thalidomide on pregnant women who, to use the archaic terminology, gave birth to monsters—children with birth defects, in this case stunted limbs. Vaccines produced from attenuated, rather than dead, viruses that caused rather than prevented the disease in question afford another example. There are several famous type I errors made by the FDA. Williams has more than once cited beta-blockers in this respect:

Which error do FDA officials have the greater incentive to make?

If a FDA official errs by approving a drug that has unanticipated, dangerous side effects, he risks congressional hearings, disgrace and termination.[62] Erring on the side of under-caution produces visible, sick victims who are represented by counsel and whose plight is hyped by the media.

Erring on the side of over-caution is another matter. A classic example was beta-blockers, which an American Heart Association study said will "lengthen the lives of people at risk of sudden death due to irregular heartbeats." The beta-blockers in question were available in Europe in 1967, yet the FDA didn't approve them for use in the U.S. until 1976. In 1979, Dr. William Wardell, a professor of pharmacology, toxicology and medicine at the University of Rochester, estimated that a single beta-blocker, alprenolol, which had already been sold for three years in Europe, but not approved for use in the U.S., could have saved more than 10,000 lives a year. The type I error, erring on the side of over-caution, has little or no cost to FDA officials. Grieving survivors of those 10,000 people who unnecessarily died each year don't know why their loved one died, and surely they don't connect the death to FDA over-caution. For FDA officials, these are the best kind of victims—invisible ones. When an FDA official holds a press conference to announce its approval of a new life-saving drug, I'd like to see just one reporter ask: How many lives would have been saved had the FDA not delayed the drug's approval?

The bottom line is, we humans are not perfect. We will make errors.

---

[61] Walter E. Williams, "Making intelligent errors". I found this online, listed with a release date of 10 August 2005. I suspect this means the article was syndicated in a number of newspapers.

[62] In a note, "FDA: friend or foe?", found at HumanEvents[TM].com, Williams quotes Alexander Schmidt, a former FDA Commissioner, "In all our FDA history, we are unable to find a single instance where a Congressional committee investigated the failure of FDA to approve a new drug. But the times when hearings have been held to criticize our approval of a new drug have been so frequent that we have not been able to count them. The message to FDA staff could not be clearer".

Rationality requires that we recognize and weigh the cost of one error against the other.[63]

Williams's remarks are a delight to read. They are forceful, to the point, backed up by numbers—but, the FDA, however much its delay may be a matter of self-protection, was following a time-honoured tradition of involving what is called the *precautionary principle*, best stated in the original words of Hippocrates: "First, do no harm". While it is true that people will die if a drug, like alprenolol, is not approved, the precautionary principle dictates that it not be approved until one has determined as much as possible if there are dangers.

There having been no Food and Drug Administration during the early years of statistics, the comparable issue under discussion—*sans* the type I / type II terminology—was inoculation. The necessity of profiting from the sale of annuities meant that tables of mortality would be drawn up early on, John Graunt's (1620 - 1674) famous *Natural and Political Observations, made upon the Bills of Mortality* first appearing in 1661 and the first real-life table of Edmond Halley (*c.* 1656 - 1743) based on work of Caspar Neumann (1648 - 1713) appearing in 1694 under the title "An estimate of the Degrees of the Mortality of Mankind, drawn from curious Tables of Births and Funerals at the City of Breslau, with an attempt to ascertain the price of Annuities upon Lives" [64]. Jakob Bernoulli's *Ars conjectandi* was perhaps too early, too original, too theoretical, or too unfinished to do so, but for the next couple of centuries it was not uncommon for treatises on probability to include tables of mortality. In the second edition of his *A Treatise of Annuities on Lives*, which was tacked onto the end of the third edition (1756) of *Doctrine of Chances*, de Moivre included four different such tables, including Halley's, along with a discussion of their relative merits. He also referred briefly to the merits of a newer table published by the Comte de Buffon in the second volume of his *Histoire naturelle*. Thus, such tables were reasonably widespread when Daniel Bernoulli, Jean le Rond d'Alembert, and Denis Diderot took up the problem of inoculation around 1760.

A precursor to modern vaccination, inoculation against smallpox as practised in the 18th century consisted in drawing some fluid from the pustules of an infected person and rubbing it into a cut or scratch on the arm of the person being inoculated. The procedure could prove fatal, but the dangers were not immediately realised and the practice was only first banned in England in 1840. The pros and cons were debated long and hard before Parliament made its decision, the most important elements of the debate offered up by Daniel Bernoulli and d'Alembert on the continent where the practice had spread to.

---

[63] Williams, "Making intelligent errors".

[64] *Philosophical Transactions of the Royal Society* 17 (1693; appeared 1694), pp. 596 - 610. Actually, bills of mortality were drawn up as a sort of early-warning system for epidemics. Graunt's purpose in analysing them is unclear, but he was the first to note statistical regularities. Halley's tables had clear actuarial implications. *Cf.* Christopher Lewin and Margaret de Valois, "History of actuarial tables", in: Martin Campbell-Kelly, Mary Croarken, Raymond Flood, and Eleanor Robson, eds., *The History of Mathematical Tables: From Sumer to Spreadsheets*, Oxford University Press, Oxford, 2003.

Those who compiled and analysed mortality tables are represented here by Jan de Witt (1625 - 1672) and Edmond Halley. De Witt, a mathematician in his own right, is mathematically remembered for his book on algebra, the *Elementa curvarum linearum* in which the analytic geometric treatment of the conics is worked out for the first time. Of central concern here is his *Waerdye van Lyf-renten naer Proportie van Los-renten* [*Value of life annuities compared with redemption bonds*] (1671), an application of the notion of mathematical expectation to annuities, and his subsequent correspondence with Jan Hudde on survivorship annuities. De Witt was first and foremost a statesman, and his title *Raadpensionaris* ["grand pensionary", a position roughly equivalent to the modern prime minister] is explicitly referred to at the bottom of the stamp issued in 1947 bearing his portrait.

Edmond Halley's name is a household word on account of the comet the orbit of which he calculated and return of which he predicted. With another return in the mid-1980's, the postal authorities of the world had a field day issuing hundreds of stamps in 1985 and 1986 celebrating Halley's Comet in particular and everything astronomical in general— Halley's comet, other comets, famous astronomers, even Halley and incidents from his life. Among the instances of his life so celebrated was his trip to the Southern Atlantic to chart the southern skies. In 1977, on the occasion of the tercentenary of this voyage and his visit to St. Helena, this island nation issued a set of three stamps commemorating the visit. These remain the most appropriate philatelic recognition of the man. I have, however, chosen for reproduction here a more attractive portrait from that nation's later 1986 Halley's Comet issue of 4 stamps.

In mid-1760, Daniel Bernoulli's memoir, *Essai d'une nouvelle analyse de la mortalité causée par la petite Vérole, et des avantages de l'Inoculation pour la prévenir* [*Attempt at a new analysis of the mortality caused by smallpox, and of the advantages of inoculation for its prevention*] was read before the French Academy.

400. Daniel Bernoulli's main object is to determine the mortality caused by the small-pox at various stages of age. This of course could have been determined if a long series of observations had been made; but at that time such observations had not been made. Tables of mortality had been formed, but they gave the total number of deaths at various ages

without distinguishing the causes of death. Thus it required a calculation to determine the result which Daniel Bernoulli was seeking.

401. Daniel Bernoulli made two assumptions: that in a year on an average 1 person out of 8 of all those who had not previously taken the disease, would be attacked by small-pox, and that 1 out of every 8 attacked would die. These assumptions he supported by appeal to observation; but they might not be universally admitted. Since the introduction of vaccination, the memoir of Bernoulli will have no practical value; but the mathematical theory which he based on his hypothesis is of sufficient interest to be reproduced here.[65]

Perhaps before discussing Bernoulli, I should pause to explain some terminology. The French word *vérole* of the title of Bernoulli's paper means "pox" and derives from the Latin *variola*, referring to smallpox, and from which the word "variolation" meaning inoculation arose. The modern *vaccination* is "a term that Louis Pasteur later popularized as a tribute to Jenner after adapting it from Jenner's term *vaccinae*, 'of the cow.'"[66] Jenner is, of course, Edward Jenner who, following a remark of a milkmaid that, having had the much milder and generally non-lethal cowpox, she would never come down with smallpox, experimented with the inoculation of cowpox in his own and various other children. This was done in the 1790's, to such effect that when Todhunter's *History* appeared in 1865, he could say of the debate over the merits of inoculation with smallpox, "the subject is happily no longer of the practical importance it was a century ago".[67]

Now, discussing Daniel Bernoulli's mathematics would be a digression from our discussion of hypothesis testing, which, as part of our discussion of de Moivre's comments, is a digression... The mathematics is interesting and it affords us practice on the calculator and an example of mathematical modelling that I would like to present here but which may be slightly out of place. I thus postpone the details to Appendix B, below, and discuss it here in only the most general terms.

The obvious argument is that one is more likely to die of smallpox if one is not inoculated. He assumed probabilities 1/8 of getting smallpox in a year given that one has not previously had the illness and 1/8 of dying of smallpox given that one has contracted the disease, thus 1/64 of dying of smallpox in a given year for one who has not been inoculated. For an inoculated person, he assumed a probability of 1/200 of death due to the illness. The probability of dying of smallpox within a year is thus

1/64 if one is not inoculated
1/200 if one is inoculated.

---

[65] Todhunter, *op. cit.*, pp. 224 - 225. Bernoulli's methods have additionally proven to be of practical value.

[66] Peter Radetsky, *The Invisible Invaders; The Story of the Emerging Age of Viruses*, Little, Brown and Company, Boston, 1991, p. 31.

[67] Todhunter, *op. cit.*, p. 265.

Pride of place for stamps of Edward Jenner go to his native Ireland and Great Britain, whence I picture the commemoratives of Jenner of those nations here. The stamp on the left is Great Britain's image, issued in one of the several small sets issued to celebrate the millennium by that country. While I ordinarily disapprove of the British habit of offering only abstract representations of scientists and their achievements, preferring straight portraits or cameos with some intelligible representational object, I must confess this stamp to have a certain charm. It features the milkmaid from whom Jenner learned that cowpox produces immunity to smallpox, a cow, and even Jenner's experimentation on small children. It would make a great illustration for a childrens' book.

The inscription running up the right side of the stamp reads

> Millenium 1999/40
> Jenner's Vaccination/P Brookes

The "40" indicates that it was the 40th stamp issued in 1999 and P. Brookes is the artist who designed the stamp.

The Irish stamp is more staid, issued in 1978 to mark the worldwide eradication of smallpox, as was formally declared the following year by the World Health Organisation. A reproduction of a painting by E.J.C. Hamman, it depicts Jenner inoculating an infant with cowpox. Note the cow peeking in at the doorway.

In 2010 Great Britain issued an attractive set of 10 stamps celebrating the 350th anniversary of the founding of the Royal Society. The lower portion of each stamp broke with English tradition by presenting a recognisable portrait (albeit cropped to reveal no chin or forehead) of an individual scientist, while the upper half remained true to tradition by offering a symbolic representation of the work of the featured scientist. In Jenner's case, the symbolism somewhat anachronistically anticipated viruses as the cause of smallpox.

And, ignoring other causes of death, the probability of dying of smallpox some-
time during one's life is

$$\frac{1}{64} + \frac{7}{8} \cdot \frac{1}{64} + \left(\frac{7}{8}\right)^2 \frac{1}{64} + \ldots = \frac{1}{64}\left(1 + \frac{7}{8} + \left(\frac{7}{8}\right)^2 + \ldots\right)$$

$$= \frac{1}{64} \cdot \frac{1}{1 - \frac{7}{8}} = \frac{8}{64} = \frac{1}{8},$$

if one never gets inoculated, and $1/200$ if one gets inoculated. Inoculation seems
a good idea.

However, if one is inoculated and comes down with smallpox, death, should
it come, will be within a month. There is a good chance of living a longer, fuller
life if one does not undergo inoculation. Daniel Bernoulli thus went further and
calculated the life expectancies of individuals who were and were not inoculated.

Bernoulli's treatment breaks into three parts. First, using Halley's table
giving the number ξ of survivors to a given age x out of a starting population
of 1300 births, he uses the probabilities $1/8$ of a person who has never had
smallpox being infected by it in a given year and $1/8$ of an infected person
dying of the disease to express the number s of people aged x who have never
had smallpox as a function of ξ and x,

$$s(x) = \frac{\xi(x)}{(1 - \frac{1}{8})e^{x/8} + \frac{1}{8}} = \frac{8\xi(x)}{7e^{x/8} + 1}.$$

With this he draws up a table of values of s.

Second, Bernoulli repeats his analysis on assumption that smallpox has
been eradicated to calculate a replacement table for Halley's, tabulating new,
larger values ξ^* for ξ.

Finally, assuming that only 1 in 200 of those inoculated with smallpox dies of
the disease, he calculates the average life expectancy of an inoculated person to
be three years longer than that of one who does not get inoculated. Todhunter,
who gives some of the details of the first two calculations[68], is silent on this
third calculation. I indicate how this could have gone in Appendix B. For here,
let it suffice to say that numerically the issue is cut and dried: inoculation leads
to a longer life and hence is the recommended course of action.

Karl Pearson devotes several densely typed pages to d'Alembert's criticism
of Bernoulli in *Réflexions sur l'Inoculation*, declaring

> I have found the work of extraordinary interest, not so much for the
> statistical information that it conveys, as for the manner in which every
> feature of the later controversies over vaccination seems to find its paral-
> lel, if not its parody in the discussions that took place over Inoculation.
> Without knowing something about this subject and the intense feeling
> it excited, we can hardly appreciate what a part it played in many of
> the writings of the great mathematicians who dealt with probability.[69]

[68] *Ibid.*, pp. 224 - 228.
[69] Pearson, *History...*, *op. cit.*, p. 543.

On this debate in general, Pearson quotes d'Alembert:

> The question of inoculation is more discussed than ever in France; it has become an affair of party and the object of a dispute almost as violent as have been those of Jansenism and Italian opera bouffe. It is true that it is an admission we ought to make this time to the honour of the French nation that the new topic about which it is impassioned today is somewhat more important than many others which have often agitated it; but the brochures, the personalities, the accusations of bad faith are just as frequent on both sides. The adversaries of inoculation call its partisans murderers, the latter dub their antagonists bad citizens. Little has been lacking, I am assured, for the quarrel to lead to bloody results, even among grave doctors, which would have obliged medicine to call surgery to its aid.[70]

D'Alembert begins with the mathematical argument. He more-or-less accepts Bernoulli's estimates of the probabilities involved, but not the conclusions. As there is no hope of improving on Pearson's discussion, I simply quote him:

> First inoculation does not take place before the child is four years old; from this age to the ordinary term of life smallpox kills $1/7$ to $1/8$ of the population. Inoculation destroys at most one in 300. D'Alembert accepts these results, which indicate that the risk of dying from smallpox is about 40 times as great[71] as dying from inoculation, and then the supporters use this 40 as an argument in favour of inoculation. D'Alembert says—and I think correctly—that this is absurd. The risk of dying from inoculation occurs in five days to a month after the operation, while the risk of dying from smallpox is spread throughout the whole length of life. According to the anti-inoculists we ought to compare the risk of dying of inoculation and of actual smallpox for the same month and a month only. According to Bernoulli's figures, the risk of dying in a month of smallpox is about $1/768$ [72] which is less than the $1/300$ of dying of inoculation. But this is not correct either. For if the inoculated does not die within the month he has nothing further to fear, but the uninoculated may still acquire smallpox and although his risk diminishes with age, it does not cease absolutely till he dies of something else. The problem is accordingly not so easy as either inoculists or anti-inoculists had made it out to be. Again, in each month of life a similar argument will apply, therefore in each month it is wiser not to be inoculated! Secondly, according to D'Alembert, to appreciate the actual risk summed through life from smallpox is impossible because sufficient observations do not exist to measure the risk at each

---

[70] *Ibid.*, p. 544.
[71] Note that $200/8 < 40 < 300/8$.
[72] I.e., $(1/8) \cdot (1/8) \cdot (1/12)$.

age. Further, he remarks that whatever the risk in the distance may be, it is not so influential with a man as a less risk at a near date.[73]

This last argument, about the difference in perceptions of immediate and future risks ought to remind us of Bernoulli's own notions of *moral expectation* and *utility* that he had brought to bear on the Petersburg Problem and other assessments of risk back in 1738. The extra three years of life to be expected from inoculation are less of a gain to the twenty-year-old who runs the risk of dying with probability 1/200 or 1/300 within 5 days to a month of inoculation than they would be to an eighty-year-old man.

D'Alembert also criticises the argument mathematically, something perhaps best discussed later, in Appendix B.

And, something that will not be discussed in the present book, d'Alembert presents his own mathematical treatment, concluding with Bernoulli, that inoculation is the wisest course of action. He distinguishes, however, two perspectives—advantage to the individual and advantage to the state:

> D'Alembert here notes that we may easily confuse two matters, the value of inoculation to the state, and the risk to the individual citizen. The state might be willing to sacrifice one citizen in five, that the remaining four might live healthily to 100, but could the individual citizen be expected to take this view, and if not can it really be an advantage to the state, if no individual citizen is likely to submit to it for its advantage.[74]

Today, of course, there is a third perspective to consider—the advantage to the oversight board, e.g., the FDA. What are best for the individual, the state, and the civil service administrators are not necessarily the same.

D'Alembert's cautious endorsement of inoculation displeased his colleague and friend Diderot. Diderot criticised d'Alembert harshly, accusing the latter of having written his paper "to the increase of his own reputation" rather than to be "attentive to the public good", a judgment echoed by Emmanuel Étienne Duvillard who accused d'Alembert of attacking inoculation out of jealousy of Daniel Bernoulli.[75] Again I quote Pearson:

> I can find in D'Alembert's paper nothing but a temperate account of the arguments on both sides of the problem, and a demand for unprej-udiced inquiry. Diderot had rushed to the conclusion that inoculation was a great scientific advance and therefore any one who questioned its efficiency was 'an enemy of the people', an opponent of public good, and seeking only to increase his own reputation. Had Diderot studied the history of medicine he would have known that in all the centuries there have existed fashionable remedies and treatments, and that those who asked for evidence of their efficiency were accused of opposing the

---

[73] Pearson, *History...*, *op.cit.*, p. 545.

[74] *Ibid.*, p. 548.

[75] *Ibid.*, pp. 552 - 553.

public good, although in twenty years' time these remedies and treatments were scorned by members of the very profession, whose more ancient members had proclaimed them the sole salvation of the public. D'Alembert's paper was a scientific inquiry; he demanded statistical facts before coming to a conclusion; he saw only rabid opponents and scarcely less ill-informed supporters of inoculation. Diderot's criticism is a bit of rampant journalism, written by a man who has made up his mind without any real study of the data, that inoculation is a great discovery and for the benefit of the people.[76]

Diderot *et alia* did not have our current knowledge of germs and viruses. Pearson, discussing d'Alembert, says

He now goes on to meet detailed criticisms of inoculation, but he nowhere refers to one of the main factors which led to its final condemnation. The risk of dying of inoculation may have been very small, but the inoculated if he had as a rule a mild form of smallpox, had the actual disease, and was capable of communicating it to others in malignant form. These others might suffer and did suffer occasionally from the more severe types, and the risk of dying of the inoculated was not the sole risk the community incurred; there was a serious risk of the actual spread of the disease; especially when we hear of scrapings from smallpox pustules being obtained by utterly untrained old women, carried about in envelopes and rubbed on children's arms! This is one objection D'Alembert does not meet and seems to have been unconscious of.[77]

The whole issue of vaccination is fascinating, but too complicated for final discussion here. There are too many variables for the simple sort of statistical test we started our discussion with. And, in real life, there are ethical, economic, and political considerations informing any decision.

The basic ethical questions are

should the individual be allowed vaccination?

and

should vaccination be compulsory?

The first question is primarily one of safety of the vaccine for the vaccinated individual *and* those he may come into contact with. There are two kinds of vaccines. One, like the Salk vaccine for polio, uses a dead virus and is perfectly safe and will not spread the infection. The other, like the Sabin vaccine for polio, uses a still living, but weakened virus, called an *attenuated* virus, and is safe for most people, but can kill those with weakened immune systems. Attenuated viruses can and do spread from individual to individual, and occasionally reacquire their original virulence. In the United States, most if not all cases of

---

[76] *Ibid.*, p. 553.
[77] *Ibid.*, p. 549.

polio since the introduction of polio vaccines have been caused by the Sabin vaccine.[78] Dead viruses, however, are not quite as effective as attenuated viruses in vaccines and re-vaccinations, formerly called "booster shots", are needed.

Polio presents a special problem, not whether or not to vaccinate, but which vaccine to use. Here, economic considerations come into play. Sabin's vaccine was simpler to produce and, being a pill, easier to deliver. It did not require boosters and, in underdeveloped countries, the ability of the immunity to spread like the infection meant one did not have to spend as much reaching everybody.

And, as I said, there are political considerations. In recent years two great, but unfounded, scares convinced much of the public that some of our vaccines were unsafe. First, a physician in England postulated a connexion between the combined effects of the measles, mumps and rubella vaccines, usually delivered in one dose as the MMR vaccine, and autism. If, as he affirmed, the MMR vaccine were a causal agent of autism, one would expect a greater incidence of autism among those who had been vaccinated than among those who hadn't. A statistical analysis revealed no such increase. Basically, the data collected for unvaccinated individuals would be used to determine the basic probability p of becoming autistic. Then one would form the null hypothesis that the probability is the same among those receiving the MMR vaccine. Finally, one would collect the data and check if the difference between the actual number of cases of autism following MMR vaccination and the value predicted using probability p is large beyond expectation. In this case it wasn't, and one failed to reject the null hypothesis. This did not completely end the MMR/autism controversy, the final blow being delivered by the revelation that the physician in question had had his research funded by lawyers involved in suing drug companies and had withheld this information from the editors of the medical journal in which he published his purported findings.

MMR was, however, quickly replaced as the danger *du jour* by *thimerosal*, a mercury-based preservative[79] used in many vaccines. Again the alarm was raised and, despite the lack of any other than anecdotal evidence, the claim that thimerosal caused autism was being touted by self-promoting politicians as well as well-meaning but scientifically uneducated celebrities and others. Although a number of clinical studies have failed to show any statistical connexion between thimerosal and autism, many vaccine manufacturers have removed thimerosal from their vaccines and the number of reported cases of autism has not thereby

---

[78] The successful eradication of the disease is a problem as uninformed parents assume that their children no longer need to be vaccinated against it. Only the effect, however, and not the cause has been eradicated. The virus is still around.

[79] No one who has seen William Eugene Smith's photographs of the victims of the mercury poisoning at Minamata in Japan can help but be nervous at any mention of mercury. Thimerosal is a compound containing ethylmercury which has never been proven harmful even in doses 10000 times that found in vaccines. It is the similar sounding methylmercury that poses a danger. *Cf.* Paul A. Offit, *Autism's False Prophets: Bad Science, Risky Medicine, and the Search for a Cure*, Columbia University press, New York, 2008, pp. 63 - 64.

been reduced, but has in fact increased.[80] As I write, the last word in the United States has been issued, not by any scientific body like the National Science Foundation or the Center for Disease Control nor by an oversight agency such as the FDA, but by a federal court. The court's decision, fortunately, was based on scientific statistical evidence and not emotion as it would have been had the issue been settled by the Senate.

Intelligent Design

The question of God's design figures strongly in the history of probability. Karl Pearson devotes an entire chapter to the subject in his massive history of statistics, and a short chapter on "Design" can be found in Ian Hacking's *The Emergence of Probability*. Pearson informs us that the century and a half to two centuries before Newton were dominated by theological argument.

> After two centuries of such controversy, men's minds had sickened of theological discussions; they turned to the new results of the renascence of science, which were opening wonderful doors into the workings of nature, and asked if justification of broad religious principles could not be found in these mysterious phenomena, rather than in writings which every reader interpreted in a different sense.[81]

> There you have the whole philosophy of the 18th century. The fundamental articles of religious belief were to be deduced from an examination of the laws of nature in a simple and ready manner. Religion had hitherto been involved in metaphysical subtilty; the new natural philosophy was to lead to a new natural basis for religion.[82]

One of the early spokesmen for this new approach to theology was William Derham (1657 - 1735) an English clergyman with sufficient leisure time to dabble in science and to wed this science to theology.

> Volcanoes are not vent holes produced by the pressure of subterranean fluids and gases, but vent holes provided by a benevolent Creator to hinder the explosion into innumerable fragments of an earth destined as the home of an immense number of living forms. For Derham there is nothing in the inorganic or organic universe which was not directly created by the deity for a useful purpose, and this purpose he usually interprets as a purpose useful to man.[83]

Pearson cites a number of examples, e.g., "Here the reindeer is commended not only for its suitability for the northern regions, but because it provides clothing

---

[80] This is believed due to more widespread diagnosis of the condition, not to any autism-preventative properties of thimerosal.
[81] Pearson, *History...*, *op. cit.*, p. 285.
[82] *Ibid.*, p. 286.
[83] *Ibid.*, p. 287.

and food for the Laplander and Greenlander and because its horns make a good glue and its tongue is a great delicacy".[84]

Determining the laws of nature and their divine purposes served theology in demonstrating God's infinite intelligence, thus in better appreciating Him, and, insofar as this purpose was ultimately for the benefit of man, to demonstrate our unique place in God's scheme of things. The view was without serious opposition until Darwin came along, extended the age of the Earth from 6000 to millions of years, providing time for species to evolve into their individual niches thereby obviating the need for an infinite intelligence to place them there. That man himself had evolved from earlier creatures weakened the argument for his being the divine reason for all things in creation. Biologists still, under the name *teleology*, ask the purpose of this or that organ or trait, but they seek the answer not in the benefit it affords man, but in how it benefits the survival of the species in question.

But Darwin was not to come forth for another couple of centuries and Derham would influence future generations.

> For Derham the *exact balancing of the number of individuals of each species* is necessary for the stability of creation. This numerical stability is a grand act of Divine Wisdom. The stability of statistical ratios is a divine ordinance (in the same manner as the 18th century thought of other natural laws, e.g. gravitation) to maintain the created world in its created condition. Here the original idea of Süssmilch's *Gottliche Ordnung* first developed, and hence we proceed by a continual evolution to Florence Nightingale's view that we learn the purpose of God by studying statistics....
>
> It is true that Derham did not do much to push his conception of the stability of statistical ratios, he left that to his successors, notably Süssmilch[85], but he was certainly the originator of the idea.[86]

I cannot say how much Arbuthnot and de Moivre knew of Derham before their works cited in the preceding section, but we find Arbuthnot drawing moral conclusions from an assumed such stability of ratios and de Moivre finding the stability of the ratio in question "manifestly serving to a wise purpose". In this they were children of their times and their comments, however "off the wall" they may strike us today, are fitting and proper for the period.

One thing: the stability which they took as evidence of God's work was observed in one place—London—and asserted likely to be universal. It was an assumption backed up by, as it were, the result they wished to prove by reference

---

[84] *Ibid.*, p. 288.

[85] Johann Peter Süssmilch (1707 - 1767) was a Lutheran minister whose major work, *Die göttliche Ordnung in den Veränderungen des menschlichen Geschlechts, aus der Geburt, Tod, und Fortpflantzung desselben erwiesen* [*The divine order in the changes of the human species, as demonstrated by its birth, death and propagation*], originally bearing a theological purpose transformed through three editions (1741, 1761/1762, 1765) into a major statistical work.

[86] Pearson, *History...*, *op. cit.*, pp. 289 - 290.

The most famous battle in the modern war over intelligent design with evolution, the Scopes Monkey Trial that took place in the summer of 1925 in the state of Tennessee, is depicted here in a stamp of Micronesia issued in 2000 celebrating the coming of the new millennium. It is one of a number of stamps in a sheet depicting events of the 1920's. Pictured from left to right are Clarence Darrow (chief lawyer for the defence), John T. Scopes (the high school teacher on trial for teaching evolution), and William Jennings Bryan (chief prosecuting attorney). Earlier in the year, the state of Tennessee had passed a law making illegal the teaching either of any scientific theory not agreeing with the biblical story of creation or that man had descended from a lower form of animal. Scopes, who had substituted briefly in teaching the biology class and may or may not have actually covered the topic of evolution himself, agreed to stand trial in a test case.

The events of that summer are the subject of legend—and a riveting drama, *Inherit the Wind*, by Jerome Lawrence and Robert E. Lee (Random House, New York, 1955), which was filmed twice.

Charles Darwin (1809 - 1882) has appeared on postage stamps of many nations, mostly in the 1980's in connexion with the anniversary of his death and more recently in 2009 in connexion with the anniversary of his birth. The earliest issues are an attractively engraved set issued by Ecuador in 1936, 101 years after Darwin's arrival in the Galapagos Islands, and I was sorely tempted to include one of these here. However, as I include Darwin because of the Scopes Monkey Trial, I find the recent stamp of St. Tomé e Príncipe more relevant and have consequently chosen it for reproduction here.

to it: it served God's purpose. The argument is circular, displaying a comforting consistency to the believer, but useless as evidence to skeptical science. Even the most devoutly religious scientist ought to eschew such reasoning:

> I have thought it needful to show how thoroughly De Moivre appreciated the results which flow with regard to statistical ratios from the knowledge of the normal curve and its standard-deviation. But as I have said, I do not understand his demonstration for design from the stabil-

BELGIË·BELGIQUE

Lambert Adolphe Jacques Quetelet, more commonly called Adolphe Quetelet, (1796 - 1874) was a mathematician, astronomer, and, most famously, a statistician whose *social physics* was an attempt to apply probability and statistics to the study of man. He served as President of the Commission Centrale de Statistique of Belgium, was Perpetual Secretary of the Académie des Sciences et Belles-Lettres de Bruxelles, presided over the First International Congress of Statistics, and served the cause of science in numerous other ways.

Florence Nightingale (1820 - 1910) was tutored by the famous mathematician J.J. Sylvester. Her primary interest, however, was in public health. She is most famous for revamping the nursing profession and caring for the wounded during the Crimean War, her nightly checking after her patients earning her the sobriquet "the Lady with the Lamp". Another, less widely known, nickname was "the Passionate Statistician", for her lasting contributions were her use of the graphical representation of statistical data to persuade Parliament to mandate her sanitary reforms in hospitals following the Crimean War. She was a great admirer of Quetelet and the statistical "laws" of his social physics, believing these laws to represent the will of God.

The stamp depicting Quetelet was issued by Belgium in 1974 on the centenary of his death. There are many more stamps of Nightingale to choose from, some with her lamp, but none with any statistical representation. (One that comes close appears on page 526 in Appendix C, below.) I have chosen one issued by Belgium in 1939 in a set of semi-postals designed to raise money for the Red Cross, as, with her link to Quetelet, it seemed somehow the most appropriate choice here.

ity of statistical ratios. It is an argument which Paley[87] used 50 years later; possibly he took it from De Moivre, for he was an excellent card-player. 'I find,' said De Moivre, 'a 35 sided die and infer a die-maker.' 'I find,' said Paley, 'a watch and infer a watch-maker.' Such arguments are fatal to science. They bring us up to a point, where further inquiry is held to be either unnecessary or irreverent. The old neolithic myth of creation stopped geological and biological inquiry until Darwinian evo-

---

[87] William Paley (1743 - 1805) was an English theologian who practised the new theology, most popularly in his *Natural Theology, or Evidence of the Existence and Attributes of the Deity* (1802).

lution crashed down the barrier. Assume an 18 to 17 ratio to be a wise creation and you will stop short of asking the physiological reasons why the ratio holds. Hence I prefer Bernoulli at this point to De Moivre. Statistical ratios are only stable as long as there is a fairly even balance between man and his environment; upset that and a new position of equilibrium must be reached. Does climate, does race, does relative age, does order of birth affect the sex ratio? These are all useful problems, if we consider the origin of the sex-ratio; but their investigation may be checked by De Moivre's dogma as to statistical ratios. Still, that dogma produced widesweeping effects for a century after De Moivre.[88]

Later researchers would find ratios other than 18 to 17. Laplace, who never relied on God in his theorising, would look for explanations, not of the ratios but of their differences.

Today, "intelligent design" is an argument appealed to by certain American religious groups to fight the teaching of Darwinian evolution. When my ex-brother-in-law left the Catholic Church, a religious body that has had no problem with the theory of evolution since Leo XIII's 1893 encyclical *Providentissimus Deus* [89], to become a born-again Christian, his first priority was evolution. He assured me of the co-existence of man and dinosaur! When he was gored by a buffalo on a friend's ranch and the pain lingered on, he could not handle it and committed suicide. It is a shame that his supposedly profound belief on which he was able to call for inspiration in his fight against Darwin could provide no succour in his time of need. Perhaps he never really had the belief at all and his embrace of the more extreme ideas[90] was merely an attempt to hide his doubts from himself.

In any event, for obvious reasons, nothing is to be gained by *my* saying anything else on the subject of intelligent design as it is understood today.

There is, however, another point to be made about de Moivre's diatribe. This concerns his attack on the use of the word "chance" and its meaninglessness in "atheistical writings". A fact I have not stressed thus far in this book is that, with a firm belief in determinism, bolstered since Descartes by a mechanical view of the universe, chance didn't exist for scholars of the day. Probability was not about what might happen by accident as a result of the capriciousness of nature. Hacking informs us

... there is a third and more elusive consideration that underlies Arbuthnot's paper. It has to do with the very nature of chance. On

---

[88] Pearson, *History...*, *op. cit.*, pp. 162 - 163.

[89] According to this document, truth cannot contradict truth and when science contradicts scripture and cannot be refuted, Church fathers should examine scripture to search out the correct interpretation thereof. Indeed, a century later, in 1996, John Paul II quoted *Providentissimus Deus* in accepting the theory of evolution as no longer merely an hypothesis.

[90] He uncritically accepted the result of Bishop Ussher's calculation of the date of the creation— 9 a.m. on 23 October 4004 B.C. This result is suspiciously precise for the outcome of what had to have been a long and involved computation based on imprecise data.

the continent, and especially in France, the investigation of chance phenomena was given a notable tincture of subjectivism. Everyone agrees
that there is no such thing as real chance but this fact can be explained
in different ways. On the continent, to talk of chance is to talk of lack
of knowledge. In England, chance is lack of skill.[91]

Determinism continued to hold sway in scientific thought until the emergence of
quantum theory in the early decades of the twentieth century. Einstein's famous
dictum that God does not play dice with the universe is a modern version of
de Moivre's rejection of Bernoulli's point that the regular excess of male over
female births could be explained by chance without reference to special action
on the part of God.

A House (of Cards) of Ill Repute

It is a bit strange to find a determinist who doesn't believe in chance writing

> Yet there are Writers... who insinuate as if the *Doctrine of Probabilities*
> could have no place in any serious Enquiry; and that Studies of this
> kind, trivial and easy as they be, rather disqualify a man for reasoning
> on every other subject.[92]

That chance should not exist is not why there was no place for it in serious
enquiry. The Earth in Galileo's day was believed to be the centre of the universe,
yet the Church had no objection to his use of the Copernican system as an
hypothesis on which to base calculations. Galileo ran afoul of authorities in
contradicting scripture and asserting heliocentrism as a *fact*.

   Whether probabilities measured ignorance or lack of skill, it could be called
chance and the calculations made. The Church would not object, unless, perhaps, one started declaring the existence of chance and thus contradicted the
Church's teachings on determinism. In those pious days one had to tread carefully: Leibniz found Newton to be irreligious in believing God, having created
the sun and stars like a clockmaker creates his clock, might have to make occasional adjustments just as the clockmaker rewinds and occasionally resets
his clocks. Perhaps de Moivre was taking no chances, if you will pardon the
expression, in being so explicit on the nonexistence of chance and the fact that
everything that happens does so by God's will.

   So why the defensiveness? In Protestant England he was beyond the reach
of the Inquisition that had gone after Giordano Bruno and Galileo, and the
Catholics who had imprisoned him in France could likewise not arrest him
again. Matthew Hopkins, England's Witchfinder General, had been executed
for witchcraft a half century earlier.[93] Given his own earlier imprisonment, the

---

[91] Hacking, *op. cit.*, p. 171.

[92] De Moivre, *op. cit.*, p. 254. This is from the end of the long quotation with which
we finished section 5.

[93] This is apparently the predominant Catholic theory of Hopkins's death; Protestants
favour his dying a natural death after his downfall.

The two most famous martyrs of science: Giordano Bruno (1548 - 1600) and Galileo. Bruno was not really a scientist, but a religious philosopher who attempted a return to a magic religion predating Christianity. Somehow mixed up with this was a belief in the Copernican system and the existence of infinitely many worlds all inhabited with divine life, the last being a state of affairs reserved by orthodox Christianity for the Earth. He ran afoul of the authorities on several occasions, finally being taken prisoner by the Inquisition in Venice, where he recanted. Sent on to Rome for further investigation, he refused to recant and was burned at the stake. It is not exactly clear what the grounds against him were.

Galileo was also tried, explicitly for his Copernicanism. He was treated more kindly: The inquisitors were allowed to "show him the instruments" of torture, but not actually use them, and upon recanting he was sentenced to a fairly comfortable house arrest.

I'm not sure what, if any, is the Bulgarian connexion, but in 1998 Bulgaria issued a lovely colourful stamp in honour of Bruno. The black and white reproduction above left does not do it justice. Two years later Italy issued an attractive stamp commemorating the 400th anniversary of Bruno's death.

Of the many Galileo stamps in existence, I know of only a few making reference to Galileo's trial. The first to appear was that of Yemen issued in 1970 and depicted on page 519, below. The second stamp, pictured above right, was issued by the Independent Postal Systems of America in 1976 and contains a sketch based on Cristiano Banti's famous 1857 painting of "Galileo Facing the Roman Inquisition". This painting is also depicted in the margin of a souvenir sheet reproduced on page 520, below, and on a more recent stamp of Chad, not pictured here, issued in 2009 celebrating an International Year of Astronomy.

danger may have been greater than I imagine, and, being a mathematician, he may have been reading John Dee's (1527 - 1608) preface to the Billingsley translation of Euclid[94]:

> AND for these, and such like marueilous Actes and Feates, Naturally, Mathematically, and Mechanically, wrought and contriued: ought any honest Student, and Modest Christian Philosopher, be counted, &

---

[94] John Dee was a most interesting character—sorcerer, courtier, mathematician, he is celebrated today by occultists as one of the Illuminati; in his day more reviled as a black magician. Sir Henry Billingsley (1551 - 1606) made the first English translation of Euclid's *Elements*, published in 1570 with Dee's preface.

called a Coniurer? Shall the folly of Idiotes, and the Mallice of the Scornfull, so much preuaile, that He, who seeketh no worldly gaine or glory at their handes: But onely, of God, the treasor of heauenly wisedome, & knowledge of pure veritie: Shall he (I say) in the meane space, be robbed and spoiled of his honest name and fame? He that seketh (by S. Paules aduertisement) in the Creatures Properties, and wonderfull vertues, to finde iuste cause, to glorifie the Æternal, and Almightie Creator by: Shall that man, be (in hugger mugger) condemned, as a Companion of the Helhoundes, and a Caller, and Coniurer of wicked and damned Spirites?... Well: I thanke God and our Lorde Iesus Christ, for the Comfort which I haue by the Examples of other men, before my time: To whom, neither in godliness of life, nor in perfection of learning, I am worthy to be compared: and yet, they sustained the very like Iniuries, that I do: or rather, greater. Pacient *Socrates*...[95]

Mathematics does have its own martyrs—Pythagoras, Hypatia, Gerbert, Dee—but of these the closest to having been martyred for his mathematics was Gerbert of Aurillac (*c.* 940 - 1003), who, before becoming Pope Sylvester II, had such a reputation for learning that it was commonly believed he had signed a pact with the devil, who eventually came to claim his prize. The mathematical element, in the narrow sense, would be his mastery of arabic numerals and devising his own tables for calculation from which the name Abacus of his familiar derived. His skill at clockmaking, thus as a maker of *mathematical instruments*, would be a broader mathematical connexion to the devil.[96]

A century or two earlier and a belief in chance might have got one into major trouble, but the real danger here was not of being accused of heresy but

---

[95] Allen G. Debus, ed., *John Dee, The Mathematical Praeface to the Elements of Geometry of Euclid of Megara*, Neale Watson Academic Publications, Inc., New York, 1975. This is a facsimile reprint with no numerical pagination. There is, however, a sort of alphanumeric numbering at the bottoms of alternate pages. The full quote begins on the reverse of page Aj and ends on the reverse of page Aij. The quote contains a lot more, with far less familiar spellynges than appear here. (I suppose I should also mention that *The Elements* was written by Euclid of Alexandria, not Euclid of Megara. The misidentification was common at the time.)

[96] Much as I would love to repeat the story here, I suppose it is not that relevant, so I will have to make do citing a few references:

Richard Erdoes, *A.D. 1000: Living on the Brink of Apocalypse*, Harper & Row, Publishers, San Francisco, 1988, pp. 1 - 9.

Elizabeth M. Butler, *The Myth of the Magus*, Cambridge University Press, Cambridge, 1948, pp. 94 - 99. This book also contains information on John Dee's magical activities.

Hans Joachim Gernentz, ed., *Der Pakt mit dem Teufel*, Müller & Kiepenheuer, Hanau, 1988, pp. 159 - 171.

The last item is a 13th century poem, redone in modern German, on Gerbert's damnation. Walther von der Vogelweide, famous *Minnesänger*, also mentions Gerbert's damnation in one of his compositions.

of guilt by association with the vice of gambling. Studying the mathematics of it may have been deemed an unhealthy obsession. I have already quoted F.N. David's remark (page 77, above) on the importance for the field of Montmort's position as canon of the Church, and on page 99 I cited de Morgan's very moral justification for discussing the problem of Gambler's Ruin:

> ...I have therefore digested the following demonstration, that no one who bears such a character may be able to weaken the evidence for the *necessity* of the pernicious results of gambling which that chapter is intended to afford.

Even Cardano, whose reputation as a gambler is such that his most famous biography[97] alludes to this fact in its subtitle, takes the high moral ground, including chapters

"Who should play and when"

"Why I have dealt with gambling"

"Why gambling was condemned by Aristotle".

The first of these begins

> So, if a person be renowned for wisdom, or if he be old and dignified by a magistracy or any other civil honor or by a priesthood, it is all the worse for him to play; on the other hand, gambling is proportionately less of a reproach to boys, young men, and soldiers. The greater the amount of money involved, the greater the disgrace; *thus a certain holder of a very high priestly office (namely, a cardinal)* was severely blamed because he played after dinner with the Duke of Milan for a stake of five thousand crowns.[98]

In their historical essay of their work on Probability, John Lubbock and John Drinkwater-Bethune cite another instance from a couple of decades earlier than de Moivre:

> It has been the misfortune of the science of probability, in consequence of the ready application made of its principles to games of cards and dice, that a prejudice has from the first existed against it as if ministering only gambling and immorality, and available for no other purpose: accordingly the anonymous writer, who in 1692, published the first English essay "On the Laws of Chance," thought it necessary to protest in his preface that the design of his book was "not to teach the art of playing at dice, but to deal with them as with other epidemic distempers, and perhaps persuade a raw squire to keep his money in his pocket."[99]

---

[97] Oystein Ore, *Cardano; the Gambling Scholar*, Princeton University Press, Princeton, 1953.

[98] *Ibid.*, pp. 186 - 187. Ore's book includes a translation of the entire *Liber de ludo aleæ*.

[99] John Lubbock and John Drinkwater-Bethune, *On Probability*, Baldwin & Chadock, London, 1830, p. 43.

Jakob and Nikolaus Bernoulli's attempts to apply probability to law, in the analysis of the credibility of witnesses, and the growing use of probability in dealing with annuities were poised to lend a respectability to probability theory that the analysis of common games of chance could not. Yet there were new dangers ahead:

> Why did Laplace leave probability alone from 1783 to 1812? I think that from 1783 to 1795, the year of his lectures, he was restrained largely by the state of affairs in France. Probability has very mundane interests, and its results too often touch social and political institutions. Condorcet, the enemy of the Jacobins, had been driven to death in 1794, the year before Laplace lectured on probability. It was an unsafe topic, and all the men of science were under suspicion, especially those that dwelt on probability. Had not Condorcet questioned the right judgment in the votes of large assemblies of untrained judges? And was not this the product of his dealing with such questions mathematically? It is scarcely to be wondered at that Laplace did not refer in his lectures to Condorcet; in 1795 he had enough to do to preserve his own stability. It was far safer to deal with the heavens, than with the more mundane scope of probability.[100]

Today probability and statistics have so many legitimate applications in science and industry that one can pursue their study and application in relative safety. What low opinion is had of statistics is ordinarily generated by its misapplication: witness the adage, "There are lies, damned lies, and statistics". There may be further danger to the fields as the results of statistical analyses fail to support popular beliefs. I refer to the ongoing debate on the supposed connexion between thimerosal and autism. Autism tends to strike at the age at which children receive vaccination and the coincidence of the onset of the illness shortly after a child's vaccination is enough to convince the parents of a causal relationship. Emotions run high and no amount of clinical studies and statistical analyses can convince them that there is no relation between thimerosal and autism. Defenders of vaccination have been harassed and even had their lives threatened. This has not yet affected the statisticians much, but as belief in the conspiracy among drug companies, the federal oversight agencies, and physicians grows, they too run the risk of being targeted.

Lest I sound too alarmist here, let me remind the reader, not of Giordano Bruno and Galileo, who lived centuries ago, nor of John T. Scopes, who was a volunteer in a legal test case and not the victim of a modern-day witch hunt, but of Nikolai Ivanovich Vavilov (1887 - 1943), whose commitment to the truth in science resulted in his death in Stalin's Russia. Vavilov was one of the leading geneticists in the first half of the 20th century. He specialised in plant genetics, travelled widely collecting seeds, and served the Soviet Union under Lenin in improving agriculture. Unfortunately for him, Lenin died and was followed by

---

[100] Pearson, *History...*, *op. cit.*, p. 651. In 1796 Laplace published his popular account of celestial mechanics, and in 1802 the technical account; his massive volume on probability, based on his work prior to 1783, did not appear until 1812.

5ᴷ ПОЧТА СССР 1987

АКАДЕМИК
Н.И. ВАВИЛОВ 1887 1943

N.I. Vavilov, the most recent martyr for science. A couple of decades after his death in prison, Vavilov was officially cleared of all charges. In 1977 the Soviet Union paid philatelic homage to Vavilov by issuing a stamp featuring his portrait. A decade later, in 1987, on the occasion of the centenary of his birth, another such stamp was issued. I present the second one here.

Stalin, who was taken in by the false promises of Trofim D. Lysenko (1898 - 1976) whose theory of the inheritance of acquired characteristics contradicted orthodox genetic theory. On the basis of single experiments Lysenko reported progress that would have major agricultural implications, while Vavilov relied on large-scale experimentation and statistical analysis. Additionally, Lysenko was able to cloak his theories in Marxist lingo and his "socialist" biology won the day in its battle with the "bourgeois" biology of Vavilov. Lysenko denounced Vavilov and in 1940 Vavilov was arrested. Initially sentenced to death by firing squad, Vavilov received a temporary reprieve when his sentence was commuted to twenty years' imprisonment, a sentence cut short by his death by starvation in 1943. Lysenkoism ruled supreme in Soviet agriculture until after Stalin's death when an intellectual thaw allowed open comparison of the failings of Soviet agriculture under Lysenko with the success of agriculture guided by scientifically based genetics in the West.[101]

Closing Remarks

Finishing the last subsection with Vavilov and the conflict between science and politics in the former Soviet Union under the extremism of Stalin might be misleading. One should not be lulled into a false sense of security. The fight between science and theology continues to this day in the courts of the land as state legislatures continue to attack the teaching of evolution:

> 1928/1968: Following the Scopes trial, Arkansas outlawed the teaching of evolution in public schools in 1928. When a textbook covering the subject was adopted by the school board in 1964, biology teacher Susan Epperson sued the state. In 1968, in the case Epperson *v.* Arkansas, the U.S. Supreme Court struck down the law.

---

[101] *Cf.* the chapter "Stalinist ideology and the Lysenko affair" in: Loren R. Graham, *Science in Russia and the Soviet Union: A Short History*, Cambridge University Press, Cambridge, 1993.

1973/1975: In 1973 Tennessee passed an "equal time" law mandating the teaching of the Biblical theory of creation on an equal basis with that of evolution in public schools. In 1975, in the case Daniel v. Waters, the U.S. Court of Appeals struck this law down.

1981/1982: Arkansas passed the "Balanced Treatment for Creation-Science and Evolution-Science Act" similar to Tennessee's "equal time" law in 1981. A lawsuit, McLean v. Arkansas Board of Education, filed by a number of religious leaders, parents, friends of underage students, and a high school biology teacher resulted in the law being declared unconstitutional by the U.S. District Court in 1982.

1982/1987: Louisiana passed the "Balanced Treatment for Creation-Science and Evolution-Science in Public School Instruction Act" in 1982. It was declared unconstitutional in 1987 by the U.S. Supreme Court in the suit Edwards v. Aguillard. 72 Nobel Prize winning scientists filed a friends-of-the-court brief in the case.

And in our current century:

2002/2005: In the case Selman v. Cobb County School District in 2005, the U.S. District Court ordered the removal of disclaimer stickers that had been affixed to textbooks discussing the theory of evolution by an act passed by the state of Georgia in 2002.

2002/2005: In 2002 the Dover Area School District in Dover, Pennsylvania chose *Of Pandas and People* as a textbook. This book attempted to pretend to be science not religion by introducing "intelligent design" as a proper name for a type of study not to be mistaken for the religious belief the phrase has always stood for. In the case Kitzmiller v. Dover Area School District, in a trial variously called Scopes II or the Dover Panda Trial, the U.S. District Court ruled that "intelligent design" is not science. The members of the board who chose the textbook were later voted out of office.

At first sight the main danger of the fundamentalist's clinging to the Bible as literal truth would seem to be his laying himself and his amenable legislature open to ridicule. But there is also a great danger in failing to accept some parts of Darwin's theory—natural selection and survival of the fittest, however repugnant they may be to one's moral sensibility, do warn us about drug resistant bacteria and against the misuse of antibiotics. Without these Darwinian concepts we are left in the dark as to the cause of the newer untreatable bacterial infections and are thus left without a clue as to how to combat them.

The greater immediate danger to science and society, however, comes not from the warfare between science and extremist theology, but from the sphere of politicised pseudo-science. Children, whose parents have not had them vaccinated against measles, mumps, and rubella, are getting ill and dying of preventable diseases. And, following Lysenko's denunciation of Vavilov, campaigns against those decrying the quackery of supposed cures for autism and against the skeptics on global warming are under way. One case that should be mentioned here, one of the most famous in the battle over global warming, is that of Bjørn Lomborg.

A Danish academic, Lomborg took his degree in political science, but is variously referred to as an economist or a statistician, the latter because he teaches statistics in the political sciences department at the University of Aarhus. In 1997 Lomborg read an article by skeptic Julian Simon declaring that environmental quality had been improving, not getting worse. Finding this hard to believe, he decided to check the facts for himself. Using data from the United Nations Environment Program, the World Health Organization, the World Bank, the Harvard Center for Risk Analysis, etc., he and his students ran the figures and discovered that, for the most part, Simon was correct. When the English edition of Lomborg's subsequent book, *The Skeptical Environmentalist: Measuring the Real State of the World* [102], was published, a storm broke over his head. He was immediately attacked in four articles in the popular magazine *Scientific American* "by four leading academic environmental alarmists in January 2002" [103]. This was followed by the publication in February 2003 of a report by the Danish Committees on Scientific Dishonesty, a branch of the Danish Research Agency (which Ebell says is composed mainly of nonscientists). Based primarily on the articles in *Scientific American*, the Committees' verdict was that Lomborg's book was "objectively dishonest" and "clearly contrary to the standards of good scientific practice".

Exactly why the Danish Ministry of Science, Technology, and Innovation next got involved is not known to me. A charge of scientific dishonesty by a government agency could easily lose an academic his position. Whether Ministry involvement was an obvious next step, whether it was the result of a request from the University in aiding any decision on disciplinary action on its part, or whether it was a response to a petition for redress from Lomborg himself, I confess not to have looked deeply enough into the affair to determine. The reason for Ministry involvement is not important. What is important is that the Ministry looked into the matter and on 17 December 2003,

> The Ministry's 70-page report found the DCSD's handling of the case "dissatisfactory" and its conclusion of scientific dishonesty "completely void of argumentation." Actually, it was worse than that. As *The Economist* noted, "The panel's ruling—objectively speaking—is incompetent and shameful." [104]

---

[102] Cambridge University Press, Cambridge, 2001.

[103] This description is from an article by Myron Ebell entitled "*The Skeptical Environmentalist* **Vindicated**: Radical Greens' Attempt to Destroy Bjørn Lomborg Fails" and published online at www.cei.org. Ebell is Director of Global Warming and International Environmental Policy at CEI, the Competitive Enterprise Institute. The CEI definitely has a political agenda, but it does seem to do the research before publishing its opinion. Checking anti-Lomborgian websites as well, I find agreement on the basic facts of the affair, if not on the interpretation, and have decided to trust Ebell's narration.

[104] Ebell, *op. cit.*, p. 4.

The Danish Ministry of Science, Technology, and Innovation did not decide the scientific issue of whether the environment has been steadily worsening or, as Simon and Lomborg found, improving. It merely ruled, in effect, that the Committees had based their decision on extra-scientific—political, if you will—considerations. There are those who believe they can dictate the laws of nature by controlling public opinion. They cannot. They can only occasionally dictate the passage of human laws, sometimes to deleterious effect.

The interesting thing is that Lomborg believes global warming is occurring, and that it is in part man-made, due to increased CO_2 emissions. What he does not believe in are the demands for the immediate and extreme reforms advocated by the alarmists. The attacks on him are thus objectively adjudged politically and not scientifically based.

As teachers it is not our place to take positions on the issues of the day and preach to our students. And one cannot take the time to look into all the evidence on every topic that comes along. But when it comes to an issue, be it evolution, vaccination, or global warming, where there is substantial disagreement, we owe it to our students not to declare the truth of one side or the other without discussing the evidence. Evolution is only opposed on religious grounds and, although there are some problems in the theory yet to be resolved—as there generally are in science—the scientific evidence overwhelmingly favours it and there is no opposing theory with comparable explanatory power. Hence scientists accept it. Vaccinations do pose dangers and are not completely safe, the decision on whether or not to vaccinate is one of risk assessment. This is something that can profitably be discussed in one's probability class. It requires a trip to a good research library as the readily accessible papers do not run the numbers. And, alas, the running of the numbers in the technical reports accessible only via the research libraries tends to involve many variables and use more advanced statistical tests than are considered here. Depending on the level of one's students, one may be able to offer a simplified analysis and discuss what conclusions are suggested by it. As for global warming, even if we accept that it is going on, the verdict is still out on whether man is the chief culprit and on what the long term effects will be. The fact that the most vocal proponents of the theory are not scientists, though it makes one want to oppose the theory, is not disproof. And, as much fear as they may instil through their witch hunts, this too is not proof. If one is going to discuss the issue in class at all, one should attempt an agnostic stand and cite and critique the evidence. I do recommend also discussing the notorious unreliability of long term predictions in a world of changing circumstances (like accelerated technological innovation). A simple example, that of predicting population growth, is discussed in Appendix B.

I fear my discussion of these matters has gone a bit beyond the intended scope of the present book, whence I will rather abruptly break off from it here. But I do find the subjects fascinating and one will want to look into these or similar issues even if nothing suitable for the classroom turns up.

6

Beyond Bernoulli's Theorem

1 De Moivre's Theorem

As we saw in the preceding chapter, the first improvement on Bernoulli's Theorem was given by Jakob Bernoulli's nephew Nikolaus. It looks to be a small, merely technical improvement on the original. However, the Danish statistician and historian of statistics Anders Hald declared

> Nicholas Bernoulli's result has been overlooked, perhaps because two otherwise reliable witnesses, namely de Moivre and Todhunter, did not grasp the significance of his result. De Moivre (1730, pp. 96 - 99)[1] gave a precise account of both theorems with the original examples but without proofs. He did not compare his own result with Nicholas Bernoulli's...
> We shall give a summary and a comparison of the two proofs and show that Nicholas Bernoulli's result is the 'missing link' between the result of James Bernoulli and de Moivre's derivation of the normal distribution as an approximation to the binomial.[2]

That the normal distribution is implicit in the proof by Nikolaus Bernoulli was only first noticed by another statistician and historian of statistics, Oscar B. Sheynin, in the late 1960's. In 1968, Sheynin published a paper, "On the early history of the law of large numbers" in the ongoing series of "Studies in the History of Probability and Statistics" in the journal *Biometrika* [3]. The new observation on the significance of Nikolaus's proof is not made therein, but it appears in the anthologised version which appeared two years later[4] and again in his history of probability:

---

[1] The reference is to de Moivre's *Miscellanea analytica*, London, 1730.

[2] Anders Hald, "Nicholas Bernoulli's Theorem", *International Statistical Review* (1984), pp. 93 - 99; here: p. 93.

[3] O.B. Sheynin, "Studies in the history of probability and statistics. XXI. On the early history of the law of large numbers", *Biometrika* 55 (1968), pp. 459 - 467.

[4] O.B. Sheynin, "On the early history of the law of large numbers", in: E.S. Pearson and M. Kendall, eds., *Studies in the History of Statistics and Probability*, Griffin, London, 1970.

I only dwell on Bernoulli's indirect derivation of the normal distribution. Let the sex ratio be m/f, n, the total yearly number of births, and μ and $|n - \mu|$ the numbers of male and female births in a year. Denote

$$n/(m + f) = r, \quad m/(m + f) = p, \quad f/(m + f) = q, \quad p + q = 1,$$

and let $s = O(\sqrt{n})$. Then Bernoulli's derivation can be presented as follows:

$$P(|\mu - rm| \leq s) \approx (t - 1)/t, \tag{1}$$
$$t \approx [1 + s(m + f)/mfr]^{s/2} \approx exp(s^2(m + f)^2/2mfn), \tag{2}$$
$$P(|\mu - rm| \leq s) \approx 1 - exp(s^2/2pqn), \tag{3}$$
$$P[|\mu - np|/\sqrt{npq} \leq s] \approx 1 - exp(-s^2/2). \tag{4}$$

This result does not however lead to an integral theorem since s is restricted (see above) and neither is it a local theorem: for one thing, it lacks the factor $\sqrt{2/\pi}$.[5]

The first thing to notice is that Sheynin's variables n, m, f, r, s, p, q are not quite in agreement with those of Nikolaus Bernoulli. The proper correspondence is listed in *Table 1*, below. We should think of m, f as $18, 17$ and n as 14000.

Sheynin	n	m	f	r	s	p	q
N. Bernoulli	n	m	f	p	l	$\frac{m}{m+f}$	$\frac{f}{m+f}$

Table 1

The variable r is not particularly interesting in itself. What is of interest is $rm = (14000/35) \cdot 18 = 7200$. Thus $P(|\mu - rm| \leq s)$ is the probability that the actual number μ of male births differs from the expected number 7200 by at most s.

The letter s corresponds to Bernoulli's variable l, the distance to the left or right of the maximum term in the binary expansion $(m + f)^n$ we wish to include. Sheynin's restriction $s = O(\sqrt{n})$ indicates that s is bounded by a constant multiple of $\sqrt{n}$. If we choose s as a function of n, say $s(n)$, this means that there is some constant c such that, for all n, $s(n) \leq c\sqrt{n}$. This condition is not mentioned anywhere in Bernoulli's letter to Montmort. Approximation (1) and the first approximation of (2) are algebraically derivable from Bernoulli's work. The second approximation of (2), however, requires the extra condition and the generalisation,

---

[5] Oscar Sheynin, *Theory of Probability. An Historical Essay*, 2nd Revised and enlarged edition, Berlin, 2009. I quote from a .pdf file that was available online and in which the pages are unnumbered.
Note the grouping: multiplication takes precedence over $/$. I have numbered the formulæ for later reference.

$$\lim_{k\to\infty}\left(1+\frac{a}{k}\right)^k = e^a,\tag{5}$$

of Jakob Bernoulli's limit (5) of Chapter 5: If $s(n) = c\sqrt{n}$, for some constant c, we successively obtain

$$\left(1+\frac{s(m+f)}{mfr}\right)^{s/2} = \left(1+s\cdot\frac{m+f}{mf}\cdot\frac{m+f}{n}\right)^{s/2}, \text{ since } r = \frac{n}{m+f}$$

$$= \left(1+s\cdot\frac{(m+f)^2}{mfn}\right)^{s/2}$$

$$= \left(1+c\sqrt{n}\cdot\frac{(m+f)^2}{mfn}\right)^{c\sqrt{n}/2}, \text{ since } s = c\sqrt{n}$$

$$= \left(\left(1+c\cdot\frac{(m+f)^2}{mf\sqrt{n}}\right)^{\sqrt{n}}\right)^{c/2}$$

$$\approx \exp(c(m+f)^2/(mf))^{c/2}, \text{ by (5)}$$

where we write $\exp(x)$ for e^x as a typographical convenience, whence

$$\approx \exp(c^2(m+f)^2/(2mf)),$$

which gives us (2),

$$\approx \exp(c^2 n(m+f)^2/(2mfn))$$
$$\approx \exp(s^2(m+f)^2/(2mfn)), \text{ re-introducing } s$$
$$\approx \exp(s^2/(2pqn)),$$

this last following from $p = m/(m+f)$, $q = f/(m+f)$. From here simple algebra yields (3) and (4).

From (4) the normal distribution can be derived via standard techniques of the Calculus. I refer the reader to Hald's paper for the details.[6]

It is not so easy for the amateur historian to discuss de Moivre's proof. The final step in his proof has been anthologised and is easily accessible[7], but the earlier part, though reported on, is not as readily available.

De Moivre's relevant works are

---

[6] Hald, *op. cit.*, pp. 97 - 98. I note that Hald reviews the two proofs of Bernoulli's Theorem by Jakob and Nikolaus Bernoulli, a variant of Sheynin's observation, and de Moivre's result. He does this with a uniform notation and I recommend the paper as a nice review for the reader who has concerned himself with the details of the proofs. Hald expands on all of this in greater detail in Chapters 2 and 3 of his *A History of Parametric Statistical Inference from Bernoulli to Fisher, 1713 - 1935*, Springer-Verlag, New York, 2007.

[7] It occupies pp. 243 - 254 of the third edition of *The Doctrine of Chances*, which can be found in: David Eugene Smith, *A Source Book in Mathematics*, Cambridge University Press, Cambridge, 1929 (reprinted by Dover Publications, New York, 1959); F.N. David, *Games, Gods and Gambling; A History of Probability and Statistical Ideas*, Charles Griffin & Co., Ltd., London, 1962 (reprinted by Dover Publications,

De mensura sortis, seu de probabilitate eventuum in ludis a casu for-tuito pendentibus (1711/1712[8]). An English translation by Bruce Mc-Clintock accompanied by commentary by Anders Hald has been pub-lished: A. de Moivre, " 'De mensura sortis' or 'On the measurement of chance' ", *International Statistical Review* 52 (1984), pp. 229 - 262. The book, like Montmort and Huyghens before, was a compendium of problems, but with a bit more generality. The binomial distribution is derived in the introduction and used repeatedly throughout.

Doctrine of Chances (1718). David's description of it reads, "it is an enlarged version of *De mensura sortis*, written in English[9], with arith-metical illustrations added for those gentlemen who do not understand algebra".[10]

Miscellanea analytica de seriebus et quadraturis or *Miscellanea analyt-ica* for short (1730). Pearson describes it thus:

> This is the most important of De Moivre's works from the standpoint of mathematics. Namely here we find these series associated with his name in Trigonometry, i.e. the expansions for sine and cosine based on
>
> $$(\cos\theta + \sqrt{(-1)}\sin\theta)^n = e^{\sqrt{(-1)}n\theta},$$
>
> the explanation of the imaginary $\sqrt{-1}$, and the factorising of $x^{2m} - 2px^m + 1$ and recurring series.[11]

And, of course there is probability, e.g., his treatment of Gambler's Ruin. I have not seen the *Miscellanea analytica* and can only quote those who have done so as regards the contents most relevant to our present discussion:

> Well, the *Miscellanea Analytica de Seriebus et Quadraturis* was published in 1730, and there is only one edition of it. But a few copies have bound up with it a supplement and in at least one case two supplements. The first of these, *inter alia*, deals with the ratio of the mid-term of a symmetrical binomial to the sum

Minneola (New York), 1988); (in untranslated English) Ivo Schneider, *Die Entwick-lung der Wahrscheinlichkeitstheorie von den Anfängen bis 1933; Einführungen und Texte*, Wissenschaftliche Buchgesellschaft, Darmstadt, 1988; and (in abbreviated form) Jacqueline Stedall, *Mathematics Emerging: A Sourcebook 1540 - 1900*, Ox-ford University Press, Oxford, 2008.

[8] The work appeared in the volume of the *Philosophical Transactions* for 1711, and was published in 1712.

[9] Following tradition, de Moivre wrote his scientific papers in Latin. However, he had no permanent position, earning his money tutoring, consulting for insurance firms, and calculating odds for gamblers. Thus, he also published more popular works in English for a broader market.

[10] David, *op. cit.*, pp. 165 - 166.

[11] Karl Pearson, *The History of Statistics in the 17th and 18th Centuries against the changing background of intellectual, scientific and religious thought*, Charles Griffith & Co. Ltd., London, 1978, p. 145.

of its terms, and also contains a table of summed logarithms or logarithms of factorials for $n = 10$ to 900, with argument interval for n of 10, to 18 figures. What I term the second supplement is a document of November 12, 1733... My own impression is that Todhunter had a copy of the *Miscellanea Analytica* which contained the first Supplement only, and so he did not see the full importance of De Moivre's work. Of the three copies of the *Miscellanea Analytica* in our College Library only one contains the second supplement which gives the normal curve.[12]

Todhunter describes the first supplement in articles 333 - 334 of his *History*. Before this, in article 332 we read

332. The ninth problem in the seventh Chapter of the *Responsio*...[13] is to find the ratio of the sum of the largest p terms in the expansion of $(1+1)^n$ to the sum of all the terms; p being an odd number and n an even number. De Moivre expresses this ratio in terms of the chances of certain events, for which chances he had already obtained formulæ. This mode of expressing the ratio is not given in the *Doctrine of Chances*, being rendered unnecessary by the application of Stirling's Theorem...[14]

Thus, prior to the supplements, we find de Moivre following in the footsteps of the Bernoullis in finding the probability

$$\sum_{k=-[p/2]}^{[p/2]} \binom{2n}{n+k} \left(\frac{1}{2}\right)^{n+k} \left(\frac{1}{2}\right)^{n-k}.$$

The rest of the supplement is pretty much superceded by the second supplement, which is:

Approximatio ad summam terminorum binomii $(a+b)^n$ in seriem expansi (1733). This, but not the first supplement, was reprinted in the second (1738) and third (1756) editions of *Doctrine of Chances* under the English title "A Method of approximating the Sum of the Terms of the Binomial $(a+b)^n$ expanded into a Series, from which are deduced some practical Rules to estimate the Degree of Assent which is to be given to Experiments" and essentially contains the normal curve and probability integral we discussed in Chapter 4.

Doctrine of Chances (2nd edition, 1738).

Doctrine of Chances (3rd edition, 1756).

Of these works, the one that needs to be discussed here is the *Approximatio*, its derivation of the normal curve and the area under it as an approximation

---

[12] *Ibid.*, p. 157.

[13] The *Responsio ad quasdam criminationes* is a seven chapter response to remarks by Montmort who had felt slighted by earlier remarks by de Moivre—none of which is relevant here.

[14] Todhunter, *op. cit.*, p. 189.

to the given probability. I cannot present de Moivre's proof in full, but can only give a taste of it. De Moivre first treats the case $p = 1/2$. Following the Bernoullis, he estimates the ratio of the terms of highest probability to a term some distance away,

$$\binom{2m}{m}\left(\frac{1}{2}\right)^{2m} \Big/ \binom{2m}{m \pm l}\left(\frac{1}{2}\right)^{2m},$$

i.e.,

$$\binom{2m}{m} \Big/ \binom{2m}{m \pm l}, \tag{6}$$

where one assumes $n = 2m$ is even. And, finally, there is the estimate of the central binomial coefficient,

$$\binom{2m}{m} = \frac{(2m)!}{m!m!}, \tag{7}$$

which nowadays reduces to finding estimates for the individual factorials.

De Moivre's *Approximatio* leads with this last problem, citing his earlier work on (6) and relying on a result of James Stirling (1692 - 1770) to replace an unknown constant he himself had calculated to several decimal places by a known constant. In modern accounts, one either first derives or one merely quotes *Stirling's formula*, a variant of this determination of the value of the constant,

$$n! \sim (2\pi)^{\frac{1}{2}} n^{n+\frac{1}{2}} e^{-n}, \tag{8}$$

where "$\sim$" reads "is asymptotic to" and means that the ratio of the two terms gets closer and closer to 1 as n grows large:

$$\lim_{n \to \infty} \frac{n!}{(2\pi)^{\frac{1}{2}} n^{n+\frac{1}{2}} e^{-n}} = 1.$$

Assuming Stirling's formula (8), we easily approximate (7) as follows:

$$\binom{2m}{m} = \frac{(2m)!}{m!m!}$$

$$\sim \frac{\sqrt{2\pi}\,(2m)^{2m+\frac{1}{2}} e^{-2m}}{\sqrt{2\pi}\,m^{m+\frac{1}{2}} e^{-m} \sqrt{2\pi}\,m^{m+\frac{1}{2}} e^{-m}}$$

$$\sim \frac{\sqrt{2\pi}}{\sqrt{2\pi}\sqrt{2\pi}} \cdot \frac{2^{2m+\frac{1}{2}} m^{2m+\frac{1}{2}}}{m^{2m+1}} \cdot \frac{e^{-2m}}{e^{-2m}}$$

$$\sim \frac{1}{\sqrt{2\pi}} \cdot \frac{2^{2m}\sqrt{2}}{\sqrt{m}}$$

$$\sim \frac{2^{2m}}{\sqrt{m\pi}} = \frac{2^n}{\sqrt{\frac{n\pi}{2}}}, \quad \text{writing } n = 2m$$

$$\sim \sqrt{\frac{2}{\pi n}} \cdot 2^n. \tag{9}$$

Occasionally, one will see (9) derived instead from the infinite product representation for π of John Wallis (1656 - 1703):

$$\frac{4}{\pi} = \frac{3 \cdot 3 \cdot 5 \cdot 5 \cdot 7 \cdot 7 \cdots}{2 \cdot 4 \cdot 4 \cdot 6 \cdot 6 \cdot 8 \cdots}. \tag{10}$$

The derivation is not all that dissimilar. One starts with (7):

$$\begin{aligned}
\binom{2m}{m} &= \frac{2m(2m-1)\cdots(m+1)m(m-1)\cdots 1}{m(m-1)\cdots 1 \cdot m(m-1)\cdots 1} \\
&= \frac{(2m-1)(2m-3)\cdots 1}{m(m-1)\cdots 1} \cdot \frac{2m(2m-2)\cdots 2}{m(m-1)\cdots 1} \\
&= \frac{(2m-1)(2m-3)\cdots 1}{m(m-1)\cdots 1} \cdot 2^m. \tag{11}
\end{aligned}$$

Next one looks at the truncation of (10) to the first $2m - 1$ terms in the numerator and denominator,

$$\begin{aligned}
\frac{4}{\pi} \sim W_{2m-1} &= \frac{3 \cdot 3 \cdot 5 \cdot 5 \cdot 7 \cdot 7 \cdots (2m-1)(2m-1)(2m+1)}{2 \cdot 4 \cdot 4 \cdot 6 \cdot 6 \cdot 8 \cdots (2m)(2m)} \\
&= \frac{(3 \cdot 5 \cdot 7 \cdots (2m-1))^2(2m+1)}{2^{2m-1}(1 \cdot 2 \cdot 3 \cdots m)^2},
\end{aligned}$$

whence

$$\begin{aligned}
\frac{2}{\sqrt{\pi}} &\sim \frac{3 \cdot 5 \cdot 7 \cdots (2m-1)\sqrt{2m+1}}{2^m(1 \cdot 2 \cdot 3 \cdots m)/\sqrt{2}} \\
&\sim \frac{(2m-1)(2m-3)\cdots 1}{m(m-1)\cdots 1} \cdot \frac{\sqrt{4m+2}}{2^m} \\
&\sim \frac{\binom{2m}{m}}{2^m} \cdot \frac{\sqrt{4m+2}}{2^m}, \quad \text{by (11)}
\end{aligned}$$

whence

$$\begin{aligned}
\binom{2m}{m} &\sim \frac{2}{\sqrt{\pi}} \cdot 2^{2m} \cdot \frac{1}{\sqrt{4m+2}} \\
&\sim \frac{2}{\sqrt{\pi}} \cdot 2^{2m} \cdot \frac{1}{2\sqrt{m}}, \quad \text{since } \sqrt{4m+2} \sim \sqrt{4m} \\
&\sim \frac{2^n}{\sqrt{\pi m}}, \quad \text{for } n = 2m, \\
&\sim \sqrt{\frac{2}{\pi n}} \cdot 2^n
\end{aligned}$$

as before.

Now (9) means that the ratio of two terms tends to 1, i.e.,

$$\frac{\binom{2m}{m}}{\sqrt{\dfrac{2}{\pi n}} \cdot 2^n} = 1 + \varepsilon_n,$$

where $\varepsilon_n \to 0$ as $n \to \infty$. Thus

$$\binom{2m}{m} = \sqrt{\frac{2}{\pi n}} \cdot 2^n \cdot (1 + \varepsilon_n),$$

i.e.,

$$\frac{\binom{2m}{m}}{2^n} = \sqrt{\frac{2}{\pi n}} \cdot (1 + \varepsilon_n) = \sqrt{\frac{2}{\pi n}} + \sqrt{\frac{2}{\pi n}} \cdot \varepsilon_n$$

$$\approx \sqrt{\frac{2}{\pi n}}, \tag{12}$$

as both $\sqrt{2/(\pi n)}$ and ε_n are very small for large values of n. Thus we have, for $p = \frac{1}{2}$, the estimate

$$A(2m, \frac{1}{2}, m) \approx \sqrt{\frac{2}{\pi n}} = \frac{1}{\sqrt{\pi m}},$$

using the notation of Chapter 4, or

$$\text{binompdf(2M,1/2,M)} \approx 1/\sqrt{(\pi M)},$$

using the notation of the *TI-83 Plus*.

With (12) we have completed—modulo the derivation[15] of Stirling's formula (8) or Wallis's product (10)—the first step in establishing the normal approximation to the binomial distribution (perhaps more correctly: the binomial approximation to the normal distribution). The logically next step is to estimate (6). In the *Approximatio*, de Moivre refers to this, but doesn't give the details. We can fill in these details as follows. Let $n = 2m$ and consider (6) for positive l:

$$\binom{2m}{m} \Big/ \binom{2m}{m+l} = \frac{(2m)(2m-1)\cdots(m+1)}{m!} \cdot \frac{(m+l)(m+l-1)\cdots 1}{(2m)(2m-1)\cdots(m-l+1)}$$

$$= \frac{(m+l)(m+l-1)\cdots(m+1)}{m(m-1)\cdots(m-l+1)}$$

$$= \frac{m+l}{m} \cdot \frac{m+l-1}{m-1} \cdots \frac{m+1}{m-l+1}$$

$$= \left(1 + \frac{l}{m}\right)\left(1 + \frac{l}{m-1}\right) \cdots \left(1 + \frac{l}{m-l+1}\right).$$

---

[15] For which see section 5 of Appendix B.

Taking logarithms,

$$\ln\left(\binom{2m}{m}\Big/\binom{2m}{m+l}\right) = \sum_{k=1}^{l} \ln\left(1 + \frac{l}{m-l+k}\right). \tag{13}$$

Now, if l is much smaller than m, say $l = O(\sqrt{m})$, then for large m, $l/(m-l+k)$ is very close to 0, and for small x,

$$\ln(1+x) \approx x.$$

Making the obvious replacements in (13), we have

$$\ln\left(\binom{2m}{m}\Big/\binom{2m}{m+l}\right) \approx \sum_{k=1}^{l} \frac{l}{m-l+k}$$

$$\approx l \cdot \sum_{k=1}^{l} \frac{1}{m-l+k}. \tag{14}$$

Now

$$\frac{1}{m} < \frac{1}{m-l+k} < \frac{1}{m-l+1},$$

so

$$\frac{l}{m} < \sum_{k=1}^{l} \frac{1}{m-l+k} < \frac{l}{m-l+1}.$$

But, for $l = O(\sqrt{m})$ and m large,

$$\frac{l/m}{l/(m-l+1)} = \frac{m-l+1}{m} = 1 - \frac{l-1}{m} \approx 1,$$

and (14) becomes

$$\ln\left(\binom{2m}{m}\Big/\binom{2m}{m+l}\right) \approx l \cdot \frac{l}{m} = \frac{l^2}{m}. \tag{15}$$

By symmetry, the same estimate holds for $-l$:

$$\ln\left(\binom{2m}{m}\Big/\binom{2m}{m-l}\right) \approx \frac{l^2}{m}.$$

Exponentiating (15), we get

$$\binom{2m}{m}\Big/\binom{2m}{m\pm l} \approx e^{l^2/m},$$

whence

$$\binom{2m}{m\pm l} \approx \binom{2m}{m}e^{-l^2/m},$$

and

$$\binom{2m}{m \pm l}\left(\frac{1}{2}\right)^n \approx \sqrt{\frac{2}{\pi n}}\,e^{-2l^2/n}.$$

Setting $\mu = np = 2m \cdot \frac{1}{2} = m, \sigma^2 = npq = \frac{n}{4}$, this can be rewritten, for $x = \mu \pm l$, as

$$\binom{n}{x}\left(\frac{1}{2}\right)^n \approx \sqrt{\frac{1}{2\pi\sigma^2}}\,e^{-\frac{1}{2}((\mu-x)^2/\sigma^2)}$$

$$\approx \frac{1}{\sqrt{2\pi}\cdot\sigma}\,e^{-\frac{1}{2}((\mu-x)/\sigma)^2}, \tag{16}$$

in accordance with formula (15) of Chapter 4, for the case $p = \frac{1}{2}$ and n even.

De Moivre did two more things, which I will not go into in detail here: He approximated a sum of probabilities (16) for x lying in a symmetric interval around μ by the area under the normal curve, and he outlined how the calculation can be modified for probabilities $p \neq \frac{1}{2}$.

Jakob Bernoulli's proof of his theorem was elementary and rigorous, if not completely general. And it allowed for an easy calculation, given ε, of the value of n necessary to guarantee $|\mu/n - x/n| < \varepsilon$. Nikolaus Bernoulli's proof gave a better bound on n, but his proof was not complete—he left out the important proof that the sequence of ratios could indeed be approximated by the given geometric progression. And de Moivre, who gave the best result, left out a host of details and didn't even begin to calculate how close the approximation was or how large n had to be to achieve such closeness. But he did move probability into a new, properly analytic direction: from de Moivre forward, probability would be based on the Calculus (called Analysis in higher circles). Texts might still include problems arising from gambling, but they would have more theory and structure. And, the normal curve would assume greater and greater importance.

2 Thomas Bayes

The next important contribution to probability theory after de Moivre appeared in an almost unreadable paper of Thomas Bayes (1702 - 1761) published posthumously in 1763. The paper, "Essay towards solving a problem in the doctrine of chances", is the centrepiece of a trio of papers on probability that appeared in two successive volumes of the *Philosophical Transactions*. The first of these was a criticism of de Moivre.[16]

---

[16] Bayes published only one mathematical work in his lifetime, a monograph with a then fashionably long title, *An Introduction to the Doctrine of Fluxions, and a Defense of the Mathematicians Against the Objections of the Author of The Analyst* (London, 1736). According to Ian Hacking's article on Bayes in the *Dictionary of Scientific Biography*, it "was perhaps the soundest retort to Berkeley['s *The Analyst*, which had criticised the foundations of Newton's Calculus,] then available". His two other mathematical works are two of the three papers on probability referred to.

Todhunter[17] informs us that in the first supplement to the *Miscellanea analytica*, de Moivre had derived the representation[18]

$$\ln(m-1)! = \sum_{k=1}^{m-1} \ln k = \left(m - \frac{1}{2}\right) \ln m$$

$$- m + \frac{1}{12m} - \frac{1}{360m^3} + \frac{1}{1260m^5} - \frac{1}{1680m^7} + \cdots$$

$$+ 1 - \frac{1}{12} + \frac{1}{360} - \frac{1}{1260} + \frac{1}{1680} - \cdots$$

and that Stirling had shown

$$1 - \frac{1}{12} + \frac{1}{360} - \frac{1}{1260} + \frac{1}{1680} - \cdots = \frac{1}{2} \ln 2\pi. \tag{17}$$

Before Stirling arrived on the scene and determined the sum to be $\frac{1}{2} \ln 2\pi$, de Moivre assumed it to take the form $\ln B$ and set out to calculate B:

> When I first began that inquiry, I contented myself to determine at large the Value of B, which was done by the addition of some Terms of the above-written Series; but as I perceived that it converged but slowly, and seeing at the same time that what I had done answered my purpose tolerably well, I desisted from proceeding farther till my worthy and learned Friend Mr. *James Stirling*, who had applied himself after me to that inquiry, found that the Quantity B did denote the Square-root of the Circumference of a Circle whose Radius is Unity, so that if that Circumference be called c, the Ratio of the middle Term to the Sum of all the Terms will be expressed by $\frac{2}{\sqrt{nc}}$.[19]

Bayes[20] noted that the series in question in fact diverges. The first few terms do approximate $\frac{1}{2} \ln 2\pi$ very closely, but after that the series diverges from its "limit". While Stirling's formula (8) is correct, his formula (17) from which (8) is derived does not hold and modern probability or analysis texts have to give more careful derivations.

When Bayes died, his friend Richard Price (1723 - 1791) found an essay[21] among Bayes's effects, edited it, and sent it to John Canton (1718 - 1772) to whom Bayes had written his note on the divergence of de Moivre's series.

---

[17] *Op. cit.*, p. 191.

[18] *Cf.* Appendix B, section 5, below, for a derivation of this.

[19] *Doctrine of Chances*, 3rd edition, p. 244. The words "middle Term" and "Sum of all the Terms" refer to the binomial expansion $(1+1)^n$. The ratio in question he showed to be nearly equal to $2/(B\sqrt{n})$.

[20] Thomas Bayes, "A letter from the late Reverend Mr. Thomas Bayes, F.R.S. to John Canton, M.A. and F.R.S.", *Philosophical Transactions* 53 (1763), pp. 269 - 271.

[21] Thomas Bayes, "An Essay towards solving a Problem in the Doctrine of Chances By the late Rev. Mr. Bayes, F.R.S communicated by Mr. Price, in a Letter to John Canton, A.M. F.R.S.", *Philosophical Transactions* 53 (1763), pp. 370 - 418.

Canton communicated the letter and the essay to the Royal Society on 24
November and 23 December 1763, respectively, and the papers were published
in the same volume of the *Philosophical Transactions*.[22]

The paper begins with a letter of 10 November 1763 from Price to Canton
describing the contents of the essay:

> Dear Sir,
>
> I Now send you an essay which I have found among the papers of
> our deceased friend Mr. Bayes, and which, in my opinion, has great
> merit, and well deserves to be preserved. Experimental philosophy, you
> will find, is nearly interested in the subject of it; and on this account
> there seems to be particular reason for thinking that a communication
> of it to the Royal Society cannot be improper.
>
> He had, you know, the honour of being a member of that illustrious
> Society, and was much esteemed by many in it as a very able mathe-
> matician. In an introduction which he has writ to this Essay, he says,
> that his design at first in thinking on the subject of it was, to find out
> a method by which we might judge concerning the probability that an
> event has to happen, in given circumstances, upon supposition that we
> know nothing concerning it but that, under the same circumstances,
> it has happened a certain number of times, and failed a certain other
> number of times. He adds, that he soon perceived that it would not be
> very difficult to do this, provided some rule could be found according
> to which we ought to estimate the chance that the probability for the
> happening of an event perfectly unknown, should lie between any two
> named degrees of probability, antecedently to any experiments made
> about it; and that it appeared to him that the rule must be to suppose
> the chance the same that it should lie between any two equidifferent de-
> grees; which, if it were allowed, all the rest might be easily calculated in
> the common method of proceeding in the doctrine of chances. Accord-
> ingly, I find among his papers a very ingenious solution of this problem
> in this way. But he afterwards considered that the *postulate* on which
> he had argued might not perhaps be looked upon by all as reasonable;
> and therefore he chose to lay down in another form the proposition in
> which he thought the solution of the problem is contained, and in a
> *scholium* to subjoin the reasons why he thought so, rather than to take
> into his mathematical reasoning any thing that might admit dispute.
> This, you will observe, is the method which he has pursued in this es-
> say.
>
> Every judicious person will be sensible that the problem now men-
> tioned is by no means merely a curious speculation in the doctrine
> of chances, but necessary to be solved in order to a sure foundation
> for all our reasonings concerning past facts, and what is likely to be

---

[22] Price himself only became a Fellow of the Royal Society two years later in 1765,
thus the necessity of relying on Canton as an intermediary in publishing the essay
in the *Transactions*.

hereafter. Common sense is indeed sufficient to shew us that, from the observation of what has in former instances been the consequence of a certain cause or action, one may make a judgment what is likely to be the consequence of it another time, and that the larger number of experiments we have to support a conclusion, so much the more reason we have to take it for granted. But it is certain that we cannot determine, at least not to any nicety, in what degree repeated experiments confirm a conclusion, without the particular discussion of the before mentioned problem; which, therefore, is necessary to be considered by any one who would give a clear account of the strength of *analogical* or *inductive reasoning*; concerning, which at present, we seem to know little more than that it does sometimes in fact convince us, and at other times not; and that, as it is the means of [a]cquainting us with many truths, of which otherwise we must have been ignorant; so it is, in all probability, the source of many errors, which perhaps might in some measure be avoided, if the force that this sort of reasoning ought to have with us were more distinctly and clearly understood.

These observations prove that the problem enquired after in this essay is no less important than it is curious. It may be safely added, I fancy, that it is also a problem that has never before been solved. Mr. De Moivre, indeed, the great improver of this part of mathematics, has in his *Laws of chance*[23], after Bernoulli, and to a greater degree of exactness, given rules to find the probability there is, that if a very great number of trials be made concerning any event, the proportion of the number of times it will happen, to the number of times it will fail in those trials, should differ less than by small assigned limits from the proportion of the probability of its happening to the probability of its failing in one single trial. But I know of no person who has shewn how to deduce the solution of the converse problem to this; namely, "the number of times an unknown event has happened and failed being given, to find the chance that the probability of its happening should lie somewhere between any two named degrees of probability." What Mr. De Moivre has done therefore cannot be thought sufficient to make the consideration of this point unnecessary: especially, as the rules he has given are not pretended to be rigorously exact, except on supposition that the number of trials made are infinite[24]; from whence it is not obvious how large the number of trials must be in order to make them exact enough to be depended on in practice.

Mr. De Moivre calls the problem he has thus solved, the hardest that can be proposed on the subject of chance. His solution he has applied to a very important purpose, and thereby shewn that those are much

---

[23] Price adds the footnote: See Mr. De Moivre's *Doctrine of Chances*, p. 243, & c. He has omitted the demonstration of his rules, but these have been since supplied by Mr. Simpson at the conclusion of his treatise on *The Nature and Laws of Chance*.
[24] This is literally true. It was the habit in those days of speaking directly of infinite integers, and not with our modern circumlocution of limits and limiting processes.

mistaken who have insinuated that the Doctrine of Chances in mathematics is of trivial consequence, and cannot have a place in any serious enquiry. The purpose I mean is, to shew what reason we have for believing that there are in the constitution of things fixt laws according to which events happen, and that, therefore, the frame of the world must be the effect of the wisdom and power of an intelligent cause; and thus to confirm the argument taken from final causes for the existence of the Deity.[25] It will be easy to see that the converse problem solved in this essay is more directly applicable to this purpose; for it shews us, with distinctness and precision, in every case of any particular order or recurrency of events, what reason there is to think that such recurrency or order is derived from stable causes or regulations in nature, and not from any of the irregularities of chance.

The two last rules in this essay are given without the deductions of them. I have chosen to do this because these deductions, taking up a good deal of room, would swell the essay too much; and also because these rules, though of considerable use, do not answer the purpose for which they are given as perfectly as could be wished. They are however ready to be produced, if a communication of them should be thought proper. I have in some places writ short notes, and to the whole I have added an application of the rules in the essay to some particular cases, in order to convey a clearer idea of the nature of the problem, and to shew how far the solution of it has been carried.

. . .

Mr. Bayes has thought fit to begin his work with a brief demonstration of the general laws of chance. His reason for doing this, as he says in his introduction, was not merely that his reader might not have the trouble of searching elsewhere for the principles on which he has argued, but because he did not know whither to refer him for a clear demonstration of them. He has also made an apology for the peculiar definition he has given of the word *chance* or *probability*. His design herein was to cut off all dispute about the meaning of the word, which in common language is used in different senses by persons of different opinions, and according as it is applied to *past* or *future* facts. But whatever different senses it may have, all (he observes) will allow that an expectation depending on the truth of any *past* fact, or the happening of any *future* event, ought to be estimated so much the more valuable as the fact is more likely to be true, or the event more likely to happen. Instead therefore, of the proper sense of the word *probability*, he has given that which all will allow to be its proper measure in every case where the word is used.[26]

---

[25] Recall the discussion of intelligent design, pp 211 - 216.

[26] Price, *op. cit.*, pp. 370 - 375. I have made two types of changes in reproducing these remarks: I have not repeated opening quotation marks at the beginnings of successive lines in the quotation from de Moivre, and I have separated some run-on words that I took to be typographical errors

This letter stands in lieu of Bayes's own introduction to the paper, as the introduction mentioned is not to be found in the final published paper. Presumably Canton preferred Price's description. Bayes's paper is not easy to read. There are several reasons for this. I quote Todhunter:

> Bayes begins, as we have said, with a brief demonstration of the general laws of the Theory of Probability; this part of his essay is excessively obscure, and contrasts most unfavourably with the treatment of the same subject by De Moivre.[27]

On another point, Todhunter says

> Price himself gives a note which shews a clearer appreciation of the proposition than Bayes had.[28]

And, amusingly,

> We pass on now to the remarkable part of the essay. Imagine a rectangular billiard table $ABCD$. Let a ball be rolled on it at random, and when the ball comes to rest let its perpendicular distance from AB be measured; denote this by x. Let a denote the distance between AB and CD. Then the probability that the value of x lies between two assigned values b and c is $\frac{c-b}{a}$. This we should assume as obvious; Bayes, however, demonstrates it very elaborately.[29]

Writing in 2007, Anders Hald added

> Bayes's mathematics is correct, but his verbal comments are obscure and have caused much discussion, which recently has led to a new interpretation of his criterion for the application of his rule for inductive inference.[30]

So we see that one reason Bayes is difficult to understand is the obscurity of his writing, possibly a combination of forces—simple inability to express himself clearly, fuzziness of his own conception of the material, and, as the last line of Todhunter's third remark might suggest, a pedantic bent.

There is another reason a modern reader would find Bayes heavy going: although the Calculus had been around for some time and Bayes was familiar with it[31], Bayes chose to give a highly geometric presentation, arguing in Euclidean fashion about areas instead of simply performing calculations. And here, in referring to "Euclidean fashion", I mean that not only did he argue geometrically, but that he also adopted the Euclidean style of beginning with Definitions and deriving Propositions, Corollaries, and Lemmas—section II even starts with a couple of Postulates.

---

[27] Todhunter, *op. cit.*, p. 295.
[28] *Ibid.*, p. 296.
[29] *Ibid.*, p. 296.
[30] Hald, *A History...*, *op.cit.*, p. 26.
[31] As remarked in footnote 16, his first mathematical publication concerned the Calculus.

Following Price's introduction, Bayes begins as follows:

PROBLEM.

Given the number of times in which an unknown event has happened
and failed: *Required* the chance that the probability of its happening in a single trial lies somewhere between any two degrees
of probability that can be named.

SECTION 1.

DEFINITION 1. Several events are *inconsistent*, when if one of
them happens, none of the rest can.

2. Two events are *contrary* when one, or other of them must; and
both together cannot happen.

3. An event is said to *fail*, when it cannot happen; or, which comes
to the same thing, when its contrary has happened.

4. An event is said to be determined when it has either happened
or failed.

5. The *probability of any event* is the ratio between the value at
which an expectation depending on the happening of the event ought to
be computed, and the value of the thing expected upon it's happening.

6. By *chance* I mean the same as probability.

7. Events are independent when the happening of any one of them
does neither increase nor abate the probability of the rest.

PROP. 1.

When several events are inconsistent the probability of the happening
of one or other of them is the sum of the probabilities of each of them.[32]

Giving credit where it is due, I must say that the statement of the problem is wonderfully clear. Most of the definitions are reasonably clear and serve
the purpose to the modern reader of explaining his terminology. Today we
say "disjoint" or "pairwise mutually exclusive" instead of "inconsistent", and
"complementary" instead of "contrary"; so the language differs but the concepts are familiar. We have not yet discussed "independent events", which are
usually discussed along with conditional probability; we will get to it in the
next section. The definition of "probability" is correctly deemed peculiar in
Price's introduction. It does not at first sight seem to make much sense. Hald
explains:

De Moivre had defined the expectation E of a game or a contract as
the value V of the sum expected times the probability P of obtaining
it, so $P = E/V$. Bayes chooses the value of an expectation as his primitive concept and defines probability as E/V. This is a generalization of

---

[32] Bayes, "An Essay...", *op. cit.*, p. 376.

the classical concept because an expectation may be evaluated objectively or subjectively. He then shows how the usual rules of probability calculus can be derived from this concept.[33]

The unfamiliar thing here, making the definition of probability unintelligible to the modern reader, is the different usage of the word "expectation". Given an event, A, today we follow Huyghens in referring to the sum

$$E(A) = PV + (1 - P)V' \qquad (18)$$

as the expectation, where V' is the value expected should A not occur, and in the de Moivre-Bayes usage the latter term is dropped. We might clarify this by defining a sum like (18) to be the *expectation of the game* and PV to be the *expectation of the event* yielding a payoff V with probability P.

The first section of Bayes's paper is a quick development of the elements of probability theory up to and including the introduction of binomial probability and the proposition that the probability of exactly k successes in n trials with probability p of success in a given trial is (in modern notation)

$$\binom{n}{k} p^k (1 - p)^{n-k}.$$

There is no need for us to discuss this development in any detail.

Section II is the main part of the paper and it is what this section contains and doesn't contain that must be discussed. A central result of probability theory that goes under the name "Bayes's Theorem" (or, sometimes, "Bayes' Theorem") concerns the reversal of conditional probability and we used it, without labelling it as such, back in Chapter 3, section 8. It has a relatively simple formal statement:

1 Theorem (Bayes's Theorem). *Let* $A_0, A_1, \ldots, A_{n-1}$ *partition the sample space* S, *i.e.,* $S = A_0 \cup A_1 \cup \ldots \cup A_{n-1}$ *with each pair* A_i, A_j *disjoint for* $i \neq j$. *Let* $E \subseteq S$ *be an event. Let* A *be one of* $A_0, A_1, \ldots, A_{n-1}$. *Then*

$$P(A|E) = \frac{P(E|A) \cdot P(A)}{\sum_{i=0}^{n-1} P(E|A_i) \cdot P(A_i)}. \qquad (19)$$

Proof. As we saw before,

$$P(A|E) = \frac{P(A \cap E)}{P(E)}$$
$$= \frac{P(E \cap A)}{P(E \cap (A_0 \cup \ldots \cup A_{n-1}))}$$
$$= \frac{P(E|A) \cdot P(A)}{\sum_{i=0}^{n-1} P(E \cap A_i)}$$
$$= \frac{P(E|A) \cdot P(A)}{\sum_{i=0}^{n-1} P(E|A_i) \cdot P(A_i)}. \qquad \square$$

[33] Hald, *A History...*, *op. cit.*, p. 26.

The proof really is that easy, but the application of the result is often not so for beginning students who believe in memorising formulæ and plugging numbers into them. All the difficulty vanishes, however, if one draws the trees as we did in Chapter 3: Students who take the trouble of drawing the trees and filling in the probabilities, unlike those who rely on formulæ, have no difficulty in applying Bayes's Theorem.

One will not find Bayes's Theorem stated explicitly in his paper. One finds it implicitly in a continuous analogue in Section 2.

It is important to remember that Bayes's Theorem 1 was not the goal of his paper but something that falls out of it on careful reading. His goal was to carry the Bernoulli-de Moivre programme a step further. Jakob Bernoulli had shown that the probability of success in a single trial could be estimated by the relative frequency of successes in a large number of trials. How close the approximation came depended on p. As we saw in discussing Arbuthnot *vs.* Nikolaus Bernoulli, the result could be good enough for showing p was not a certain number, but it could not tell us how much confidence we could place in a given estimate of p. Bayes set out to determine the probability from a number of trials that p was within certain fixed limits. This meant a sample space of probabilities, no longer a discrete set, but now the continuous interval $[0, 1]$. Hence he dealt with a continuous analogue to Theorem 1.

Bayes considered a two-stage procedure. The first was a simple, but overly elaborate mechanism for choosing a probability p at random. The second stage consisted of conducting a number of Bernoulli trials based on p. The choice of p in the first stage was to be performed in accordance with the *principle of insufficient reason*, referred to by him as Postulate 1:

> Postulate. 1. I Suppose the square table or plane $ABCD$ to be so made and levelled, that if either of the balls o or W be thrown upon it, there shall be the same probability that it rests upon any one equal part of the plane as another, and that it must rest somewhere upon it.[34]

His design is that a ball W is thrown upon the table and when it is brought to rest, a line perpendicular to two opposite edges and passing through the resting point is drawn. This partitions the square into two regions. When another ball is tossed, it either comes to rest in one region (success) or in the other (failure). The probability p of success is the ratio of the area of the region of success to the area of the whole square. Note that this ratio is that of the length of the side of the region lying on the side of the square to which the perpendicular has been drawn to the length of the side of the square, and the 2-dimensionality of the figure is irrelevant. Indeed, all that is needed is the assumption that p has been chosen randomly from a uniform distribution of probabilities, the supposition, as Price put it (page 236, above), that "the chance [be] the same that it should lie between any two equidifferent degrees". Thus, for $0 \leq a < b \leq 1$, one assumes an *a priori* probability:

$$P(a < p < b) = b - a.$$

---

[34] Bayes, "An Essay...", *op. cit.*, p. 385.

As already remarked, the second stage of Bayes's experiment is to run n Bernoulli trials with the given probability p of success—which he describes as tossing a ball n additional times onto his square and counting the number, say S_n, of successes. For any $k = 0, 1, \ldots, n$, the conditional probability of exactly k successes given that p is the probability of success on a single trial is

$$P(S_n = k \mid p) = \binom{n}{k} p^k (1-p)^{n-k}. \tag{20}$$

If there were only finitely many equally likely probabilities to choose from, with $p_0, p_1, \ldots, p_{m-1}$ among them, we would calculate

$$P(S_n = k \mid p_0 \text{ or } p_1 \text{ or } \ldots \text{ or } p_{m-1}),$$

which turns out to be

$$\sum_{i=0}^{m-1} P(S_n = k \mid p_i) \cdot P(p_i \mid p_0 \text{ or } p_1 \text{ or } \ldots \text{ or } p_{m-1}) = \frac{1}{m} \sum_{i=0}^{m-1} P(S_n = k \mid p_i).$$

Given a continuous distribution of p's over an interval $[c, d]$, the analogous calculation is integration:

$$P(S_n = k \mid c < p < d) = \frac{1}{d-c} \int_c^d \binom{n}{k} p^k (1-p)^{n-k} dp, \text{ by (20)}$$

$$= \frac{\binom{n}{k}}{d-c} \int_c^d x^k (1-x)^{n-k} dx, \tag{21}$$

using the more customary variable x as the variable of integration.

Now suppose we have k successes in n trials and $0 \le a < b \le 1$ are given. The probability that p lies between a and b given that $S_n = k$ is

$$P(a < p < b \mid S_n = k) = \frac{P(a < p < b \,\&\, S_n = k)}{P(S_n = k)}$$

$$= \frac{P(S_n = k \mid a < p < b) \cdot P(a < p < b)}{P(S_n = k)}$$

$$= \frac{\frac{1}{b-a} \binom{n}{k} \int_a^b x^k (1-x)^{n-k} dx \cdot (b-a)}{\binom{n}{k} \int_0^1 x^k (1-x)^{n-k} dx}, \text{ by (21)}$$

$$= \frac{\int_a^b x^k (1-x)^{n-k} dx}{\int_0^1 x^k (1-x)^{n-k} dx}. \tag{22}$$

Bayes did not express his result in these terms, but in terms of the ratio of the areas under the curve $y = x^k (1-x)^{n-k}$ over the intervals $[a, b]$ and $[0, 1]$, all expressed in the pre-Cartesian, Euclidean fashion of describing the regions whose areas are to be calculated by labelling points on the paths forming the perimeters of the respective regions.

In theory, the areas in question are easy to calculate by anyone who has taken a first course in the Calculus, as the curve in question, $y = x^k(1-x)^{n-k}$, is a polynomial. In practice, for large n, the area calculation is a bit laborious. Bayes thus added some "Rules" for approximating the areas in question fairly quickly. The integral

$$\int_0^p x^k(1-x)^{n-k}dx$$

is called the *incomplete beta function* and is sufficiently important in statistics that Karl Pearson published tables[35] of its values to 7 decimals for n, k up to 50 and p increasing from .01 in increments of .01.

2 Example. For small enough n, the integral can be calculated directly by anyone who knows Calculus. For those not familiar with the Calculus, there is the calculator. Pick n, k and enter $y = x^k(1-x)^{n-k}$ in the equation editor:

$\quad$ Y$_1$=X^3(1−X)^4 .

One can then choose values for a, b, say .4 and .6, respectively, and enter

$\quad$ fnInt(Y$_1$(X),X,.4,.6)/fnInt(Y$_1$(X),X,0,1) .

The result is .420416: For 3 successes in 7 trials the odds are slightly less than even that the probability of success lies between .4 and .6. But for $n = 70, k = 30$, the probability that p lies between .4 and .6 is raised to .6955193473 or just about 7/10. For $n = 700, k = 300$, the numerator and denominator of the fraction are too small and one gets a division by 0 error message.

3 Remark. For those unfamiliar with the Calculus, I should explain[36] that fnInt(, which is found in the MATH menu accessed by the MATH button, performs function integration, i.e., it calculates the area beneath a curve given by an expression *expr* depending on a *variable*, and trapped between the lines $x = $ *lower_bound* and $x = $ *upper_bound*. The syntax is

$\quad$ fnInt(*expr, variable, lower_bound, upper_bound*[,*tolerance*]) ,

the optional variable specifying the desired level of accuracy in the computation. If the expression is, say, Y$_1$(X), one can also perform the integration by graphing Y$_1$(X) over the interval, pressing the CALC button, choosing the item $\int$f(x)dx from the menu, and entering the lower and upper bounds when prompted in the graphics screen. The region will be shaded on the graph and the value of the area will be displayed on the bottom of the graphics screen. To evaluate the ratio (22), one enters Y$_1$(X) in the equation editor, performs the above using a, b as lower and upper bounds, respectively, then exits the graphics screen to the home screen to store the area in a variable, e.g., N for "numerator":

---

[35] Karl Pearson, *Tables of the incomplete beta function*, Biometrica Office, London, 1934.

[36] A more detailed discussion appears in Appendix B.

Ans→ N .

One then repeats the area calculation using $0, 1$ as lower and upper bounds before exiting and calculating

N/Ans.

Once again, if n is very large the calculator will be unable to distinguish the numerator and denominator from 0, and the attempt to calculate the ratio will fail. [It is perhaps worth mentioning that this second method of evaluating the area actually uses fnInt(with the optional fifth variable specifying a modest tolerance level for the calculation. Its advantages are the interactive nature of the calculation, which frees one of the responsibility of listing the variables in the correct order, and the visual display in which the area being calculated is shaded in the graph.]

Following the discussion of three "Rules" for approximating the probability in question, Price added "an appendix containing an application of the foregoing rules to some particular cases" [37]. One of these applications is worth considering here.

4 Example. Suppose we have no knowledge about an event other than that it has occurred n times in n trials. What conclusion can we draw about the probability of the event's happening on the next trial? The calculations involved require a minimum knowledge of the Calculus, and, in the special case $n = 1$, not even that. The part that might not be easy is choosing the interval $[a, b]$. We shall find the probability of there being a more than even chance of the event occurring:

$$P\left(\frac{1}{2} < p \le 1 \,\middle|\, S_n = n\right) = \frac{\int_{1/2}^1 x^n (1-x)^{n-n} dx}{\int_0^1 x^n (1-x)^{n-n} dx} = \frac{\int_{1/2}^1 x^n dx}{\int_0^1 x^n dx}.$$

For $n = 1$ the areas in question are the areas of a trapezium and a triangle, respectively, as in *Figure 1*, below. The area of the trapezium is the difference

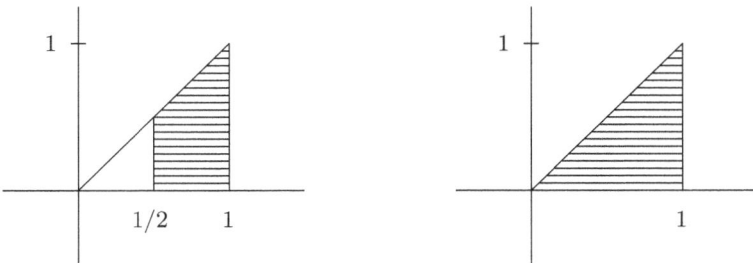

Fig. 1.

---

[37] Bayes, "An Essay...", *op. cit.*, p. 404. In citing the title of the appendix I have suppressed the typographical variation.

of the areas of the two triangles

$$\frac{1}{2} \cdot 1 \cdot 1 - \frac{1}{2} \cdot \frac{1}{2} \cdot \frac{1}{2} = \frac{1}{2} - \frac{1}{8} = \frac{3}{8}$$

and that of the triangle is $\frac{1}{2} \cdot 1 \cdot 1 = \frac{1}{2}$, whence the ratio is

$$\frac{3/8}{1/2} = \frac{3}{4}.$$

That is, "the answer is that there would be an odds of three to one for somewhat more than an even chance that it would happen on a second trial".[38]

For general n, one learns in the first semester of Calculus that

$$\int_c^d x^n\, dx = \frac{d^{n+1}}{n+1} - \frac{c^{n+1}}{n+1},\qquad\qquad (23)$$

whence

$$P\left(\frac{1}{2} < p \le 1 \,\middle|\, S_n = n\right) = \frac{\dfrac{1^{n+1}}{n+1} - \dfrac{\left(\frac{1}{2}\right)^{n+1}}{n+1}}{\dfrac{1^{n+1}}{n+1} - \dfrac{0^{n+1}}{n+1}} = 1 - \left(\frac{1}{2}\right)^{n+1}.$$

Thus, if the event has happened twice, the odds are 7 to 1 that the probability of its happening in a single trial is greater than $1/2$.

One of the more imaginative applications of Bayes's work discussed by Price concerns the question of whether or not the sun will rise tomorrow:

> One example here it will not be amiss to give.
> Let us imagine to ourselves the case of a person just brought forth into this, world and left to collect from his observation of the order and course of events what powers and causes take place in it. The Sun would, probably, be the first object that would engage his attention; but after losing it the first night he would be entirely ignorant whether he should ever see it again. He would therefore be in the condition of a person making a first experiment about an event entirely unknown to him. But let him see a second appearance or one *return* of the Sun, and an expectation would be raised in him of a second return, and he might know that there was an odds of 3 to 1 for *some* probability of this. This odds would increase, as before represented, with the number of returns to which he was witness. But no finite number of returns would be sufficient to produce absolute or physical certainty. For let it be supposed that he has seen it return at regular and stated intervals a million of times. The conclusions this would warrant would be such as follow—There would be the odds of the millioneth power of 2, to one, that it was likely that it would return again at the end of the usual

---
[38] *Ibid.*, p. 405.

interval. There would be the probability expressed by .5352, that the odds for this was not *greater* than 1.600,000 to 1; And the probability expressed by .5105, that it was not *less* than 1.400,000 to 1.[39]

The computations referred to are fairly straightforward applications of (23):

$$P\left(\frac{1}{2} < p \leq 1 \,\middle|\, S_{1000000} = 1000000\right) = 1 - \left(\frac{1}{2}\right)^{1000001}$$

$$P\left(\frac{1600000}{1600001} < p \leq 1 \,\middle|\, S_{1000000} = 1000000\right) = P(.999999375 < p \leq 1)$$

$$= 1 - .999999375^{1000001}.$$

Entering

1−.999999375^1000001

directly into the *TI-83 Plus* we get .4647390106 and a probability that p does not exceed 1600000/1600001 of $1 - .4647390106 = .5352609894 \approx .5353$. Similarly,

$$P\left(0 \leq p < \frac{1400000}{1400001} \,\middle|\, S_{1000000} = 1000000\right) = P(0 \leq p < .9999992857)$$

$$= .9999992857^{1000001}$$

$$\approx .4895414368,$$

yielding $1 - .4895414368 = .5104585632 \approx .5105$ as the probability that p is not less than 1400000/1400001.

Price follows this with some general remarks, most importantly

> What has been said seems sufficient to shew us what conclusions to draw from *uniform* experience. It demonstrates, particularly, that instead of proving that events will *always* happen agreeably to it, there will be always reason against this conclusion. In other words, where the course of nature has been the most constant, we can have only reason to reckon upon a recurrency of events proportioned to the degree of this constancy; but we can have no reason for thinking that there are no causes in nature which will *ever* interfere with the operations of the causes from which this constancy is derived, or no circumstances of the world in which it will fail.[40]

Indeed, it is not too hard to imagine cases where the inductive inference would be wrong: A stranger from another world, unfamiliar with human society, lands in a small village and observes people hard at work 6 days in a row. He considers it 98.4375% probable that it is more likely than not that the inhabitants of the village will be industrious the next day. However, this is a religious community

---

[39] *Ibid.*, pp. 409 - 410.
[40] *Ibid.*, pp. 410 - 411.

and the 7th day is a day of rest. Waldemar Daninsky's new neighbours enjoy 27 straight nights of quiet rest and, not knowing he is a werewolf, will be surprised to find themselves awakened in the middle of the night by howling next door as it happens to be a full moon that night. Etc.

5 Exercise. (For those who know the Calculus.) It is now the sabbath in the village and our visitor from another world has observed industrious activity 6 times and failed to observe it once. How much confidence should he have in his feeling that it is more likely than not that the villagers will be hard at work on the following day?

I could give the formula analogous to (23) that applies in this last Exercise for the benefit of those readers who know no Calculus, but I'd rather not, so as to emphasise the fact that as the difference $n - k$ grows larger (22) becomes progressively more difficult to calculate. Price preceded his applications with rules Bayes had devised for the approximate calculation of these probabilities, but he omitted Bayes's derivation of the second rule. A year later, in the next volume of the *Transactions*, he published "A Demonstration of the Second Rule in the Essay towards the Solution of a Problem in the Doctrine of Chances, published in the Philosophical Transactions, Vol. LIII. Communicated by the Rev. Mr. Richard Price, in a Letter to Mr. John Canton, M.A. F.R.S." in which he presented Bayes's deduction along with some improvements. I forego the pleasure of discussing such technicalities here as we shall not be making these calculations. What we should discuss instead are some generalities.

Probability is one of the conceptually most difficult to understand areas of mathematics. Some areas, like Number Theory, are difficult in the sense that the results are difficult to prove—the proofs are deep and rely on the mysterious introduction of seemingly irrelevant concepts. But even the most difficult to prove results are easy to understand: while one cannot begin to explain the proof of the Prime Number Theorem to the proverbial man in the street, the statement of the result can be communicated fairly easily. Even more easily conveyed is the, as yet unsolved, problem of whether or not there are infinitely many twin primes, i.e., pairs of prime numbers of the form $p, p + 2$ like $3, 5$ or $5, 7$ or $11, 13$, or...

But with Probability Theory, alongside the combinatorial difficulty of counting the numbers of elements of various sets and the computational difficulty of calculating well-defined expressions (like binomial probabilities and the new Bayesian probabilities), there is the added difficulty of explaining what a result means.

Bernoulli's Theorem is already somewhat abstract. It does not say that if n is large enough and n Bernoulli trials are made, the fraction of successful trials will be close to the probability p of success on a single trial; the Theorem says only that the fraction will "probably" be close to p, and more probably so as n grows larger. Yet this has an objective meaning. We can think of the sample space of all possible sequences of n Bernoulli trials. The collection of those sequences for which the proportion of successful cases is close to p will include a large percentage of those sequences, a percentage that grows nearer to 100% as n grows larger.

De Moivre's Theorem also has clear meaning. Over and above its sharing the same meaning as Bernoulli's, it offers a computational shortcut to estimating binomial probabilities as we discussed in Chapter 4. This is concrete and instantly intelligible.

But now we have Bayes's probability calculation. What does it mean to say that, for example, having observed 2 successes and 5 failures in 7 trials, the probability that the probability p of success in a single trial lies between .2 and .4 is approximately .48? That with probability approximately .65 the value of p lies between .1 and .4? We can probably safely infer that we should not bet even money that the next trial will result in a success, but does it tell us what odds to demand?

Bayes's probabilities that probabilities lie within certain bounds are more abstract and harder to get a feel for than the other probabilities we've thus far encountered. One gets some feel for their significance by introducing the notion of subjectivity and "degrees of belief", "confidence levels", and the like. Thus, in our Bayesian example, if I have witnessed 2 successes and 5 failures in 7 trials, I am 65% confident that the probability of success on a given trial lies between .1 and .4.

In the early days of probability the difficulty was computational rather than interpretational. One thought of the cases and calculated the ratio of favourable to unfavourable ones (odds) or the ratio of favourable to all cases (probability) or one thought in terms of frequency of occurrence. Galileo cleared up an apparent paradox when the two notions seemed to disagree, but other than that there was no problem in understanding what probability was. Once the theory developed, it became natural to apply it in situations where there were no cases to enumerate and no frequencies to refer to. The probability of God's existence in Pascal's Wager and degrees of credibility in law are examples. Jakob Bernoulli had envisioned social applications of probability in his *Ars conjectandi*, but his work being incomplete, this was a desideratum, not a *fait accompli*. Nikolaus Bernoulli had written his dissertation on law and the application of his uncle's work on probability to it. I haven't discussed law because the application of probability to legal issues such as weighing evidence and testimony have not had a substantial impact on the development of the theory. Problems from gambling had—the problem of points gave birth to the mathematical theory, problems like Gambler's Ruin led to the introduction of finite difference methods and more advanced techniques we haven't discussed. And the physical sciences, astronomy in particular, were soon to influence the field through the problem of reconciling errors of observation. But law would lead to no exciting new problems requiring novel solutions and developments. What law could do, in the absence yet of applications to science, was to lend an air of respectability to a field mired in gambling. With the birth of statistics, however, this rôle seems to have been taken over by the application of probabilistic techniques to annuities and insurance. Possibly the only thing law had to offer was an area where probability was not objectively measurable either *a priori* through the enumeration of cases, or *a posteriori* through the reporting of frequencies, but was entirely *subjective*.

At this point, let me pause to quote John Maynard Keynes, whose prose is an absolute delight:

> 3. But in the meantime the subject had fallen into the hands of the mathematicians, and an entirely new method of approach was in course of development. It had become obvious that many of the judgments of probability which we in fact make do not depend upon past experience in a way which satisfies the canons laid down by the Port Royal Logicians[41] or by Locke[42]. In particular, alternatives are judged equally probable, without there being necessarily any actual experience of their approximately equal frequency of occurrence in the past. And, apart from this, it is evident that judgments based on a somewhat indefinite experience of the past do not easily lend themselves to precise numerical appraisement. Accordingly James Bernoulli, the real founder of the classical school of mathematical probability, while not repudiating the old test of experience, had based many of his conclusions on a quite different criterion—the rule which I have named the Principle of Indifference[43]. The traditional method of the mathematical school essentially depends upon reducing all the possible conclusions to a number of 'equi-probable cases.' And, according to the Principle of Indifference, 'cases' are held to be equi-probable when there is no reason for preferring any one to any other, when there is nothing, as with Buridan's ass[44], to determine the mind in any one of the several possible directions. To take Czuber's example of dice, this principle permits us to assume that each face is equally likely to fall, if there is no reason to suppose any particular irregularity, and it does not require that we should *know* that the construction is regular, or that each face has, as a matter of fact, fallen equally often in the past.
>
> On this Principle, extended by Bernoulli beyond those problems of gaming in which by its tacit assumption Pascal and Huyghens had worked out a few simple exercises, the whole fabric of mathematical probability was soon allowed to rest. The older criterion of experience, never repudiated, was soon subsumed under the new doctrine. First,

---

[41] Recall the *Port-Royal Logic*, which Bernoulli favourably cited for its description of the empirical method.

[42] John Locke (1632 - 1704) was an English empiricist philosopher best known for his *An Essay Concerning Human Understanding* (1690), a work strongly influenced by the *Port-Royal Logic*.

[43] Meant here is the principle of insufficient reason, which Keynes finds implicit in the *Ars conjectandi* in the passage cited in Chapter 5, section 1, above (page 154).

[44] Jean Buridan (*c.* 1300 - 1358) was an Aristotelian scholar most famous for his example of a donkey which, equidistant between two stacks of hay, died of starvation unable to decide which stack to feed from.

in virtue of Bernoulli's famous Law of Great Numbers[45], the fractions representing the probabilities of events were thought to represent also the actual proportion of their occurrences, so that experience, if it were considerable, could be translated into the cyphers of arithmetic. And next, by the aid of the Principle of Indifference, Laplace established his Law of Succession[46] by which the influence of any experience, *however limited*, could be numerically measured, and which purported to prove that, if B has been seen to accompany A twice, it is two to one that B will again accompany A on A's next appearance. No other formula in the alchemy of logic has exerted more astonishing powers. For it has established the existence of God from the premiss of total ignorance[47]; and it has measured with numerical precision the probability that the sun will rise to-morrow.[48]

These remarks are from the chapter "Historical Retrospect" of Keynes's treatise and follow two chapters later than his more careful criticism of the principle of insufficient reason. Hence he allowed himself the lighter, somewhat sarcastic, tone here. To Keynes, to whom probability was a branch of logic, such an assumption as the principle of insufficient reason was simply not allowed. There is, however, a perspective under which the principle is justified. If by "probability" we mean to measure the degree of confidence we have in asserting some proposition or our degree of belief therein, then having no prior knowledge among a number of symmetrical possibilities, our degrees of belief in any of them ought not to differ from those of the others. The initial uniform distribution of *a priori* probabilities is a *subjective* probability distribution representing our lack of knowledge, not an objective measure of real-world probability (like a record of frequencies would give). And, the *a posteriori* probabilities that p lies within given intervals is thus likewise subjective, a measure of our subsequent degree of confidence in the new estimate.

It is the rule, not the exception, that a careful study of the foundations of a subject follows at a much later stage than the development of the subject. The Calculus, born in the 17th century and developed throughout the 18th, was given a firm foundation in the 19th. The real numbers themselves were first firmly grounded as a result of work done between $c.$ 1830 and 1872. The foundations of probability were first subject to serious study in the early decades of the 20th century, at first via the long familiar notion of probability as the ratio of the numbers of cases, and via the intuitive notion of probability as limits of frequencies. In 1921, in the work from which the above quotation is

---

[45] Keynes, like any good philosopher, writes in his own private language. The common designation is "Law of Large Numbers", a designation introduced by Poisson for Bernoulli's Theorem and various generalisations. As one may gather from our use of "Law of Large Numbers" as the title to this, second part of the present book, the term is widespread. We will have more to say about it later.

[46] *Cf.* section 3, below.

[47] I.e., Pascal's Wager of Chapter 2, section 3, above.

[48] John Maynard Keynes, *A Treatise on Probability*, Macmillan and Co., London, 1921; reprinted by Dover Publications, Inc., Mineola, NY, 2004; pp. 81 - 82.

taken, Keynes attempted to explain probability as a branch of logic in which *propositions* were assigned probabilities conditional on various evidences. True subjective probability was ignored until the 1930's when the Italian probability theorist Bruno de Finetti (1906 - 1985) developed a probability theory on subjective foundations. Perhaps it is time to quote an expert:

> Bernoulli undoubtedly imported the word 'subjective' into probability theory but by now the word in this connection has become equivocal. Several distinct modern theories of probability have been called subjective, and since there has been so much idle controversy about them it is worth getting the terminology straight. For my part I prefer to avoid the word 'subjective' altogether, so as to avoid potential ambiguity, but since Bernoulli started the trend we cannot evade it here. In recent writing three different kinds of probability have been called subjective. The most extreme subjectivism is that of Bruno de Finetti. L.J. Savage has felicitously called it 'personalism'. Then there are the theories of logical or inductive probability developed by J.M. Keynes and others, and which, though insisting on a measure of objectivity, are often called 'subjective' by their detractors. Thirdly there is yet another concept of subjectivity current among many present philosophers of quantum physics.[49]

The weak point behind Bayes's derivation of (22), as Bayes himself apparently realised[50], is his assumption of a uniform *a priori* distribution of probabilities in line with the principle of insufficient reason. Although this assumption can be justified on a subjective reading of probability, *philosophes* of the day would have wanted something more objective:

> Price attempts to remedy this defect in his commentary. As examples he discusses the drawings from a lottery and the probability of a sunrise tomorrow. Recognizing that Bayes's criterion of ignorance cannot be applied to himself regarding sunrises he invents "a person just brought forward into this world, knowing nothing at all about this phenomena." [51] He concludes that "It should be carefully remembered that these deductions suppose a previous total ignorance of nature." This implies that in the natural sciences "unknown events" are the exception

---

[49] Ian Hacking, *The Emergence of Probability; A Philosophical Study of Early Ideas About Probability, Induction and Statistical Thinking*, Cambridge University Press, Cambridge, 1975, pp. 146 - 147. The final chapter of Jan von Plato, *Creating Modern Probability: Its Mathematics, Physics and Philosophy in Historical Perspective*, Cambridge University Press, Cambridge, 1994, offers a more detailed discussion of de Finetti's work, if not any terse quotations I can use here.

[50] Price: "But he afterwards considered that the *postulate* on which he had argued might not be looked upon by all as reasonable..." (cited in context on page 236, above).

[51] The ending quotation marks should follow "world", not "phenomena". The full context of "the person just brought forth into this, world" (Bayes, "An Essay...", p. 409) is cited on page 246, above.

rather than the rule. Usually we know something about the probability of a phenomenon under investigation and Bayes's rule is therefore seldom applicable. On this background it is no wonder that the Essay did not evoke any response from British mathematicians and natural scientists.[52]

Indeed, it wasn't until Laplace independently rediscovered Bayes's Theorem and the Bayesian analysis, publishing them in a memoir of 1774, that these ideas were widely adopted:

> The influence of this memoir was immense. It was from here that "Bayesian" ideas first spread throughout the mathematical world, as Bayes's own article was ignored until after 1780 and played no important role in scientific debate until the twentieth century. It was also this article of Laplace's that introduced the mathematical techniques for the asymptotic analysis of posterior distributions that are still employed today... After more than two centuries, we mathematical statisticians cannot only recognize our roots in this masterpiece of our science, we can still learn from it.[53]

All of this brings us to the most important figure in the history of probability.

3 Laplace and the Central Limit Theorem

Todhunter begins his chapter on Laplace with the words

> 860. LAPLACE was born in 1749, and died in 1827. He wrote elaborate memoirs on our subject, which he afterwards embodied in his great work the *Théorie analytique des Probabilités*, and on the whole the Theory of Probability is more indebted to him than to any other mathematician.[54]

This opinion of Laplace's importance is borne out by a simple page count. Ignoring front and back matter, 150 of the 618 pages of Todhunter's history of probability, roughly 24% of the book, is taken up by the chapter on Laplace. Karl Pearson's later history of statistics devotes 99 out of 734 pages, or roughly $13\frac{1}{2}\%$, to him.

---

[52] Hald, *A History...*, *op. cit.*, p. 28. I have taken two liberties in citing Hald: I have omitted one insertion he made in his second quote from Bayes/Price, and I have omitted the bibliographic references.

[53] The memoir of Laplace referred to is his "Mémoire sur la Probabilité des causes par les évènemens", *Mémoires de l'Académie royale des sciences presentés par divers savans* 6 (1774), pp. 621 - 656. An English translation exists: S.M. Stigler, "Laplace's 1774 memoir on inverse probability", *Statistical Science* 1 (1986), pp. 359 - 378. The quotation, cited by Hald, *A History...*, *ibid.*, pp. 34 - 35, is taken from Stigler's introduction to his translation. As in the preceding quotation, I have omitted a couple of bibliographic references.

[54] Todhunter, *op. cit.*, p. 464.

From 1774 on, Laplace wrote voluminously on probability, covering just about every aspect of the subject. He collected the material in 1812 in his massive *Théorie analytique des probabilités*, which saw two further editions (1814 and 1820) during his lifetime, the second with a long introduction and the third with supplements. The importance of Laplace and this book is testified to by everyone writing on the subject, for example Hans Fischer:

> To the end of the 19th century, Laplace's *Théorie analytique* remained the most influential book of mathematical probability theory, which was considered less a part of mathematics in the narrower sense, but a discipline of "mathesis mixta." Reduced as it was by a major field of application of the classical theory, the moral sciences, and augmented only by problems which could be mastered within the framework of simple stochastic techniques, such as the kinetic theory of gases, hardly any probabilistic concepts were put forward which were new with regard to Laplace's.
>
> In the field of statistics, Laplace had mainly presented theoretical concepts in a rather unsystematic way in his *Théorie analytique*. His analytical deductions were written in a very difficult style, and his mode of reasoning within error theory became far less popular in comparison with Gauss', which was easier to understand and to apply. The general relevance for statistics of Laplacian error theory was appreciated only by the end of the 19th century. However, it influenced the further development of a largely analytically oriented probability theory; limit distributions of sums of independent random variables became a basis of modern probability theory.[55]

The cited importance and difficulty of the *Théorie analytique* had earlier been noted[56]:

> In a contemporary review of Laplace's *Théorie*, De Morgan described it as
>> the Mont Blanc of mathematical analysis; but the mountain has this advantage over the book, that there are guides always ready near the former, whereas the student has been left to his own method of encountering the latter.

It was in the same year that his review appeared that de Morgan's English exposition of Laplace appeared. Here he explains

> No one was more sure of giving the result of an analytical process correctly, and no one took so little care to point out the various small corrections on which correctness depends. His *Théorie des Probabilités* is by very much the most difficult mathematical work we have ever met with, and principally from this circumstance; the *Mécanique Céleste*

---

[55] Hans Fischer, "Pierre-Simon Marquis de Laplace", in: C.C. Heyde and E. Seneta, eds., *Statisticians of the Centuries*, Springer-Verlag New York, Inc., New York, 2001, p. 99. The explanation of the last line will be given shortly.

[56] Adrian Rice, "Augustus De Morgan", in: Heyde and Seneta, *op. cit.*, p. 160.

has its full share of the same sort of difficulty, but the analysis is less intricate.[57]

Laplace's *Théorie analytique* is formidable on various accounts. The careless formulations of results referred to by de Morgan constitute one reason. Another is expositional: it is reported[58] that it is not a coherent piece so much as a compilation of his earlier works. And for the historian, of course, there is the annoying habit of the age of not citing one's sources, making it difficult to sort out exactly what Laplace's contributions were. However, it is clear even to the novice from the esteem in which he is held that these contributions were great. I cannot hope to do him justice in a work such as this and shall limit my comments to brief remarks on his extensions of the results of Bernoulli, de Moivre and Bayes.

Law of Succession and Bayes's Theorem

Laplace was initially unaware of Bayes's work and independently rediscovered the results, stating Bayes's Theorem explicitly, deriving (22), and deriving his own Law of Succession mentioned by Keynes in our quotation on page 251, above. In describing Laplace's fundamental memoir of 1774, the "Mémoire sur la Probabilité. . ." [59], Todhunter writes

> 870. Laplace first takes the standard problem in this part of our subject: Suppose that an urn contains an infinite number of white tickets and black tickets in an unknown ratio; $p + q$ tickets are drawn of which p are white and q are black: required the probability of drawing m white tickets and n black tickets in the next $m + n$ drawings.
> Laplace gives for the required probability
> $$\frac{\int_0^1 x^{p+m}(1-x)^{q+n}dx}{\int_0^1 x^p(1-x)^q dx},$$
> so that of course the m white tickets and n black tickets are supposed to be drawn in an assigned order.[60]

Todhunter gives no hint as to the proof, but he does tell us that Laplace proceeds to derive (22) and to consider next the problem of points. There is more in this memoir but we have enough already to consider.

We begin by considering Laplace's *Law of Succession*, cited in Todhunter's paragraph 870, for the case $m = 1, n = q = 0$ and p now called n:

6 Theorem (Law of Succession). *Suppose an event has occurred n times in n trials. The probability of its occurrence in the $(n+1)$-th trial is approximately $(n + 1)/(n + 2)$.*

---

[57] Quoted by Pearson, *History of Statistics, op. cit.*, p. 714.
[58] I have not read the work, which is in French and runs several hundreds of pages.
[59] Cited in the footnote 53 at the end of the preceding section.
[60] Todhunter, *op. cit.*, pp. 466 - 467.

Armed with knowledge of Bayes's work, we might try to prove this by appeal to (23). The probability that the actual probability of success on the $(n+1)$-th trial after n straight successes lies in the interval

$$\left[\frac{n+1}{n+2} - \frac{1}{n+2}, \frac{n+1}{n+2} + \frac{1}{n+2}\right].$$

i.e., in the interval $[n/(n+2), 1]$ is

$$\frac{\int_{n/(n+2)}^{1} x^n\, dx}{\int_{0}^{1} x^n\, dx} = \frac{\dfrac{1^{n+1}}{n+1} - \dfrac{(n/(n+2))^{n+1}}{n+1}}{\dfrac{1^{n+1}}{n+1} - \dfrac{0^{n+1}}{n+1}}, \text{ by (23)}$$

$$= 1 - \left(\frac{n}{n+2}\right)^{n+1} = 1 - \left(1 - \frac{2}{n+2}\right)^{n+1}$$

$$= 1 - \left(1 - \frac{2}{n+2}\right)^{n+2} \Big/ \left(1 - \frac{2}{n+2}\right)$$

$$\sim 1 - e^{-2}/1 = .8646647168.$$

Thus it is highly probable that the probability lies between $n/(n+2)$ and 1, and is thus approximately $(n+1)/(n+2)$.

Of course, since Laplace established the Law of Succession before establishing (22), this would not be how he proved the result. In his classic probability textbook of the twentieth century, William Feller proves the result as follows:

The following example is famous and illustrative but somewhat artificial. Imagine a population of $N+1$ urns, each containing N red and white balls; the urn number k contains k red and $N-k$ white balls ($n = 0, 1, 2, \ldots, N$). An urn is chosen at random and n random drawings are made from it, the ball drawn being replaced each time. Suppose that all n balls turn out to be red (event A). We seek the (conditional) probability that the next drawing will also yield a red ball (event B). If the first choice falls on urn number k, then the probability of extracting in succession n red balls is $(k/N)^n$. Hence

$$\mathbf{P}\{A\} = \frac{1^n + 2^n + \ldots + N^n}{N^n(N+1)}. \tag{24}$$

The event AB means that $n+1$ drawings yield red balls, and therefore

$$\mathbf{P}\{AB\} = \mathbf{P}\{B\} = \frac{1^{n+1} + 2^{n+1} + \ldots + N^{n+1}}{N^{n+1}(N+1)}. \tag{25}$$

The required probability is $\mathbf{P}\{B|A\} = \mathbf{P}\{B\}/\mathbf{P}\{A\}$.
The sums (24) and (25) can be considered Riemann sums approximating integrals[61], so that when N is large

---

[61] *Cf.* Appendix B.

$$N^{-1} \sum_{k=1}^{N} \left(\frac{k}{N} \right)^n \sim \int_0^1 x^n \, dx = \frac{1}{n+1}. \tag{26}$$

We have therefore for large N approximately

$$\mathbf{P}\{B|A\} \approx \frac{n+1}{n+2}. \tag{27}$$

...This is the so-called law of succession of Laplace (1812).[62]

The positing of $N+1$ urns, like Bayes's first toss of a ball, is simply a means of choosing a probability of success in a given trial with each probability being equally likely. Bayes had opted for a continuous distribution of probabilities p between 0 and 1, while Feller chooses a discrete listing, $0 = 0/N, 1/N, 2/N, \ldots, N/N = 1$.

The calculation of (24) is a straightforward application of the Sequential and Addition Principles. One chooses the urn giving probability k/N of success with probability $1/(N+1)$ and then one chooses n successive red balls with probability $(k/N)^n$, thus having a combined probability

$$\frac{1}{N+1} \left(\frac{k}{N} \right)^n$$

of choosing the urn with k red balls and then choosing n successive red balls. Adding these up gives

$$P(A) = \frac{1}{N+1} \left(\frac{0}{N} \right)^n + \frac{1}{N+1} \left(\frac{1}{N} \right)^n + \ldots + \frac{1}{N+1} \left(\frac{N}{N} \right)^n$$

$$= \frac{1}{N+1} \cdot \frac{0^n + 1^n + \ldots + N^n}{N^n} = \frac{1^n + 2^n + \ldots + N^n}{(N+1)N^n},$$

as asserted in (24). Formula (25) is established in the same way.

There are now a couple of ways we can reach the final result. In the second part of his *Ars conjectandi*, Jakob Bernoulli had established a number of combinatorial results, among which were formulæ for summing the n-th powers of the first N integers:

$$1^n + 2^n + \ldots + N^n = \frac{N^{n+1}}{n+1} + \text{terms of lower degree in } N.$$

The reader will already be familiar with

$$1 + 2 + \ldots + N = \frac{N(N+1)}{2} = \frac{N^2}{2} + \frac{N}{2}$$

$$1^2 + 2^2 + \ldots + N^2 = \frac{(2N+1)N(N+1)}{6} = \frac{N^3}{3} + \frac{N^2}{2} + \frac{N}{6},$$

---

[62] William Feller, *An Introduction to Probability Theory and Its Applications*, vol. 1, 2nd edition, John Wiley & Sons, Inc., New York, 1957, p. 113. I have taken the liberty of renumbering his formulæ to conform with the present numbering.

and possibly with

$$1^3 + 2^3 + \ldots + N^3 = \frac{N^2(N+1)^2}{4} = \frac{N^4}{4} + \frac{N^3}{2} + \frac{N^2}{4}.$$

For n fixed and N large,

$$\frac{1^n + 2^n + \ldots + N^n}{N^{n+1}} = \frac{1}{n+1} + \text{terms in } \frac{1}{N}$$

$$\sim \frac{1}{n+1}.$$

Thus, taking the ratio of (24) and (25),

$$P(B|A) = P(B)/P(A) = \frac{1^{n+1} + 2^{n+1} + \ldots + N^{n+1}}{N^{n+1}(N+1)} \Big/ \frac{1^n + 2^n + \ldots + N^n}{N^n(N+1)}$$

$$= \frac{1^{n+1} + 2^{n+1} + \ldots + N^{n+1}}{N^{n+1}} \Big/ \frac{1^n + 2^n + \ldots + N^n}{N^n}$$

$$\sim \frac{1}{n+2} \Big/ \frac{1}{n+1} = \frac{n+1}{n+2},$$

as promised by (27).

Feller's proof starts with (24) and rewrites it as

$$\frac{1}{N+1} \sum_{k=1}^{N} \left(\frac{k}{N}\right)^n \sim \frac{1}{N} \sum_{k=1}^{N} \left(\frac{k}{N}\right)^n$$

and notes that this sum is approximately the area under the curve $y = x^n$ between $x = 0$ and $x = 1$, i.e., it is approximately

$$\int_0^1 x^n dx = \frac{1^{n+1}}{n+1} - \frac{0^{n+1}}{n+1}, \text{ by (23)}$$

$$= \frac{1}{n+1},$$

thus establishing (26). It follows that (24) is approximately $1/(n+1)$ and (25) approximately $1/(n+2)$, whence follows (27).

7 Exercise. For k fixed, consider the polynomial

$$P_k(X) = X^k + (X-1)^k + \ldots + (X-k+1)^k.$$

i. Show: $P_k(X) = (k+1)X^k +$ terms of lower degree.
ii. Show: As $x \to \infty$, $P_k(x) \sim (k+1)x^k$, whence

$$\frac{P_k(x)}{x^k} \sim k+1.$$

iii. Use ii for $k = n, n+1$ to derive (27).

Feller follows his proof of Laplace's Law of Succession with an illustration and a pertinent remark:

> Before the ascendence of the modern theory, the notion of equal probabilities was often used as synonymous for "no advance knowledge." Laplace himself has illustrated the use of (27) by computing the probability that the sun will rise tomorrow, given that it has risen daily for 5000 years or $n = 1,826,213$ days[63]. It is said that Laplace was ready to bet 1,826,214 to 1 in favor of regular habits of the sun, and we should be in a position to better the odds since regular service has followed for another century. A historical study would be necessary to render justice to Laplace and to understand his intentions. His successors, however, used similar arguments in routine work and recommended methods of this kind to physicists and engineers in cases where the formulas have no operational meaning. We should have to reject the method even if, for sake of argument, we were to concede that our universe was chosen at random from a collection in which all conceivable possibilities were equally likely. In fact, it pretends to judge the chances of the sun's rising tomorrow from the *assumed* risings in the past. But the assumed rising of the sun on February 5, 3123 B.C., is by no means more certain than that the sun will rise tomorrow. We believe in both for the same reasons.[64]

Like a number of Hungarian mathematicians of the 20th century, Georg Pólya moved to Germany for a period before settling in the United States. He criticised Laplace's Law of Succession thus:

> A boy is 10 years old today. The Rule says that, having lived 10 years, he has probability 11/12 to live one more year. The boy's grandfather attained 70. The Rule says that he has probability 71/72 to live one more year.
>
> These applications look silly, but none is sillier than the following due to Laplace himself. "Assume," he says, "that history goes back 5,000 years, that is 1,826,213 days. The sun rose each day, and so you can bet 1,826,214 against 1 that the sun will rise tomorrow." I would certainly be careful not to offer such a bet to a Norwegian colleague who could arrange air transportation for both of us to some place within the Arctic Circle.
>
> Yet the rule can even beat this absurdity. Let us apply it to the case $m = 0$: the Rule's derivation is as valid for this case as for any other case. Yet for $m = 0$ the Rule asserts that any conjecture without any verification has the probability 1/2. Anybody can invent examples to show that such an assertion is monstrous.[65]

---

[63] *Exercise.* 5000 · 365 would yield 1,825,000 days, not 1,826,213. Explain *fully* the discrepancy.

[64] Feller, *op. cit.*, pp. 113 - 114.

[65] George Polya, *Patterns of Plausible Inference*, Princeton University Press, Princeton, 1954, p. 136. This is the second volume of a pair of books collectively titled

Pólya cheats a bit here. He does not present a proof of Laplace's Law and so does not reveal the assumptions upon which its application rests. The assumption is that we have no idea of the value of p, the probability of a person's living another year, and thus we have a uniform distribution of the probabilities, whereas we actually know that the distribution of p's is not uniform. A 10 year old is very likely to live to be 11—with probability approximately .99 according to Halley's table reproduced in Appendix D—while a 70 year old is less likely, not more likely, to live another year—with probability only .92.[66] The Law of Succession does not apply because we know that it is much more likely than not that someone living today will still be alive next year. Should Price's "person just brought into this, world" suddenly appear, having no prior knowledge of human longevity, his *a priori* subjective probability would be uniform and *he* could apply the Law of Succession to determine the subjective degree of credibility to give the hypothesis that this 10 year old or that 70 year old will live another year. Pólya's point, however, is that *we* cannot apply the Law in this case. It has no objective validity and no actuary would be caught using it.

On the issue of the sun's rise, Feller makes a good point and Pólya disingenuously evades it to make a mild joke: Laplace could simply express himself more clearly on the geographic location at which he expects to see a sunrise tomorrow to escape that criticism.

What is important in these quotations is the applied mathematician's warning against the mindless application of a formula. In teaching mathematics the temptation always is to concentrate on the formalism and let the application fend for itself. The danger is that students learn how to use the formulæ, but not when[67], and certainly not when not to. Probability is particularly susceptible of misapplication and one should be ever-vigilant in applying it. There was the example back in Chapter 2 of Pascal's Wager, there is the doubtful assumption of the principle of insufficient reason hinted at here in the opening line of Feller's quotation, and now the numerically exaggerated assumption of 5000 years of observations of risings of the sun. As the student learning probability in the middle school, high school, or a freshman liberal arts course in college is unlikely to go on to be a statistician, an actuary, or some other practitioner of probability, but may very well be a consumer of probability, I deem it less important to teach him or her the details of performing complicated probability calculations as is generally done in such courses than to teach the overall principles, a healthy but not mindless skepticism of the application of proba-

---

Mathematics and Plausible Reasoning. Pólya's writings are a must for anyone involved in teaching mathematics at any level. For probability in particular, I recommend the two chapters of *Patterns of Plausible Inference* devoted to the subject.

[66] *Exercise.* Look up a more current table and determine the probabilities in question.

[67] One of the things students learn in the Calculus is how to find the maximum and minimum values of a function. Later in Differential Equations, students learn that under certain conditions the equations under consideration have unique solutions in some interval which can be determined by solving for a maximum. The relevance of the familiar technique is often overlooked.

bility theory, and an alertness to the assumptions made in any application. I think this most important in the earlier grades. By the time they reach college, students have an expectation that they will be taught technique in mathematics courses and are very resistant to anything different: They expect to apply formulæ and simple algorithms and not to have to think deeply about the nature of what they are doing. In college, this is particularly true of juniors and seniors who have delayed taking their required freshman mathematics courses as long as possible.

Central Limit Theorem: Introduction

Following his Law of Succession, I wish to describe Laplace's contributions to de Moivre's Theorem, a task that I fear I can only carry out here at the most superficial level. For the most part, prior to Laplace probability had been combinatorial, occasionally difficult, but yet elementary in character. This begins to change with Bernoulli's Theorem, the meaning of which does not present itself immediately to the senses. And de Moivre introduced new techniques— finite difference methods, recurrent series, etc. Joseph Louis Lagrange further developed these techniques, and Laplace completed them, transforming the entire theory of probability into a branch of Analysis, an area of application of advanced techniques of the Calculus.

Laplace's contributions to de Moivre's Theorem are of three types: computation, refinement, and generalisation.

As regards computation, we have already seen in Chapter IV, in quoting Thomas Galloway, how Laplace calculated

$$\int_0^\tau e^{-t^2} dt$$

by means of infinite series (*cf.* pp. 126 - 130, above). Additionally, he gave the first calculation of the area,

$$\int_{-\infty}^\infty e^{-t^2} dt = \sqrt{\pi}$$

under the curve $y = e^{-t^2}$.[68] This calculation requires the Calculus and I defer it to Appendix B for the reader with sufficient background.

Laplace not only supplemented de Moivre's Theorem with such computations, but gave it a new derivation and, in doing so, provided an error term estimating how closely the binomial distribution was approximated by its limiting normal distribution. The derivation is highly analytic and goes beyond the standard Calculus course and thus beyond the scope of this book. The same holds, alas, of his derivation of his most important result, the *Central Limit Theorem*.

Here is how one authority introduces the Central Limit Theorem:

---

[68] As cited on page 126, Galloway more-or-less credits the result to Leonhard Euler because an obvious substitution reduced the integral to another tricky one which Euler had earlier evaluated.

Laplace's major result in probability theory is now called the central limit theorem, where central is to be understood as meaning fundamental. It was read to the Academy on 9 April 1810...and can be most simply described as a major generalization of De Moivre's limit theorem: Any sum or mean (not merely the total number of successes in n trials) will, if the number of terms is large, be approximately normally distributed. Regularity conditions and exceptional cases would come later, but even without such refinements the achievement was major and the analysis that produced it a triumph.[69]

The Central Limit Theorem opened up whole new vistas in probability theory, and is indeed a central result of the theory. For more than a century after Laplace proved his Central Limit Theorem, much of the work in probability theory was devoted to clarifying the theorem—determining the conditions under which it was valid, revealing the exceptions, and providing rigorous proofs. The Central Limit Theorem becomes, thus, not a single theorem, but a family of theorems—a common theme as it were.

The term "Central Limit Theorem" did not originate with Laplace. The name traces back only as far as Georg Pólya who used it in the title of a paper that appeared in 1920: "Über den zentralen Grenzwertsatz der Wahrscheinlichkeitsrechnung und das Momentenproblem" [On the central limit theorem in the probability calculus and the problem of moments][70].

One of those generalising Laplace's result was Simeon Denis Poisson, who coined the term *Law of Large Numbers* to describe the result. Today we take the Law of Large Numbers to be any result of the flavour of Bernoulli's Theorem asserting that the probability that some statistic does not stray too far from its expected value tends to 1 as the number of trials increases without bound. We will be more explicit on this shortly; for now I only note that Bernoulli's Theorem itself is often called the *Weak Law of Large Numbers*, there being a much sharper version due to Émile Borel (1871 - 1956) called the *Strong Law of Large Numbers*. The term "Central Limit Theorem" is reserved for certain results of the flavour of de Moivre's Theorem asserting normality and offering the probability integral as an approximation. De Moivre's Theorem itself seems to be an exception to this rule despite the fact that it is the prototype of such limit theorems.

---

[69] Stephen M. Stigler, *The History of Statistics; The Measurement of Uncertainty before 1900*, Harvard University Press, Cambridge (Mass.), 1986, pp. 136 - 137.

[70] *Mathematische Zeitschrift* 8 (1920), pp. 171 - 181. H.A. David has attempted to trace the first uses of many terms in probability theory in an appendix, "First (?) occurrence of common terms in statistics and probability", in: H.A. David and A.W.F. Edwards, eds., *Annotated Readings in the History of Statistics*, Springer-Verlag New York, Inc., New York, 2001. In addition to Pólya's German, he finds the first English use of the term in a textbook by statistician Harald Cramér (1893 - 1985) published in 1937.

Some Definitions

To describe the Central Limit Theorem properly, it will be convenient to make a few formal definitions. We begin by recalling, with slightly modified notation, the slightly abstract definition of probability in terms of distribution functions, as introduced in Chapter 4 on page 125, above. We assume given a set Ω of *outcomes*, and a *distribution function* $f : \Omega \to [0, 1]$. If $\Omega = \{O_0, \ldots, O_{k-1}\}$ is a finite set, we assume $f(O_0) + f(O_1) + \ldots + f(O_{k-1}) = 1$. A subset $E \subseteq \Omega$ is called an *event* and its probability is

$$P(E) = \sum_{O_i \in E} f(O_i).$$

If Ω is the real line or some interval I, we assume f takes on nonnegative real numbers as values and the area under the curve $y = f(x)$ over the interval I is 1:

$$\int_I f(x)dx = 1,$$

i.e.,

$$\int_{-\infty}^{\infty} f(x)dx = 1$$

if Ω is the full real line, and

$$\int_{\alpha}^{\beta} f(x)dx = 1,$$

if Ω is the interval (α, β) or $[\alpha, \beta]$. An event is any subset $E \subseteq \Omega$ over which the area in question exists and the probability of E is this area. Thus, if, say, $E = [\gamma, \delta]$, we have

$$P(E) = \int_{\gamma}^{\delta} f(x)dx.$$

These two examples are not exhaustive. Indeed, the problem of Gambler's Ruin discussed back in Chapter 3 does not assume either form. Fitting probabilities on finite spaces, geometric probabilities determined by area ratios, and Gambler's Ruin into a common framework would require the kind of abstraction in mathematics that had its roots in the late 19th century and matured in the 20th. Laplace, of course, worked a century earlier and his Central Limit Theorem can be discussed in the less rarified atmosphere of these two examples.

We do have to introduce two more notions, the first sounding rather more abstract than it is. This is the notion of a *random variable*. For a finite space $\Omega = \{O_0, \ldots, O_{k-1}\}$, a random variable is just some function $X : \Omega \to \mathbb{R}$ assigning a real number to each outcome. $X(O_i)$ could be 1 or 0 according to whether or not O_i is a winning outcome, or, it could be the payoff in ducats, dollars, euros, etc., that a given player will receive if outcome O_i occurs. If Ω is the set of all repetitions of n trials of a game, $X(O_i)$ could be the number of wins O_i represents for a given player.

In the continuous case, in which Ω is some interval, to be a random variable a function $X : \Omega \to \mathbb{R}$ must also be *integrable*, i.e., the area

$$y = \int_I X(t)dt$$

must exist for all intervals $I \subseteq \Omega$. The reason for this is that one is interested in the *expected value* and *variance* of X, which are defined to be

$$E(X) = \sum_{O_i \in \Omega} X(O_i)f(O_i) \text{ and } \sigma^2(X) = \sum_{O_i \in \Omega} (X(O_i) - E(X))^2 f(O_i),$$

respectively, in the finite case, and

$$E(X) = \int_\Omega X(t)f(t)dt \text{ and } \sigma^2(X) = \int_\Omega (X(t) - E(X))^2 f(t)dt,$$

respectively, in the continuous case.

For example, if we toss a coin twice, then $\Omega = \{HH, HT, TH, TT\}$, $f(O) = 1/4$ for all four outcomes O, and $X(O) =$ the number of heads in outcome O. We have

$$E(X) = X(HH) \cdot f(HH) + X(HT) \cdot f(HT) + X(TH) \cdot f(TH)$$
$$+ X(TT) \cdot f(TT)$$
$$= 2 \cdot \frac{1}{4} + 1 \cdot \frac{1}{4} + 1 \cdot \frac{1}{4} + 0 \cdot \frac{1}{4} = \frac{2+1+1}{4} = 1,$$
$$\sigma^2(X) = (2-1)^2 \cdot \frac{1}{4} + (1-1)^2 \cdot \frac{1}{4} + (1-1)^2 \cdot \frac{1}{4} + (0-1)^2 \cdot \frac{1}{4}$$
$$= \frac{1}{4} + \frac{1}{4} = \frac{1}{2},$$

the same values we would have got back in Chapter 2.

For a continuous example, let $\Omega = [0, 1]$, $f(t) = 1$, and $X(t) = t$, then[71]

$$E(X) = \int_0^1 t \cdot 1 \, dt = \frac{t^2}{2} \Big|_0^1 = \frac{1}{2}$$

is the average value of t in the interval $[0, 1]$. For $f(t) = 2t$ and $X(t) = t$, we have[72]

$$E(X) = \int_0^1 t \cdot 2t \, dt = \frac{2t^3}{3} \Big|_0^1 = \frac{2}{3}$$

as a weighted average of values of t in the interval $[0, 1]$.

---

[71] For those unfamiliar with the Calculus, the area is that of a triangle and is easily computed by the familiar formula.

[72] For those unfamiliar with the Calculus, use the option $\int f(x)dx$ found in the CAL-CULATE menu accessed via the CALC button after graphing $y = 2x^2$ to evaluate the integral.

I leave it to the more ambitious reader to calculate the variances in these examples.

One often expresses events, probabilities, and expectations in terms of random variables. In the example of the coin tosses, we might write $X = 1$ to denote the event $E = \{$a single head occurs$\} = \{O|X(O) = 1\}$ and

$$P(X = 1) = f(HT) + f(TH) = \frac{1}{4} + \frac{1}{4} = \frac{1}{2},$$

since the outcomes O for which $X(O) = 1$ are just HT and TH. In our second continuous example, in which $\Omega = [0, 1]$, $f(t) = 2t$, and $X(t) = t$, we might write

$$\frac{1}{2} \leq X \leq \frac{2}{3} \quad \text{for} \quad \{t|1/2 \leq X(t) \leq 2/3\} = \left[\frac{1}{2}, \frac{2}{3}\right],$$

and

$$P\left(\frac{1}{2} \leq X \leq \frac{2}{3}\right) = \int_{1/2}^{2/3} 2t\, dt = \frac{2t^2}{2}\Big|_{1/2}^{2/3} = t^2\Big|_{1/2}^{2/3} = \frac{4}{9} - \frac{1}{4} = \frac{7}{36},$$

as we can check on our calculators.

It is perhaps most natural to refer to the values of X and their distribution. That is, we can consider the sample space

$$\Omega_X = \{X(O) \,|\, O \in \Omega\}, \quad f_X(x) = P(X = x) \text{ for } x \in \Omega_X.$$

With this notation,

$$\mu_X = E(X) = \sum_{x \in \Omega_X} x \cdot f_X(x), \quad \sigma^2(X) = \sum_{x \in \Omega_X} (x - \mu)^2 f_X(x)$$

in the finite case, and

$$\mu_X = E(X) = \int_{\Omega_X} x f_X(x)\, dx, \quad \sigma^2(X) = \int_{\Omega_X} (x - \mu_X)^2 f_X(x)\, dx,$$

in the continuous case.[73] Notationally, this saves us the trouble of writing the O's, and if X is understood it too can be suppressed. More importantly, it allows us to represent the distribution graphically by plotting the points $\langle x, f_X(x) \rangle$.

For example, in n Bernoulli trials with probability p of success, we might think of an outcome as a sequence of S's (success) and F's (failure). Each sequence of letters, $L_0 \ldots L_{n-1}$ will have a probability,

$$f(L_0 \ldots L_{n-1}) = p^k (1 - p)^{n-k},$$

where k is the number of L_i's that are S. Introducing the random variable X,

$$X(L_0 \ldots L_{n-1}) = \text{the number } k \text{ of } S\text{'s in } L_0 \ldots L_{n-1},$$

---

[73] I refer the interested reader to Appendix B for the details of the proof that the computations of μ_X and σ_X using Ω and Ω_X agree in the finite case.

we would have

$$P(X = k) = \binom{n}{k} p^k (1-p)^{n-k}$$

and the usual binomial distribution which we could graph as an xyLine statistical plot on the calculator.

The Central Limit Theorem of Laplace is the calculation in the simplest case of the limits of probabilities connected with sequences of random variables. Laplace's result assumes *independence*—the last notion we need to define before we can state his result. There are, in fact, three notions of independence to sort out—an informal intuitive notion, and two formal notions of independence of events and of random variables, respectively.

The intuitive notion of independence of events is nicely summarised by Definition 7 of Bayes, cited on page 240, above: Events are independent if the occurrence of one of them has no effect on the probability of the rest. The intuition behind independence comes from situations in which a series of tasks are performed. For example, a person might toss a coin, then roll a die, and finally draw a ball from an urn. Assuming the coin, die, and urn are fixed in advance, none of the three actions has any bearing on the others. The result of the coin toss does not affect the outcome of the roll of the die and neither of these will influence the drawing of a ball. The events tossing heads, rolling a 5, drawing a white ball are all independent of each other. There would be a dependence if, for example, we used the coin toss to determine which of two unequally weighted dice to use for the die roll, and the outcome of this would determine which of 6 urns with varying compositions of white and black balls to draw from for the final experiment.

If two events A, B are independent, then neither event affects the probability of the other:

$$P(A|B) = P(A), \quad P(B|A) = P(B).$$

From either of these and the formulæ

$$P(A \cap B) = P(A) \cdot P(B|A) = P(B) \cdot P(A|B),$$

it follows that

$$P(A \cap B) = P(A) \cdot P(B).$$

Provided neither $P(A)$ nor $P(B)$ is 0, the argument can be reversed:

$$P(A) \cdot P(B) = P(A \cap B) = P(A) \cdot P(B|A),$$

whence division by $P(A)$ yields $P(B) = P(B|A)$, and similarly division by $P(B)$ yields $P(A) = P(A|B)$. Hence we formally define two events A, B to be *independent*, or *stochastically independent* [74], if $P(A \cap B) = P(A) \cdot P(B)$.

For more than two events one must define the notion of independence more carefully and less ambiguously than Bayes did. Formally, we define a family $\mathcal{E}$

---

[74] The word "stochastic" derives from a Greek word for guessing and is used to emphasise that it is the probability of an event that is unaffected by the occurrence of the other event.

of events to be *independent* if, for any finite subset $\{E_0, E_1, \ldots, E_{k-1}\} \subseteq \mathcal{E}$, with $k \geq 2$, we have

$$P(E_0 \cap E_1 \cap \ldots \cap E_{k-1}) = P(E_0) \cdot P(E_1) \cdots P(E_{k-1}).$$

This is not implied by the assumption that each pair E_i, E_j is independent or that the probability of the intersection of all the elements of $\mathcal{E}$ is the product of all the individual probabilities.

8 Exercise. Suppose $\Omega = A \cup B \cup C$ is determined by the diagram of *Figure 2*, below. Show:

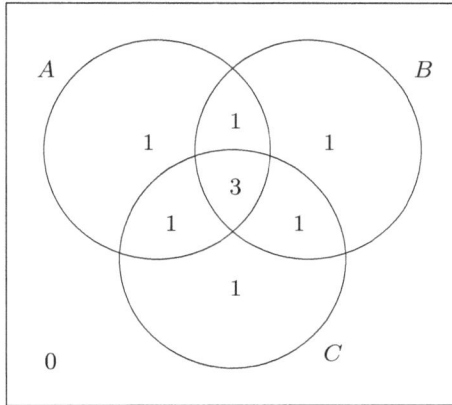

Fig. 2.

i. A, B are independent: $P(A \cap B) = P(A) \cdot P(B)$
ii. A, C are independent
iii. B, C are independent
iv. A, B, C are not independent: $P(A \cap B \cap C) \neq P(A) \cdot P(B) \cdot P(C)$.

9 Exercise. Construct an example for which $P(A \cap B \cap C) = P(A) \cdot P(B) \cdot P(C)$, but $P(A \cap B) \neq P(A) \cdot P(B)$.

The notion of the independence of events is generally introduced quite early in elementary courses in probability theory, right along with conditional probability. I leave it to the reader to decide whether I have not done so accidentally, having simply overlooked it until now because I haven't had to use it earlier, or I have done so deliberately for precisely this reason. Due to its affinity to conditional probability, it is logically natural to introduce the notion at that point; however, at that stage one need not even be aware of the notion. It is convenient to introduce it now.

Actually, it is not the independence of events, but the independence of random variables that we need.

Random variables $X_0, X_1, \ldots$ are *independent* if, for any values $x_0, x_1, \ldots$, the events $X_0 = x_0, X_1 = x_1, \ldots$ are independent.

Statements of the Central Limit Theorem

With all of this, we can pretty much state the Law of Large Numbers and the Central Limit Theorem as follows:

10 Theorem (Law of Large Numbers). *Let* $X_0, X_1, \ldots$ *be a sequence of pairwise independent random variables, and let* $S_n = X_0 + X_1 + \ldots + X_{n-1}$. *For any* $\epsilon > 0$, *as* $n \to \infty$,

$$P\left(\left| \frac{S_n}{n} - \mu \right| < \epsilon \right) \to 1,$$

provided appropriate conditions are satisfied.

11 Theorem (Central Limit Theorem). *Let* $X_0, X_1, \ldots$ *be a sequence of pairwise independent random variables, and let* $S_n = X_0 + X_1 + \ldots + X_{n-1}$. *As* $n \to \infty$, *the distribution of values of* S_n/n *approaches a normal distribution, provided appropriate conditions are satisfied.*

These are marvellously vague formulations. The "appropriate conditions" are conditions under which a given proof is valid. Because limits are involved, the conditions are those guaranteeing convergence and can be quite technical. As de Morgan has already warned us, Laplace is generally not very explicit on points such as this. Even modern textbooks will vary in their statements of the conditions, which are usually expressed as some restrictions on the sequences $E(X_0), E(X_1), \ldots$ and $\sigma^2(X_0), \sigma^2(X_1), \ldots$ Sitting on the floor next to me is a stack of books in which I find a number of distinct explicit formulations of the Central Limit Theorem. Several of them can be found in Feller's standard textbooks of the 1950's and 1960's. His first statement of the result is presented without proof in the first volume of his text and reads there as follows:

> **Central Limit Theorem.** *Let* $\{\mathbf{X}_k\}$ *be a sequence of mutually independent random variables with a common distribution. Suppose that* $\mu = \mathbf{E}(\mathbf{X}_k)$ *and* $\sigma^2 = \mathrm{Var}(\mathbf{X}_k)$ *exist and let* $\mathbf{S}_n = \mathbf{X}_1 + \ldots + \mathbf{X}_n$. *Then for every fixed* β
>
> $$P\left\{ \frac{\mathbf{S}_n - n\mu}{\sigma n^{\frac{1}{2}}} < \beta \right\} \to \Phi(\beta) \qquad (28)$$
>
> where $\Phi(\beta)$ is the normal distribution.[75]

Here $\Phi(\beta)$ is the cumulative normal distribution function,

$$\Phi(\beta) = \frac{1}{\sqrt{2\pi}} \int_{-\infty}^{\beta} e^{-\frac{1}{2}t^2} \, dt,$$

and the rest of the notation should be self-explanatory. Replacing $\mathbf{S}_n$ by $\overline{\mathbf{X}} = \mathbf{S}_n/n$, (28) reads

---

[75] Feller, *op. cit.*, vol. 1, p. 229.

$$P\left(\frac{\overline{\mathbf{X}} - \mu}{\sigma/\sqrt{n}} < \beta\right) \to \Phi(\beta),$$

i.e., the values of $\overline{\mathbf{X}}$ are approximately normally distributed with mean μ and standard deviation $\sigma/\sqrt{n}$, i.e., with variance σ^2/n.

Feller offers the following explicit statement of the Law of Large Numbers:

> **Law of Large Numbers.** *Let $\{\mathbf{X}_k\}$ be a sequence of mutually independent random variables with a common distribution. If the expectation $\mu = \mathbf{E}(\mathbf{X}_k)$ exists, then for every $\epsilon > 0$ as $n \to \infty$*
>
> $$P\left\{\left|\frac{\mathbf{X}_1 + \ldots + \mathbf{X}_n}{n} - \mu\right| > \epsilon\right\} \to 0;$$
>
> in words, the probability that the average S_n/n will differ from the expectation by less than an arbitrarily prescribed $\epsilon > 0$ tends to one.[76]

Feller notes that, in this generality, the result was first published in 1929 by the Russian mathematician Aleksandr Khinchin (1894 - 1959).

Returning to the Central Limit Theorem I note that this statement of the Law of Large Numbers makes no assumption about the existence of the variance $\sigma^2(X_k)$.

Some pages further on Feller cites a more general version for independent random variables $X_0, X_1, \ldots$ which do not have a common distribution. He does not single the result out as a labelled theorem, but spreads its statement across two pages. He starts

> To fix ideas we shall imagine that an infinite sequence of probability distributions is given so that for each n we have n mutually independent variables $\mathbf{X}_1, \ldots, \mathbf{X}_n$ with the prescribed distributions. We shall assume that the means and variances exist and put
>
> $$\mu_k = \mathbf{E}(\mathbf{X}_k), \quad \sigma_k^2 = \mathrm{Var}(\mathbf{X}_k).$$
>
> The sum $\mathbf{S}_n = \mathbf{X}_1 + \ldots + \mathbf{X}_n$ has also finite mean and variance
>
> $$m_n = \mathbf{E}(\mathbf{S}_n), \quad s_n^2 = \mathrm{Var}(\mathbf{S}_n)$$
>
> given by
>
> $$m_n = \mu_1 + \ldots + \mu_n, \quad s_n^2 = \sigma_1^2 + \ldots + \sigma_n^2.^{[77]}$$

Let me interject to note that this last pair of equations asserting the mean and variance of the sum of a finite number of independent random variables to equal the sums of the means and variances of the variables will be proven rigorously in Appendix B, below. Feller continues, coming after a couple of lines to:

---

[76] *Ibid.*, vol. 1, p. 228.

[77] *Ibid.*, vol. 1, p. 238.

The sequence $\{\mathbf{X}_k\}$ is said to obey the central limit theorem if for every fixed $\alpha < \beta$

$$P\left\{\alpha < \frac{\mathbf{S}_n - m_n}{s_n} < \beta\right\} \to \Phi(\beta) - \Phi(\alpha).\,^{78}$$

This, of course, only states what it is for a sequence $X_0, X_1, \ldots$ of independent random variables to obey the Central Limit Theorem. The Theorem itself is not stated until the conditions under which the sequence "obeys" the theorem are specified. Feller then offers some very general conditions due to Jarl Waldemar Lindeberg (1876 - 1932), noting that they are met provided the variables X_k are *uniformly bounded* (i.e., there is some bound B such that $|X_k| < B$ for all k) and $s_n \to \infty$.[79] This gives us a very readable general, but not most general, statement:

12 Theorem. [80] *Let $\{X_i\}$ be a uniformly bounded sequence of mutually independent random variables. Assume for each i,*

$$\mu_i = E(X_i) \text{ and } \sigma_i^2 = \sigma^2(X_i)$$

exist, and let $S_n = X_0 + \ldots + X_{n-1}$. Define

$$m_n = E(S_n) = \mu_0 + \ldots + \mu_{n-1}, \quad s_n^2 = \sigma^2(S_n) = \sigma_0^2 + \ldots + \sigma_{n-1}^2.$$

If $\lim_{n\to\infty} s_n = \infty$, then for every fixed $\alpha < \beta$,

$$P\left(\alpha < \frac{S_n - m_n}{s_n} < \beta\right) \to \Phi(\beta) - \Phi(\alpha).$$

Again, we can rewrite the conclusion of the Theorem as

$$P\left(\alpha < \frac{\overline{X}_n - m_n/n}{s_n/n} < \beta\right) \to \Phi(\beta) - \Phi(\alpha).$$

And, if the limits

$$\mu = \lim_{n\to\infty} \frac{m_n}{n} = \lim_{n\to\infty} \frac{\mu_0 + \ldots + \mu_{n-1}}{n}$$

$$\sigma^2 = \lim_{n\to\infty} \frac{s_n}{n} = \lim_{n\to\infty} \frac{\sigma_0^2 + \ldots + \sigma_{n-1}^2}{n}$$

exist, the distributions of the means, $\overline{X}_n$ approach the normal distribution with mean μ and standard deviation σ.

In the second volume, where he proves the Theorem, Feller first restates the result in a normalised version:

---

[78] *Ibid.*

[79] *Ibid.*, vol. 1, p. 239.

[80] I hope the slight change in notation isn't confusing. Feller begins his enumerations at 1, while I, being a logician, start at 0. And in the present context I prefer i for the index, reserving k for the number of successes in n Bernoulli trials—as will be done in the next subsection.

Theorem 1. (*Identical distributions on* $\mathcal{R}^1$.) *Let* $\mathbf{X}_1, \mathbf{X}_2, \ldots$ *be mutually independent random variables with a common distribution* F. *Assume*

$$\mathbf{E}(\mathbf{X}_k) = 0, \quad \mathrm{Var}(\mathbf{X}_k) = 1.$$

As $n \to \infty$ *the distribution of the normalised sums*

$$\mathbf{S}_n^* = \frac{\mathbf{X}_1 + \cdots + \mathbf{X}_n}{\sqrt{n}}$$

tends to the normal distribution $\mathfrak{N}$ *with density* $\mathfrak{n}(x) = e^{-\frac{1}{2}x^2}/\sqrt{2\pi}$.[81]

The terminology differs slightly but the result, modulo the normalisation, is recognisably the same. A few pages later we find the formulation:

Theorem 3. (*Lindeberg*). *Let* $\mathbf{X}_1, \mathbf{X}_2, \ldots$ *be mutually independent random variables with distributions* $F_1, F_2, \ldots$. *Assume*

$$\mathbf{E}(\mathbf{X}_k) = 0, \quad \mathrm{Var}(\mathbf{X}_k) = \sigma_k^2$$

and put

$$s_n^2 = \sigma_1^2 + \cdots + \sigma_n^2.$$

Assume that for each $t > 0$

$$\frac{1}{s_n^2} \sum_{k=1}^{n} \int_{|y| \geq t s_n} y^2 F_k\{dy\} \to 0$$

or, what amounts to the same, that

$$\frac{1}{s_n^2} \sum_{k=1}^{n} \int_{|y| < t s_n} y^2 F_k\{dy\} \to 1.$$

Then the distribution of the normalized sum

$$\mathbf{S}_n^* = \frac{\mathbf{X}_1 + \cdots + \mathbf{X}_n}{s_n} \tag{29}$$

tends to the normal distribution $\mathfrak{N}$ *with zero expectation and unit variance.*[82]

This is unreadable without some explanation of the notation. Even with such an explanation, it is barely readable and requires some mathematical skill to see that his Theorem 1 and our Theorem 12 are special cases. In introducing it, Feller himself says

---

[81] Feller, *op. cit.*, vol. 2. (1966), p. 253.
[82] *Ibid.*, vol. 2, p. 256.

We turn to a generalization of theorem 1 to variable distributions. The conditions give the impression that they are introduced artificially with the sole purpose of making the same proof work. Actually it turns out that the conditions are also necessary for the validity of the central limit theorem with the classical norming used in (29).[83]

As to the matter of proof, Laplace, building on earlier work of Lagrange on *generating functions*, proved his version of the Central Limit Theorem using Fourier series, an analytic tool one learns in post-Calculus courses. Pafnuty Lvovich Chebyshev pioneered a new method which he applied to the problem, full success being achieved by his student Andrei Andreevich Markov. Another student Aleksandr Mikhailovich Lyapunov applied a new technique and he, in turn, influenced Jarl Waldemar Lindeberg (1876 - 1932), who, in 1922, first proved the versions cited by Feller. Feller, commenting on the proof of Lindeberg's general result given in his textbook, says in a footnote

> Lindeberg's method appeared intricate and was in practice replaced by the method of characteristic functions developed by P. Lévy[84]. That streamlined modern techniques permit presenting Lindeberg's method in a simple and intuitive manner was shown by H.F. Trotter, Archiv. der Mathematik, vol. 9 (1959), pp. 226 - 234.[85]

Feller also notes that, as he himself showed in 1935, Lindeberg's general conditions are necessary as well as sufficient.

For a statement of the Central Limit Theorem that is not instantly recognisable as being related to any of the statements offered thus far, I refer the reader to Michel Loève's[86] massive *Probability Theory*[87]. I do not repeat it here

---

[83] *Ibid.* I have, of course, taken the liberty of changing the numbering of formulæ in these quotations.

[84] Paul Lévy (1886 - 1971) was a major contributor to the modern theory of probability from the 1930's on till his death.

[85] Feller, *op. cit..*, vol. 2, p. 256.

[86] Michel Loève (1907 - 1979) was a French-American probability theorist and statistician.

[87] D. van Nostrand Company, Inc., Princeton, 1955. I have the third (1963) edition in which the statements of the Central Limit Theorem and an Extended Central Limit Theorem appear on pages 309 and 310, respectively. A review (*Bulletin of the American Math. Society*, 67, (1961), pp. 446 - 447) of the second edition of his text by J.L. Doob (1910 - 2004), another important 20th century probability theorist, warns the reader of the depth of the treatment:

> According to the first edition preface, calculus was a prerequisite. The second edition prerequisite is "honest" calculus. A student whose calculus course has been honest enough to prepare him for this difficult book is indeed fortunate... Loève's first edition marked the end of a golden research era in probability theory, in which the subject had been advancing too fast to be treated by anything but research articles, specialized or research books, and elementary texts. He was the first to attempt a text covering the most significant aspects of the subject in a rigorous manner. Even the over 600 pages of the second edition have of course necessarily omitted many things of interest,

because I just don't have the heart to face half a page of daunting typesetting, nor do I look forward to trying to explain the notation and terminology in a few short pages. Perhaps were I younger...

More recognisable is the following statement drawn from a statistics text-book more-or-less contemporary with Feller and Loève:

THEOREM 3: *If x has a distribution with mean μ and standard deviation σ for which the moment generating function exists, then the variable $t = (\overline{x} - \mu)\sqrt{n}/\sigma$ has a distribution that approaches the standard normal distribution as n becomes infinite.*[88]

This statement follows the proof, about which the author comments:

This theorem is known as a *central limit theorem*. Such theorems have been studied a great deal by mathematicians interested in probability. Although the preceding proof required the existence of the moment generating function of x, a proof very similar to the preceding proof can be constructed that requires only the existence of the first two moments; however it requires a knowledge of complex variables. From a practical point of view, this theorem is exceedingly important because it permits the use of normal curve methods on problems related to means of the type illustrated in the preceding section even when the basic variable x has a distribution that differs considerably from normality. Of course the more the distribution differs from normality, the larger n must become to guarantee approximate normality for $\overline{x}$.[89]

A few words of explanation are in order. The *moments* of a random variable X are the sums

$$\mu'_k = \sum_{x \in \Omega_X} x^k f_X(x)$$

and the *moments about the mean* are the sums

$$\mu_k = \sum_{x \in \Omega_X} (x - \mu)^k f_X(x).$$

---

although the exceedingly compact style sometimes goes beyond letting the reader see beauty bare, in fact forcing him to guess at the beauty from the skeleton. But he has shown it possible to write a reasonably complete text as a central core, from which specialized works can radiate... This reader feels that Loève's attempt to be so complete in a book of normal length would have been more successful if about 100 more pages had been allotted, and devoted to discussion and examples, but the book is an excellent pioneering text which will have an enormous influence.

Publication of the text was taken over by Springer-Verlag in 1977/78 as a two volume 4th edition, and is still in print.

[88] Paul G. Hoel, *Introduction to Mathematical Statistics*, 3rd edition, John Wiley & Sons, Inc., New York, 1962, p. 145.

[89] *Ibid.*, pp. 145 - 146.

Thus the expectation $E(X)$ is the first moment and the variance $\sigma^2(X)$ the second moment about the mean, which (as is shown in Appendix B) equals the second moment. The other moments have no fixed interpretations, but they can be grouped together to form a power series called the *moment generating function*,

$$\mu_X(\theta) = 1 + \mu_1'\theta + \frac{\mu_2'}{2!}\theta^2 + \frac{\mu_3'}{3!}\theta^3 + \ldots$$

If, for example, the moments are bounded by some number M, the series defining the moment generating function will converge, i.e., the moment generating function exists. The theorem then tells us that $\overline{X}$ is approximately normal for large n. Lindeberg's result is sharper, requiring only some conditions on $\mu_1' = E(X)$ and $\mu_2' = \sigma^2(X)$.

The Meaning of the Central Limit Theorem

There are three ways we can get a feel for the meaning of a theorem. If we understand the terms in the statement of the theorem, we supposedly understand the theorem. This isn't always the case, however, and when it isn't, we must look elsewhere for understanding the meaning and significance of the result. The proof of a theorem, insofar as it reveals why the result proven is true, can be of great help. But, again, it may be insufficient. The various proofs, or parts thereof, that we have seen of Bernoulli's Theorem—Jakob Bernoulli's, Nikolaus Bernoulli's, de Moivre's—were all rather technical and probably less revealing than Bernoulli's initial remarks in the *Meditationes* or the pictures created by the program BERNULLI. And, as to the Central Limit Theorem, we haven't seen a hint of the proof, which lies well beyond the scope of the present work. This leaves us only with one thing to fall back on—the use to which a theorem is put.

An obvious first application of the Central Limit Theorem should be de Moivre's Theorem, and, indeed, his theorem is an easy corollary. The derivation of de Moivre's result from Laplace's generalisation is fairly simple and direct, with some minor algebraic grubbiness occasioned by the normalisation process. Suppose we are to run a sequence of n Bernoulli trials with fixed probability p of success on a given trial. The outcome O of a given trial is either S (success) or F (failure); the outcome of a sequence of trials is thus a sequence $O_0O_1\ldots O_{n-1}$ of S's and F's. On this space define the random variables $X_0, X_1, \ldots, X_{n-1}$ by

$$X_i(O_0O_1\ldots O_{n-1}) = \begin{cases} 1, & O_i \text{ is } S \\ 0, & O_i \text{ is } F. \end{cases}$$

Thus, $S_n = X_0 + X_1 + \ldots + X_{n-1}$ is given by

$$S_n(O_0O_1\ldots O_{n-1}) = \text{ the number of } S\text{'s among } O_0, O_1, \ldots, O_{n-1}.$$

Now we can calculate

$$E(X_i) = p \cdot 1 + (1 - p) \cdot 0 = p,$$

since $O_i = S$ with probability p, and

$$
\begin{aligned}
\sigma^2(X_i) &= p(1-p)^2 + (1-p)(0-p)^2 \\
&= pq^2 + qp^2 = pq(q+p) = pq \cdot 1 \\
&= pq.
\end{aligned}
$$

It follows from Theorem 12 that

$$
\begin{aligned}
\mu &= E(S_n) = p + p + \ldots + p = np \\
\sigma^2 &= \sigma^2(S_n) = pq + pq + \ldots + pq = npq,
\end{aligned}
$$

as announced in Chapter 4 and as will be shown directly without appeal to the Central Limit Theorem in Appendix B, below. The conclusion of Theorem 12 thus yields

$$
P\left(\alpha < \frac{S_n - \mu}{\sigma} < \beta\right) = P\left(\alpha < \frac{S_n - np}{\sqrt{npq}} < \beta\right) = \Phi(\beta) - \Phi(\alpha).
$$

Now de Moivre's Theorem concerns sums,

$$
\sum_{k=-l+1}^{l-1} \binom{n}{[\mu]+k} p^{[\mu]+k} q^{n-[\mu]-k},
$$

arranged almost symmetrically around the mean. This will translate to an expression

$$
P([\mu] - l < S_n < [\mu] + l),
$$

which de Moivre's theorem would express as $\Phi(\ldots [\mu] + l \ldots) - \Phi(\ldots [\mu] - l \ldots)$. In the case where μ is integral, i.e., $[\mu] = \mu$, simple algebra suffices: S_n takes on values $\mu + k$ and

$$
\mu - l < S_n < \mu + l \Leftrightarrow -l < S_n - \mu < l
$$
$$
\Leftrightarrow \frac{-l}{\sigma} < \frac{S_n - \mu}{\sigma} < \frac{l}{\sigma},
$$

and we have

$$
P(\mu - l < S_n < \mu + l) = P\left(\frac{-l}{\sigma} < \frac{S_n - \mu}{\sigma} < \frac{l}{\sigma}\right) = \Phi\left(\frac{l}{\sigma}\right) - \Phi\left(\frac{-l}{\sigma}\right).
$$

When μ is not integral, the argument requires greater attention to detail, but is not much more complicated.

The next application one would find in a textbook in pure theory would likely be that the Poisson distribution tends to normality. Applications oriented textbooks tend to discuss *sampling*. Sampling with replacement is unproblematic. The idea is that one starts with an initial population, chooses an element O_0, takes its measure and replaces it, then chooses an element O_1, takes its measure, etc. That is, we think of a sample space Ω which equals a population P. A sampling of n elements with replacement would be an element $O_1O_2 \ldots O_{n-1}$ of the space $\Omega^n = \Omega \times \Omega \times \cdots \times \Omega$. If the measure in question (height, weight, etc.) is given by the variable X on Ω, then the sequence $X_0, X_1, \ldots, X_{n-1}$ of random variables defined by

$$X_i(O_0 O_1 \ldots O_{n-1}) = X(O_i)$$

is independent, $S_n = X_0 + X_1 + \ldots + X_{n-1}$ is the sum of the measurements of the elements of the sample and $\overline{X} = S_n/n$ is their average. In this case, the Central Limit Theorem says that, regardless of the shape of the original distribution, the distribution of the variable $\overline{X}$ has mean $E(X)$ and variance $\sigma^2(X)/n$, and, for large values of n, is close to the normal distribution with mean $E(X)$ and variance σ^2/n.

Because the variance σ^2/n tends to 0 as $n \to \infty$, the means of the samples cluster ever closer to the mean of the population as the samples increase in size. This is reminiscent of Bernoulli's Theorem asserting that long term frequencies are probably close to the actual probabilities, more probably so when the number of trials increases. It is, however, more general, applying to any independent variables—average winnings, average heights, average weights, etc.

Somewhat more impressive, perhaps, is a not exactly convincing but heuristically suggestive application cited by Feller:

(c) An application to the *theory of inheritance* will illustrate the great variety of conclusions based on the central limit theorem. In chapter V, section 5, we have studied traits which depend essentially only on one pair of genes (alleles). We conceive of other characters (like height) as the cumulative effect of many pairs of genes. For simplicity, suppose that for each particular pair of genes there exist three genotypes *AA,Aa* or *aa*. Let x_1, x_2, and x_3 be the corresponding contributions. The genotype of an individual is a random event, and the contribution of a particular pair of genes to the height is a random variable $\mathbf{X}$, assuming the three values x_1, x_2, x_3 with certain probabilities. The height is the cumulative effect of many such random variables $\mathbf{X}_1$, $\mathbf{X}_2$, …, $\mathbf{X}_n$, and since the contribution of each is small, we may in first approximation assume that the height is the *sum* $\mathbf{X}_1 + \ldots + \mathbf{X}_n$. It is true that not *all* the $\mathbf{X}_k$ are mutually independent. However, the central limit theorem holds also for large classes of dependent variables, and, besides, it is plausible that the great majority of the $\mathbf{X}_k$ can be treated as independent. These considerations can be rendered more precise; here they serve only as indication of how the central limit theorem explains why many biometric characters, like height, exhibit an empirical distribution close to the normal distribution. This theory permits also the prediction of properties of inheritance, e.g., the dependence of the mean height of children on the height of their parents. Such biometric investigations were initiated by F. Galton and Karl Pearson.[90]

---

[90] Feller, *op. cit.*, vol. 1, pp. 240- 241. The overall form of the argument is borrowed from Laplace; *cf.* page 293, page 309, and pages 477 - 482, below.

Feller's reasoning is far from mathematically rigorous, but it is intuitive and suggestive. To see why credence should be given to its suggestions, we should examine what the Central Limit Theorem does *not* say, and how close it comes to saying it.

Let P denote the population of the United States. Suppose, for the sake of argument, that height depends on n gene pairs, each with 3 alleles, and the alleles of each gene are independent of the alleles of the other genes (e.g., they lie on different chromosomes). Each person has a sequence $a_0 a_1 \ldots a_{n-1}$ of alleles of the genes in question and we think of random variables $X_0, X_1, \ldots, X_{n-1}$, each $X_i(a_0 a_1 \ldots a_{n-1})$ being the contribution to height made by allele a_i. $S_n = X_0 + X_1 + \ldots + X_{n-1}$ is the total genetic contribution to an individual's height. Each X_i and S_n induces a function on P, namely

$$\hat{X}_i(\text{person}) = X_i(a_0 a_1 \ldots a_{n-1}), \quad \hat{S}_n(\text{person}) = S_n(a_0 a_1 \ldots a_{n-1}),$$

if the person in question has allele sequence $a_0 a_1 \ldots a_{n-1}$. The domain of the X_i's and S_n themselves is not P, but the space Ω of *all possible* allele sequences and the Central Limit Theorem refers to Ω or, more precisely, to the values of S_n on Ω and the frequencies of occurrence of these values on Ω—not on P. In short, the Central Limit Theorem does not immediately apply to explain why heights in P are approximately normally distributed.

P and Ω differ quite a bit. It is 2010 as I write these words and the current census is not complete, but the population has been estimated to be around 300000000. If we assume $n = 18$, the cardinality of Ω is $3^{18} = 387420489$, somewhat larger than that of P. And if we assume $n = 20$, Ω is more than 11 1/2 times as large as P. P does not include all possible allele sequences. Moreover, P may contain duplicates among the allele sequences in question. We may thus imagine P as a sample of 300000000 elements of Ω, drawn with replacement.

Now the Central Limit Theorem tells us that the means of samples of size 300000000 from Ω will cluster around the mean of Ω with, given the largeness of the number 300000000, a very small standard deviation. Thus, if we imagine P chosen randomly from Ω, its mean should be close to that of Ω. This is as much as we can firmly conclude right now. But we can imagine, and it is the task of higher theory to quantify and prove such, that P should resemble Ω, i.e., it should also approximate a normal distribution.

One must be careful about reasoning like this. The argument depends on the measured characteristic depending on a large number of independent random variables to conclude the mean values on Ω to be fairly normally distributed, and then on P being a large random sample from Ω to conclude P to resemble Ω and thus be nearly normal.

In many applications one takes samples *without* replacement from a large but finite set. Indeed, the following statement of the Central Limit Theorem is taken from a good statistics textbook written for undergraduate business majors:

The Central Limit Theorem
Suppose a random variable X has population mean μ and standard

deviation σ and that a random sample of size n is taken from this population. Then the sampling distribution of $\bar{x}$ becomes approximately normal as the sample size n increases. The mean of the distribution is $\mu_{\bar{x}} = \mu$ and the standard deviation is $\sigma_{\bar{x}} = \dfrac{\sigma}{\sqrt{n}}$.[91]

This statement omits the important caveat that the size n of the sample must be much smaller than the size N of the population. The point is that in sampling without replacement, the probabilities of obtaining values x of $X_i(O)$ change. If, for example, there are K successful outcomes in a space of N elements, in j trials one has had k successes and $j - k$ failures, the probability of success on the $(j+1)$-th trial is

$$\frac{K - k}{N - j}, \tag{30}$$

which differs from, e.g.,

$$\frac{K - k - 1}{N - j} \quad \text{or} \quad \frac{K - k + 1}{N - j},$$

i.e., the probability of successes on the $(j+1)$-th trial depends on what has happened before. The random variables

$$X_i(O_0 O_1 \ldots O_{n-1}) = \begin{cases} 1, & O_i \text{ is a success} \\ 0, & \text{otherwise,} \end{cases}$$

are not independent. If K and N are extremely large, much larger than $n > j \geq k$, however, then (30) is barely distinguishable from K/N and $X_0, X_1, \ldots, X_{n-1}$ will practically be independent and the Central Limit Theorem will apply. Indeed, with probability close to 1, in drawing a small sample with replacement, no replacements will actually have been made.[92] If, however, n is the same magnitude as N or K, the result as stated can fail miserably.

13 Example. Let $\Omega = \{0, 1, \ldots, 20\}$, $f(i) = 1/21$. Then

$$\mu = \frac{\sum\limits_{i=0}^{20} i}{21} = \frac{1}{21} \cdot \frac{20 \cdot 21}{2} = 10$$

$$\sigma^2 = \frac{\sum\limits_{i=0}^{20} (i - 10)^2}{21} = \frac{1}{21} \left[\sum_{i=0}^{20} i^2 - 20 \sum_{i=0}^{20} i + \sum_{i=0}^{20} 10^2 \right]$$

---

[91] Michael Sullivan, III, *Statistics; Informed Decisions Using Data*, Pearson Education, Inc., Upper Saddle River (NJ), 2004, p. 432.

[92] In view of the Birthday Problem, this may be surprising, but if K, N are sufficiently bigger than n, it is a fact. I refer the reader to the final section of Appendix A for details (page 393).

$$= \frac{1}{21}\left[\frac{20^3}{3} + \frac{20^2}{2} + \frac{20}{6} - 20\cdot\frac{20\cdot 21}{2} + 21\cdot 100\right]$$

$$= \frac{1}{21}\left[\frac{8000}{3} + \frac{400}{2} + \frac{20}{6} - 4200 + 2100\right]$$

$$= \frac{1}{21}[2870 - 4200 + 2100] = \frac{770}{21} = 36\frac{2}{3}$$

$$\sigma = \sqrt{36\frac{2}{3}} \approx 6.05530708.$$

Now $N = 21$ and if we set $n = 20$, the sampling of means is given by

$$\Omega' = \{9.5, 9.55, 9.6, \ldots, 10.5\}, \quad f(i) = \frac{1}{21},$$

which is nowhere near normal. If we store the values of Ω' in the calculator as L_1,

seq(9.5+.05X,X,0,20)→L_1,

we can enter

1-Var Stats L_1

and read off $\bar{x} = 10, \sigma x = .3027650354$ as the mean and standard deviation of the distribution. The application of the Central Limit Theorem correctly predicts the mean to be 10, but incorrectly predicts

$$\frac{\sigma}{\sqrt{20}} = \frac{6.055300708}{\sqrt{20}} \approx 1.354006401$$

as the standard deviation.

What happens here is that the distributions of means of n-element samples become progressively more normal as n increases from 1 to 2 to 3...up to 10 and 11, and then flatten out more and more until $n = 20$. For $n = 10$, the distribution is quite close to normal. There are $_{21}C_{10}$ 10-element subsets of Ω, so we cannot graph the full distribution to see this, but we can approximate it through random sampling. I defer this to the end of the final section of Appendix A (cf. page 393).

Caveat

A normal distribution is a nice thing to have. It is completely determined by the mean and standard deviation and these can be used to determine probabilities and percentiles. Many distributions are so close to being normal that one can replace the actual distribution by the normal one with the given mean and standard deviation. For example, if I give an exam to a large class of students and find a mean of 68 and standard deviation of 15, and want a grade distribution of 5% A's, 15% B's, 60% C's, 15% D's, and 5% F's.[93], the cut-offs for the A's, B's, C's, and D's would thus be the 95th, 80th, 20th, and 5th percentiles and I would thus want to find α's such that

$$\Phi(\alpha) = \frac{1}{\sqrt{2\pi}} \int_{-\infty}^{\alpha} e^{-\frac{1}{2}(t-\mu)^2/\sigma^2}\, dt,$$

equals .95, .80, .20, and .05, respectively.

In the old days we would have done this by looking α up in a table. Today we use the calculator. Without introducing new functions, we could do this by graphing

   Y_1=normalcdf(0,X,68,15)[94]

and

   Y_2=.95, .8, .2, .05,

successively, and use the intersect option of the CALCULATE menu accessed by the CALC button on the *TI-83 Plus*. (The ZDecimal window is certainly large enough for all the intersections to be visible.) There is, of course, a shortcut: The third item on the DISTR menu accessed by the DISTR button is the function invNorm(inverse to normalcdf(. One can successively enter[95]

   invNorm(.95,68,15)
   invNorm(.80,68,15)
   invNorm(.20,68,15)
   invNorm(.05,68,15)

---

[93] Grade scales differ from country to country. I've seen transcripts from Asian schools with numerical grades ranging from 0 to 100. When I taught in the Netherlands, the numerical grades ranged from 1 to 10, with 1 through 5 being unnecessarily refined degrees of failure, 6 through 9 successively better passing grades, and 10 a rarely given high score. In the United States letter grades A, B, C, D, F (or, occasionally, A, B, C, D, E) are assigned, sometimes accompanied by $+$ and $-$ as further refinement. The letters were supposed to mean: superior (A), above average (B), average (C), below average (D), and failure (F). Three decades ago the average grade at the college I was teaching at was $B-$; the average grade given in the mathematics department was $C+$. Given three additional decades of grade inflation, I'd guess that the current meanings are: better than average to superior (A), average (B), below average (C), and failure (D and F).

[94] The theoretical lower bound would be $-\infty$, not 0. However, 0 is just over $4\,1/2$ standard deviations below the mean, whence the area below the bell curve to the left of 0 is negligible.

[95] For some reason one cannot use the full list $\{.95, .80, .20, .05\}$ as input in the first argument of invNorm(. One can use lists by successively entering

   $\{.95,.80,.20,.05\}\to L_1$
   seq(invNorm(L_1(X),68,15),X,1,4)$\to L_2$,

but this is hardly any convenience in the present situation.

to get the cut-off grades

$$92.67\ldots,\quad 80.62\ldots,\quad 55.37\ldots,\quad 43.32\ldots,$$

or, say,

93 - 100	A	
81 - 92	B	
56 - 80	C	
44 - 55	D	
0 - 43	F .	

I've not checked any of them, but there are programs for one's computer that will allow one to enter the scores and will calculate the mean and standard deviation, and, I assume, will calculate letter grades for one, most usefully for any chosen breakdown in percentages, e.g., 10% A's, 20% B's, 40% C's, 20% D's, 10% F's.

For grading purposes this does seem a bit silly. If one is teaching a large lecture class of 150 students in college, then 10% of 150 is 15 and one can assign A's to the top 15 scores, B's to the next 30, etc. Or, going back to the 5, 15, 60, 15, 5 distribution, 7 or 8 A's, 21 to 24 B's, etc.

Moreover, using the mean and standard deviation to determine letter grades only makes sense if the distribution is close to normal, which is not likely to be the case for a small class. The calculations via mean and standard deviation are conveniences when the class size is so great that sorting the exam scores is time-consuming and the distribution is normal; it is not any sort of sacred ritual giving more reliable results than simply ordering the scores, giving the top so-many percent A's, the next so-many B's, etc.

I can, perhaps, make a more convincing case with an hypothetical example involving a subject less emotion-laden than grading. Suppose a hat manufacturer wishes to make a decent profit. The costs of making hats of a given size are such that he deems it profitable only to make hats of common sizes. If head sizes are normally distributed he might use the percentages to decide that the market for hats more than 2 standard deviations from the average hat size is too small for his firm to find affordable to make. He can use normalcdf(to determine what percentage of hats to make for people in each size range. If, on the other hand, his potential market consisted of mainly small-headed and large-headed people[96], with few people of moderately sized heads, i.e., if the distribution were *bi-modal*, knowledge of the mean and standard deviation would not be of much use.

A simple illustration is given by combining two barely overlapping normal distributions. In the calculator enter

Y$_1$=.5normalpdf(X,5,1)+.5normalpdf(X,10,1),

---

[96] As could be the case, going back to Feller's example, were one gene the main factor determining headsize and the others minor contributors.

simulating a 50 - 50 mixture of a small-headed population of individuals with size 5 on average hats and big-head individuals wearing size 10 on average. Graph this using the window with settings

Xmin=−1
Xmax=15
Xscl=1
Ymin=−.1
Ymax=.5
Yscl=1 .

Visual inspection should be enough to convince you that the mean of the bimodal distribution is 7.5. Using the $\int f(x)dx$ option of the CALCULATE menu accessed via the CALC button and choosing 7, 8 as the lower and upper bounds, respectively, yields an area .0214002339, i.e., 2.14% of the population having hat sizes between 7 and 8 inclusive. The standard deviation of the distribution graphed is approximately 2.7,[97] whence the normal estimate would be

normalcdf(7,8,7.5,2.7),

which is .14691691, or approximately 14.7%. Using the normal approximation on which to base his production strategy would result in our manufacturer making a disproportionately large number of hats for a small percentage of the market.

The lesson is clear: Before applying a normal approximation to a distribution of data, check to make sure the data are approximately normally distributed. One can graph the data. If the graph appears to be bell-shaped, determine the mean and standard deviation and then superimpose the bell curve on the graphed data as we did in Chapter 4 in the programs BELL1 and BELL2 (and as will be done again in Appendix A in CENTRAL1 and CENTRAL2). Modern statisticians have, in fact, devised tests to determine how normal an empirical distribution is and thus how much confidence one can place in applying the normal distribution to one's data. Such tests are given in college level statistics courses and are not far beyond the scope of the present book, but, nonetheless, do lie beyond this scope and I do not discuss them here.

Sampling

Before the *caveat* we had started to discuss *sampling*, the process of determining some characteristic of the population by collecting data on the individuals in a manageable subset of the population. Both the Law of Large Numbers and the Central Limit Theorem provide theoretical justification for this procedure. Each law says, the Central Limit Theorem a bit more quantitatively, that if the

---

[97] To see this, enter $Y_2=(X-7.5)^2Y_1(X)$ in the equation editor, graph this using Zoom-Fit, and integrate between 0 and 15 to get a good estimate of σ^2. Then take the square root. Likewise, integrating $Y_3=XY_1(X)$ between 0 and 15 would have approximated the mean 7.5.

size of the sample is large enough, then the estimate obtained from the sample is *probably* close to the true value one is looking for.

The word "probably" is very important here. The theorems say that the probability of drawing a sample yielding a numerical value not too far from the true value is nearly 1 if the sample is large enough. This is not a statement about a specific sample drawn, but about the collection of all such samples. Not every sample will resemble the general population, but the "average" one, the "typical" one, the one drawn at *random*, will. It is important in choosing the sample as randomly as possible, to make the choices depend as much on chance as possible. Modern Statistics has devised a number of techniques for doing this that we cannot go into here, other than to note that a key probabilistic element to choosing samples is the generation of random sequences of numbers, techniques and applications of which I defer to Section 3 of Appendix A (and to a very small extent the first section of Appendix B).

What I can do here is take the lead from Howard Wainer who wrote an interesting little article[98] on the often overlooked consequences of these laws, in particular the consequence that smaller samples have larger variances. Two of his examples concern education, a topic of particular interest to the originally intended readership of this book, and I will discuss them.

The first was the cause of a lot of wasteful spending. It is the matter of school size. Wainer states, "In the 1990s, it became popular to champion reductions in the size of schools. Numerous philanthropic organizations[99] and government agencies funded the division of larger school[s] based on the fact that students at small schools are over-represented in groups with high test scores".[100] When Wainer and his colleague Harris Zwerling looked into the matter, however, they found not only over-representation of small schools among the high-scoring schools in exams given to 5th graders in 1662 Pennsylvanian schools, but also the over-representation of small schools among the low-scoring schools. The fact is, if one thinks of a school as a sample of students more-or-less randomly selected from a large population of students and compares the distributions of averages for the large and small schools, one should expect the scores for the larger schools to cluster more closely together around the population average, and the scores for the smaller schools to spread out more in *both* directions. If the average score for the population is μ and the standard deviation is σ, the standard deviation of the means of samples of size n is $\sigma/\sqrt{n}$.

Taking a numerical example, consider samples of size 300 and 3000, respectively. The standard deviation of the first distribution is $\sigma/\sqrt{300}$, while that of the second is $\sigma/\sqrt{3000}$. The former is $\sqrt{10} \approx 3.16$ times as large as the latter. A deviation from the mean of $\sigma/\sqrt{300}$ among the larger samples is thus more than 3 standard deviations from the mean and fewer than 1% of the samples

---

[98] Howard Wainer, "The most dangerous equation", *American Scientist* 95 (2007), pp. 249 - 256.

[99] Wainer (*Ibid.*, p. 252.) cites, for example, the Bill and Melinda Gates Foundation having spent 1.7 billion dollars.

[100] *Ibid.*, p. 253.

will differ by that amount, whereas over 30% of the smaller samples will differ by that amount.

Thus, the high performance of some small schools can as easily be attributed to chance—a lucky sample—and not to the size of the schools. Indeed, this is what Wainer and Zwerling found. In graphing the regression line of average scores against school size, they found a horizontal line—for 5th graders: when test scores for 11th graders were compared, they found average scores rose with school size. This, Wainer points out, can be explained simply by the fact that larger schools have more teachers with a broader spectrum of specialties, allowing the schools to do a better job at teaching the more advanced subjects.

Along these same lines, one ought to consider the appropriateness of grading on a curve—giving a fixed percentage of A's, B's, etc.—for small classes. In a large lecture course, one can be fairly confident that average performance is near the national average. With smaller class sizes, one will occasionally have above average students who on standardised exams would mostly get A's and B's. And one will occasionally get students who should be encouraged, via D's and F's, to pursue other interests. In my own personal experience, the best and worst undergraduate classes I ever had were small classes of engineering students learning Differential Equations. The good class probably all deserved much higher grades than I, then young and inexperienced, assigned. By the time I had the bad class I was more experienced and assigned F's to a third of them without qualms. There was no soul-searching or self-doubt on my part: these students not only could not do simple algebra (distribute a minus sign across a parenthetical expression), but they couldn't follow such an operation in class unless every step was explained. A teacher should expect to encounter a few such students occasionally in Calculus I, but not a couple of courses later in Differential Equations.

Wainer's other educational example concerns a subject that is controversial, and rightly so:

> For many years it has been well-established that there is an over-abundance of males at the high end of academic test-score distributions. About twice as many males as females received National Merit Scholarships and other highly competitive awards. Historically, some observers used such results to make inferences about differences in intelligence between the sexes. Over the past few decades, however, most enlightened investigators have seen that it is not necessarily a difference in level but a difference in variance that separates the sexes. Public observation of this fact has not always been greeted gently, witness the recent outcry when Harvard (now ex-) President Lawrence Summers pointed this out. Among other comments, he said:
>> It does appear that on many, many, different human attributes—height, weight, propensity for criminality, overall IQ, mathematical ability, scientific ability—there is relatively clear evidence that whatever the difference in means—which can be debated—there is a difference in standard deviation/variability

of a male and female population. And it is true with respect to
attributes that are and are not plausibly culturally determined.
The males' score distributions are almost always characterized by
greater variance than the females'. Thus while there are more males
at the high end, there are also more at the low end.[101]

Wainer supports this by citing data from the *National Assessment of Educational Progress (NAEP)*, a series of standardised exams in various academic
subjects given at irregular intervals since 1990. The exams are given to 4th, 8th
and 12th graders and cover mathematics, general science, reading, geography,
and U.S. history. In all cases, for each year and each subject, the ratio of the
standard deviations of the scores for males to that for females was between 1.03
and 1.09, and the ratios for the 6 mathematics exams ranged between 1.03 and
1.06. In mathematics, the means themselves differed by at most 2 points (out of
500 points) with the male scores being the higher in 5 out of 6 administrations
of the exam. With standard deviations in the mid- to upper 30's, the difference
in means has no practical significance. The difference in standard deviations
can be significant.

Wainer does not give us all the information we need, so I will have to invent
some figures.[102] For the year 2005, he reports

$$\mu_{\text{male}} = 280, \ \sigma_{\text{male}} = 37$$
$$\mu_{\text{female}} = 278, \ \sigma_{\text{female}} = 35.$$

He does not tell us how many of each gender took the exam nor the mean
and standard deviations of the group taken as a whole. So let us assume the
numbers of male and female students to be equal, Arbuthnot's experience to
the contrary, thus making the overall mean 279. And let us simply guess the
standard deviation of the group as a whole to be 36.[103] Supposing all distributions are normal, a score of 0 is 7.75 standard deviations below the mean and
we can take it as $-\infty$ in the cumulative distribution.

Suppose now we wish to reward academic achievement by granting scholarships to the top 5 percent of students. The minimum score for this is determined
by entering

invNorm(.95,279,36),

which yields 338.2147305. In other words, any student with a score of 339 or
higher will get a scholarship. Now enter

normalcdf(0,338.2147305,280,37)

to get .9421835362, whence 5.78% of the males will get a scholarship. Entering

normalcdf(0,338.2147305,278,35)

---

[101] *Ibid.*, pp. 253 - 254.
[102] I have gone to the NAEP website, but it is not a convenient place for getting raw
data and I am impatient when on the computer.
[103] We have insufficient reason to assume it closer to 37 than to 35 or closer to 35 than
37.

yields .9573220534, whence 4.27% of the females will get a scholarship. $57\frac{1}{2}\%$ of the scholarships will go to males, while the females will get only $42\frac{1}{2}\%$ of the scholarships.

Wainer does not tell us whether these scores are for 4th, 8th, or 12th graders. I did check the NAEP site and seem to recall his figures being for 8th graders, hardly the likely recipients of scholarships. Also, the NAEP is far from universal and of questionable randomness in the selection of schools at which the exam is given. A more realistic model might be the *Scholastic Aptitude Test* (*SAT*), which is given to practically all high school students in the U.S. who plan to attend college. It too is not universal, as the weaker students will, one would hope, not be planning to attend college and will thus not be taking the SAT. As there is, by the greater spread, a greater proportion of potentially low scoring males opting out of taking the SAT, the increase in average obtained by dropping the poorly performing students ought to be greater among males than females. Assuming equal averages of ability in the general population, we would nonetheless expect the male averages on the SAT to be higher than that for the females. And, indeed, it is—and has been every year since 1972 at least.

The SAT website offers a very convenient presentation of the data with a number of tables for the year 2010 all brought together in a single pdf file.[104] Indeed, there is so much I almost do not know where to begin. From their *Table 5: Score Distributions* I copy the mathematics results in *Table 2*, below.

Score range	200 - 290	300 - 390	400 - 490	500 - 590	600 - 690	700 - 800
Male	14485	72103	186613	223156	158830	65606
Female	21820	121865	272397	244699	127688	38728

Table 2: Distribution of Math Scores, SAT 2010

I cannot explain the 10 point gaps between the intervals. There is a note to the table referring one to www.collegeboard.com/sat-skills for an explanation of how each interval can be interpreted by the skills typical students in those ranges possess. Being lazy, I will simply interpret each interval as including the missing points at the end, thus 200 - 299, 300 - 399, ..., with all the scores concentrated at the midpoints 250, 350, ...

14 Exercise. Create lists as follows:

seq(250+100X,X,0,5)$\rightarrow$L$_1$
{14485,72103,...,65606}$\rightarrow$L$_2$
{21280,121865,...,38728}$\rightarrow$L$_3$.

Plot xyPlots using L$_1$,L$_2$ and L$_1$,L$_3$, respectively, and examine the graphs for signs of normality: are the graphs bell-shaped? Use L$_2$ and L$_3$ as frequency lists to estimate the mean scores and standard deviations of the male and female score distributions using 1-Var Stats. Store L$_2$ + L$_3$ in L$_4$ and examine the overall distribution.

---

[104] "2010 College-Bound Seniors Total Group Profile Report".

The actual data are collected in *Table 3*, below. The difference in means is

	Test-takers	μ	σ
Male	720793	534	118
Female	827197	500	112
Total	1547990	516	116

Table 3: SAT 2010 Summary of Math Scores

quite a bit larger in this example. On the NAEP, with a difference of 2 points in the 2005 exam, the average score for males was only $.0\overline{5}$ standard deviations above that for females. On the SAT, the 34 point difference is $34/116 = .293\ldots$ standard deviations, over 5 times as large. If we took one boy and one girl and compared their test scores, getting 534 and 500, we might not be inclined to think much of it. After all, a score of 534 is only 18 points above the mean. The probability that a student chosen at random has a score at least that high would be (using 10000 for $+\infty$)

normalcdf(534,10000,516,116),

which is .4383426825. And the probability of getting a score of 500 or below is at least

normalcdf(0,500,516,116),

which is .4451430994. The probability of both events occurring in two independent draws is thus at least $.438 \times .445 = .19491$, i.e., it should occur at least 19% of the time, making it an unremarkable occurrence.

However, we did not choose one boy and one girl, but over half a million of each in two samples and got averages of 534 and 500.

A simple way to test if the difference between male and female means can be attributed to chance is to assume there is no difference, the males and females having equal mathematical ability. With nothing else to go on, we assume this common average to be 516 and the standard deviation to be 116. Now the mean among all samples of size 720793 is 516 and the standard deviation of these samples is $116/\sqrt{720793} \approx .1366320892 \approx .137$. Now $(534 - 516)/.137 = 131.3868613$, i.e., 534 is over 131 standard deviations above the mean, an event of such low probability as to register 0 on the calculator:

1−normalcdf(498,534,516,.1366320892)

returns 0.

The calculation given assumes the normality or near normality of the distribution of means, which should certainly hold for a sample of 720793 taken from an infinite population of all possible high school students present or future, which we could easily assume if we believed the mean 516 and standard deviation 116 were immutable laws of nature. They are not and we should take into account that the population from which we drew our massive sample is only about 2 times as large as the sample. More careful calculation shows that the standard deviation of the sample means, often called the *standard error*[105], is better estimated by

$$\frac{\sigma}{\sqrt{n}} \cdot \sqrt{\frac{N-n}{N-1}},$$

where n is the sample size and N the population size. In the present case we have $N = 1547990$ and, again, $n = 720793$, $\sigma = 116$, yielding a standard error

$$\sigma_{\text{se}} = \frac{116}{\sqrt{720793}} \cdot \sqrt{\frac{1547990 - 720793}{1547990 - 1}} \approx .098787847 \approx .1.$$

And, $(534 - 516)/.1 = 180$, making the mean score for the males around 180 standard deviations above the expected value, which is again a most improbable occurrence.

Likewise, drawing a sample of 827197 individuals yielding a mean of 500 is highly improbable. I conclude that the difference between male and female scores on this exam is not due to chance. What the difference is due to— naturally greater mathematical ability among boys, cultural factors, or merely the combined effect of greater variance and the loss of the potentially lowest scores—I leave to the reader to determine. To this end, I note that the breakdown by ethnicity shows the 84856 females who described themselves as "Asian, Asian American, or Pacific Islander" had a mean score of 577, much better than the mean for males. Similarly the mean of those 86509 males describing themselves as "Black or African American" was 436, well below the mean score for females. The correlation of these results with known cultural factors (respect for learning in the Japanese-American community, peer pressure among young black males not to succeed academically) together with the near identical performance of males and females on the NAEP exams, lead me to favour cultural differences. But that is mere opinion not backed up by any hard data.

4 Gauss, the Error Distribution, and the Method of Least Squares

Ivo Schneider begins his discussion of the application of probability theory to the calculus of errors with the words, "The roots for the development of an error concept in the exact natural sciences and in mathematics have up to now not been exposed".[106] He offers a tentative beginning:

> The period around 1700 was particularly suitable too for the development of measuring instruments, which moved the idea of measuring errors into the foreground. The reason for this is, that at this time considerable improvements with regard to precision were achieved. With

---

[105] *Cf.* also page 469, below.
[106] Schneider, *op. cit.*, p. 215.

the offer of continually improved instruments not only would a consciousness of the relative accuracy of the therewith achieved measurements [arise] but also the need for a quantitative determination of the error [would] arise.[107]

With repeated observation inconsistent data were collected and the problem of reconciling the data arose. The common sense solution would have been to toss out any measurements that were totally out of line with the others and then to average the remaining values. The earliest declaration of a principle seems to have been given posthumously by Roger Cotes (1682 - 1716) in 1722:

> Let p be the place of some object determined on the basis of a first observation, q, r, s the positions of the same object [determined] by succeeding observations; let in addition P, Q, R, S be weights which are inversely proportional to the extent of the domain over which the errors arising from the individual observations can extend... One imagines the points p, q, r, s assigned the weights P, Q, R, S and finds their centre of gravity Z; I claim that Z is the most probable place of the object, and can be viewed with greatest certainty as the true place.[108]

The most obvious interpretation of this is that a weighted sum of the disparate values obtained from repeated measurements gives the most probable estimate for the true value (on condition that the given values have been observed).

Histories of probability record the first attempt at justifying the choice of the mean as the best guess at the "true value" of the measure being made as occurring in mid-century. This was done by the Englishman Thomas Simpson (1710 - 1761), first in a paper published in 1755 in the *Transactions*[109] and in greater detail in 1757[110]. In this work he sets out to determine in a couple of cases the probability that the mean does not deviate from the "true value" by more than a given quantity. The weakness of his approach is the utter simplicity and implausibility of the distributions examined.

Laplace considered the problem in the 1770's:

> Earlier writers had confronted the choice of an error curve by making an arbitrary selection. Simpson, for example, had based his choice of the uniform and triangular distributions upon mathematical expediency.

---

[107] *Ibid.*, pp. 215 - 216.

[108] *Ibid.*, p. 216. I have translated from Schneider's German translation from the Latin. A direct English translation of the remark from the Latin can be found in Stigler, *A History...*, *op. cit.*, p. 16.

[109] Thomas Simpson, "A letter to the Right Honorable George Earl of Macclesfield, President of the Royal Society, on the advantage of taking the mean of a number of observations in practical astronomy", *Philosophical Transactions of the Royal Society of London* 49 (1755), pp. 82 - 93.

[110] Thomas Simpson, "An attempt to shew the advantages arising by taking the mean of a number of observations, in practical astronomy", in: Thomas Simpson, *Miscellaneous Tracts on Some Curious, and Very Interesting Subjects in Mechanics, Physical-Astronomy, and Speculative Mathematics*, London, 1757. Excerpts (in the original English) can be found in Schneider, *op. cit.*

For his limited aim (showing that an arithmetic mean was better than a single observation) this arbitrariness was not a major drawback—his audience only needed to accept the distribution as qualitatively correct in order to believe his qualitative conclusion—that averaging increased accuracy. Laplace, however, was attempting an exact computation, and he needed both a convincing case for any choice of $\phi(x)$ he might make *and* a mathematical analysis equal to the task of calculating the best mean for that curve.[111]

The distribution we are discussing is the *error distribution*, more properly: the probability distribution function for errors. When one takes a measurement of some object, the result will not quite be correct, but will be off by some amount we call an error. Ignoring such subtleties as the existence of analysable and correctible systematic errors, we can see that the errors themselves form a distribution with its own probability distribution function. Now measurements are real numbers coming from some, finite or infinite, interval, whence the distribution will be some integrable function with 1 being its integral over the interval in question. Simpson considered two distributions,

the uniform one: $f(x) = 1/(2a)$ on $[-a, a]$

the triangular one: $f(x) = 1/a - |x|/a^2$ on $[-a, a]$.

[In these cases, if we define $f(x) = 0$ for $x \notin [-a, a]$, we can graph these on the calculator by entering

Y$_1$=(−A≤X)(X≤A)/(2A)
Y$_2$=(−A≤X)(X≤A)(1/A−abs(X)/A^2) ,

in the equation editor—provided some non-zero value of a has been stored in the variable A.]

In a paper published in 1778, Daniel Bernoulli used the parabola

$$f(x) = a^2 - x^2 \text{ on } [-a, a],$$

normalised to yield area 1, as his candidate for an error distribution.

These selections were indeed arbitrary and one is reminded of the sort of thrashing around for a solution exhibited by Pacioli, Cardano, and Tartaglia in tackling the problem of points. Ignoring the uniform distribution, it seems most distributions satisfied the common-sense adequacy conditions that the zero error would be the likeliest one, the distribution was symmetrical around the maximum, and it decreased to 0 as one moved away from the maximum. Other than that, there wasn't much justification for the actual choice as authors attempted to use theirs to justify the practice of taking the average of the observed values.

Laplace considered a number of candidates for an error distribution in the 1770's. He rejected the uniform distribution as he considered large errors less likely than small ones. In 1774 he assumed

---

[111] Stigler, *A History...*, *op. cit.*, p. 110.

Now, as we have no reason to suppose a different law for the ordinates than for their differences, it follows that we must, subject to the rules of probabilities[112], suppose the ratio of two infinitely small consecutive differences to be equal to that of the corresponding ordinates.[113]

The argument here is completely nonrigorous. Assuming dx infinitesimal he writes, letting $\phi(x)$ denote the distribution function,

$$\frac{d\phi(x+dx)}{d\phi(x)} = \frac{\phi(x+dx)}{\phi(x)}$$

from which one can conclude

$$\frac{d\phi(x+dx)}{dx} \cdot \frac{1}{\phi(x+dx)} = \frac{d\phi(x)}{dx} \cdot \frac{1}{\phi(x)}.$$

Supposedly this makes the ratio $\phi'(x)/\phi(x)$ constant throughout on either half of the x-axis[114] and not merely in an infinitesimal neighborhood, and we can conclude

$$\ln \phi(x) = C_1 x + C_2$$

for some constants C_1, C_2. Thus

$$\phi(x) = e^{C_2} \cdot e^{C_1 x}.$$

Actually, since $\phi(x)$ goes to 0 as $|x|$ goes to infinity, this has to be adjusted: x is replaced by $|x|$ and C_1 by $-m$ for some positive number m. The requirement that the area under the entire curve is 1 allows him to determine by a simple calculation (well, simple for those who know the Calculus) e^{C_2} to be $m/2$. Thus he concludes

$$\phi(x) = \frac{m}{2} e^{-m|x|}, \quad \text{for some positive real } m.$$

This derivation is hardly convincing. He did worse in 1777 when he decided on[115]

$$\phi(x) = \frac{1}{2a} \ln\left(\frac{a}{|x|}\right) \text{ on } [-a, a].$$

Again Stigler:

Here a represented the upper limit of the possible errors. The argument leading to this curve was as difficult as any mathematical argument Laplace attempted in this period, and it seems reasonable to speculate that it was his success in the mathematics of this intricate investigation that led him to emphasize the result. The curve itself was more difficult to work with than his earlier exponential distribution had been.[116]

---

[112] Stigler explains these "rules of probabilities" to be the principle of insufficient reason.

[113] Stigler, *A History...*, *op.cit.*, p. 111.

[114] The graph of the final solution has a cusp at $x = 0$, whence $\phi(x)$ is not differentiable there and ϕ'/ϕ is defined separately on $(-\infty, 0)$ and $(0, \infty)$.

[115] Note that this ϕ is undefined at $x = 0$, not merely not differentiable there.

[116] Stigler, *A History...*, *op. cit.*, p. 120.

This was the state of affairs when Carl Friedrich Gauss jumped in:

> It fell to Gauss' lot to create a theory of errors that immediately found numerous followers. In the second book of his "Theory of motion of heavenly bodies rotating around the Sun in conic sections" (*Theoria motus corporum cœlestium in sectionibus conicis Solem ambientum.* Hamburg 1809) Gauss proved that among unimodal, symmetric and differentiable distributions $\phi(x - x_0)$ there is exactly one normal distribution for which the maximum likelihood estimator of the location parameter x_0 coincides with the arithmetical mean.[117]

Gauss's result was a tiny bit less general than this. In common with his predecessor, Gauss assumed given a distribution function $\phi(x)$ which had a maximum at 0, was symmetric (i.e., $\phi(-x) = \phi(x)$ for all x), and decreased to the limit 0 as x moved progressively farther from the origin. Unlike Laplace, whose main distribution functions failed to be differentiable at $x = 0$, Gauss implicitly assumed ϕ to be continuously differentiable everywhere, that is he assumed ϕ to be a continuous function with no corners or sudden changes in acceleration. There are infinitely many curves of this sort, but only the normal curve has the additional property assumed by Gauss that the mean is the "maximum likelihood estimator".

From Simpson on, the task had been to justify the use of the mean to reconcile disparate measurements. One postulated an error curve and demonstrated the choice of mean as optimal. Gauss reversed the process—he assumed the optimality of the mean and concluded the distribution function to be normal:

> ... we will, approaching the subject from another point of view, inquire upon what function, tacitly, as it were, assumed as a base, the common principle, the excellence of which is generally acknowledged, depends. It has been customary certainly to regard as an axiom the hypothesis that if any quantity has been determined by several direct observations, made under the same circumstances and with equal care, the arithmetical mean of the observed values affords the most probable value, if not rigorously, yet very nearly at least, so that it is always most safe to adhere to it.[118]

Gauss's proof depends on the Calculus, but, modulo this appeal to that subject, is fairly simple. Gauss himself omits one small argument that the average Calculus student might not be able to supply, and the historical accounts are usually even sketchier. Thus, I present his proof with a few more details pedantically included in Appendix B, below, for the reader with a background in the Calculus.

---

[117] Gnedenko and Sheĭnin, *op. cit.*, pp. 226 - 227.

[118] Charles Henry Davis (translator), *Motion of the Heavenly Bodies Moving About the Sun in Conic Sections: A Translation of Gauss's "Theoria Motus."*, Little, Brown and Company, Boston, 1857, p. 258. The relevant passages from Gauss's book are paragraphs 175 - 179. Schneider, *op. cit.*, includes a German translation of 1887 of paragraphs 175 - 178 and the opening lines of 179.

There is a bit more to Gauss's contribution than this, but one can already see how his name might get attached to de Moivre's normal distribution. Astronomers might or might not have taken an interest in probability and might or might not have appreciated the Central Limit Theorem, but they were definitely aware of observational errors, and Gauss's recognition that such were normally distributed would definitely have caught their attention. And it caught Laplace's attention. Laplace read Gauss's monograph about the time he first proved the Central Limit Theorem and realised the relevance of his own work: If the phenomenon under observation were the result of a large number of independent random causes, each having only a small effect on the final outcome, then, by the Central Limit Theorem, the distribution would of necessity be close to normal.

One more step needed to be taken. In 1818 Friedrich Wilhelm Bessel (1784 - 1846) published the first empirical investigation of the subject, examining some 60000 observations performed during the years 1750 - 1762 by James Bradley at Greenwich Observatory. Bessel compared the distributions of deviations from the accepted declinations and right ascensions in three groups of stars with the estimates calculated via the normal curve.[119] Bessel concluded

> The agreement between observation and calculation is overall as good as one could generally only expect with this number of observations. However the most serious errors, those which far exceed the usual limits, are in each of the three sorts of observations a little more frequent than seems to correspond to theory. Can this perhaps be ascribed to particular causes for the largest errors, such as an error in the tube, an accidental movement of the instrument, or some similar matter? In any event the difference is too insignificant that it cannot also be ascribed to an insufficiently large number of observations.[120]

Stigler sums the situation up nicely:

> Bessel's investigation was a lonely and unrepeated first look at an empirical distribution. If it caught public notice at all, it was as added support for the assumption of Gauss's special form of error distribution. It seems to have occurred to no one, not even Bessel, that such an exercise would be a useful part of many statistical studies. No one at that early date seems to have realized that comparisons of this type could have implications far beyond those they held for the theory of errors...[121]

Looking backward instead of forward, Anders Hald, another authority on the history of statistics, tells us

---

[119] Friedrich Wilhelm Bessel, *Fundamenta astronomiae pro anno 1755 deducta ex observationibus viri incomparabilis James Bradley in specula astronomica Grenovicensi per annos 1750 - 1762 institutis*, Königsberg, 1818. An excerpt in German translation appears in Schneider, *op. cit.*, pp. 277 - 279, and his table of comparisons can also be found in Stigler, *A History...*, *op. cit.*, p. 204.
[120] Schneider, *op. cit.*, p. 279.
[121] Stigler, *A History...*, *op. cit.*, p. 203.

Portraits of Gauss and Bessel on West German stamps. The former was issued in 1955 on the centenary of Gauss's death, the latter in 1984 in celebration of the bicentenary of Bessel's birth. (The background of the Bessel stamp depicts the graphs of a couple of functions named after Bessel and, unfortunately, are of no special significance here.)

It is a surprising fact that nobody before Bessel studied the form of empirical error distributions, based on the many astronomical observations at hand. Presumably, they would then have realized that the error distributions on which they spent so much mathematical labor (the rectangular, triangular, and quadratic) were poor representatives of the real world.[122]

This state of affairs, that it would be some 60 years from the onset of interest in the form of the error distribution and someone's actually examining such a distribution, is indeed surprising. But it shouldn't be. The old Baconian description of the scientific method as the inductive inference of universal laws from a plethora of observational data never was a good description of scientific practice. In general, science is governed by basic principles, speculations on them, speculations on speculations, ... We saw this in our discussion of Intelligent Design in Chapter 5. Numerical data are usually tests of these speculations. Perhaps no one reported on an empirical distribution before Bessel because, prior to Gauss's determination and Laplace's subsequent theoretical justification, there was no plausible theoretical distribution to compare the data with. Moreover, distributions such as those considered by Simpson were not put forth as accurate predictions of error distributions but as oversimplified, calculation-friendly examples suggestive of what to expect in practice—namely, that the mean of a number of observations offers the best estimate of the true value obtainable from these observations. Also, before Laplace's theoretical justification of the normal curve via his Central Limit Theorem, there was no particular reason to believe there to be one common type of error distribution: there may have been quite a variety. That the normal distribution should

---

[122] Hald, *A History...*, *op. cit.*, p. 58.

indeed be common, if not universal, only became a testable hypothesis after Gauss and Laplace. The gap of 9 years between Gauss and Bessel is much less remarkable and surprising, particularly if one can avoid thinking in terms of today's faster pace, pressure on academics to publish, and greater competition to get results among a vastly greater population of scientists. And, of course, this was early in the history of statistics; what is routine today was only being developed.

Hald summarises Gauss's determination of the error curve thus:

> Gauss's invention of the normal distribution marks the beginning of a new era in statistics. Natural scientists now had a two-parameter distribution[123] which (1) led to the arithmetic mean as estimate of the true value..., (2) had an easily understandable interpretation of the parameter h in terms of the precision of the measurement method, and (3) gave a good fit to empirical distributions of observations, as shown by Bessel.[124]

I said earlier that there was a bit more to Gauss's contribution than his determination of the shape of the error curve. The deleted words from this last quotation from Hald allude to this: "and thus to a probabilistic justification of the method of least squares".

Finding the number best estimated by a set of numbers, i.e., finding the best estimate of the "true value" approximated by inconsistent data, is only the first, simplest problem of working with such data. When the quantity being measured depends on one or more parameters, it is not a single quantity being sought, but the equation of a curve, surface, or higher dimensional surface. Astronomy and geodesy provided problems of this sort that resulted in various solutions in the latter half of the 18th century and the early years of the 19th. Tobias Mayer (1723 - 1762) tackled the librations of the moon, Laplace the orbits of Jupiter and Saturn, Ruđer Josip Bošković (Roger Joseph Boscovich) (1711 - 1787) the shape of the Earth, Adrien Marie Legendre the orbits of comets, and Gauss the orbits of "minor planets" or "planetoids" as certain asteroids were called. Initially, these had nothing to do with probability, so I shall take the ahistorical liberty of discussing this work with a much simpler example than any of them tackled.

In Appendix D are a couple of population tables for the United States. I reproduce a portion of the table provided by the U.S. Census Bureau in *Table 2*, below, letting t denote the number of decades that have passed since 1790 and $P(t)$ the population at time t. For convenience I also include a row giving the values of $\ln P(t)$.[125] The reason for taking logarithms is simply that the early US population, as discussed in Appendix B, was essentially exponential:

---

[123] The parameters are the mean and standard deviation, μ and σ. More precisely, Gauss used $h = 1/(\sigma\sqrt{2})$ as the second parameter, a form often more convenient to use than σ.

[124] Hald, *A History...*, *op. cit.*, p. 58.

[125] The table contains an error: $\ln P(1)$ should be 15.48482. However, when I discovered my error and redid the calculations I found the data too perfect, ruining the overall

$x:t$	0	1	2
$P(t)$	3929214	5308483	7239881
$y : \ln P(t)$	15.18395	15.57483	15.79516

$x:t$	3	4	5
$P(t)$	9638453	12860702	17063353
$y : \ln P(t)$	16.08127	16.36969	16.65244

Table 2

$$P(t) = Ae^{Bt},$$

for some constants A, B, whence the logarithm would be essentially linear:

$$\ln P(t) = \ln A + Bt.$$

Thus in labelling t as x and $\ln P(t)$ as y, one is considering a linear equation $y = a + bx$ for which one wants to determine the coefficients a and b.

First things first. Create two lists in your calculator by putting $\{0, 1, \ldots, 5\}$ into L$_1$ and $\{15.18395, \ldots, 16.65244\}$ into L$_2$. Then, in the statistical plotter turn Plot1 on, choose the first pictogram as the Type, set Xlist to L$_1$, Ylist to L$_2$, and choose the small box as the Mark. If you now graph the x, y-pairs by using ZoomStat you will see that the points are very nearly linearly arranged.

We can imagine the plot representing points on a line, deviations from that line being errors of measurement, and we want to use the imperfect information to make a best guess at the equation of the line. If we had only two points this would be simple: use them to determine the line. The problem here is that different pairs determine different lines. Choosing $t = 0$ and $t = 1$ gives us the equations

$$15.18395 = a + 0 \cdot b = a$$
$$15.57483 = a + 1 \cdot b$$

from which we conclude

$$a = 15.18395, \quad b = .39088.$$

If we enter

Y$_1$=15.18395+.39088X

in the equation editor and press the GRAPH button we see that this is a poor choice. Indeed, a good rule of thumb is to choose two points far from each other—where neither is too much out of line from the others. Our choice violated both conditions as the points were adjacent, ignoring values farther out, and the point $(1, 15.57483)$ is clearly out of line from the others. The choice $t = 0$ and $t = 5$ is clearly going to be better. It gives us the equations,

---

demonstration of the improvements made by more sophisticated techniques. Hence I kept the error, an action which seemed appropriate in the present context.

$$15.18395 = a + 0 \cdot b = a$$
$$16.65244 = a + 5 \cdot b$$

from which we conclude

$$a = 15.18395, \quad b = .293698.$$

If we now enter

Y$_2$=15.18395+.293698X,

we see a much better fit. Indeed, the fit in this case is so good there is no point in considering more sophisticated methods for this data. The results will either be barely distinguishable from Y$_2$ or, perhaps, they will even be worse.

Not all linearly related data fit this well and there are methods that work better in the general case. So let us pretend we haven't seen the line given by $t = 0$ and $t = 5$ and that every line we've tried separates from the group like Y$_1$ or the line obtained from the points $t = 2$ and $t = 3$. And let us see how the method could be improved. The table gives us 6 points, from which we derive 6 equations:

$$15.18395 = a + b \cdot 0$$
$$15.57483 = a + b \cdot 1$$
$$15.79516 = a + b \cdot 2$$
$$16.08127 = a + b \cdot 3$$
$$16.36969 = a + b \cdot 4$$
$$16.65244 = a + b \cdot 5.$$

We have an inconsistent set of 6 equations in 2 unknowns and (we are pretending) no single pair of equations in 2 unknowns yields a suitable line fitting the data. What is there to do?

In a paper of 1748 Mayer came up with a simple solution; divide the equations into as many groups as there are variables, add the equations of each group, and solve the resulting system of equations. In our case, we might add the first 3 equations to obtain

$$46.55394 = 3a + 3b, \tag{31}$$

and add the final 3 to obtain

$$49.1034 = 3a + 12b. \tag{32}$$

We now solve the system (31) - (32) to get

$$a = 15.23471, \quad b = .28327.$$

If we graph this with our plotted points and our old Y$_1$, we see a vast improvement and a good fit.

Laplace wasn't born until a year after Mayer's work was published and it wasn't until 1787 that he modified Mayer's procedure. With Mayer we obtained 2 equations, each formed from half the existing equations. With Laplace each of the two final equations is an amalgam of all 6 equations. We obtain the first by adding all 6 equations:

$$95.65734 = 6a + 15b. \tag{33}$$

The next equation might be obtained by subtracting the sum of the first 3 equations from the sum of the final 3 (i.e., $(32) - (31)$):

$$2.54946 = 0a + 9b.$$

With this choice we get the same solution as Mayer:

$$a = 15.23471, \quad b = .28327.$$

(*Exercise.* Why is that?) We could vary this a bit by making a different choice of which equations are to be added and which subtracted. For example, adding the 2nd, 4th, 6th and subtracting the 1st, 3rd, 5th results in

$$.95974 = 0a + 3b. \tag{34}$$

Solving (33) - (34) yields

$$a = 15.14311, \quad b = .31991.$$

Graphing the equation resulting from this choice of a, b yields a good fit, much better than our naïvely chosen Y_1, but not as good as the first line obtained by Mayer and Laplace or even our old two-point Y_2.

The success of the Mayer and Laplace methods can be attributed to the fact that by combining the equations errors can cancel each other, thus reducing the overall error. The degree of success varies with the manner in which the equations are organised—how they are grouped in Mayer's case and which one's get subtracted in Laplace's. Bošković addresses the issue by directly seeking an optimal solution, giving along the way the first criterion for optimality.

Bošković is not as well-known among mathematicians as he should be, so I may be forgiven if I pause to say a few brief words about him. His lineage is mixed. His paternal grandfather was Serbian, his mother Italian, and he was born in Dubrovnik, now part of Croatia. In a biographical essay from 1961 one reads, "Today he can be described conveniently as a Jugoslav, for unquestionably he was born within the borders of the present Federation of the National Republics of Jugoslavia".[126] Today, of course, the Federation no longer exists, but in 1774 he was made a French subject. He was a Jesuit scholar

---

[126] Elizabeth Hill, "Roger Joseph Boscovich; a biographical essay", in: Lancelot Law White, ed., *Roger Boscovich, S.J., F.R.S., 1711 - 1787; Studies of His Life and Work on the 250th Anniversary of His Birth*, George Allen & Unwin, Ltd, London, and Fordham University Press, New York, 1961; p. 17.

of partial Italian extraction who studied and taught in Italy, his name often Italianised to Ruggiero Giuseppe Boscovich. Thus he can currently legitimately be deemed Serbian, Croatian, Italian, and French. His travels took him as well to England and he was made a Fellow of the Royal Society, an honour not quite up to the task of adding "English" to the list of his nationalities.

Portraits of Bošković from Croatia and Jugoslavia. The Croatian stamp is one of a pair of identical design issued in 1943. The Jugoslavian stamp was issued Christmas Eve in 1960 in honour of the imminent 250th anniversary of Bošković's birthday; another stamp celebrating the bicentennial of his death was issued by that country in 1987.

As was typical in the day, Bošković was a polymath: astronomer, mathematician, physicist, geodesist, hydrographer, philosopher, and even a poet. He is best-known for his physical atomism which strongly influenced Lord Kelvin and, in mathematics, for his work on the combination of observations[127].

> Boscovich was the first to devise a completely objective procedure for uniquely determining the coefficients of a two-parameter line $y = \alpha + \beta x$ from a set of three or more observational points. Like the *median*, Boscovich's procedure is comparatively insensitive to the more extreme of a set of observations, and is especially well suited to summarizing the linear trend evidenced by a more or less heterogeneous set of data compiled from various sources, or obtained by a measurement procedure that has a tendency to yield discordant values.
>
> ...
>
> Nearly half a century before Legendre announced (1805) his Principle of

---

[127] A good phrase I steal from the title of the essay: Churchill Eisenhart, "Boscovich and the combination of observations", in: White, *op. cit.* This essay gives as well a discussion of the advantages of Bošković's procedure relative to the other approaches. More libraries, however, are likely to have Stigler, *A History...*, *op. cit.*, where Bošković's approach and Laplace's variation of it are discussed in some detail (pp. 39 - 55).

Least Squares, and twenty years before the birth of Gauss (1777), Roger Joseph Boscovich had formulated and applied the principle that, given more than two pairs of observed values of variables x and y connected by a linear functional relationship of the form $y = \alpha + \beta x$, then the values (a and b) that one should adopt for α and β in order to obtain the line ($y = a + bx$) that is most nearly in accord with all of the observations should be those determined jointly by the two conditions:

(I) *The sums of the positive and negative corrections (to the y-values) shall be equal.*

(II) *The sum of (the absolute values of) all of the corrections, positive and negative, shall be as small as possible.*

He justified these two conditions as follows: The first he considered to be required by the traditional assumption that positive and negative errors are of equal probability; and the second, to be necessary in order to bring the solution into closest possible agreement with the observations.[128]

As to what this means, we imagine for values $x_0, x_1, \ldots, x_{n-1}$ of the variable x we have observed values $y_0, y_1, \ldots, y_{n-1}$ of y and we want to fit a line $y = a + bx$ to the data. Bošković gives two conditions a and b must satisfy. First, the deviations between observed and predicted values must sum to 0:

$$\sum_{i=0}^{n-1} \left(y_i - a - bx_i \right) = 0. \tag{35}$$

Second, the sum of the absolute values of the deviations must be minimum among all such lines:

$$\sum_{i=0}^{n-1} \left| y_i - a - bx_i \right| \text{ is minimum.} \tag{36}$$

Condition (I), i.e., (35) readily yields some information:

$$\sum_{i=0}^{n-1} \left(y_i - a - bx_i \right) = \sum_{i=0}^{n-1} y_i - \sum_{i=0}^{n-1} a - b \sum_{i=0}^{n-1} x_i = 0,$$

whence

$$\bar{y} - a - b\bar{x} = \frac{1}{n}\sum_{i=0}^{n-1} y_i - \frac{1}{n} \cdot na - \frac{1}{n} \cdot b \sum_{i=0}^{n-1} x_i = \frac{1}{n} \cdot 0 = 0,$$

---

[128] Eisenhart, *op. cit.*, pp. 200 - 201. I have chosen to quote Eisenhart for the clarity of his prose. Stigler (*A History...*, *op. cit.*, p. 46) offers an English translation of Bošković's own statement which refers very specifically to the geodesic problem he was considering.

and the centre of gravity, i.e., the point with coordinates given by the mean of the x_i's and the mean of the y_i's, must lie on the line $y = a + bx$. Moreover, replacing a by $\overline{y} - b\overline{x}$ transforms the sum in (36) into

$$\sum_{i=0}^{n-1} \left| (y_i - \overline{y}) - b(x_i - \overline{x}) \right|. \tag{37}$$

Finding the value of b that minimises (37) is nowadays fairly easy on the calculator. We have but to graph the function on the calculator and use it to locate b approximately. On the calculator, of course, b will be represented by the variable X. Now the sum that we want to minimise is the sum of a list. So we will use the list summation to avoid having to write a sum of n carefully entered terms. To this end, one first enters $x_0, x_1, \ldots, x_{n-1}$ and $y_0, y_1, \ldots, y_{n-1}$ into lists L_1 and L_2, respectively. If you haven't already done so, do it now with the x_i's and y_i's of *Table 2*. Then store the deviations in L_3 and L_4:

$L_1 - \text{mean}(L_1) \rightarrow L_3$
$L_2 - \text{mean}(L_2) \rightarrow L_4$.

Now enter the function in the equation editor:

$Y_1 = \text{sum}(\text{abs}(L_4 - X*L_3))$.

The hard part about graphing, of course, is choosing the window. I do not want to go into details here, but it turns out b is to be found among the ratios

$$\frac{y_i - \overline{y}}{x_i - \overline{x}}, \quad x_i \neq \overline{x}. \tag{38}$$

So perform all the divisions:

$L_4 / L_3 \rightarrow L_5$

and find the extreme values

$\{\text{min}(L_5), \text{max}(L_5)\}$.

The result, for the data at hand, is

$\{.2453733333 \ .303576\}$,

and $Y_1(\text{Ans})$ yields

$\{.3411 \ .182724\}$,

So we can press the WINDOW button and enter the following values:

Xmin$=.24$
Xmax$=.31$
Xscl$=.01$
Ymin$=0$
Ymax$=.35$
Yscl$=.1$.

When we (turn off Plot1 and) press the GRAPH button we see a polygonal line that looks like a straight line descending to a point where it changes to a clunky ascending curve. Inspection shows the minimum to lie somewhere between .28 and .29. The TRACE button will yield a minimum Y value of .11748 at X = .28393617. We can, of course, try for a better estimate by pressing CALC and choosing minimum, selecting .28 for the left bound, .29 for the right, and, say, .284 for the guess. The calculator yields a minimum of .11748 at .28400376. If we zoom in, centering our zoom near this point, we see what appears to be a polygonal line with a tiny flat interval starting at .28393617 and ending around .28449468. The function is minimised for all b in that interval. We are free to choose any value in the interval. Having come from the pre-calculator days of hand computation, I choose .284 as the simplest.

From the equation $\overline{y} = a + b\overline{x}$, we quickly calculate a:

mean(L_2)$-$.284mean(L_1) .

The result is 15.23289. We can now go into the equation editor, turn on Plot1, turn off Y_1, and enter

Y_2=15.23289+.284X,

and graph the result using ZoomStat. The fit is quite good.

Bošković did not have a calculator and had to perform all the divisions by hand[129], but so many divisions make the method computationally intensive for large groups of data. The Method of Least Squares would alleviate this.

Credit for the introduction of the *Method of Least Squares* attaches itself to Legendre, who was the first to publish[130]; the American Robert Adrain[131], who derived, albeit incompletely, the normal distribution of errors and the Method of Least Squares which "he seems to have learned... from Legendre's memoirs" [132]; and Gauss, who first published the Method in his *Theoria motus* in 1809, claiming priority for having known and used it since 1795.

The Method of Least Squares is very easy to describe. In place of Bošković's two conditions (I) and (II), i.e., (35) and (36), there is the single condition that the sum of the squares of all the corrections shall be as small as possible, i.e.,

---

[129] His method of calculation was slightly different, but he still had to perform all the divisions of (38). *Cf.* Eisenhart, *op.cit.*, for details: pp. 202 - 204 for Bošković's approach and pp. 204 - 205 for Laplace's approach. *Cf.* also Appendix B, below.

[130] Adrien Marie Legendre, *Nouvelles méthodes pour la détermination des orbites des comètes*, Courcier, Paris, 1805. Anthologised versions of that portion of the work discussing the Method of Least Squares can be found in D.E. Smith, *op. cit.* (English), and Schneider, *op. cit.* (German).

[131] Robert Adrain, "Research concerning the probabilities of the errors which happen in making observations", *The Analyst; or Mathematical Museum* 1, no. 4, (1808), pp. 93 - 109. The journal was most obscure, lasting only a single year, and Adrain's contribution was unknown in Europe. This, along with a couple of later papers of his applying the Method, was anthologised in Stephen M. Stigler, ed., *American Contributions to Mathematical Statistics in the 19th Century*, Arno Press, New York, 1980.

[132] Gnedenko and Sheǐnin, *op. cit.*, p. 229.

$$\sum_{i=0}^{n-1}\left(y_i-a-bx_i\right)^2 \text{ is minimum.} \tag{39}$$

Using the Calculus, finding a, b minimising this reduces to solving the two linear equations in two unknowns a, b:

$$2\sum_{i=0}^{n-1}\left(y_i-a-bx_i\right)(-1)=0$$

$$2\sum_{i=0}^{n-1}\left(y_i-a-bx_i\right)(-x_i)=0.$$

The first is obviously equivalent to (35) and thus yields

$$\bar{y}=a+b\bar{x}. \tag{40}$$

The second becomes, after a little algebra,

$$\sum_{i=0}^{n-1}y_ix_i=a\sum_{i=0}^{n-1}x_i+b\sum_{i=0}^{n-1}x_i^2. \tag{41}$$

The coefficients are quickly calculated:

$$\bar{y}=\mathsf{mean}(\mathsf{L}_2),\quad \bar{x}=\mathsf{mean}(\mathsf{L}_1)$$

$$\sum_{i=0}^{n-1}y_ix_i=\mathsf{sum}(\mathsf{L}_2*\mathsf{L}_1),\quad \sum_{i=0}^{n-1}x_i=\mathsf{sum}(\mathsf{L}_1),\quad \sum_{i=0}^{n-1}x_i^2=\mathsf{sum}(\mathsf{L}_1^2),$$

and the pair of equations is now easily solved by the familiar techniques.

The Method of Least Squares has become so commonplace that it has been built into the calculator. Press the STAT button, enter the CALC submenu, and scroll down to 8: LinReg(a+bx) and hit ENTER. Add the variables L_1, L_2, Y_3:

LinReg(a+bx) $\mathsf{L}_1,\mathsf{L}_2,\mathsf{Y}_3$.

This will perform a *linear regression*[133] on the lists, solving equations (40) and (41), and entering the result in the equation editor:

Y_3=15.227665714286+.28608971428571X.

This gives quite a bit more precision than is realistically called for, but it is simpler to leave it as is than to start deleting digits. Anyway, with it now stored in a function variable, it is a simple matter to graph the line with the plotted points and verify that it gives an excellent fit.

15 Exercise. If you haven't done anything else with the function variables since our discussion of Bošković's method, you should still have

---

[133] *Cf.* the section on population growth in Appendix B for a tiny bit more on regression.

$$Y_2 = 15.23289 + .284X$$

in the editor. Compare the results:

i. Enter $\mathsf{sum(abs(L_2 - Y_i(L_1)))}$ for $i = 2, 3$ to compare the sums of the absolute deviations.

ii. Enter $\mathsf{sum((L_2 - Y_i(L_1))^2)}$ for $i = 2, 3$ to compare the sums of the squared deviations.

Thus we have the Method of Least Squares: choose the curve which minimises the sum of the squares of the deviations from the mean of the observed values. By the time Gauss published his account of it, it had already appeared in print in works of Legendre and Adrain. As already noted, Adrain's contribution was unknown to Gauss, who was, however, aware of Legendre's paper and acknowledged, if not very graciously, its existence:

> Our principle, which we have made use of since the year 1795, has lately been published by LEGENDRE in the work *Nouvelles methodes pour la determination des orbites des cometes, Paris*, 1806[134], where several other properties of this principle have been explained, which, for the sake of brevity, we here omit.[135]

Questions of authorship aside, Gauss's name merits connexion with the Method of Least Squares for another reason: he sought to justify its use on probabilistic grounds. On assumptions that the error distribution was smooth, symmetric about a maximum at 0, and that the conditional distribution given certain observed values was maximised by the average of the values, Gauss had concluded[136] the distribution function to take the form

$$\varphi(y) = \frac{h}{\sqrt{\pi}} e^{-h^2 y^2}.$$

Now, if $y_0, y_1, \ldots, y_{n-1}$ are the observed values from n independent observations,

$$\Omega(y) = \prod_{i=0}^{n-1} \frac{h}{\sqrt{\pi}} e^{-h^2 (y - y_i)^2} = \left(\frac{h}{\sqrt{\pi}}\right)^n e^{-h^2 ((y - y_0)^2 + (y - y_1)^2 \ldots + (y - y_{n-1})^2)}$$

is the distribution function for the probability that y is the correct value given that $y_0, y_1, \ldots, y_{n-1}$ have been observed. And this is maximised when the absolute value of the exponent of e is minimised, i.e., when

$$(y - y_0)^2 + (y - y_1)^2 + \ldots + (y - y_{n-1})^2 \text{ is minimum.}$$

Thinking of y as $a + bx$, we see that $\Omega(y)$ is maximised when a, b minimise (39), whence the Method of Least Squares yields a, b determining the most likely estimate of y given the observed values. In Gauss's own words,

---

[134] I imagine this refers to the 1806 reprint of Legendre's memoir which contained a supplement. A second supplement was added in 1820.

[135] Davis, *op. cit.*, p. 270.

[136] His argument is given in Appendix B.

We will now develop the conclusions from this law. It is evident, in order that the product

$$\Omega = h^\mu \pi^{-\frac{1}{2}\mu} e^{-hh(vv+v'v'+v''v''+...)}$$

may become a maximum, that the sum

$$vv + v'v' + v''v'' + \text{etc.},$$

must become a minimum. *Therefore, that will be the most probable system of values of the unknown quantities* p, q, r, s, *etc.*[137]*, in which the sum of the squares of the differences between the observed and computed values of the functions* V, V', V'', *etc. is a minimum...*[138]

Praise for Gauss's achievement is not universal. Stigler declares:

> Viewed from the perspective of more than a century and a half later, we can see much to criticize in Gauss's argument. First and foremost, the argument was a logical aberration—it was essentially both circular and non sequitur. In outline its three steps ran as follows: The arithmetic mean (a special, but major, case of the method of least squares) is only "most probable" if the errors are normally distributed; the arithmetic mean is "generally acknowledged" as an excellent way of combining observations so that errors may be taken as normally distributed (as if the "general" scientific mind had already read Gauss!); finally, the supposition that errors are normally distributed leads back to least squares. Even Gauss himself, returning to this subject twelve and thirty years later was to find this chain of reasoning unpalatable.[139]

I find this a bit strong. Gnedenko and Sheynin offer a milder criticism:

> As a simple corollary [to his derivation of the error curve], Gauss found that the density function [i.e., probability distribution function] of a given set of observations attains its maximal value when the sum of the squares of the discrepancies between the observed values of the measured constant and its 'true' value attains its minimum. This was his justification of the principle of least squares.
> Certain faults were inherent in this exceptionally elegant derivation and Gauss himself indicated them later on. First, only normally distributed errors were recognised as random errors of observations. In pointing out this restriction in §17 of part 1 of his "Theory of combinations of observations least prone to errors" (*Theoria combinationis observationum erroribus minimis obnoxiae*. Göttingen 1823) Gauss added that he will offer "a new discussion of the subject" so as to prove that "the method

---

[137] I.e., a, b. Gauss considers the general case in which y may depend linearly on any number of parameters. I add that his $v, v', v'', \ldots$ and V,V', V'', ...are our $x_0, x_1, \ldots, x_{n-1}$ and $y_0, y_1, \ldots, y_{n-1}$, respectively.

[138] Davis, *op. cit.*, p. 260; Schneider, *op. cit.*, p. 243.

[139] Stigler, *A History...*, *op. cit.*, pp. 141 - 143.

of least squares leads to the best combination of observations... for any probability law of the errors... "[140]

Gnedenko and Sheynin's remarks are not only more moderate than Stigler's, but they are a bit more measured. Gauss's argument is certainly neither circular nor an aberration. He assumes—on the basis of practice—that the arithmetic mean yields the most likely estimate and concludes the distribution to be normal. The added argument then shows the converse, that the arithmetic mean yields the most likely estimate for a normal distribution. Additionally, the argument shows that, in this case, the Method of Least Squares leads to the optimum result. As a justification of the Method of Least Squares, the argument applying only to normal distributions is a bit disappointing, but it is not as bad as Stigler finds it to be. Gauss produced a new argument, published a dozen years later in his *Theoria combinationis observationum erroribis minimis obnoxiæ* (1823), but not because he apparently found the argument unpalatable, rather because he found a more general proof:

In the "Theory of motion of heavenly bodies" we have shown, how the *most probable* values of the unknowns is to be derived when the law for the probability of observational error is known; and as the nature of this law remains hypothetical in almost all cases, we have applied to this theory the most plausible law, by which the probability of an error x is taken to be proportional to $e^{-h^2 x^2}$; and from this developed the procedure, which had long been used by us, particularly in astronomical calculations, and which is now applied by most reckoners under the name of the method of least squares.

Later LAPLACE showed, attacking the problem differently, that exactly this principle, how the law of the probability of errors can also be obtained, in preference to all others, if the number of observations is very large. If however the number of observations is moderate, the question remains undecided, so that by proposing our hypothetical law the method of least squares is only deserving to be recommended over others, because it is best suited for the simplification of calculations.

We hope therefore to prove to be of service to mathematicians in that we show by this new treatment of the matter, that the method of least squares offers the best of all combinations, and indeed is not approximately but unconditionally, whatever the probability law for errors and whatever the number of observations are, if one only takes the definition of mean error not in the sense of LAPLACE but as given by us in Articles 5 and 6.[141]

Oft cited in this respect is a letter from Gauss to Bessel which begins as follows:

---

[140] Gnedenko and Sheĭnin, *op. cit.*, p. 227.
[141] Schneider, *op. cit.*, p. 249. Gauss's definition of mean error [*mittleren Fehler*] is given explicitly in Article 7 (p. 246 in Schneider) and amounts to the standard deviation of the distribution. This quote is from Article 17.

Highly respected friend.

I have yet to send you my warmest thanks for the generous gift which you have made me of the work on your degree measurements and of the latest volume of the *Königsberger Beobachtungen*.

I have read your essay in the *Astronomische Nachrichten* on the approximation of the law for the probability of observational errors from combined sources to the formula $e^{-\frac{xx}{hh}}$ with greatest interest, though if I should speak honestly, this interest refers less to the subject matter itself than to your presentation. For it has been familiar to me for many years, whereas I have never myself got round to carrying out its development completely.[142]

That I have incidentally abandoned the metaphysics applied in the *Theoria motus corporum cœlestium* to the Method of Least Squares, has been done chiefly on a ground that I have never publicly mentioned. Namely I must hold it in all ways less important to determine the value of an unknown quantity whose probability is the greatest, which of course remains ever infinitely small, as much more [to find] that [value] at which one has attained at the least unfavourable game; or, if fa denotes the probability of the value a for the unknown x, it matters less that fa is a maximum than that $\int fx - F(x - a) \cdot dx$ [143] extended over all possible values of x becomes a minimum; where for F is chosen a function which is always positive and in a proper fashion grows always larger for larger arguments. That one chooses the square for this is purely arbitrary, and this arbitrariness lies in the nature of the subject. Without the known extraordinarily large advantages which the choice of the square offers, one could choose any other [function] respecting these conditions and it is done in wholly singular cases. I do not know, however, if I have expressed myself sufficiently clearly on this, what I actually mean with this distinction, and apologise in the contrary case.

... [144]

Perhaps a few words of clarification are in order. In earlier times the word "probability" ambiguously referred to what we now call "probability" and also to what we now call the "odds". Gauss used the word ambiguously to describe the density function or likelihood and its integral or probability. In the discrete case the likelihood of an outcome equals its probability when the outcome is considered as an event, while in the continuous case the probability of any individual outcome is 0 and the likelihood measures a density of probability and is nowadays called the density or distribution function. This ambiguity is harmless enough, but perhaps should be noted.

---

[142] Believe it or not, some people found such Gaussian claims for credit annoying. Legendre and János Bolyai are perhaps the most famous targets of this practice.

[143] Nowadays we would write $\int f(x)F(x - a)dx$.

[144] Letter from Gauss to Bessel, 28 February 1839, in: *Briefwechsel zwischen Gauss und Bessel*, Verlag von Wilhelm Engelmann, Leipzig, 1880, pp. 523 - 524.

The important bit of clarification is this. In Gauss's first justification of the Method of Least Squares he showed that the method maximised the likelihood of the value chosen, i.e., it produced a such that $f(a)$ was maximum. The second proof, for $F(x) = x^2$, showed that the method minimised the expectation of error,

$$E(a) = \int f(x)F(x - a)dx,$$

where $F(x)$ is taken to be the measure of an error x. The first justification held only for normal distributions, whereas the second was valid in more general circumstances. It is a matter of some subtlety that would only achieve greater clarity with greater experience dealing with such generalised expected values.

Sheynin sums up the situation with an impressively concise display of erudition as follows:

9.6. More about the Method of Least Squares

1) In spite of Gauss' opinion, his first justification of the MLSq became generally accepted, in particular because the observational errors were (and are) approximately normal whereas his mature contribution was extremely uninviting; and the work of **Quetelet** and **Maxwell** did much to spread the idea of normality. Examples of deviation from the normal law were however accumulating both in astronomy and in other branches of natural sciences as well as in statistics. And, independently from that fact, several authors came out against the first substantiation. **Markov**, who referred to Gauss himself (to his letter to **Bessel**) is well known in this respect but his first predecessor was **Ivory**. The second justification was sometimes denied as well. Thus **Bienaymé** declared that Gauss had provided considerations rather than proofs; see also **Poincaré's** opinion.

2) When justifying the MLSq in 1823 in an essentially different way, Gauss called the obtained estimators *most plausible (maxime plausibiles*, or, in his preliminary note (1821), *sicherste*, rather than as before, *maxime probabile, wahrscheinlichste*. For the case of the normal distribution, these are jointly effective among unbiassed regular estimators. The second substantiation of the MLSq can be accomplished by applying the notions of multidimensional geometry. Kolmogorov also believed that the formula for m^2 should, after all, be considered as its definition. Much earlier Tsinger stated that it already "concealed" the MLSq.

3) Mathematicians had not paid due attention to Gauss' work on the MLSq, and neither did statisticians... [145]

[145] Sheynin, Oscar B., *Theory of Probability. A Historical Essay*, 2nd revised and enlarged edition, 2009. The work has no page numbers listed; however, the quote is from pages 125 - 126 in the pdf file, which is available online. In reproducing these words I have omitted the bibliographic references. The attentive reader will note that a right parenthesis is missing from this quotation. I haven't quite decided whether Sheynin meant to place it after *plausibiles* or after *wahrscheinlichste*.

A modern, matter-of-fact summary of the matter is given by Freudenthal and Steiner, whom I quote for the extra insight they offer:

Linear Estimation and Related Matters

17. The estimation-calculation[146] is the first proper application of probability theory. In the 18th century the question became more pressing, how one could balance inconsistent astronomical and geodetic observations, e.g., how one should distribute the deviation of the sum of the measured angles of a triangle from 180° among the three angles. More generally: one associates each measured quantity with a point in some m-dimensional space R, described by (presumably linear or approximately linear) conditional equations, so that the observed point should lie on a linear manifold V from R. How does one bring it to V, if it deviates from V?

The answer is very natural: By the shortest path, thus through orthogonal projection. This answer is called after LEGENDRE (1805) the *method of least squares*, precisely because one measures distance with the root of the sum of squares— today we should rather say: method of the shortest distance. That this method was not so self-evident around the turn of the 18th century as it is today is shown by the search for other methods before and after LEGENDRE—only GAUSS had, independently, practised LEGENDRE's method since 1795 already. Outright unfruitful discussion over the foundation of the method of least squares extended far into the 19th century. We shrug our shoulders because in the meantime we have learned to think in n dimensions. In the 19th century the simple geometric facts were drowning in a chaotic algebraic formalism.[147]

Freudenthal and Steiner then discuss some technical points including Gauss's derivation of the error curve, finishing the discussion with

> GAUSS's essay is artificial and barely convincing. Still GAUSS had posed a problem which LAPLACE dismissed. Was it not really so, that the observations of a quantity (and not only their mean) distributed according to DE MOIVRE? There was empirical evidence that testified to this. And if one accepted on the basis of this evidence the error law being of the de-Moivre form as an empirical law, how should one explain this theoretically?
>
> 19. It is hardly amazing that the answer to this question came at once from several sides independently. It is called the *"hypothesis of elementary errors"* and would be given by THOMAS YOUNG, G. HAGEN,

---

[146] Freudenthal and Steiner use the word *Vergleichung*, meaning balance or compensation. "Estimation" seems to be more descriptive. *Cf.* Stigler, *A History...*, *op. cit.*, p. 143, for such a use of the word.

[147] Hans Freudenthal and Hans-Georg Steiner, "Aus der Geschichte der Wahrscheinlichkeitstheorie und der mathematischen Statistik", in: Heinrich Behnke, Günther Bertram and Robert Sauer, eds., *Grundzüge der Mathematik IV*, Vandenhoek & Ruprecht, Göttingen, 1966; here: p. 175

F.W. BESSEL[148], and A. QUETELET. The observational errors combine additively from a large number of "elementary errors" and thereby according to the central limit theorem obey approximately the de-Moivre law.[149]

The "Hypothesis of Elementary Errors", like the "Law of Large Numbers" and the "Central Limit Theorem", is a vague designation of a general principle, originating in this case with Laplace, given reformulations and various justifications of varying levels of rigour and simplicity, and eventually given a name. A popular formulation, cited by Hald, was given by Gotthilf Heinrich Ludwig Hagen (1797 - 1884) in a textbook (1837) for civil engineers:

> The hypothesis, which I make, says: the error in the result of a measurement is the algebraic sum of an infinitely large number of elementary errors which are all equally large, and each of which can be positive or negative with equal ease.[150]

Hagen assumed a literal version of this and derived the error curve on that basis in a proof Hald reports was often reprinted. I refer the interested reader to Appendix B for a bit more on this matter.

Bessel's paper[151], which inspired the remarks on the Method of Least Squares in Gauss's oft-cited letter to Bessel quoted from above, was another such attempt to offer theoretical justification of the principle overcoming some simplifying assumptions Laplace had based his derivation on.

5 Chebyshev's Inequality

The presentation of Chebyshev's Inequality and the simple derivation of the Law of Large Numbers from it almost belongs at the end of Chapter 5 rather than here. However, there are good reasons, historical and logical, why it belongs here. Chronologically it came after the works of Laplace and Gauss, and, indeed, arose from the effort to rigorise Laplace's derivation of the method of least squares. The logical reasons are that the result makes essential reference to the standard deviation and variance, concepts which crystallised in the works of de Moivre and Laplace. Moreover, the result, concerning random variables, is of a higher level of abstraction than Bernoulli's Theorem and is more modern in character.

Chebyshev's Inequality should properly be called Bienaymé's Inequality and is sometimes referred to as the Bienaymé-Chebyshev Inequality, as Irénée Jules

---

[148] This was done in the essay referred to by Gauss in his letter quoted from above.

[149] Freudenthal and Steiner, *op. cit.*, p. 177.

[150] Hald, *A History...*, *op. cit.*, p. 167.

[151] F.W. Bessel, "Untersuchungen über die Wahrscheinlichkeit der Beobachtungsfehler", *Astronomische Nachrichten* 15 (1839), pp. 369 - 404. The paper is reprinted in the second volume of Bessel's collected works (1876) and an excerpt can be found in Schneider, *op. cit.*

Johann Franz Encke (1791 - 1865) studied under Gauss, who arranged Encke's first astronomical post. Encke is particularly noted for applying and improving Gauss's technique of calculating orbits. His most famous cometary calculation was of the orbit of Encke's comet. First observed by Pierre François André Méchain in 1786, then by Caroline Herschel in 1795, the comet was rediscovered by Jean Louis Pons in 1805 and again observed by him in its 1818/19 re-appearance. Encke calculated its orbit and predicted its re-appearance in 1822: Encke's comet has the shortest period (3.3 years) of any known comet.

Of interest to us here is not Encke's comet, however, but his having published a long three-part exposition of the Method of Least Squares in three successive volumes (1834, 1835, 1836) of the *Berliner Astronomisches Jahrbuch*.

The stamp shown here was issued by the Comoro Islands in 1986 in a set cashing in on the worldwide philatelic celebration of the 1985 return of Halley's comet. The Comoro stamps featured a few famous comets, those for Halley's and Encke's also displaying portraits of the men after whom they were named.

Comets and probability theory have crossed paths on occasion, directly via the Method of Least Squares and the calculations of orbits, and indirectly by sharing some scientists. One of the great heroes of our story, Jakob Bernoulli, was less than successful in his cometary work. I refer the reader to pp. 165 - 168 of Donald K. Yeomans, *Comets; A Chronological History of Observation, Science, Myth, and Folklore*, John Wiley & Sons, Inc., New York, 1991.

Ernst Abbe (1840 - 1905), shown here on a stamp of the German Democratic Republic issued in 1956 is best known for his optical work for the firm of Carl Zeiss. He was also Professor of Astronomy and director of the observatory in Jena, but is important here for his statistical work. His habilitation thesis (Jena, 1863), titled "Ueber die Gesetzmässigkeit in der Vertheilung der Fehler bei Beobachtungsreihen" ["On the conformity-to-a-law of the distribution of errors in a series of observations". (A short excerpt can be found in Schneider, *op. cit.* and a longer excerpt in English translation in David and Edwards, *op. cit.*)], dealt with the *goodness of fit* of the Gaussian distribution and anticipated important 20th century developments in statistics. [The year 1846 on the stamp refers to the founding of the firm VEB Carl Zeiss.]

Bienaymé (1796 - 1878) was the first to prove and publish the result.[152] However, the inequality "was not appropriately distinguished from the general text of the work"[153] and is easily overlooked.

Pafnuty Lvovich Chebyshev (1821 - 1884) was much more explicit when he published a paper "On mean values" in 1867 in Russian[154] and in French translation[155]. Here the inequality is presented as the main result, properly labelled as a Theorem, provided a rigorous and completely elementary proof, and applied to derive the Law of Large Numbers as a corollary.

Although Bienaymé's paper was reprinted with the French version of Chebyshev's, immediately preceding it in fact, the result is more memorably presented in Chebyshev's paper which, continuing the practice de Morgan had criticised Laplace for, contained no attribution. Reading both papers, one could overlook or forget its presence in Bienaymé but not in Chebyshev. Chebyshev had, in a very real sense, made the result his own. He may or may not have independently rediscovered it, he gave it a new proof, and he applied it explicitly to give a new proof of Bernoulli's Theorem. Later, in fact, he would take a closer look at Bienaymé's proof and crystallise the method of moments using it to give an almost complete, new and rigorous proof of the Central Limit Theorem itself.[156] It is thus not too surprising that Chebyshev's name would attach itself to Bienaymé's inequality and Bienaymé's connexion with it de-emphasised or even forgotten. Reinforcing this, of course, is the *Matthew Effect*: "Them that's got shall get, Them that's not shall lose"[157]. In science this refers to the fact

---

[152] Irénée Jules Bienaymé, "Considérations à l'appui de la découverte de Laplace sur la loi de probabilité dans la méthode des moindres carrés" ["Reasons in support of the discovery of Laplace over the law of probability in the method of least squares"], *Compte rendus hebdomadaires des séances de l'Academie des sciences* 37 (1853), pp. 309 - 325; reprinted in *Journal de mathématiques pures et appliquées* 12, ser. 2, (1867), pp. 158 - 176. The extract containing the result in question appears in German translation in Schneider, *op. cit.*, pp. 152 - 153. Moreover, his derivation of the inequality is essentially that given in textbooks today (see below).

[153] B.V. Gnedenko and O.B. Sheĭnin, "Chapter Four. The Theory of Probability", in: A.N. Kolmogorov and A.P. Yushkevich, eds., *Mathematics of the 19th Century: vol. 1, Mathematical Logic–Algebra–Number Theory–Probability Theory*, 2nd revised edition, Birkhäuser Verlag, Basel, 2001; p. 262. The translation of the volume is by A. Shenitzer, H. Grant, and O.B. Sheynin.

[154] *Matematicheskii sbornik* 2 (1867), pp. 1 - 9.

[155] Pafnouty Lvovitsch Tchebychef, "Des valeurs moyennes", *Journal de mathématiques pures et appliquées* 12, ser. 2, (1867), pp. 177 - 184. The paper was reprinted in the first volume of his collected works (pp. 685 - 694) and a complete German translation from the French appears in Schneider, *op. cit.* The English translation in David Eugene Smith, ed., *A Source Book in Mathematics* (Cambridge University Press, Cambridge, 1929, reprinted by Dover Publishing Company, New York, 1959) is complete but drops a number of ellipses where in midproof a sequence $x, y, z, \ldots$ of indeterminate length becomes the shorter x, y, z.

[156] This new proof would be completed by his student Andrei Andreevich Markov.

[157] As formulated in the lyrics of the tune "God Bless the Child" by Billie Holiday and Arthur Herzog, Jr.

that the greater portion of shared credit tends to accrue to the more famous of the claimants for credit. Bienaymé was no slouch, but

> Bienaymé was far ahead of his time in the depth of his statistical ideas. Because of this and the other characteristics of his work, and his being overshadowed by the greatest figures of his time, his name and contributions are little known today.[158]

One of those overshadowing him, particularly in probability theory, was Chebyshev who not only contributed to the development of probability theory through his own research but also founded the Russian school of probability which produced fundamental results in the field at the end of the 19th and beginning of the 20th centuries.

With due attention having been paid to its authorship, I present the following:

16 Theorem (Chebyshev's Inequality). *Let X be a random variable, $\mu = E(X), \sigma = \sigma(X)$. For any $t > 0$,*

$$P(|X - \mu| \geq t) \leq \frac{\sigma^2}{t^2}.$$

Proof. Let Ω denote the sample space, f its distribution function, and let

$$\Omega_t = \{x \in \Omega \mid |x - \mu| \geq t\}.$$

By definition

$$\sigma^2(X) = \sum_{x \in \Omega} (x - \mu)^2 f(x)$$
$$\geq \sum_{x \in \Omega_t} (x - \mu)^2 f(x)$$
$$\geq \sum_{x \in \Omega_t} t^2 f(x) = t^2 \sum_{x \in \Omega_t} f(x)$$
$$\geq t^2 P(\Omega_t) = t^2 P(|X - \mu| \geq t),$$

whence

$$P(|X - \mu| \geq t) \leq \frac{\sigma^2}{t^2}. \qquad \square$$

The proof really is this simple, and this is the reason I have not included a passage from either author: Explaining Bienaymé's notation alone would take up more space than required by the above proof, and Chebyshev's proof is rather more complicated and less revealing. To include something of an historical extract, I quote William Feller's remarks:

---

[158] C.C. Heyde and E. Seneta, "Bienaymé, Irénée-Jules", in: Charles Coulston Gillispie, ed., *Dictionary of Scientific Biography*, vol. 15, Charles Scribner's Sons, New York, p. 32. One contribution that is occasionally explicitly credited to Bienaymé is Bienaymé's Identity, which the reader can find as Corollary 28 on page 467 in Appendix B, below.

Chebyshev's inequality must be regarded as a theoretical tool rather than a practical method of estimation. Its importance is due to its universality, but no statement of great generality can be expected to yield sharp results in individual cases.

Examples. (a) If $\mathbf{X}$ is the number scored in a throw of a true die, then... $\mu = \frac{7}{2}, \sigma^2 = \frac{35}{12}$. The maximum deviation of $\mathbf{X}$ from μ is $2.5 \approx 3\sigma/2$. The probability of greater deviations is zero, whereas Chebyshev's inequality only asserts that this probability is smaller than 0.47.

(b) For the binomial distribution $\{b(n, p)\}$ we have... $\mu = np, \sigma^2 = npq$. For large n we know that

$$\mathbf{P}\{|\mathbf{S}_n - np| > x(npq)^{1/2}\} \approx 1 - \Phi(x) + \Phi(-x).$$

Chebyshev's inequality states only that the left side is less than $1/x^2$; this is obviously a much poorer estimate.[159]

On the other hand, we can apply Chebyshev's Inequality to the problem of male and female SAT scores considered in section 3 (*cf.* page 287, above). Our calculation presupposed the normality of the various distributions considered. Using Chebyshev's Inequality, we do not need this assumption. Let the random variable $X(S)$ be the mean of a sample S of the correct size, which was 720793, the number of male high school students taking the exam. The population mean and standard deviation were taken to be 516 and 116, respectively, with the mean for males being 534, 18 points above the mean. Now Chebyshev's Inequality yields

$$P(|X - 516| \geq 18) \leq \frac{\sigma^2/720793}{18^2} \approx .000576\ldots,$$

where we use $\sigma/\sqrt{n}$ for the standard deviation of the distribution of means[160]. Multiplying $\sigma/\sqrt{n}$ by the correction factor $\sqrt{(N - n)/N - 1}$ for the student population $N = 1547990$ will merely reduce the probability estimate. Generally, one would reject the null hypothesis, in this case that $\mu_{\text{male}} = 516$, if the probability of its being true was less than .05 or .01, the cut-off to be chosen in advance. The reason for choosing the cut-off in advance is to keep the statistician honest and free of any charge of manipulating the results should the probability estimate turn out to lie between .01 and .05. As it is much less than either of these, we can confidently assert that the difference in mean scores for male and female test-takers is not attributable to chance and that the difference between these means is *statistically significant*.

As mentioned, the Law of Large Numbers is a corollary to Chebyshev's Inequality.

17 Corollary (Law of Large Numbers). *Let $X_0, X_1, \ldots$ be a sequence of pairwise independent random variables with a common distribution and $\mu = E(X_i)$ the common expected value. Let $S_n = X_0 + \ldots + X_{n-1}$. For every $t > 0$,*

---

[159] Feller, *op. cit.*, vol. 1, pp. 219 - 220.
[160] *Cf.* page 468 for the derivation of this value.

$$\lim_{n\to\infty} P\left(\left|\frac{S_n}{n} - \mu\right| > t\right) = 0.$$

The result in this generality, which includes infinite sample spaces for which the variance is infinite, is due to Aleksandr Khinchin, who published the proof in 1929. The reduction of the result in this generality to Chebyshev's Theorem requires an additional trick and for that reason I do not include the proof here, but refer the interested reader to Feller's textbook[161]. What I will do is give the proof in the Bernoulli case, which is more immediate.

18 Corollary (Bernoulli's Theorem). *Let* $X_0, X_1, \ldots$ *be a sequence of pairwise independent random variables sharing common mean and variance:* $\mu = E(X_i), \sigma^2 = \sigma^2(X_i)$. *Let* $S_n = X_0 + \ldots + X_{n-1}$. *For every* $t > 0$,

$$\lim_{n\to\infty} P\left(\left|\frac{S_n}{n} - \mu\right| > t\right) = 0.$$

Proof. S_n/n is itself a random variable and, as noted in section 3 above (and as will be proven in Appendix B),

$$E(S_n/n) = \mu, \quad \sigma^2(S_n/n) = \sigma^2/n.$$

One simply applies Chebyshev's Inequality to the random variable S_n/n:

$$P\left(\left|\frac{S_n}{n} - \mu\right| > t\right) \leq P\left(\left|\frac{S_n}{n} - \mu\right| \geq t\right)$$
$$\leq \frac{\sigma^2/n}{t^2} = \frac{\sigma^2}{nt^2}.$$

$t > 0$ and σ^2 are constants, whence $\sigma^2/(nt^2)$ goes to 0 as n tends to ∞. □

At the risk of belabouring the obvious, thoughts of the criticism given Bernoulli's Theorem reported on in the preceding chapter and of Feller's remark above on the poorer bounds afforded by Chebyshev's Inequality spur me on to see what the present proof does by way of bounds for Bernoulli's example. Recall the problem: how many trials n are needed to insure a probability $1000/1001$ that $|S_n/n - \mu|$ is less than $1/50$ given $p = 3/5$? Here, as shown in Appendix B, section 5, below, we have $\mu = p$ and $\sigma^2 = pq$. We want

$$P\left(\left|\frac{S_n}{n} - \frac{3}{5}\right| < \frac{1}{50}\right) \geq \frac{1000}{1001},$$

i.e.,

$$P\left(\left|\frac{S_n}{n} - \frac{3}{5}\right| \geq \frac{1}{50}\right) \leq 1 - \frac{1000}{1001} = \frac{1}{1001}. \tag{42}$$

We know by Chebyshev's Inequality that

$$P\left(\left|\frac{S_n}{n} - \frac{3}{5}\right| \geq \frac{1}{50}\right) \leq \frac{\sigma^2/n}{(1/50)^2},$$

---
[161] Feller, *op. cit.*, pp. 231 - 233.

Russian Probabilists

Russian probability started slowly with expository works by Viktor Yakovlevich Bun-yakovsky (1804 - 1889), Mikhail Vasilievich Ostrogradsky (1801 - 1862), and August Yul'evich Davidov (1823 - 1885). It quickly reached maturity with the work of Pafnuty Lvovich Chebyshev (1821 - 1894), who shared with Nikolai Ivanovich Lobachevsky (1792 - 1856) the honour of being the top Russian mathematicians of the day. Lobachevsky made only a minor excursion into probability theory, just enough to justify including his portrait here, Chebyshev published widely in mathematics, only a few of his main works being on probability, but these few were fundamental. He gave new proofs of the Law of Large Numbers, proved the Bienaymé-Chebyshev inequality, generalised the Law of Large Numbers and introduced a technique called the method of moments, a description of which must lie beyond the scope of this book.

The stamp bearing Ostrogradsky's portrait was issued in 1951 in celebration of his 150th birthday.

Lobachevsky's portrait appeared on one of a set of 16 stamps issued in 1951 honouring Russian scientists and the likeness was repeated with a slightly different border in 1956 on the occasion of the centenary of his death. Pictured here is the second issue.

The stamp bearing Chebyshev's portrait is one of two issued in 1946 by the Soviet Union in honour of Chebyshev's 125th birthday. The two stamps have face values of 30 kopecks and 60 kopecks, but have identical designs; they have slightly different colours and the 30 kopeck stamp comes in at least two colour variations.

Chebyshev started a whole school of mathematics and had numerous illustrious students, grandstudents, etc. His students who made contributions to probability theory include Aleksandr Mikhailovich Lyapunov (1857 - 1918), Aleksei Nikolaevich Krylov (1863 - 1945), and, above all, Andrei Andreevich Markov (1856 - 1922).

Lyapunov's 100th birthday was celebrated philatelically in 1957 with the issue of a very tiny stamp pictured here, and later a very attractive stamp issued by the Ukraine in 2010 in a set celebrating the 125th anniversary of the university of Kharkov, where he taught. A stamp commemorating the 10th anniversary of Krylov's death was issued in January 1956 and one celebrating his 100th birthday was issued in 1963.

To the right and slightly to the rear of his fellow mathematicians Henri Poincaré and Kurt Gödel on a Portuguese stamp issued in 2000 in celebration of 20th century achievements is Andrei Nikolaevich Kolmogorov (1903 - 1987). Kolmogorov is one of the great Russian probability theorists of the early 20th century. He is best remembered in the field for his work with Aleksandr Khinchin (1894 - 1959) on Markov processes and his axiomatic foundation for the theory of probability, *Grundbegriffe der Wahrscheinlichkeitsrechnung* [*Basic concepts of probability calculation*] (Springer-Verlag, Berlin, 1933; Russian translation 1936; and English translation by Nathan Morrison, *Foundations of the Theory of Probability*, Chelsea Publishing Company, New York, 1950). But for the presence of formulæ like the Addition Principle for Probabilities and some of the terminology, there is almost no overlap between this slim volume of abstract theory and the intensive combinatorial analysis of individual problems that filled probability textbooks of the 19th century and treatises of the 18th, and which still forms the subject matter of elementary courses in probability theory today.

Ostrogradsky

Russian
Probabilists

Lobachevsky

Chebyshev

Lyapunov

Krylov

Kolmogorov (far right)

whence (42) will hold if

$$\frac{\sigma^2/n}{(1/50)^2} = \frac{2500\sigma^2}{n} \le \frac{1}{1001},$$

i.e., if

$$n \ge 1001 \cdot 2500\sigma^2 = 1001 \cdot 2500 \cdot \frac{3}{5} \cdot \frac{2}{5}$$
$$\ge 1001 \cdot 100 \cdot 6 = 600600,$$

which is quite a bit larger than Bernoulli's much maligned lower estimate for n of 25550.

This brings us to the end of our story if not of this book. Throughout its history, Probability has undergone major changes. Up through Bernoulli it had been a combinatorial endeavour, the art of counting. Under de Moivre it became more analytic, relying on the Calculus of Finite Differences and the Calculus itself, reaching a peak under Laplace. Laplace marks the end of a mathematical era. The new era, stretching from the publication in 1817 of Bernard Bolzano's (1781 - 1848) proof of the Intermediate Value Theorem until around 1930, was largely concerned with foundations as mathematicians strove for rigour. Chebyshev introduced rigour into the analytic theory of Probability, a practice followed by his students, leading for example to the first completely rigorous proof, by Aleksandr Mikhailovich Lyapunov (1857 - 1918), of Laplace's version of the Central Limit Theorem in the early decades of the 20th century. The mathematical prerequisites for this discussion go beyond what I intended to assume for this book. The same must be said for the later developments, the introduction of abstract measure theory begun by Emile Borel in a paper of 1898, a publication which Jan von Plato singles out as the beginning of Modern Probability Theory[162], and brought to completion by Andrei Nikolaevich Kolmogorov (1903 - 1987) with the publication of his *Grundbegriffe der Wahrscheinlichkeitsrechnung* in 1933. Today Probability Theory is a branch of abstract measure theory bearing little resemblance to the subject as taught in middle school, high school, or service courses in college.

There are developments to the story other than the transformation of the field of Probability from applied mathematics to pure theory. The three that come most readily to mind are the applications of Probability in physics, beginning in earnest in the latter half of the 19th century[163]; the philosophers' attempts to explain what probability is, encompassing such issues as subjective probability; and the development of statistical theory and the creation of statistical tools. I deem the first and third of these as other stories for other writers and refer the reader to the book of von Plato[164] for the physics and the oft-cited books of Hald, Pearson, and Stigler for the statistics. The philosphical issues

---

[162] Jan von Plato, *Creating Modern Probability*, Cambridge University Press, Cambridge, 1994, p. 8.

[163] I discuss a simple example in the section on Maxwell in Appendix B, below.

[164] Von Plato, *op.cit.*

are something for which I have no particular talent in discussing and I simply refer the reader to von Plato[165] and Ian Hacking[166] as good starting points. In doing so I acknowledge that I leave my narrative never having explained what Probability is, i.e, what the book is about.

[165] *Ibid.*
[166] Ian Hacking, *The Emergence of Probability; A Philosophical Study of Early Ideas About Probability*, Cambridge University Press, Cambridge, 1975.

Part III

Appendices

A

The Calculator; Beyond the Basics

I assume the reader already to have some basic familiarity with the *TI-83 Plus* and its use in calculating and graphing numerical functions. What I do not assume is familiarity with the non-numerical data types the calculator was designed to handle, nor do I assume the reader has familiarised himself with programming on the calculator. The dual purpose of this Appendix is to explain the rudiments of programming on the *TI-83 Plus* and to introduce the data types used sufficiently to allow one to understand the applications we have made of the calculator in the present book.

1 Basic Programming

The modern calculator is a complex device with so many built-in functions and operations that most of its buttons must have multiple uses determined by shift keys[1], and, this being insufficient, some of the buttons are used just to access menus. Even with all of this, one quickly runs into computational tasks that are not handled directly by single buttons or menu items. When this happens one must break the calculation down into steps and either perform the steps one at a time by hand pressing buttons and choosing menu items one after another, or enter a program into the calculator to automate the calculator's performance of the sequence of tasks prescribed by the aforesaid button presses and menu selections.

There are advantages to such automation—speed, the ability to go back and edit the sequence of instructions to correct any errors, the guarantee that the steps will be followed without the sort of error humans make when performing repetitive and boring tasks, and the ability to quickly and easily perform the calculation anew on other sets of data. This last advantage is particularly useful in that it allows fairly effortless exploration and therewith the discovery of

---

[1] Initially, after evolving beyond the basic four-function calculator, the calculator had one shift key INV allowing a function and its inverse to share a single button. The *TI-83 Plus* has two shift keys: 2nd and ALPHA. The more advanced *TI-89 Titanium* has four: 2ND, ↑, ◇, and ALPHA.

patterns one might not see in a single worked out example. The programs BERNULLI and RUIN to be discussed later offer nice illustrations of this.

Programming is easily described. It is the translation into a highly formal language of a set of instructions giving the steps of an algorithm for performing a task. This is a very broad definition that covers not only programming of the kind one does in the program editor, but also the storing of explicitly definable functions in the equation (Y=) editor. An expression,

$$Y_1 = \text{int}(X/3) - (X/3 = \text{int}(X/3)),\qquad(1)$$

may not look like a program, but on the calculator it is used exactly the same way one would expect to use a program: It is a set of instructions stored in the calculator for later repeated use on various inputs. The instructions are these: Take whatever number is stored in the location denoted by X, divide it by three, find the greatest integer less than or equal to the result and store it somewhere, say in location a. Now take the number stored in the location denoted by X, divide it again by three, and store it in some location, say, b. Again take (the number stored in) X, divide it by 3, take the greatest integer in the result and compare it to the number stored in location b. If the two numbers are the same, subtract 1 from the number stored in location a, and, if not, subtract 0. When read this way, we see that the expression (1) is indeed a program, and a mildly complicated one at that. We might not have recognised it as such because it is written in ordinary algebraic symbolism and we are quite familiar with this symbolism.

What we usually consider to be a program is written in a more extensive, more powerful programming language, and has a number of lines. Indeed, we can almost define a program to be a list of instructions, called *command lines*, ordering the calculator to perform a sequence of tasks in step-by-step fashion. There are actually four types of commands in TI-BASIC, the programming language of the *TI-83 Plus*. Only one, the most common, is the direct command entered from somewhere in the keypad for the calculator to perform some computation it already knows how to. It can be complicated, like ordering the calculator to evaluate (1) and storing the result in some variable:

:int$(X/3) - (X/3 = \text{int}(X/3)) \rightarrow Y$,

or, if (1) has already been entered in the equation editor,

:$Y_1(X) \rightarrow Y$ or   :$Y_1 \rightarrow Y$,

the latter because Y_1 will automatically be interpreted as $Y_1(X)$.

The other types of commands are *control* commands, *interface* commands, and *calls* to other programs, and are accessed from special menus that appear when one presses the PRGM button from inside the program editor. All programs are created in the program editor.

On pressing the PRGM button in the Home screen, one is faced with a choice of three menus labelled EXEC, EDIT, and NEW. If there are any programs in existence, they will be displayed in the EXEC and EDIT menus; if no programs exist, these menus are empty and all that is left is to create a new program.

Choosing NEW gives the single option of creating a new program. On choosing this option one is asked to create a name for the new program. The name can be up to eight characters in length, must start with one of the letters A to Z or θ, and the remaining characters can be letters from A to Z, θ, or numerals from 0 to 9.

When you have chosen a name for the program and pressed ENTER, you are in the program editor and can write a program. The screen looks like

PROGRAM:*program_name*
:

The last colon specifies the beginning of a command. All commands begin with colons. Generally, you needn't worry about them as the editor inserts them automatically whenever you press ENTER to start a new line. With them, however, one can include more than one command on a single line. I generally prefer one command per line for readability's sake, but when memory is in short supply it is recommended to occasionally double up, especially if the commands are short.

Let us consider how we might program (1). First, before we pick a name and enter the program editor, we should be aware that one cannot program functions on the *TI-83 Plus* like one can on a computer or on some other calculators like the *TI-89 Titanium*. So we have to decide how the program will acquire the value of X for which we want to calculate $Y_1(X)$ and then what we want the program to do with the value of $Y_1(X)$ once it has calculated this value. There are two solutions to the first problem. The program can assume that the desired value of X has been stored beforehand in the variable X, or it can interactively ask for a value supplied by whoever is running the program. The approach chosen is largely a matter of taste. The former method might be preferred if the program is to be called by another program, especially if it is to be called a number of times with a succession of values of X. But the latter is nice if we are writing a program to be used by someone else who is not familiar with the program and might not know which values to store or the names of the variables to store them in.

As for the value of $Y_1(X)$, one can store it in a variable, say Y, to be used later, display it on the home screen, or do both.

The simplest program would assume some number stored in X and place the result in a variable Y:

PROGRAM:CALCY1
:int(X/3)−(X/3=int(X/3))→Y .

If we were calculating this by hand, we would not bother to divide by 3 each time the expression X/3 occurred, but would divide once and plug the result into the expression three times. Or, we might even plug it once into int($\cdot$) and plug this result into the given expression twice and plug the value for X/3 into it once:

PROGRAM:CALCY2
:X/3→Q

:int(Q)→I
:I−(Q=I)→Y .

Once one is in the program editor, one can enter more than just the direct commands of the sort one might enter in succession from the home screen. Pressing the PRGM button allows access to three menus. CTL lists the *control* commands, I/O lists the *interface* commands, and EXEC lists the programs available for *call* commands.

The control commands allow one to control the execution of subsequent commands. Ordinarily, we think of a program as a list of instructions that the machine will follow line by line. Control commands allow one to alter the order of execution of succeeding instructions. There are 18 such commands, not all of which we use in this book. The ones we need, as labelled in the CTL menu are:

1: If
2: Then
3: Else
4: For(
5: While
6: Repeat
7: End
8: Pause
9: Lbl
0: Goto
D: prgm
G: DelVar .

Commands 1, 2, 3, 7 are used in *branching*—executing different sets of instructions according as certain conditions are or are not met. These commands are more naturally discussed later, in section 2, where we discuss truth values, and I postpone their discussion until then.

Commands 4, 5, 6, 7 are used in *looping*—repeating the execution of sets of commands a number of times. While and Repeat rely on tests to escape the loop and, as with the branching commands, I postpone their discussion until the next section.

The command For(has the following syntax[2]:

:For(*variable,start,finish*[,*increment*])
:*commands*
:End .

For example we might use the following bit of code to sum the first N positive integers:

---

[2] The square brackets indicate that the argument *increment* is optional. If no increment is given, an increment of 1 is assumed as default. Thus, For(X,1,N) means the same as For(X,1,N,1).

```
:0→S
:For(X,1,N,1)
:X+S→S
:End.
```

What this does is the following. The first line of code sets the value of S to 0, thus initialising it for use as a running sum. The For(command says that the command X+S→S will be carried out for X =1, 2, ..., N, i.e., that one will successively add 1, 2, ..., N to S, thus resulting in $0 + 1 + \ldots + N$.

If we wanted to add the even integers ≤ N, we would use 2 as the increment:

```
:0→S
:For(X,2,N,2)
:X+S→S
:End.
```

In such a case, should N be odd, X will never equal N. The loop does not run on forever, however, but will stop as soon as X exceeds N. Thus, for N = 5, for example, after adding 4 to S, X becomes $6 > 5$ and the command X+S→S is not executed. The program then either moves on to execute the first command following the End command, or, if there is no such command, stops execution entirely. The End command closes the loop, signifying its end, not the end of the program.

Increments can be negative, thus allowing one to count down instead of counting up. This is occasionally convenient. I use negative increments in the programs RANDPERM and TREIZE found on pages 391 and 392, respectively, below.

Command 8 is the Pause command. It is useful in programs that display a lot of information, like a graph or a table, during their executions. Using a Pause command directly after the display will pause the program, allowing the user to examine the display before continuing by pressing ENTER. The only use I make of it in this book is on page 368 below in an enhanced version of the program BERNULLI of page 164.

Commands 9 and 0 are *jump* commands, traditionally used in connexion with branching commands to tell the machine which commands to go to and execute next following the outcome of some test. It is also commonly used to exit a loop early should some condition be met. I illustrate this in the program PRIME on page 336, in section 2, below. Jumps can also be used to create loops as in the following

```
:Lbl 1
:commands
:Goto 1.
```

Here Lbl is used to create a label at some point in the program. It does nothing else. The calculator continues to execute the succeeding *commands* until the Goto command sends it back to the labelled line and *commands* are executed again, until the Goto command is reached, ... This will continue forever unless there is among the *commands* a branching command which, when some condition is met, takes the calculator out of the loop via another pair of jump commands, a STOP command (which stops execution entirely), or a RETURN command (which, in case the program in question was called from another program, stops execution from the called program and returns to the calling program). STOP and RETURN are items F and E, respectively, in the CTL menu; I have not used them in this book.

I can give one example here of the use of the jump commands that does not rely on some test to either decide to jump to a labelled command line or to escape a loop. This is to skip over some lines one *never* wants executed:

```
PROGRAM:NULL
:Goto A1
:"THIS PROGRAM"
:"DOES NOTHING."
:Lbl A1 .
```

This is indeed a useless program in that it does nothing. The skipped lines serve merely as *comments* and can be used in writing a program to remind ourselves later, should we need to read it, what various parts of the program do. Probably because of the limited amount of memory in the earlier models of the *TI-83* and its predecessor the *TI-82*, no provision for adding comments to one's programs was made in the programming language. Thus, one might want to keep the number of comments low. Documentation of programs, however, is important and I have not emphasised it in this book only because each of my programs has been embedded in discussion and embedding comments in programs seemed unnecessary. My programs are rather small, but for longer programs you might consider beginning them with a list of variables:

```
:Goto A1
:"STORED VARIABLES:"
:"R=RATE"
:"T=TIME"
:"FINAL RESULT:"
:"DISTANCE=D"
:Lbl A1
:R*T→D .
```

Being able to determine quickly the variables you need to store numbers in is handy when you come back to a program later after forgetting how it works. Additionally, you might add simple comments explaining the tasks performed by the different parts of the program.

Labels can be one or two characters long, the individual characters being letters A to Z, θ, or numerals, 0 to 9. Thus there are $37 + 37 \cdot 37 = 1406$ possible labels and one is in no danger of running out of them. The only deterrent to their use for embedding comments as above would be limitations of memory: I've seen advice on where closing parentheses are not needed and how their omission will save a few bytes, and on how placing two commands on one line will likewise save memory. It may be necessary, if you are dealing with a long program, to store your fully documented program in a text file on the computer or even on paper and only enter an undocumented version in the calculator.

Command D, prgm, is a program call. One way of documenting a complex program is to break its execution into numerous subtasks and write programs for each of these, giving them names suggestive of their tasks. Program names can be up to 8 characters long, so this should be quite doable—provided you don't mind the occasional simplified spelling. Once you have the subordinate programs in place, you can write a master program that calls each one as necessary. The command prgm offers one way of doing this. The program BUFFON of Project 3 of Chapter 2 calls the program PETE with the line

   :prgmPETE.

This line can be entered in one of three ways: 1. Press the PRGM button to access the CTL menu and choose prgm, then type the name of the program you wish to call; 2. press the CATALOG button and scroll down to choose prgm, then type the name of the program you wish to call; and 3. press the PRGM button, open the EXEC menu, and then choose the program you wish to call. In the third option, the letters prgm will automatically be prefixed to the name of the called program. Note that the third alternative is only available if you have already stored the program to be called in the calculator.

The last control command we have to mention is G: DelVar. This command is a housecleaning command and is used to delete variables to free up memory and unclutter the list of list names in the NAMES menu accessed by pressing the LIST button[3]. It is the most used command in this book; it could easily have been the least used or even unused.

There are fewer interface commands listed in the I/O menu. The ones used in this book are

   1: Input
   3: Disp
   4: DispGraph
   6: Output(
   8: ClrHome.

The simplest of these is ClrHome, which clears the home screen, making it less cluttered and more readable when the program has to display some information.

The basic display command is Disp and its syntax is

   :Disp [*value1,value2,...,value n*].

All the arguments are optional and if they are all missing the program will merely display the home screen. If values *value1*, *value2*, ... are present, the program will cause these values to be displayed on successive lines on the home screen. Thus

   :Disp "PLEASE ENTER","A VALUE."

and

---

[3] Lists are discussed in detail in the next section.

```
:Disp "PLEASE ENTER"
:Disp "A VALUE." .
```

will both result in the display

```
PLEASE ENTER
A VALUE.
```

If *value* is a numerical expression entered between quotation marks, Disp "*value*" will display the expression literally on the left side of the home screen; if there are no quotation marks, the expression is evaluated and the result entered on the right. Thus, Disp 0 will result in a 0 on the right side of the screen, while Disp "0" will result in a 0 on the left side of the screen. As will be discussed in the subsection on strings in the next section, the two 0's are different types of entities, one a number and one a string.

The Output(command is similar but allows more control. Its syntax is

```
:Output(row,column,value) ,
```

where *value* may again be an expression in quotes or an expression to be evaluated. If one has a variable X and one wants to display its value on the home screen,

```
:Disp "X=",X
```

will display "X=" on one line and the value of X on the next, while

```
:ClrHome
:Output(1,1,"X=")
:Output(1,3,X)
```

will display "X=" and its value on the first line. If the value is, say, 5 the first line of the home screen will read

```
X=5 .
```

The command DispGraph opens the graphics screen and displays those function graphs and statistical plots that are turned on.

The DispGraph command is used to show something to the user. The Disp and Output(commands may also be used to display information, such as a calculated value. Additionally, the Disp command is often paired with the Input command to request inputs. The syntax for the Input command is

```
:Input [text,variable] .
```

With no argument, Input opens the graphics screen and waits for the user to position the cursor on a point and press enter. The x-value of the cursor is then stored in the variable X. If *variable* is present, Input causes the program to pause until the user enters a value, which is then stored in the given variable. The argument *text* is usually some semi-explanatory text string. Additional explanation would be provided by Disp commands:

```
PROGRAM:DISTANCE
:ClrHome
:Disp "PLEASE ENTER","THE VALUES:"
:Input "RATE=",R
:Input "TIME=",T
:R∗T→D
:Output(5,1,"DISTANCE=")
:Output(5,10,D) .
```

And, of course, you might want to finish this up with the housecleaning commands:

```
:DelVar R
:DelVar T
:DelVar D .
```

This works nicely. If you run the program choosing, say, 2, 5, for R, T, you will get the readout:

```
PLEASE ENTER
THE VALUES:
RATE=2
TIME=5
DISTANCE=10 .
```

There will be a slight problem as the cursor will be flashing over the D. Anything you type will print over the line "DISTANCE=10". Pressing Enter will not move you to the next line, but will run the program again, as it was the last entry made from the home screen. Going back into the editor and appending the line

```
:Disp "" ,
```

with no space between the quotation marks, will remedy the situation. The Output(command does not create a new line like Disp does. Thus, after clearing the screen, only four line returns have been made and the cursor rests on the first spot in the fifth line, where, coincidentally, the program has placed the line "DISTANCE=10". The Disp "" command adds a fifth line return without altering the readout on the home screen.

The reader will probably have noticed that the interactivity of the program, necessitating explanatory instructions to the user, supplies some documentation of the program, in this case explaining what numbers are to be entered where. For a program with long execution times, one can display additional information to appease the impatient user. For example,

```
:Disp "WE WILL NOW","CALCULATE THE","AVERAGE."
```

Another thing one can do in a program that runs a long time to keep the user informed is to display a progress report onscreen. For example, in the BUFFON program, 2048 calls to PETE are made in the loop:

```
:For(K,1,2048)
:prgmPETE
:more_commands
:End .
```

One could replace these lines by

```
:ClrHome
:Output(3,3,"CURRENTLY ON")
:Output(4,3,"SIMULATION")
:Output(6,3,"OUT OF 2048")
:For(K,1,2048)
:If K/20=int(K/20)
:Output(5,7,K)
:prgmPETE
:more commands
:End .
```

Note that this anticipates our understanding of the If command, which will be explained in the next section. For now, it suffices to note that this bit of code will result in a counter noting the start of every 20th execution of the program PETE. You can view a simpler example on the calculator by entering the following program in the program editor and running it:

```
PROGRAM:COUNT20
:ClrHome
:Output(3,3,"CURRENTLY ON")
:Output(4,3,"ITERATION")
:Output(6,3,"OUT OF 500")
:For(K,1,500)
:If K/20=int(K/20)
:Output(5,7,K)
:End .
```

2 The Data Types

The *TI-83 Plus* computes with essentially five distinct types of data: numbers, truth values, lists, matrices, and strings. We are, of course, quite familiar with numbers as objects we compute with, and the other types of data, if not the computations with them, would not be unfamiliar to us: *truth values* are just 1 and 0 standing for "true" and "false", respectively; *lists* are just lists of numbers; *matrices* (singular: *matrix*) are rectangular arrays of numbers; and *strings* are just words in an alphabet consisting of letters, numbers, matrix and function names, etc.—just about any symbol that can be displayed on the calculator screen.

Truth Values, Conditionals, and Loops

In the 19th century George Boole noted that the truth values "true" and "false" could be manipulated algebraically just like numbers. Today we commonly, but not universally, represent them by the numbers 1 for "true" and 0 for "false", and this is what is done on the calculator. The TEST button gives one access to two menus. One called TEST provides relational tests that yield truth values— 1 if the relation holds for the values tested and 0 otherwise; the LOGIC menu provides four functions for combining truth values: and, or, xor, and not(. The first three are infix binary operators and the last is a prefix unary function.

These logical functions, as implemented on the *TI-83 Plus*, can take any numbers as input, treating all nonzero numbers as true, and 0 as false. The output is always 1 or 0 according as the result should be true or false. If both inputs to and are true (i.e., nonzero) the result is 1; otherwise it is 0. Thus

  1 and 1 yields 1
  2 and π yields 1
  2 and 0 yields 0.

The function or is the *inclusive or* yielding 1 if any argument is true (i.e., nonzero) and 0 only if both arguments are 0:

  1 or 0 yields 1
  2 or π yields 1
  0 or 0 yields 0.

The function xor is the *exclusive or* yielding 1 if one or the other of its arguments is true, but not both, and 0 if both are true or both false:

  π xor e yields 0 because π, e are both true
  π xor 0 yields 1 because only π is true
  0 xor 0 yields 0.

And not(maps true to false and false to true:

  not(1) yields 0
  not(3) yields 0
  not(0) yields 1.

The main use of the truth functions would be to combine simple tests to make more complicated tests in branching commands in programs. Because the tests produce truth values, which are in fact numerical values, they can also be used in numerical expressions which, at first sight, appear to be meaningless. On page 163, for example, we find the definition

  $Y_1 = \text{int}(X/3) - (X/3 = \text{int}(X/3))$.

Here, the subexpression $X/3 = \text{int}(X/3)$ is a test yielding 1 if the values of $X/3$ and $\text{int}(X/3)$ are equal and 0 otherwise. The function int(is the *greatest integer function* yielding the greatest integer less than or equal to its input. For $X = 5$, for example, $\text{int}(5/3)$ is 1, the greatest integer $\leq 5/3$. Hence

$$Y_1(5)=\text{int}(5/3) - (5/3=\text{int}(5/3)) = 1 - (5/3=1) = 1 - 0 = 1,$$

since the test $5/3=1$ evaluates to 0 (i.e., false).

A similar use of tests is to define functions piecewise in the Equation Editor without programming. For example, if the absolute value function were not already available as abs(in the NUM menu accessed by the MATH button, we could define it as follows:

$Y_1=(-X)*(X<0)+X*(0\leq X)$.

For definitions on bounded intervals, we can use and. For example:

$Y_1=(X+1)*(X<0)+1*(0<X \text{ and } X< 1)+X*(1\leq X)$.

However, as I stated, the main use of tests is in providing decisions for use in branching commands in programs. On the *TI-83 Plus* these are variants of two forms—straightforward conditionals and criteria for ending loops.

TI-BASIC, the programming language of the *TI-83 Plus* offers three variants of the conditional. The simplest of these consists of an If command, followed by a command to be executed if the condition is true, and then followed by other commands. The syntax is:

```
:commands
:If condition
:command   (if true)
:commands .
```

The program BUFFON of Project 3 of Chapter 2 uses lists and will be discussed below. It runs the PETE program of page 48 2048 times and compiles a list of results. It starts with a list of 25 0's and when the result of PETE is a number I, adds 1 to the I-th entry in the list. At one point it checks to see if the list is long enough and, if not, extends it. Then, regardless of whether the extension is made, adds 1 to the I-th entry:

:If I>dim(L$_1$)   checks if I is greater than the length of L$_1$
:I→dim(L$_1$)   extends L$_1$ if the condition is true
:L$_1$(I)+1→L$_1$(I) adds 1 to the I-th element no matter what.

The simple If offers one occasion where the program may become more readable if two commands are written in a single line:

:If condition:command
:commands .

For the example at hand, this would be

:If I>dim(L$_1$):I→dim(L$_1$)
:L$_1$(I)+1→L$_1$(I) .

The second variant is the If-Then pairing. It consists of an If statement testing the truth of some condition, followed by a Then, followed by one or more commands to be executed if the condition is true, and finally completed by an End statement. All subsequent commands are executed regardless of the truth of the condition. Its syntax is

```
:If condition
:Then
:commands  (if true)
:End
:commands .
```

Both of these types of conditionals can be found in the program SUBST1 on page 341, below. Lines 2 through 7 of that program read

```
:If N=0
:Goto 1
:If N=1
:Then
:Str2+sub(Str1,2,length(Str1)−1)→Str1
:Goto 1
:End .
```

Here, a number N has been determined and some actions are to be performed depending on the value of N. If N = 0, nothing is to be done, so a simple If construction tests to see if N = 0 and, if it is, the program jumps to another line in the program that has been labelled 1 using a Lbl 1 command. The next conditional is an If-Then grouping that tests if N = 1 and if this is the case performs an action to be described later and then jumps to the line labelled 1. The End statement closes the conditional and tells the calculator that subsequent commands are to be executed regardless of the truth of the condition.

Finally, there is the If-Then-Else constellation, with the syntax:

```
:If condition
:Then
:commands₁  (if true)
:Else
:commands₂  (if false)
:End
:commands .
```

The If-Then-Else command structure is no more powerful than the If-Then structure, as it can be replaced by two If-Then structures:

```
:If condition₁
:Then
:commands₁  (if true)
:end
:If not(condition₁)
:Then
:commands₂  (if true, i.e., if condition₁ is false)
:End
:commands ,
```

or, with the aid of jump commands, by two simple If's:

```
:If condition₁
:Goto 1
:If not(condition₁)
:Goto 2
:Lbl 1
:commands₁
:Goto 3
:Lbl 2
:commands₂
:Lbl 3
:commands .
```

The If-Then and If Then-Else conditionals are more convenient and more efficient to use than their simulations by simple If's and, more importantly, yield more readable programs. Thus they are used whenever possible.

Looping structures allow one to repeat a set of actions multiple times. The simplest looping structure is the For(loop, the syntax of which is

```
:For(variable,start,finish[,increment])
:commands   (while variable does not exceed finish)
:End .
```

The variable *increment* written in square brackets may be absent, in which case it has an implied value of 1. For each value of *variable* from *start* to *finish*, with increments of *increment*, the *commands* are executed. The End statement ends the grouping of *commands* to be iterated. The BUFFON program mentioned above uses a For(loop to iterate the program PETE 2048 times and build a record in the shape of a list L_1 as it goes:

```
:For(K,1,2048)
:prgmPETE
:If I>dim(L₁)
:I→dim(L₁)
:L₁(I)+1→L₁(I)
:End .
```

Note that the If command is a simple If, whence End closes off the For(loop and not the If. Conditionals and loops can be nested. To figure out what a given End statement applies to, think of If-Then, If-Then-Else, and For(as left parentheses and End as a right parenthesis and match the parentheses.

The For(loop repeats the actions specified by its commands a fixed number of times. Occasionally, one might want to break out of the loop early because some condition has been met. One can use a conditional to do this. For example, we have the following program to test if a positive integer stored in the variable N is a prime:

```
PROGRAM:PRIME
:For(I,2,N−1)
:If N/I=int(N/I)
```

```
:Then
:Disp "NOT PRIME"
:Goto 1
:End
:End
:Disp "PRIME"
:Lbl 1
:DelVar I .
```

This is not a particularly efficient program. In practice one would stop the loop when I exceeded the square root of N, but I thought I would go for the simpler program which simply tests for each I = 2, 3, 4, ..., N − 1 if it is a divisor of N or not. If a divisor is found, the message "NOT PRIME" is displayed and the Goto command takes the computation out of the loop. If after testing all possible divisors none is found, the message "PRIME" is displayed. Note that the label is placed after this latter display command so that no mixed message is sent.

Digression. The last command of PRIME is the delete variable command, DelVar. As mentioned earlier, it is found in the CTL (for *control*) menu (item G) accessed by the PRGM button while writing or editing a program. It can also be found in the CATALOG and used in the home screen outside the confines of a program.

Because all variables in TI-BASIC are *global*, undeleted variables maintain their values when a program stops execution. This can be useful. In the snippet of code from BUFFON cited above, the command prgmPETE ran the program PETE, which produced a value for the variable I, a value that was then used by BUFFON. There are 27 real variables in the *TI-83 Plus*, namely A, B, ..., Z, and θ, all with default values of 0. Each variable used in a program is assigned a value, with nonzero values tending to occupy more memory than the default 0. This is true even of the variables used as counters in loops. When execution of a loop is finished, the counter variable has as its value the last value assigned in the iteration. As such a value is unlikely to be needed outside the loop, it is not a bad idea to routinely delete such variables at the end of a program. This frees up a little memory.

The fact that variables are global, shared by the home screen and all programs, means that one must be careful in choosing one's variables in writing a program. Execution can overwrite a variable the value of which one had intended to use elsewhere. For example, while separated loops can usually safely be assigned the same variable as counters, nested loops should be given different counter variables. And, in the case where one program calls another, special care should be taken to guarantee the called program does not change the value of any variables being used in the master program— the one doing the calling. For example, BUFFON calls the program PETE and each program uses variables. If you read the programs[4] you will see that they have no vari-

---

[4] The program PETE is listed on page 48, above; BUFFON can be found on pages 353 - 354, below.

ables other than I in common and there is no danger of a crucial variable in BUFFON having its value changed by PETE. The programs in the present book are fairly simple and there is not much danger of running into any problems here, but if one is going to get one's money's worth out of the calculator, one will be doing a lot more programming and this is one of the points that should be borne in mind when writing programs.

There are two other loop structures in TI-BASIC. These are the While and Repeat structures. They are designed to handle loops for which one does not know in advance how many times an action is to be repeated: the looping stops as soon as some condition is met or is no longer met.

The While command iterates the execution of a set of commands as long as some condition holds. Its syntax is:

```
:While condition
:commands₁
:End
:commands₂ .
```

In running, While tests if *condition* is true and, if so, *commands₁* are executed, the *condition* is tested again and if it is still true, *commands₁* are executed again; and this pattern is repeated until the *condition* is false, at which point *commands₁* are skipped over and the program executes *commands₂*. Here is a simple program using While:

```
PROGRAM:TEST
:1→I
:While A=0
:Disp I
:3→A
:I+1→I
:End .
```

Store 0 in A and run the program. Then store 1 in A and run the program. At the end of each program execution, check the current value of I.

The Repeat command iterates execution of a set of commands until some condition is met. Its syntax is:

```
:Repeat condition
:commands₁
:End
:commands₂ .
```

Unlike the If commands and the While command, in a Repeat loop the *condition* is not checked until after the *commands₁* have been executed and the End command is encountered.

The program PETE on page 48 used a Repeat command. That program was designed to simulate the tossing of a coin until heads showed up.

There is a danger in using While and Repeat loops instead of a For(loop. If you are not careful, the exit conditions might never be met and the program will run forever, or until the calculator runs out of memory. For example, replace the While in the program TEST by Repeat, store 1 in A and run the program. If you get tired of watching an increasing sequence of numbers scrolling down the screen, press the On button to interrupt execution and then select the Quit option.

The program PETE, by the way, comes with no guarantee that it will not likewise run forever. However, assuming randInt(is truly random, it is extremely unlikely its iteration will be repeated more than a couple of dozen times. To put it another way, it is morally certain, if not necessarily so, that PETE will stop in a reasonable number of steps.

Strings

Following truth values, strings form the simplest data types the calculator is designed to handle. They are also the least used, but they do have their uses.

The most basic thing one can do with strings is to attach them to list and function names. The usual method of defining a function is to pick a function variable, say Y_1, go into the equation editor (called the Y= editor in the calculator manual), and enter an expression. For example,

$Y_1=e^\wedge(-1/X^2)$.

One can also do this from the home screen by placing the expression on the right within quotation marks and storing it in Y_1, i.e., entering

$"e^\wedge(-1/X^2)" \rightarrow Y_1$.

If you do this and press the Y= button to enter the equation editor, you will see that you have indeed defined Y_1 to be the desired function. Now this is nothing to celebrate in itself. The first method requires fewer keystrokes and is more intuitive. However, the second method can be used within a program. Examples of this are given by the programs BELL1 (page 123, above) and BELL2 (page 137, above).

I defer discussion of the analogous attachment of an expression to lists to our discussion of lists (Exercise 6, below).

Another straightforward use of strings is to print instructions that should appear on the screen during the execution of a program, as in the following piece of code:

```
:Disp "PLEASE ENTER"
:Disp "YOUR NAME:"
:Input Str1
```

when the program reaches these lines of code during execution, the following lines will appear on the screen

```
PLEASE ENTER
YOUR NAME:
```

and a flashing cursor will appear below them waiting for the user to spell out a name before hitting the ENTER button. When this is done, the name is stored in the string variable Str1.

The *TI-83 Plus* is not very generous when it comes to naming strings. There are only 10 names allowed, all built-in and accessible through the String... submenu of the VARS menu accessed by the VARS button. They are Str1, Str2, ..., Str9, Str0. On the other hand, strings can be thousands of characters long and, if one reserves one character as a delimiter, any number of strings can be strung together into a single string coding them as a string of strings.

Functions for handling strings are built into the calculator. The most accessible of these is concatenation, which is given by the + button of the calculator. If, for example, Str1 is the string 123 and Str2 is 456, then Str1+Str2 is the string 123456, while Str2+Str1 is 456123. Concatenation is not commutative. It is, however, associative and we thus need not use parentheses in grouping when concatenating more than two strings.

The crew at Texas Instruments must not have envisioned that people would need to use strings all that much, for, not only did they limit the number of names one could use to name a string, but they also chose not to supply them with their own menu. All the rest of the string functions are accessed only via the CATALOG. The most straightforward of these are length(, inString(, and sub(.

The functions length(, inString(, and sub(all deal with the individual characters of the string, e.g., length(applied to a string or string variable tells you how many characters are in a string: length("ABC") returns the value 3. It is important here to know that a character is not determined by what we see on the screen, but by how it was entered. Expressions accessed by buttons and menus (variable and function names, commands, etc.) count as single characters[5]:

length("sin(X)") is 3 as it consists of the three characters sin(, X, and)

length("[A]") is 1 if [A] is a matrix name entered via the menu accessed by pressing the MATRX button, and 3 if [A] is entered by pressing the [, A, and] keys.

1 Exercise. One can test strings for equality using the = test. Verify that the test [A]=[A] can result in different answers depending on how [A] is entered in its two occurrences.

The second of these operators, inString(, takes two strings as arguments and gives the position of the first character of the first occurrence of the second string as a substring of the first string if the second string occurs as a substring of the first, and it returns the value 0 if the second string does not occur as a substring. Position numbering begins with 1. Thus

inString("DOG","D") returns 1

---

[5] *Warning.* This is specific to the *TI-83 Plus* and kindred calculators. It no longer holds for the *TI-89 Titanium*, for which dim("sin(x)") is 6 as the string has 6 characters s, i, n, (, x, and) regardless of how it was entered.

inString("DOG","G") returns 3

inString("DOG","OG") returns 2

inString("DOG","GO") returns 0.

Using 1 and 0 as the logical values "true" and "false", respectively, the expression

not(inString($string_1$,$string_2$)=0)

results in a test on whether or not $string_2$ occurs as a substring of $string_1$:

not(inString("DOG","X")=0) returns 0

not(inString("DOG","G")=0) returns 1.

From this batch of functions we finally come to sub(, which takes 3 arguments—a string, a starting position, and a length. sub($string$,$begin$,$length$) produces the substring of $string$ of length $length$ starting at position $begin$. Thus,

sub("CROCODILE",4,3) returns "COD".

We can illustrate the use of these functions by writing the following simple program that takes two strings Str1 and Str2 as inputs and replaces the first occurrence of the letter X in Str1 by the string Str2:

```
PROGRAM:SUBST1
:inString(Str1,"X") → N
:If N=0
:Goto 1
:If N=1
:Then
:Str2+sub(Str1,2,length(Str1)−1)→Str1
:Goto 1
:End
:If N=length(Str1)
:Then
:sub(Str1,1,N−1)+Str2→Str1
:Goto 1
:End
:sub(Str1,1,N−1)+Str2+sub(Str1,N+1,length(Str1)−N)→Str1
:Lbl 1
:DelVar N .
```

The idea of the program is fairly simple. Str1 should look something like

$a_1 \ldots a_{N-1} X a_{N+1} \ldots a_m$.

One finds N and concatenates

$a_1 \ldots a_{N-1} + Str2 + a_{N+1} \ldots a_m$.

Now $a_1 \ldots a_{N-1}$ is the string sub(Str1,1,N$-$1) and $a_{N+1} \ldots a_m$ is the string sub(Str1,N+1,length(Str1)$-$N). The program is so long because one has to treat separately the cases N $=$ 0 (X does not occur in Str1), N $=$ 1 ($a_1 \ldots a_{N-1}$ is the empty string), and N $=$ m ($a_{N+1} \ldots a_m$ is the empty string), the latter two cases because the calculator doesn't deal with empty strings. One can create an empty string, e.g., by entering

   " "→Str1 ,

and even apply the length(function to it to get 0, but this seems to be the limit of what one can do with the empty string. For example, instead of returning an empty string, sub(Str1,N+1,1) will result in an error message when N is length(Str1). I hasten to add that SUBST1 is only guaranteed to work properly if neither Str1 nor Str2 is empty.

    SUBST1 is a bit like the Find-and-Replace function of word processors. The difference is that SUBST1 only replaces a single occurrence of X by Str2. Word processors usually allow the option of replacing all occurrences of the string being searched for by the replacement string. For strings Str2 not containing the character X, this can be accomplished by a simple iteration of SUBST1:

```
PROGRAM:SUBST2
:Repeat inString(Str1,"X")=0
:prgmSUBST1
:End .
```

    SUBST2 does not work if X occurs in Str1 and Str2, but will run forever. For, after each call to SUBST1, there is still at least one X in Str1. This can be seen by storing any two strings containing X in the variables Str1 and Str2, respectively, and running the program. The program will run on forever, seemingly doing nothing, until you quit it by pressing the On button. You can watch the program's "progress" by inserting the command

   :Disp Str1

just before the :End command. Try this with the pairs

   Str1: X, Str2: X and Str1: X, Str2: XX.

2 Project. i. Write a program SUBST3 that will replace all occurrences of the letter X in a nonempty string Str1 by a nonempty string Str2.
ii. Do the same, but where, instead of X, all occurrences of a string Str3 are replaced.
[N.B. i. Instead of concatenating

   $a_1 \ldots a_{N-1}$+Str2+$a_{N+K} \ldots a_m$,

as in SUBST1 (where STR1 has length K), create two strings

   $a_1 \ldots a_{N-1}$+Str2 and $a_{N+K} \ldots a_m$,

then apply this again to $a_{N+K} \ldots a_m$ and tack the first string obtained from this onto the end of $a_1 \ldots a_{N-1}$+Str2 and proceed to work on the new tail. How many string variables do you need for this program?

ii. The task of this program is not unambiguously defined if occurrences of Str3 overlap each other in Str1. Let, for example, Str1 be BOBOBOB, Str2 be TOM, and Str3 be BOB. Should the result be TOMOTOM or BOTOMOB? What does your favourite word processor yield in this case? Your program needs only to guarantee a correct answer when there is no such overlap.]

The next batch of commands String▶Equ(, Equ▶String(, and expr(concern the use of strings as expressions for functions. We have already seen how we can attach an expression to a function variable outside the equation editor, e.g., in a program, via an assignment,

$$"5X+1" \to Y_1.$$

The command

$$\text{String▶Equ}("5X+1",Y_1),$$

or the pair of commands

$$"5X+1" \to \text{Str1}$$
$$\text{String▶Equ}(\text{Str1},Y_1)$$

will both do the same thing, namely result in the line

$$Y_1=5X+1$$

appearing in the equation editor.

The command Equ▶String(is the inverse operation. If a function variable Y_n has an expression, this expression can be converted to a string. If, for example, one enters $Y_1=5X+1$ in the equation editor, the command

$$\text{Equ▶String}(Y_1,\text{Str3})$$

stores the expression 5X+1 in the string named Str3.

The final string function expr(will take a string, convert it to an expression, and then evaluate it. Thus, if, say, 5 has been stored in the variable X,

$$\text{expr}("X") \text{ returns } 5$$
$$\text{expr}("5X+1") \text{ returns } 26$$
$$\text{expr}(\text{Str3}) \text{ returns } 26,$$

where we assume 5X+1 has been stored in Str3 as in the preceding paragraph. Likewise, if $Y_1=5X+1$ has been entered in the equation editor, expr$("Y_1")$ will also evaluate to 26.

If a string *string* stores a numerical expression, we can achieve the same result via the pair of commands

$$\text{String▶Equ}(\textit{string},Y_1)$$
$$Y_1(X).$$

The function expr(also applies, however, to other data types. For example, if we enter expr("Str3") the evaluation will be Str3. The expression expr(expr("Str3")), however, will evaluate to the value of the expression given by Str3, which would be 26 in the example of the previous paragraph.

It is important to realise that strings consisting of ordinary digits are not the same as the numbers they represent, even though the two may look the same on screen: A 2 on screen may be the number 2 or the string with the single character 2. On entering a 2 on the keypad, a number is entered and a 2 appears on the screen. Putting the 2 in quotes creates a string and when one presses ENTER a 2 appears on the screen. The two look alike, but are not the same. One quick way to check this[6] is to enter

Ans+0

immediately after entering the string or the number and see what happens. The number will yield 2 as a new answer and the other will result in an error message. Occasionally one wants to go back and forth between a number and the string of its digits. The function expr(performs half of this task, producing the number from the expression. If, for example, one has stored the number 23 in the string Str1, then expr(Str1) will produce the number 23. The reverse conversion is trickier. For any *fixed* number, like 23, we can create the string and store it in a string variable by entering the simple command

"23"→Str1 .

But the general function taking one from a number to a string whose characters are the digits of the number, a function that might be needed in a program, is more difficult to produce.

One can convert a number to a string by writing a program that converts each digit of the number using a bunch of commands like

:If X=5
:"5"→Str1 .

Such a program, though straightforward, seems a bit too complicated for such a basic task. There are simpler programs—simpler in complexity of programming if not in conception. The calculator does have one built-in application that will do the conversion in a special case, namely its application for doing statistical regression. Some clever person exploited this and posted the following program on the Texas Instruments web site:

PROGRAM:NUM2STR1
:{1,2}→L₁
:{N,N}→L₂
:LinReg(ax+b) L₁,L₂,Y₁
:Equ►String(Y₁,Str2)

---

[6] Another way is to look *where* the answer is. When the calculator returns the 2, it will appear on the left side of the screen if the 2 is a string and on the right side if it is a number.

```
:sub(Str2,4,length(Str2)−3)→Str1
:DelVar Y₁
:DelVar Str2
:ClrList L₁
:ClrList L₂ .
```

This program, which uses lists and list functions soon to be described, will convert whatever number that is stored in the variable N into a string representing that number. The program works as follows. Suppose some value, say 455, is stored in N. Lists L_1 of the x-coordinates $\{1,2\}$ and L_2 of the y-coordinates $\{455, 455\}$ of the points $\langle 1, 455\rangle, \langle 2, 455\rangle$ are created. The command

LinReg(ax+b) L_1, L_2, Y_1

calculates the coefficients of the line best fitting these data, in this case the constant line $y = 455$, and places the equation of this line in the equation editor, thus:

$Y_1 = 0X + 455$.

The Equ▶String(command produces the string 0X+455, of which the desired string 455 is the substring of length 3 starting at position 4.

A less clever, but even simpler program for single digit numbers stored in the variable X is the following:

```
PROGRAM:NUM2STR2
:"0123456789"→Str2
:sub(Str2,X+1,1)→Str1
:DelVar Str2 .
```

For multidigit numbers one would perform an iteration, perhaps as follows:

```
PROGRAM:NUM2STR3
:"0123456789"→Str2 .
```

We could call the program NUM2STR2 to convert each digit into a string as it is found. This would mean repeating the storing of the string "0123456789" in the variable Str2 multiple times. Instead, we perform this task only once, and iterate only the second command line of NUM2STR2. The iteration will depend on the number of digits in a number x. This is approximately the logarithm of x to the base 10, and, for positive integers x, is in fact $[\log x] + 1$. 0 must be treated separately, as $\log 0$ is undefined. The next few lines of the program are thus,

```
:If X=0
:Then
:"0"→Str1
:Goto 1
:End
:int(log(X))+1→L .
```

If $x = a_0 \ldots a_{l-1}$ is an l-digit number, then $a_0 = [x/10^{l-1}]$ and $a_1 \ldots a_{l-1} = x - a_0 \cdot 10^{l-1}$. We can thus generate $a_0, a_1, \ldots, a_{l-1}$ one after another, converting each to a string "a_0", "a_1", $\ldots$, "a_{l-1}", and, via concatenation, forming successively the substrings "a_0", "a_0a_1", $\ldots$, "$a_0a_1 \ldots a_{l-1}$". We do this in the program as follows:

```
:int(X/10^(L−1))→Y
:sub(Str2,Y+1,1)→Str1
:X−Y*10^(L−1)→X
:For(I,L−1,1,−1)
:int(X/10^(I−1))→Y
:Str1+sub(Str2,Y+1,1)→Str1
:X−Y*10^(I−1)→X
:End .
```

Alternatively, one could work one's way backwards by noting that $a_{l-1} = x - 10[x/10]$ and $a_0 \ldots a_{l-2} = [x/10]$. *Cf.* page 355, below.

One now has just some cleaning up to do: a Lbl command in case 0 is stored in X, variable deletion, and a command to display Str1 to show us that the program has indeed performed the conversion:

```
:Lbl 1
:DelVar I
:DelVar L
:DelVar X
:DelVar Y
:DelVar Str2
:Disp Str1 .
```

Note that this program assumes one has no use for X itself after program execution. To recover X, one has but to enter expr(Str1).

We now move on to our next data type.

Lists

The second most important data type for elementary probability and statistics, after that of numbers, is that of lists. A list in the calculator is just a list of numbers, something like the way a string is a list, in the informal sense, of characters. The implementation differs, but one can do much the same with lists as one did with strings: one can concatenate them, generate sublists, etc. Insofar as lists on the *TI-83 Plus* are lists of numbers, they are subject to arithmetic operations as well.

A list is entered in the calculator by entering the left brace ({), then the elements of the list—in order—separated by commas, and finally a right brace (}). Thus, for example, one would enter

$$\{1,2,3,4\}.$$

When the calculator displays a list, the commas are replaced by spaces:

{1 2 3 4}.

Once we have created a list, we can store it in a list variable using a variable name. The *TI-83 Plus* is quite generous with names for list variables. While it directly provides only six names $L_1, \ldots, L_6$ for list variables, each accessed by a button, it allows one to generate a large number of one's own names for lists. One can use any word up to 5 letters long in which the first character is one of A, B, C, ..., Z, θ and the remaining characters are letters A, B, C, ..., Z, θ or digits 0, 1, ..., 9. This allows for 52007967 possible 1-, 2-, ... or 5-character names that can be used in addition to $L_1, \ldots, L_6$ to name lists. Some statistical operations use one of these names, RESID, for a special list of residuals generated during their computations, so it is best to regard that name as reserved and not to use it. This still leaves one with quite a wide choice.

A good list name should be mnemonic. For example, one could use EVENS to name a list of even numbers, and create it by entering

$$\{2,4,6,8\} \rightarrow \text{EVENS}. \tag{2}$$

Accessing the list to subtract 1 from each of its elements and create a list ODDS is trickier. It happens we can subtract 1 from all the elements of a list by subtracting 1 from the list. Thus

{2,4,6,8}−1

will result in {1 3 5 7}. And

{2,4,6,8}−1→ODDS

will create the same list and give it the name ODDS. However, entering

$$\text{EVENS}-1 \rightarrow \text{ODDS} \tag{3}$$

simply yields an error message. Entering

EVENS−1

does not result in an error, but it doesn't produce the list {1 3 5 7} either. It will produce a number, probably −1. What happens is that the calculator interprets EVENS as the product of the numerical variables E, V, E, N, and S, and in all likelihood these have not all been assigned values other than the default 0, resulting in a product 0, whence $0 - 1 = -1$ as the answer. The error message of (3) reflects this: The instruction says to store a number in a list variable, which the calculator will not do. One must inform the calculator that one wants EVENS to represent a list, not a product of numbers. One does this by prefixing the character ∟, which is the last item on the OPS submenu accessed via the LIST button, to the name EVENS. Thus one should enter

$$\text{∟EVENS}-1 \rightarrow \text{ODDS}. \tag{4}$$

in place of (3). Personally, I always use the little corner even when it is not needed. Thus, in place of (4), I use

∟EVENS−1→∟ODDS.

The extra key presses are a nuisance, but making it a habit to use them precludes leaving them out at the wrong time.[7]

Most operations on lists are to be found in the menus accessed via the LIST button. Two exceptions are the access function and the arithmetic ones. The access function allows one to access particular elements of a list. Its syntax is

listname(element_number).

Thus, if one has stored {1 2 3} in the list L_1, $L_1(2)$ will return 2. It is important that one use the list name or variable and not the list itself. {1 2 3}(2) will return {2 4 6} for reasons soon to be made clear. It is also important that one use the parentheses: $L_1 2$ will not return 2, but the list {2 4 6}.

Before explaining the seemingly odd values obtained for {1 2 3}(2) and $L_1 2$, I quickly interject that the access function works not only to get the i-th element of a list, but allows one to replace it by storing a new value in that element: The command

$-2 \rightarrow L_1(2)$

will change L_1 to {1 −2 3}. We can also extend the list by one element:

$4 \rightarrow L_1(4)$

converts L_1 to {1 −2 3 4}. This only works for the very next element. The additional command

$6 \rightarrow L_1(6)$

will not result in {1 −2 3 4 0 6}, with 0 or some other default value in the fifth position, but in an error message.

The simplest use of lists is to combine several alternatives into a single expression. The calculator will evaluate an expression involving a list for each element of the list in turn and output the list of results. For example, if one wants both solutions to the quadratic equation,

$$6X^2 - 7X - 3 = 0,$$

one can enter

$(-(-7)+\{1,-1\}\sqrt{((-7)^2 - 4*6*(-3))})/(2*6),$

and the calculator will display

{1.5 −.3333333333}.

This explains the results obtained previously when we entered {1 2 3}(2) and $L_1 2$: The calculator read them as multiplications and not as function evaluations.

Similarly, in graphing the circle, instead of entering two functions, e.g.,

---

[7] Once a list has been created, it appears in the NAME menu accessed by the LIST button and can be pasted directly without worrying about the corner. However, in writing programs, the list might not already exist and one will have to spell it out—using the corner.

$$Y_1 = \sqrt{(4-X^2)}$$
$$Y_2 = -\sqrt{(4-X^2)},$$

we can use the single expression

$$Y_1 = \{1, -1\}\sqrt{(4-X^2)}.$$

More than one list can occur in such expressions, but they must all be of the same length and the expression will evaluate to a list obtained by evaluating the terms with the lists replaced successively by their first, second, ... elements.[8] For example,

$$\{1,2,3\}*5 + \{4,6,7\}*2$$

is treated as $\{1*5+4*2, 2*5+6*2, 3*5+7*2\}$ and results in

$$\{13 \ 22 \ 29\}.$$

For any fixed length $n \geq 1$, the lists of length n inherit arithmetic operations:

$$\{a_0, \ldots, a_{n-1}\} + \{b_0, \ldots, b_{n-1}\} = \{a_0 + b_0, \ldots, a_{n-1} + b_{n-1}\}$$
$$\{a_0, \ldots, a_{n-1}\} - \{b_0, \ldots, b_{n-1}\} = \{a_0 - b_0, \ldots, a_{n-1} - b_{n-1}\}$$
$$\{a_0, \ldots, a_{n-1}\} * \{b_0, \ldots, b_{n-1}\} = \{a_0 * b_0, \ldots, a_{n-1} * b_{n-1}\}$$
$$\{a_0, \ldots, a_{n-1}\}/\{b_0, \ldots, b_{n-1}\} = \{a_0/b_0, \ldots, a_{n-1}/b_{n-1}\},$$

and even

$$\{a_0, \ldots, a_{n-1}\}^{\{b_0, \ldots, b_{n-1}\}} = \{a_0^{b_0}, \ldots, a_{n-1}^{b_{n-1}}\},$$

where, in the cases of division and exponentiation, all the individual elements a_i/b_i and $a_i^{b_i}$ must be defined. Note that this applies to tests as well. For example,

$$\{a_0, \ldots, a_{n-1}\} = \{b_0, \ldots, b_{n-1}\} \quad \text{is} \quad \{a_0 = b_0, \ldots, a_{n-1} = b_{n-1}\}.$$

That is, the equality test does not give 1 if two lists are identical and 0 if they differ, but gives a list of 1's and 0's telling where the lists agree or disagree. This assumes the lists have the same length; if the lengths disagree an error message appears onscreen. The test for equality of lists requires some additional operations.

Other arithmetical operations on lists can be found in the MATH submenu accessed by pressing the LIST button. Several of these are self-explanatory: min(and max(applied to a list yield its minimum and maximum elements, respectively. sum(and prod(respectively add and multiply all elements of the list.

A quick demonstration of the use of list arithmetic on the calculator is afforded by the calculation, alluded to above, of the number of possible names

---

[8] There are exceptions. For example, binompdf(and binomcdf(will not accept lists as inputs to their first arguments.

for lists. By the Addition and Sequential Principles of Chapter 2, Section 1, the number is

$$27 + 27 \cdot 37 + 27 \cdot 37^2 + 27 \cdot 37^3 + 27 \cdot 37^4.$$

We would normally evaluate this on the calculator by entering

27+27*37+27*37^2+27*37^3+27*37^4 ,

or, perhaps,

27*(1+37+37^2+37^3+37^4).   (α)

With list arithmetic, we would enter

sum(27*37^{0,1,2,3,4})   or   27*sum(37^{0,1,2,3,4}).   (β)

The number of possible program names could be calculated by entering

27*sum(37^{0,1,2,3,4,5,6,7}),   (γ)

which results in 2.63435959E12. An additional list operation seq(to be discussed shortly allows us to express (β) and (γ) as

27*sum(37^seq(X,X,0,4))   and   27*sum(37^seq(X,X,0,7)),

respectively. For this particular example, we can use our knowledge of geometric progressions to get the answer with even fewer keystrokes and without the use of lists by entering

27*(37^5−1)/36   and   27*(37^8−1)/36,

respectively. In general, however, the use of lists in such situations can be very convenient.

We can use prod(to devise an equality test for two lists of equal length:

$$\mathsf{prod}(\{a_0,\ldots,a_{n-1}\} = \{b_0,\ldots,b_{n-1}\}) = \prod_{i=0}^{n-1} a_i = b_i$$

$$= \begin{cases} 1, & \text{each } a_i \text{ equals } b_i, \\ 0, & \text{otherwise.} \end{cases}$$

Thus the product is 1 if the lists are equal and 0 otherwise. So if $\mathsf{L_1}$ and $\mathsf{L_2}$ are names of lists of equal length, we can enter

prod(L$_1$=L$_2$)=1   or even   prod(L$_1$=L$_2$)

to test for equality.

The function mean(returns the arithmetic average of the elements of the list, and median(returns the middle element when the elements are listed in order, if the list has an odd number of elements, and the midpoint between the middle two elements if the list has an even length. The remaining two functions, stdDev(and variance(, like median(, are statistical functions.

The LIST button allows access to two additional menus: NAMES and OPS The first of these merely lists the variables currently being used to name lists, or, to put it another way, it lists the names of currently existing lists. The third I have already discussed before the Digression. What may be new are the operations in the OPS menu. These are, for the most part, more-or-less obvious things that one can do with lists that one might not have thought of as algebraic operations or as computational in character.

The first two operations sort the list—SortA in ascending order and SortD in descending order. If one wants to sort a list, say {1,2,1,3,2}, one must first assign it a name, say

{1,2,1,3,2}→L$_1$.

Then one can enter

SortA(L$_1$)

and the calculator will sort the list, displaying Done when it is finished. Entering L$_1$ will now result in {1 1 2 2 3}. The sort operators are not functions that produce lists as output, but operations that change the inputs themselves.

The sorting operations can be applied to more than one list at a time. In doing this, they do not sort the lists individually, but sort the first list and permute the elements of the other lists accordingly. All the lists must have the same length. One can imagine, for example, two lists ∟XLIST and ∟YLIST of the x- and y-coordinates of points $\langle x_0, y_0 \rangle, \ldots, \langle x_{n-1}, y_{n-1} \rangle$. The command SortA(∟XLIST,∟YLIST) will sort the x_i's in order and sort the y_i's to match their corresponding x_i's. In other words, it will sort the pairs $\langle x_0, y_0 \rangle, \ldots, \langle x_{n-1}, y_{n-1} \rangle$, ordering them according to their first coordinates. SortA(∟YLIST,∟XLIST) will sort them by their second coordinates.

More generally useful is the next operator, dim(, which gives the length or *dimension* of the list. Thus, entering

dim({1,2,3,4})

into the calculator will result in the number 4. Note that, unlike the sort commands, dim(accepts lists as well as their names as inputs. dim(has another use. Storing a number $n > 0$ to the dimension of a list will create a list of n 0's if no list of that name exists, extend a list of length less than n to one of length n by adding 0's at the end, and truncate a long list to one consisting of the first n elements of the list.

In theory, the empty list of length 0 is no different from any other list. One can store 0 to the dimension of a list to create an empty list, or to remove all the elements of a list. Thus,

0→dim(∟AB)

will return 0, and subsequently entering dim(∟AB) will also receive 0 as an answer. But entering ∟AB results in the error message INVALID DIM.

Before continuing on to the next command I pause to note that the dim(command can be used to test if lists L$_1$, L$_2$ are equal:

```
PROGRAM:LISTEQ
:If not(dim(L₁)=dim(L₂))
:Then
:0→ E
:Else
:prod(L₁=L₂) → E
:End
:Disp E .
```

Here, the program assumes one wants to test if lists L_1, L_2 are equal. If the dimensions are not equal, it stores 0, the numerical value for "false", in the variable E. If the dimensions are equal, it stores the result of the test cited earlier in E.

The Fill(command takes two inputs, a number that one is to put into every position of a list and the name of an existing list of positive length. Thus, if, say, ∟SCORE is the list $\{3,7,5\}$ representing the scores of three players, one can reset the scores to 0 by simply entering Fill(0,∟SCORE). Here, again, one cannot substitute the list itself, but must use the name of a list. Fill(is an operation on the list, not a function, and it cannot be used as an argument of a function. Thus, for example,

Fill(0,∟SCORE) → L₁

is declared a syntax error. To get the desired result requires two commands:

Fill(0,∟SCORE)
∟SCORE→ L₁ ,

or, if one wants to keep the original list,

∟SCORE→ L₁
Fill(0,L₁) .

The Fill(command is more of a programming convenience than a necessity. The task performed by the command Fill(X,∟List) can also be accomplished by using the arithmetic operations via the command,

X+0∗∟LIST→∟LIST .

Indeed, the arithmetic expression X+0∗∟LIST is functional and, what we required two commands to do what Fill(0,∟SCORE) →∟ L₁ was intended to do, we can arithmetically do in the single command

0∗∟SCORE→ L₁ .

The program BUFFON uses the Fill(command, as well as the next command in the OPS menu. This is the seq(function, the syntax of which is

seq(*expression, variable, start, end*[, *increment*]) .

This function produces a list by successively evaluating *expression* for values of the *variable* beginning at *start* and increasing (or decreasing) by *increment* (implicitly assumed to be 1 if it is left unspecified) until *end* is reached. We used

it twice in the program BINGRAPH of Chapter 4 (page 120), first to generate
the list $\{0\ 1\ \ldots\ 10\}$ with the command

$\quad$ seq(X,X,0,10) $\rightarrow L_1$,

and then to simulate binompdf(10,.5) via the command

$\quad$ seq((10 $_nC_r$ X)*(.5^10),X,0,10) $\rightarrow L_2$.

We can also use seq to give a functional test for equality of lists:

$\quad$ prod(seq(L_1(I)=L_2(I),I,1,min({dim(L_1),dim(L_2)})))*(dim(L_1)=dim(L_2)) .

And I use seq(in my version of BUFFON, which was assigned earlier in
Chapter 2 (in Project 3, page 48) as an exercise for the reader already familiar
with programming the *TI-83 Plus*. Here follows an annotated copy of my version
of the program:

$\quad$ PROGRAM:BUFFON
$\quad$:25$\rightarrow$dim(L_1)
$\quad$:Fill(0,L_1) .

BUFFON will call PETE 2048 times and keep track for $I = 1, 2, \ldots, 25$ how
many games a head was first achieved on the I-th toss of a coin. L_1 will keep
a running tally of these numbers. The first two lines of BUFFON set the 25
counters of L_1 to 0.

$\quad$:For(K,1,2048)
$\quad$:prgmPETE
$\quad$:If I>dim(L_1)
$\quad$:I$\rightarrow$dim(L_1)
$\quad$:L_1(I)+1$\rightarrow L_1$(I)
$\quad$:End .

PETE is the simulation of tossing a coin until it turns up heads. The number
of tosses required is stored by PETE in the variable I. It could happen that I
is so large we don't have a counter for it, i.e., there is no I-th element of L_1.
When that happens, we simply extend L_1 by appending 0's at the end until the
length, i.e., dimension, of L_1 is I. We then add 1 to the I-th counter.

$\quad$:dim(L_1) $\rightarrow$ D
$\quad$:seq(2^(X−1),X,1,D) $\rightarrow L_2$.

These lines create the list of payoffs

$\quad$ $\{1\ 2\ \ldots\ 16777216\ (\ldots)\}$.

(Assuming, as is highly probable, one never has to extend L_1, the last element
of the list will be $2^{24} = 16777216$.)

$\quad$:L_1*$L_2\rightarrow L_3$
$\quad$:sum(L_3)/2048$\rightarrow$ B
$\quad$:Disp B .

This now calculates the average winnings of the 2048 games. To avoid cluttering up the memory with useless variables, one might want to do a little cleaning up before displaying B:

```
:ClrList L₁,L₂,L₃
:DelVar K
:DelVar I
:DelVar D
:DelVar X.
```

Do not delete B before the Disp command or the Disp B command will result in a 0 appearing on the screen. If you want to delete B, you may safely do so after displaying it on the screen.

The use of seq(to generate short sequences as was done in the programs BINGRAPH and BUFFON is the most obvious application of this function, but it is not the only one. We can use it to simulate the sub(function on strings. Given a list

$$L_1 = \{a_0\ a_1\ \ldots\ a_{k-1}\ a_k\ \ldots\ a_{k+m-1}\ \ldots\ a_{n-1}\},$$

we can extract $\{a_k\ \ldots\ a_{k+m-1}\}$ by entering

```
seq(L₁(I),I,K,K+M−1).
```

where the numbers k, m are stored in the variables K, M, respectively.

In a similar fashion, we can reverse a list:

```
seq(L₁(dim(L₁)−I),I,0,dim(L₁)−1),
```

or, remembering that we can choose our own increment and that it can be negative:

```
seq(L₁(I),I,dim(L₁),1,−1).
```

If L_1 is the list $\{a_0, a_1, \ldots, a_{n-1}\}$ and L_2 is a list $\{i_0, i_1, \ldots, i_{k-1}\}$ of positive integers between 1 and $n-1$, we can construct the list $\{a_{i_0}, a_{i_1}, \ldots, a_{i_{k-1}}\}$ using seq(:

```
seq(L₁(L₂(I)),I,1,dim(L₂)).
```

[Note that, just as we can replace a number variable by a list variable in an arithmetic expression to produce a list of values of the expression, we might expect that we could enter $L_1(L_2)$ to obtain the same list as this last expression using seq(, e.g, that $L_1(\{1,2\})$ would result in the list $\{L_1(1)\ L_1(2)\}$. Attempts to compose lists in this way merely result in error messages.]

Concatenation was another basic operation on strings, one we made use of in the programs SUBST1, SUBST2, and SUBST3. Concatenation is easily programmed via the list operations we already have:

```
PROGRAM:CONCAT1
:dim(L₁) →A
:dim(L₂) →B
:For(I,1,A+B)
```

:If I≤A
:Then
:L_1(I) →$_L$CNCT1(I)
:Else
:L_2(I−A) →$_L$CNCT1(I)
:End
:End[9]
:DelVar I
:DelVar A
:Delvar B .

The program constructs the concatenated list $_L$CNCT1 one element at a time, first listing the elements of L_1 and then those of L_2. We could shorten the runtime by starting with L_1 and then appending the elements of L_2:

PROGRAM:CONCAT2
:dim(L_1) →A
:dim(L_2) →B
:L_1→$_L$CNCT2
:For(I,1,B)
:L_2(I) →$_L$CNCT2(A+I)
:End
:DelVar I
:DelVar A
:Delvar B .

Both programs work fine, as one may check by storing nonempty lists in L_1 and L_2 and running the programs. Unfortunately, programs on the *TI-83 Plus* are not functions. One might try using seq(to simulate CONCAT1 by entering

seq(L_1(I)∗(I≤dim(L_1))+L_2(I−dim(L_1))∗(I>dim(L_1)),I,1,dim(L_1)+dim(L_2)) .

This will give an error message. (*Exercise.* Explain the error and the message.) Fortunately, the folks at Texas Instruments have thoughtfully included the function augment(as the 9th item in the OPS menu: augment(L_1,L_2) will produce the desired concatenated list functionally and without the need for a new program.

The rest of the functions in the OPS menu are not necessary for the discussion of the present book and I shall not discuss them, other than to note that two of them allow for conversion between lists and matrices—the remaining data type we wish to consider. Before introducing matrices, however, I digress to discuss calculations with statistical lists.

Before considering these statistical lists, however, let me consider a more mundane example, that of obtaining the list of digits of a number. If we store a number in the variable X, we can obtain the last digit by finding the remainder

---

[9] Note the necessity of two End commands. The first closes the conditional and the second closes the loop. If only one End were given, the program would fail to iterate the loop and $_L$CNCT1 would have length 1, its sole element being L_1(1).

on dividing by 10. If we subtract this digit and divide by 10, the last digit of the result is the next to the last digit of X. And so on:

```
PROGRAM:DIGITS
:If X=0
:Then
:{0}→L₁
:Goto 1
:End
:int(log(X))+1→L
:L→dim(L₁)
:X−10∗int(X/10)→L₁(L)
:(X−L₁(L))/10→X
:For(I,L−1,1,−1)
:X−10∗int(X/10)→L₁(I)
:(X−L₁(I))/10→X
:End
:Lbl 1
:DelVar I
:DelVar L
:DelVar X
:Disp L₁ .
```

Once one has the program DIGITS, one can give a slightly more transparent version of our program NUM2STR3 converting a number to a string of digits:

```
PROGRAM:NUM2STR4
:"0123456789"→Str2
:prgmDIGITS
:dim(L₁)→L
:For(I,1,L)
:If I=1
:Then
:sub(Str2,L₁(I)+1,1)→Str1
:Else
:Str1+sub(Str2,L₁(I)+1,1)→Str1
:End
:End
:DelVar I
:DelVar L
:DelVar X
:DelVar Str2
:ClrList L₁
:Disp Str1 .
```

On the *TI-89*, which allows lists of strings as well as lists of numbers, the program can be made a bit more transparent by simply replacing L₁ by a list of the digits represented as strings and writing a program concatenating a list of strings.

Some Statistical Calculations

That the DRAW and PRGM menus are accessed by the same button on the *TI-83 Plus* or that the DISTR and VARS menus are accessed by a single button appear to be accidental. There is no strong connexion between the groups of these two pairs. However, that the LIST menus and STAT menus share a common button is a matter of significance. The basic data of statistics are lists and the manipulation of statistical data affords a rich collection of applications of lists and list operations.

In olden days, before the advent of the pocket calculator, we did most computation by hand. There were mechanical devices such as adding machines to reduce the drudgery, and although they were small enough to sit on one's desk without getting in the way when not in use, they were hardly portable. Students did not carry them around between classes, taking them out of their backpacks in study halls or libraries to do their mathematics homework. Indeed, although not very loud individually, the cacophony of the mass usage of adding machines in such sanctums of silence would have been deemed sufficiently distracting by the authorities to have their use banned in places where peace and quiet are essential. Computational labour was spared by and large, not by machine, but by the efficient organisation and transformation of data. Statistical computations offer a dramatic demonstration of this practice.

In statistics, the mean of a collection $x_0, x_1, \ldots, x_{n-1}$ of numbers is usually denoted $\overline{x}$ or μ or, where there are also y_i's around as well, μ_x (and the mean of the y_i's being $\overline{y}$ or μ_y). In the present context, $\overline{x}$ is often used:

$$\overline{x} = \frac{x_0 + \ldots + x_{n-1}}{n} = \frac{1}{n} \sum_{i=0}^{n-1} x_i.$$

Conceptually, the calculation of $\overline{x}$ couldn't be any simpler: add up all the x_i's and divide the sum by their number. On the calculator we would naturally simulate this by storing $x_0, x_1, \ldots, x_{n-1}$ in a list, say, ∟X and then entering

sum(∟X)/dim(∟X) .

Or, of course, we could enter the expression mean(∟X), which yields the same result with just under half the keystrokes and without having to know what goes into its calculation.

In calculating by hand, we would demonstrate that we know what the mean is by simply adding the numbers and dividing by their quantity. However, we can actually cut down on the labour by complicating the procedure slightly: Instead of calculating $\overline{x}$ directly, make a first rough estimate of the mean and then average the differences of the x_i's and the estimate, finally adding this average to the estimate to obtain $\overline{x}$. A little algebra justifies the procedure: Let m be an initial estimate of the mean of the data and observe

$$m + \frac{1}{n} \sum_{i=0}^{n-1} (x_i - m) = m + \frac{1}{n} \left(\sum_{i=0}^{n-1} x_i - \sum_{i=0}^{n-1} m \right)$$

$$= m + \frac{1}{n} \left(\sum_{i=0}^{n-1} x_i \right) - \frac{1}{n} \cdot nm$$

$$= m + \overline{x} - m = \overline{x}.$$

As we see, we will get the correct result for any choice of m. A value somewhere in the midst of the x_i's will yield numbers $x_i - m$ smaller than the numbers x_i themselves and which will thus be easier to handle.

3 Example. Nine men appear at a ritual weighing in ceremony and their weights (in pounds), in no particular order, are

$$210, 203, 197, 203, 185, 206, 212, 199, 197.$$

Using the definition, we would find the mean by storing the elements of the list in a list variable,

$\{210,203,\ldots,197\} \rightarrow \llcorner X$,

and then calculating $\mathsf{sum}(\llcorner X)/\mathsf{dim}(\llcorner X)$ as before. The computational shortcut is simulated on the calculator by first storing the elements of the list in $\llcorner X$, and then choosing an estimate for the mean. We can make any choice we please, the closer to the mean the better. We could note that some values lie above 200 and some below, and thus choose 200 as our initial estimate. Or, to demonstrate the use of the commands, we could first sort the list

$\mathsf{SortA}(\llcorner X)$,

then find the middlemost element of the list,

$\mathsf{median}(\llcorner X)$,

which turns out to be 203. This is a good choice—about half the numbers $x_i - 203$ are positive and half are negative, and all are much smaller than the individual entries. We would next enter

$\mathsf{sum}(\llcorner X{-}203)/\mathsf{dim}(\llcorner X){+}203$.

Doing all of this on the calculator should not impress you with the advantages of the new method. All we have done is add two steps to the procedure—subtracting 203 from L_1 and adding 203 to the calculated answer. Now, however, store $\llcorner X$ in L_1 and $L_1{-}203$ in L_2, enter the List Editor accessed by the STAT button and compare the two lists L_1 and L_2. If you had to add the elements of one of these two lists by hand, which list would you choose? If you are indifferent to the choice, imagine comparable lists with, say, 50 elements in each list and ask yourself again. Or, carry out both computations by hand and see which computation you found more pleasant (less unpleasant?).

Notice that a couple of numbers used in the this example repeat. Quite often data is presented in two lists—the list of values and a *frequency* list telling how many times the individual values occur.

4 Example. Suppose there was a larger group of men weighing in this year and we recorded the weights,

$$173, 175, 182, 187, 192, 193, 196, 205, 207,$$

occurring with the frequencies,

$$1, 3, 2, 7, 13, 9, 4, 2, 3,$$

respectively. Enter these as lists,

{173,175,...,207}→∟X
{1,3,...,3}→∟FREQ .

In the past we would have set this up by making a table like that of *Table 1*, below, entering the x_i's in the first column, the frequencies f_i in the second

x_i	f_i	$f_i * x_i$	$x_i - 192$	$f_i * (x_i - 192)$
173	1	173	-19	-19
175	3	525	-17	-51
182	2	364	-10	-20
187	7	1309	-5	-35
192	13	2496	0	0
193	9	1737	1	9
196	4	784	4	16
205	2	410	13	26
207	3	621	15	45
Σ:	44	8419		-19

Table 1.

column, and the products of the corresponding elements in the third column. Summing the second column gives us the total number of x-values counting multiplicities (in this example, the total number of men weighed) and summing the third column gives the combined total weight of all the men. Simple division yields the average weight. On the calculator, we enter the weights and frequencies

{173,175,...,207}→∟X
{1,3,...,3}→∟FREQ ,

and then enter

sum(∟FREQ∗∟X)/sum(∟FREQ) ,

to find the mean. The computational shortcut begins, after entering ∟X and ∟FREQ, with our making an initial guess at the mean. The midpoint of the range, 173 - 207, is

$(\mathsf{max}(\llcorner\mathsf{X})+\mathsf{min}(\llcorner\mathsf{X}))/2$,

or 190. The median, $\mathsf{median}(\llcorner\mathsf{X})$, is 192, which is also the *mode*, or most frequently occurring score. Either 190 or 192 would be a good choice. I chose 192.
Returning to hand calculation, we would extend our three columns of *Table 1*
by adding two new columns: In column 4 we tabulate all results of subtracting
192 from the first column, and in column 5 we list the results of multiplying
column 4 by column 2. Summing column 5 and dividing by the result of summing column 2 yields the average deviations from the mean, whence restoring
the 192 yields the mean:

$\mathsf{sum}(\llcorner\mathsf{FREQ}*(\llcorner\mathsf{X}-192))/\mathsf{sum}(\llcorner\mathsf{FREQ})+192$.

Note that the third column is not used at all in the shortcut calculation.

Once again, look at the figures in *Table 1*. If you had to find the mean by
hand, would you rather have to multiply columns 1 and 2 and add up column
3 or multiply columns 2 and 4 and add up column 5?

Let us pause for a couple of exercises:

5 Exercise. There is no button or menu item yielding the mode. Write a simple
program MODE which, given a list $\llcorner\mathsf{X}$ of values and a list $\llcorner\mathsf{FREQ}$ of frequencies,
will find the mode and store it in a variable M.

6 Exercise. Attaching expressions to lists allows automatic updates of the lists
defined by expressions. Enter the lists $\llcorner\mathsf{X}$ and $\llcorner\mathsf{FREQ}$ as above, store 192 in
the real variable M, and then attach expressions to lists as follows:

   "$\llcorner\mathsf{FREQ}*\llcorner\mathsf{X}$"→$\llcorner\mathsf{FX}$
   "$\llcorner\mathsf{X}-\mathsf{M}$"→$\llcorner\mathsf{XM}$
   "$\llcorner\mathsf{FREQ}*\llcorner\mathsf{XM}$"→$\llcorner\mathsf{FXM}$.

Then, choosing SetUpEditor from the EDIT menu accessed via the STAT button,
enter

   SetUpEditor $\llcorner\mathsf{X},\llcorner\mathsf{FREQ},\llcorner\mathsf{FX},\llcorner\mathsf{XM},\llcorner\mathsf{FXM}$.

Then go into the List Editor. All but the bottom row of *Table 1* should be there.
Exit the editor, store 190 in M, and re-examine the table. Exit the editor and
calculate $\mathsf{sum}(\llcorner\mathsf{FXM})$, $\mathsf{M}+\mathsf{sum}(\llcorner\mathsf{FXM})/\mathsf{dim}(\llcorner\mathsf{FREQ})$. Repeat with 191 stored in
M.

Calculating the variance and standard deviation offers a couple of new wrin-
kles, one computational and one a subtle question of meaning. By definition,
the variance σ^2 is just the average of the squares of the deviations of the values
from their mean:

$$\sigma^2 = \frac{1}{n}\sum_{i=0}^{n-1}(x_i - \overline{x})^2, \tag{5}$$

for $x_0, \ldots, x_{n-1}$ as in Example 3, and

$$\sigma^2 = \frac{1}{N} \sum_{i=0}^{n-1} f_i \cdot (x_i - \overline{x})^2, \tag{6}$$

for $x_0, \ldots, x_{n-1}$ as in Example 4, where $f_0, \ldots, f_{n-1}$ are the frequencies with which the values $x_0, \ldots, x_{n-1}$ occur and $N = f_0 + f_1 + \ldots + f_{n-1}$ is the total number of weights collected.

In calculating the variance by hand, one often does a little algebra first:

$$\sigma^2 = \frac{1}{N} \sum_{i=0}^{n-1} f_i \cdot (x_i - \overline{x})^2 = \frac{1}{N} \sum_{i=0}^{n-1} \left(f_i x_i^2 - 2 f_i x_i \overline{x} + f_i \overline{x}^2 \right)$$

$$= \frac{1}{N} \left(\sum_{i=0}^{n-1} f_i x_i^2 - 2\overline{x} \sum_{i=0}^{n-1} f_i x_i + \overline{x}^2 \sum_{i=0}^{n-1} f_i \right)$$

$$= \frac{1}{N} \sum_{i=0}^{n-1} f_i x_i^2 - \frac{1}{N} 2\overline{x} \sum_{i=0}^{n-1} f_i x_i + \frac{1}{N} \overline{x}^2 N$$

$$= \frac{1}{N} \sum_{i=0}^{n-1} f_i x_i^2 - 2\overline{x}^2 + \overline{x}^2$$

$$= \frac{1}{N} \left(\sum_{i=0}^{n-1} f_i x_i^2 \right) - \overline{x}^2, \tag{7}$$

which can also be written in the form

$$\sigma^2 = \frac{1}{N} \left(\sum_{i=0}^{n-1} f_i x_i^2 - \frac{1}{N} \left(\sum_{i=0}^{n-1} f_i x_i \right)^2 \right) \tag{8}$$

or

$$\sigma^2 = \frac{1}{N^2} \left(N \left(\sum_{i=0}^{n-1} f_i x_i^2 \right) - \left(\sum_{i=0}^{n-1} f_i x_i \right)^2 \right). \tag{9}$$

Statistics manuals and textbooks for students outside the exact sciences, e.g., college textbooks for freshman level business or liberal arts majors might follow the definition of variance with one of these formulæ unjustified by any derivation. Textbooks at a higher level may derive one of these variants and offer no further comment. The classic textbook of the 20th century by William Feller[10] is an example. And occasionally an author will offer some explanation:

> The calculation of the standard deviation from its definition (6) becomes inaccurate unless an accurate value of $\overline{x}$ is used, and then the computations usually become tedious

. . .

---

[10] William Feller, *An Introduction to Probability Theory and Its Applications*, vol. 1, 2nd edition, John Wiley & Sons, Inc., New York, 1957, p. 213.

This form is often more convenient than (6)... particularly when the x_i contain at most two digits each.[11]

Of course, where no frequency list is involved, the calculator's variance(function can be applied, thus by-passing the tabular calculations altogether. Here, however, is where the subtlety briefly mentioned above comes in. The functions stdDev(and variance(do not calculate the standard deviation and variance as given by the definition (5) of the variance, but instead variance(calculates

$$S^2 = \frac{1}{n-1} \sum_{i=0}^{n-1} (x_i - \overline{x})^2,$$

and stdDev(calculates, not σ, but

$$S = \sqrt{\frac{1}{n-1} \sum_{i=0}^{n-1} (x_i - \overline{x})^2}.$$

The quantities σ and σ^2 are called the *standard deviation* and the *variance of the population*, while S and S^2 are termed the *standard deviation* and the *variance of the sample*. Briefly, one uses σ and σ^2 if the list $x_0, x_1, \ldots, x_{n-1}$ exhausts the entire population, and one uses S and S^2 as estimates of the standard deviation and variance of the population if $x_0, x_1, \ldots, x_{n-1}$ is just a sample taken from the population. Note that

$$S^2 = \frac{n}{n-1} \sigma^2, \quad S = \sqrt{\frac{n}{n-1}} \, \sigma,$$

and for large n the values are close. Two explanations for this dichotomy are offered in the literature, but what it boils down to is that S offers a better estimate of the σ of the entire population than does the σ value of the sample— unless n is large and there is not much difference between the values of S and σ calculated for the sample. *Cf.* Appendix B, page 468, below.

Now it is not my intention to turn this book into an introduction to Statistics, but Statistics is closely allied to Probability and it offers many of the most natural applications of lists. Thus I have gone into some detail here. And I shall go into one more statistical detail: The *TI-83 Plus* offers one more useful feature here. If we are given a list of values of the variable x, say, ∟XLIST, and a frequency list, ∟FREQ, we can press the STAT button, go to the CALC submenu, and choose the first item 1-Var Stats and enter

1-Var Stats ∟XLIST,∟FREQ .

The result will be a scrolling display. For the lists of Example 4 the first screen of the display will read

---

[11] Paul G. Hoel, *Introduction to Mathematical Statistics*, 3rd ed., John Wiley & Sons, Inc., New York, 1962, pp. 74 - 75. I have changed the reference number to the formula (6) to agree with the numbering of the present text.

1-Var Stats
  x̄=191.3409091
  Σx=8419
  Σx²=1613569
  Sx=7.879744955
  σx=7.789687769
  ↓n=44 ,

the arrow indicating that one can scroll down for more information (information which we will not concern ourselves with here). These values are also stored in the calculator under these names in the Statistics... submenu accessed via the VARS button. The sums Σx and Σx^2 are useful for other reasons, but hardly seem necessary for statistical purposes as Sx and σx have already been calculated. Nonetheless, one might like to use them in (7), (8), or (9) to see that these formulæ do indeed yield $(\sigma x)^2$.

If one is given a list $x_0, x_1, \ldots, x_{n-1}$ without an associated frequency list and possible repetitions of values, one can use $1, 1, \ldots, 1$ as a frequency list, but it is unnecessary: The command 1-Var Stats can in this case be used in the form

1-Var Stats ∟XLIST .

Once one knows what the mean and two standard deviations are, 1-Var Stats is a convenient time saver. It does not itself have much educative value other than as something to aim for in learning to program the calculator. Here are a couple of projects for those who wish to become proficient at programming the calculator:

7 Project. Write programs ONEVARS1 and ONEVARS2 which do the following: ONEVARS1 assumes given a list ∟XLIST and ONEVARS2 assumes ∟XLIST and ∟FREQ. Each program will produce a list ∟INFO containing in order, the mean of the x-values in ∟XLIST, the sum of the x's, the sum of the squares of the x's, the standard deviation of the x's as a sample, the standard deviation of the x's as a population, and the total number of x's (taking the frequencies into account in ONEVARS2). You are not allowed to use the functions mean(, stdDev(, variance(, 1-Var Stats, or 2-Var Stats in either program.

8 Project. To better emulate 1-Var Stats, it would be nice to display the results in a similar fashion. Complete the following ending for ONEVARS1 and ONEVARS2:

```
:ClrHome
:Disp "x̄="
:Output(1,3,∟INFO(1))
:Disp "Σx="
:Output(2,4,∟INFO(2)) .
```

Note that Output(controls the exact location that its third input appears on the screen. The characters x̄, Σx, Σx², etc. are found in the Statistics... menus accessed by the VARS button.

9 Project. When data comes in pairs $\langle x, y \rangle$, there are, in addition to the means, variances, and standard deviations of the x's and y's, a *correlation coefficient* (with all sums ranging over $i = 0, \ldots, n-1$):

$$r = \frac{1}{n} \sum \frac{(x_i - \overline{x})}{\sigma_x} \frac{(y_i - \overline{y})}{\sigma_y}$$

$$= \frac{n \sum x_i y_i - \left(\sum x_i\right)\left(\sum y_i\right)}{\sqrt{\left(n \sum x_i^2 - \left(\sum x_i\right)^2\right)\left(n \sum y_i^2 - \left(\sum y_i\right)^2\right)}}.$$

Write a program TWOVARST which assumes given two lists ∟XLIST and ∟YLIST and produces three lists, ∟XINFO, ∟YINFO, and ∟MISC. ∟XINFO should list $\overline{x}$, Σx, Σx^2, Sx, and σx in order; ∟YINFO should list $\overline{y}$, Σy, Σy^2, Sy, and σy in order; and ∟MISC should list n (the number of $\langle x, y \rangle$-pairs), Σxy, r in order. You are again not allowed to use mean(, stdDev(, variance(, 1-Var Stats, or 2-Var Stats, but you may call or copy ONEVARS1.

Here are a couple of additional programs dealing with statistical lists. The first takes a given list ∟X of numbers and splits it into an ordered list of values occurring in the list and a list of their corresponding frequencies.

```
PROGRAM:FREQLIST
:SortA(∟X)
:dim(∟X)→N
:{∟X(1)}→∟VALS
:{1}→∟FREQ
:For(I,2,N)
:dim(∟VALS)→K
:If ∟X(I)=∟VALS(K)
:Then
:∟FREQ(K)+1→∟FREQ(K)
:Else
:augment(∟VALS,{∟X(I)})→∟VALS
:augment(∟FREQ,{1})→∟FREQ
:End
:End
:DelVar I
:DelVar K
:DelVar ∟X.
```

The variables I and K being useless I have deleted them. Also, I doubt there is any further need for the original list ∟X and have deleted it. The variable N, being the number of elements of the original list and now the sum of ∟FREQ may still be of some use, e.g., in calculating the mean or variance. It can be recovered by summing ∟FREQ, but since we already have it, why calculate it again?

Note that the augment(commands only extend ∟VALS and ∟FREQ by single elements and could thus be replaced by simple store commands, e.g.,

:∟X(I)→∟VALS(K+1)

in place of

:augment(∟VALS,{∟X(I)})→∟VALS .

If we have not previously run the program and lists ∟VALS and ∟FREQ do not already exist, we could similarly create the lists with such commands, e.g.,

:∟X(1)→∟VALS(1)

in place of

:{∟X(1)}→∟VALS .

However, if these lists already exist, such commands merely change the first elements of the lists and the remaining elements will be kept. Thus we do not want to make such changes to the third and fourth commands of the program to match any possible replacements of the augment(commands.

If the number of elements in the list ∟X is very large and the frequencies are very small, one can vary the procedure, breaking the range of the x's into intervals of equal length and generating a list of the frequencies with which the elements occur in the given intervals. For hand calculation, this offers the human calculator a smaller set of data which results in estimates of the mean and standard deviation wih errors no larger than half the length of an interval. If ∟X is not too long[12], we can program the calculator to do this for us by modifying the program FREQLIST. The following program assumes given a list ∟X of x-values to be analysed, and a number N of intervals one wishes to split the range of the x's into:

```
PROGRAM:INTVLFRQ
:min(∟X)→A
:max(∟X)→B
:(B−A)/N→L
:L/2→H
:seq(A+IL,I,0,N)→∟ENDPT
:seq(∟ENDPT(I)+H,I,1,N)→∟MIDS .
```

This is the setting up portion of the program. A and B are the minimum and maximum elements of the list ∟X. Dividing B−A by N gives us the common length L of each of the intervals. Further dividing by 2 gives us half the length of the intervals. ∟ENDPT is the sequence of endpoints of the intervals we partition [A,B] into, and ∟MIDS is the list of midpoints of these intervals. For statistical calculations, all the x's are identified with the midpoints of the intervals in which they lie. The error introduced is at most H. Note that, by first calculating H and then taking ∟ENDPT(I)+H to obtain ∟MIDS(I)

---

[12] The maximum number of elements a list on the *TI-83 Plus* can have is 999.

requires only a single division by 2, while not pre-calculating H and taking $(\llcorner ENDPT(I)+\llcorner ENDPT(I+1))/2$ to obtain $\llcorner MIDS(I)$ requires N such divisions.

$\llcorner MIDS$ serves the same purpose here that $\llcorner VALS$ did in the program FREQLIST. I have used the new name simply because variable names should convey as much information as they can in five or fewer letters.

We know that to each interval there will correspond a frequency counting the number of x's in the interval. Thus the dimension of $\llcorner FREQ$ will equal N, the number of intervals. We can thus create a frequency list with all of its elements initially set to 0:

```
:0→dim(∟FREQ)
:N→dim(∟FREQ) .
```

The first command empties the list $\llcorner FREQ$ should it already exist and creates one otherwise; the second command fills $\llcorner FREQ$ with N 0's.

The next step is to tally the distribution of the x's. In FREQLIST, I used the variable N to denote the dimension of $\llcorner X$. Here I have used N to denote the number of intervals. Thus I introduce the not particularly mnemonic M to denote the dimension of $\llcorner X$:

```
:dim(∟(X)→ M .
```

The tallying is very simple. For each element x_k of the list $\llcorner X$, one finds I such that

$$A+LI \le x_k < A+L(I+1)$$

and increments $\llcorner FREQ(I+1)$ if $I < N$ and $\llcorner FREQ(N)$ if $I = N$. If $\llcorner X(K)$ happens to be an endpoint other than B, this places $\llcorner X(K)$ in the interval to its right; B is placed in the last interval.

```
:For(K,1,M)
:int((∟X(K)−A)/L)→I
:If I≠N
:Then
:∟FREQ(I+1)+1→∟FREQ(I+1)
:Else
:∟FREQ(N)+1→∟FREQ(N)
:End
:End .
```

This portion of the program would then be followed by some cleaning up commands to remove unneeded variables:

```
:DelVar I
:DelVar K
:DelVar ∟ENDPT .
```

Deletion of the remaining variables depends on what one wants to do with the lists. One can, for example, finish off the program with the following commands which draw a histogram displaying the data:

```
:A−H→Xmin
:B+H→Xmax
:L→Xscl
:max(∟FREQ)/5→Y
:−Y→Ymin
:max(∟FREQ)+Y→Ymax
:PlotsOff
:FnOff
:ClrDraw
:Plot1(Histogram,∟MIDS,∟FREQ)
:DispGraph .
```

And, again, we can now delete unnecessary variables—definitely Y, probably H, L, A, B, N—and, if one uses 1-Var Stats to find the mean, variance, and standard deviation, M is no longer needed and can be deleted as well.

This finishes the examples I have to offer from Statistics. However, I have one more program using statistical graphing that I would like to present here. In Project 1 of Chapter 5, on page 164, the reader was asked to write a program BERNULLI illustrating the monotonic decrease of the sequences $q_k, q_{k+t}, q_{k+2t}, \ldots$ of left tails of the binomial expansions $(p+q)^{nt}$ by graphing a scatter plot of the terms $q_1, q_2, q_3, \ldots, q_{nt}$ for various values of r, s, t. Here is my initial version of the program:

```
PROGRAM:BERNULLI
:Disp "ENTER"
:Disp "PARAMETERS:"
:Input "R=",R
:Input "S=",S
:Input "N=",N
:R+S→T
:R/T→P
:seq(X,X,1,N)→L₂
:"int((R−1)X/T−((R−1)X/T=int((R−1)X/T))"→Y₁
:seq(binomialcdf(X,P,Y₁(X)),X,1,N)→L₁
:PlotsOff
:FnOff
:Plot1(Scatter,L₂,L₁,□)
:DelVar R
:DelVar S
:DelVar T
:DelVar N
:DelVar P
:ZoomStat .
```

If you haven't already worked out your own version of BERNULLI and run it with the choices of values for R, S, T suggested in Project 1, enter the above program now and run it supplying the given values, carefully examining the graphs. The points graphed oscillate, but decrease and one should see a pattern making it evident that for whatever $k \leq t$ one starts with, the sequence $q_k, q_{k+t}, q_{k+2t}, \ldots$ decreases monotonically. To make this more visually evident, replace the last six lines of the program by the following, which "connect the dots" $q_k, q_{k+t}, q_{k+2t}, \ldots$ for $k = 1, \ldots, t$:

```
:DelVar R
:DelVar S
:DelVar P
:For(I,1,T)
:Plot1(Scatter,L₂,L₁,□)
:For(J,1,int(N/T)−1)
:Line(L₂(I+(J−1)T),L₁(I+(J−1)T),L₂(I+JT),L₁(I+JT))
:End
:Pause
:End
:DelVar T
:DelVar I
:DelVar J
:DelVar N
:ZoomStat.
```

When you run the new program, it will display the scatter plot and lines connecting the points $\langle 1, q_1 \rangle, \langle 1+t, q_{1+t} \rangle, \langle 1+2t, q_{1+2t} \rangle, \ldots$ and then pause for you to examine the graph. When you hit ENTER, it redraws the plot, this time with lines connecting $\langle 2, q_2 \rangle, \langle 2+t, q_{2+t} \rangle, \langle 2+2t, q_{2+2t} \rangle, \ldots$, and then pauses again. Each time you hit ENTER it repeats with the next subsequence singled out for special treatment, until all the subsequences have been displayed. The final ENTER displays the scatter plot without any of the special lines.

Matrices

Matrices are rectangular arrays of numbers and as such share some of the same functions and properties as lists and strings—they have dimensions, they can be augmented, their individual elements can be accessed, etc. Also, like lists, they can be added, subtracted, multiplied, and even multiplied by numbers. Matrix arithmetic, however, is fundamentally different from the arithmetic of lists. Whereas the latter arises more-or-less as a convenience, matrix multiplication arises from applications. In the present book we have not yet fully encountered the arithmetic of matrices, having used them only in Chapter 1 in the program POINTS (page 20), where they were used simply for storage.

The *TI-83 Plus* has ten built-in names for matrices, [A], [B], $\ldots$, [J], and these are listed in the NAMES menu accessed by pressing the MATRX button. This is the only way of accessing the names: the name [A] is different from the string [A] entered by using the [, A, and] buttons. The horizontal lines in a matrix are called *rows* and the vertical ones are *columns*. Thus, for example, in

$$\begin{bmatrix} 1 & 2 & 3 \\ 4 & 5 & 6 \end{bmatrix}, \tag{10}$$

1, 2, 3 and 4, 5, 6 are rows, and 1, 4; 2, 5; and 3, 6 are columns. A matrix consisting of m rows and n columns is said to have *dimension* $m \times n$, which is represented in the calculator as the list $\{m\ n\}$.

The function dim(applies to matrices as well as to lists: if [A] represents the matrix (10), then entering

dim([A])

results in

$\{2\ 3\}$.

One can use an assignment command,

$\{3,4\} \rightarrow$dim([B]),

to create a matrix or re-dimension an already existing matrix. 0's are placed where previous elements did not exist. Thus, for example, for [A] as above,

$\{3,4\} \rightarrow$dim([A])

will result in the matrix,

$$\begin{bmatrix} 1 & 2 & 3 & 0 \\ 4 & 5 & 6 & 0 \\ 0 & 0 & 0 & 0 \end{bmatrix},$$

which is displayed in the calculator as

[[1 2 3 0]
 [4 5 6 0]
 [0 0 0 0]],

while

$\{2,2\} \rightarrow$dim([A])

yields

$$\begin{bmatrix} 1 & 2 \\ 4 & 5 \end{bmatrix}.$$

Individual elements are accessed by entering the *name* of a matrix followed by the coordinates of the desired element in parentheses and separated by a comma, first the row number and then the column number. For our standard [A] of (10),

[A](1,2) yields 2
[A](2,3) yields 6.

As with lists, this access function can be used to replace an element of a matrix. For our customary [A],

$7 \rightarrow$[A](2,2)

results in

$$\begin{bmatrix} 1 & 2 & 3 \\ 4 & 7 & 6 \end{bmatrix}.$$

Unlike lists, we cannot use this to extend either dimension by 1. The commands

$7 \rightarrow [A](3,1)$ and $7 \rightarrow [A](1,4)$

both result in ERR: INVALID DIM.

To extend a matrix there is the augment(command which takes two matrices with equal numbers of rows and creates a new matrix consisting of the columns of the first matrix followed by the columns of the other. Thus, if one creates and stores the matrices

$$\begin{bmatrix} 1 & 2 \\ 4 & 5 \end{bmatrix} \text{ in } [A] \quad \text{and} \quad \begin{bmatrix} 3 \\ 6 \end{bmatrix} \text{ in } [B],$$

then augment([A],[B]) results in

$$\begin{bmatrix} 1 & 2 & 3 \\ 4 & 5 & 6 \end{bmatrix}.$$

One can think of a matrix as a list of columns of a fixed length and augment(as working the same on such lists as it did on lists of numbers. Indeed, there are operations Matr▶list(which will extract columns from a matrix as lists and List▶matr(which will combine several lists into the columns of a matrix. Other than in one exercise (page 419, below), we do not use this capability in the present work.

And, finally, before discussing matrix arithmetic, I mention one last operation familiar from our discussion of lists. This is Fill(, which again takes a number and a name, this time a matrix name, and fills all entries of the matrix with the given number. Filling a matrix with 0's is a convenient initialisation procedure. For our usual [A], Fill(0,[A]) returns

$$\begin{bmatrix} 0 & 0 & 0 \\ 0 & 0 & 0 \end{bmatrix}.$$

The important arithmetical operations on matrices, in order of increasing complexity, are multiplication by a constant and negation, addition and subtraction, and multiplication. And, after defining multiplication, we can exponentiate, raising a matrix to a positive integral power. There is no division, although there are special matrices that behave where they should like identities, and some matrices will have inverses, leading to a sort of occasional division. We we will make no use of this in the present work.

Multiplication by a constant and negation are performed elementwise:

$$5 \begin{bmatrix} 1 & 2 \\ 3 & 4 \end{bmatrix} = \begin{bmatrix} 5 & 10 \\ 15 & 20 \end{bmatrix}, \quad -\begin{bmatrix} 1 & 2 \\ 3 & 4 \end{bmatrix} = \begin{bmatrix} -1 & -2 \\ -3 & -4 \end{bmatrix}.$$

Both operations apply to any matrix. To be added or subtracted, as with the addition and subtraction of lists, two matrices must have the same dimensions. For, these operations are again computed componentwise:

$$\begin{bmatrix} 1 & 2 & 3 \\ 4 & 5 & 6 \end{bmatrix} + \begin{bmatrix} 1 & -1 & 1 \\ -2 & 2 & -2 \end{bmatrix} = \begin{bmatrix} 1+1 & 2-1 & 3+1 \\ 4-2 & 5+2 & 6-2 \end{bmatrix} = \begin{bmatrix} 2 & 1 & 4 \\ 2 & 7 & 4 \end{bmatrix}.$$

One enters these operations in the calculator in the usual way one does with numbers and lists:

5[A], 5∗[A], [A]5, [A]+[B], [A]−[B].

The same is true of multiplication: one enters

[A][B]   or   [A]∗[B]

via the keypad and the product will appear if it is defined and the calculator has the necessary memory. A simple 25-by-25 matrix occupies over 5000 bytes of RAM—in a device holding only 24K RAM. Even the less computationally intense operation of addition of two such matrices may require one to delete some programs and variables before the operation can be performed. (We will indeed shortly be considering a 25-by-25 matrix in connexion with one of the problems discussed in Chapter 3, so this is not just a theoretical possibility but one we will actually encounter.)

As with addition, to be able to multiply two matrices, their dimensions must match. This does not mean the dimensions must equal, as multiplication is not elementwise, but that the number of rows of the second matrix must equal the number of columns of the first. That is, a product [A][B] can be formed if there are positive integers m, n, p such that [A] has dimension $m \times n$ and [B] has dimension $n \times p$. Their product will have dimension $m \times p$ and the element of the i-th row, j-th column of the product will be obtained by taking the elements of the i-th row of [A], multiplying them by the corresponding elements of the j-th column of [B], and adding the individual products. Thus, for example,

$$\begin{bmatrix} 1 & 2 & 3 \\ 4 & 5 & 6 \end{bmatrix} \begin{bmatrix} 2 & 1 \\ 2 & 1 \\ 1 & 2 \end{bmatrix} = \begin{bmatrix} 1 \cdot 2 + 2 \cdot 2 + 3 \cdot 1 & 1 \cdot 1 + 2 \cdot 1 + 3 \cdot 2 \\ 4 \cdot 2 + 5 \cdot 2 + 6 \cdot 1 & 4 \cdot 1 + 5 \cdot 1 + 6 \cdot 2 \end{bmatrix}$$

$$= \begin{bmatrix} 9 & 9 \\ 24 & 21 \end{bmatrix},$$

while

$$\begin{bmatrix} 2 & 1 \\ 2 & 1 \\ 1 & 2 \end{bmatrix} \begin{bmatrix} 1 & 2 & 3 \\ 4 & 5 & 6 \end{bmatrix} = \begin{bmatrix} 2 \cdot 1 + 1 \cdot 4 & 2 \cdot 2 + 1 \cdot 5 & 2 \cdot 3 + 1 \cdot 6 \\ 2 \cdot 1 + 1 \cdot 4 & 2 \cdot 2 + 1 \cdot 5 & 2 \cdot 3 + 1 \cdot 6 \\ 1 \cdot 1 + 2 \cdot 4 & 1 \cdot 2 + 2 \cdot 5 & 1 \cdot 3 + 2 \cdot 6 \end{bmatrix}$$

$$= \begin{bmatrix} 6 & 9 & 12 \\ 6 & 9 & 12 \\ 9 & 12 & 15 \end{bmatrix},$$

as one can check on the calculator by entering the matrices as [A] and [B] in the calculator[13] and successively multiplying them by entering [A][B] and [B][A] in the home screen. Notice that matrix multiplication is not commutative, yielding here not only different matrices for the two products, but matrices of different

---

[13] The simplest way to do this is to press the MATRX button, navigate to the EDIT menu and then choose the matrix you want to define. Once you have done that, you simply enter the dimensions and matrix elements before pressing the QUIT button to exit. Remember to hit ENTER before quitting to register the last change made.

dimensions. Even for *square* matrices ($m = n$ and $n = p$), where the two products have the same dimensions, these products can fail to be equal. And it can happen (if $p \neq m$) that one can form the product [A][B] but not [B][A].

The mechanics of matrix multiplication are variously motivated by simple examples. If we have a $1 \times n$ *row matrix* [A] and an $n \times 1$ *column matrix* [B], we can think of the elements of [A] as quantities of items to be purchased and the elements of [B] as the unit prices of these items, whence [A][B] will be (a 1×1 matrix, the sole element of which is) the total amount spent. In physics, we might think of the elements of [A] as being weights and those of [B] as being heights, in which case (the element of) [A][B] will represent the total work done raising all the given weights to their designated heights.

Elementary probability also offers a simple motivating example of the multiplication of a row and a column vector. Given a game with outcomes $O_0, O_1, \ldots, O_{n-1}$ having probabilities $p_0, p_1, \ldots, p_{n-1}$ and payoffs $x_0, x_1, \ldots, x_{n-1}$ for a given player A, the expected value of the game for A is

$$E = x_0 p_0 + x_1 p_1 + \ldots + x_{n-1} p_{n-1},$$

and

$$[E] = \begin{bmatrix} x_0 & x_1 & \cdots & x_{n-1} \end{bmatrix} \begin{bmatrix} p_0 \\ p_1 \\ \vdots \\ p_{n-1} \end{bmatrix}.$$

Markov Chains

The application of matrices in Probability Theory we need to discuss here deals with *Markov chains* or *Markov processes*, first introduced in 1906 by Andrei Andreevich Markov (1856 - 1922), sometimes referred to as Markov the Elder, his identically named son having also been an important mathematician. In its simplest form a Markov process consists of a closed system which can assume one of finitely many *states* $s_0, s_1, \ldots, s_{n-1}$, and a set of fixed probabilities p_{ij} of making the transitions from states s_i to states s_j in a single discrete step. Concrete examples abound. The Problem of Points, for example, deals with a Markov process. The states would be represented by the various possible distributions of points and the *transition probabilities* p_{ij} being determined by the probabilities the individual players have of winning a round of the game.

The first of Pacioli's problems involves two teams playing for 6 points, one has made 5 and the other 2. They are equally likely to score an individual point. The possible states are the pairs $\langle m, n \rangle$ of numbers of points the two teams can have. Considering only what is possible from the present state $\langle 5, 2 \rangle$ on, the possible states are

$$\langle 5, 2 \rangle, \langle 5, 3 \rangle, \langle 5, 4 \rangle, \langle 5, 5 \rangle, \langle 5, 6 \rangle, \langle 6, 2 \rangle, \langle 6, 3 \rangle, \langle 6, 4 \rangle, \langle 6, 5 \rangle.$$

There are thus 9 possible states $s_0, s_1, \ldots, s_8$. We can list the transition probabilities in a matrix, which, for intelligibility's sake I give in the form of a table with the states as row and column headers (*Table 2*, below).

	$\langle 5,2 \rangle$	$\langle 5,3 \rangle$	$\langle 5,4 \rangle$	$\langle 5,5 \rangle$	$\langle 5,6 \rangle$	$\langle 6,2 \rangle$	$\langle 6,3 \rangle$	$\langle 6,4 \rangle$	$\langle 6,5 \rangle$
$\langle 5,2 \rangle$	0	1/2	0	0	0	1/2	0	0	0
$\langle 5,3 \rangle$	0	0	1/2	0	0	0	1/2	0	0
$\langle 5,4 \rangle$	0	0	0	1/2	0	0	0	1/2	0
$\langle 5,5 \rangle$	0	0	0	0	1/2	0	0	0	1/2
$\langle 5,6 \rangle$	0	0	0	0	1	0	0	0	0
$\langle 6,2 \rangle$	0	0	0	0	0	1	0	0	0
$\langle 6,3 \rangle$	0	0	0	0	0	0	1	0	0
$\langle 6,4 \rangle$	0	0	0	0	0	0	0	1	0
$\langle 6,5 \rangle$	0	0	0	0	0	0	0	0	1

Table 2. Transition Probabilities

If we start with the matrix

$$[A] = [1\ 0\ 0\ 0\ 0\ 0\ 0\ 0\ 0]$$

representing the probabilities of being in states $s_0 = \langle 5,2 \rangle$, $s_1 = \langle 5,3 \rangle, \ldots$, and let [B] be the 9×9 *transition matrix* of 0's, 1/2's, and 1's embedded in the *Table*, then

$$[A][B] = [0\ 1/2\ 0\ 0\ 0\ 1/2\ 0\ 0\ 0]$$

represents the probabilities of being in the given states after one more play of the game.

$$[A][B]^2 = [0\ 0\ 1/4\ 0\ 0\ 1/2\ 1/4\ 0\ 0]$$

represents the probabilities of being in the given states after two plays of the game. Note that, no matter how many plays are made, once a player has made 6 points, the state remains constant and at least one player will have 6 points after 4 games. Thus, the final probabilities are given by

$$[A][B]^4 = [0\ 0\ 0\ 0\ 1/16\ 1/2\ 1/4\ 1/8\ 1/16].$$

The first team wins in the states $\langle 6,m \rangle$ (i.e., s_5, s_6, s_7, or s_8), thus with probability

$$\frac{1}{2} + \frac{1}{4} + \frac{1}{8} + \frac{1}{16} = \frac{15}{16},$$

which agrees with the matrix on page 21 obtained by running the program POINTS. On that page we also mentioned the variant of the problem discussed by Tartaglia, one with an initial state of $\langle 5,3 \rangle$. For this case, we can simply choose

$$[A] = [0\ 1\ 0\ 0\ 0\ 0\ 0\ 0\ 0]$$

and use the same [B]. The game is over in 3 plays:

$$[A][B]^3 = [0\ 0\ 0\ 0\ 1/8\ 0\ 1/2\ 1/4\ 1/8]$$

and the probability of the first team's winning is

$$\frac{1}{2} + \frac{1}{4} + \frac{1}{8} = \frac{7}{8},$$

as previously calculated.

In theory, Pacioli's second problem involving 3 players in a game that will be decided in 7 rounds would ostensibly have 2187 states and a 2187×2187 transition matrix. Now most of these states can be ignored. Recall the problem: The first player has 4 points, the second 3, and the third 2. They are playing for 6. Thus, we need only consider states $\langle m, n, p \rangle$ where $m \in \{4, 5, 6\}, n \in \{3, 4, 5, 6\}$, and $p \in \{2, 3, 4, 5, 6\}$, there being $3 \cdot 4 \cdot 5 = 60$ of these. Of these, we can further eliminate those triples with more than one 6 among m, n, p. There are 10 of these, leaving us with only 50 states and a 50×50 transition matrix to deal with. The calculator allows matrices of dimensions up to 99×99, but one had better clear as much as possible from RAM to be able to do anything with such. Entering

{50,50}→dim([B])

simply to create a 50×50 matrix consisting of nothing but 0's results in an entity occupying 22511 bytes of RAM—and this in a calculator containing only 24K RAM. One cannot multiply [B] by itself in calculating $[A][B]^7$, but must multiply

[A][B], ([A][B])[B], (([A][B])[B])[B], $\cdots$

until the desired product occurs.

I definitely prefer solving Pacioli's second problem by the technique used in Chapter 3.

However, Markov chains are ideally suited to exploring the Gambler's Ruin problem of Chapter 3 (pages 92 - 105, above). Recall the problem:

A and B take each twelve counters and play with three dice on this condition, that if eleven is thrown A gives a counter to B, and if fourteen is thrown B gives a counter to A; and he wins the game who first obtains all the counters. Shew that A's chance is to B's as 244140625 is to 282429536481.

There are 24 counters in all, whence 25 states as A can have $0, 1, \ldots, 24$ counters. A loses a counter to B with probability $9/14$ and wins one from B with probability $5/14$, unless A has 0 or 24 counters. When either of these is the case, the game is over and the state never changes thereafter. This gives us the transitional probabilities p_{ij} for $i, j = 0, 1, \ldots, 24$,

$$p_{ij} = \begin{cases} 1 & i = j = 0 \text{ or } i = j = 24, \\ 9/14, & j = i - 1 \text{ and } i \neq 0, 24, \\ 5/14, & j = i + 1 \text{ and } i \neq 0, 24, \\ 0, & \text{otherwise.} \end{cases}$$

The 25×25 transition matrix [B] looks like

$$\begin{bmatrix} 1 & 0 & 0 & 0 & \ldots & 0 & 0 \\ 9/14 & 0 & 5/14 & 0 & \ldots & 0 & 0 \\ 0 & 9/14 & 0 & 5/14 & \ldots & 0 & 0 \\ \vdots & & & & & & \\ 0 & 0 & 0 & 0 & \ldots & 5/14 & 0 \\ 0 & 0 & 0 & 0 & \ldots & 0 & 5/14 \\ 0 & 0 & 0 & 0 & \ldots & 0 & 1 \end{bmatrix}.$$

And the starting state can be represented by the matrix [A],

$$[0\ 0\ \ldots\ 0\ 1\ 0\ \ldots\ 0],$$

with the sole nonzero element occurring in the 13th position representing A's possession of 12 counters. [The possible states start at 0 counters for A, whence all the indices are shifted up 1.] Successively multiplying [A] by [B] will yield the sequence [A], [A][B], ([A][B])[B], ... with $[A][B]^n$ listing the probabilities of the possible states after n tosses of the dice, i.e., the probabilities that A will possess the various numbers of counters.

Now, calculating $[A][B]^n$ by hand is a bit of a laborious procedure, especially as n will have to be quite large. There is no way either player can win in as few as 10 tosses, yet entering

[A][B]^10

in the calculator results, for me, in an error message: the calculator ran out of memory. Granted, I have cluttered up my calculator with a number of programs and have a few too many undeleted variables, while the reader may have an uncluttered calculator and maybe even a *TI-84* with more memory, and thus may not encounter this problem. But it will require less memory if one associates one's multiplications to the left and does not try to use matrix exponentiation. One can enter [A], then Ans*[B], and then keep pressing ENTER, watching for changes in the results. This gets boring fast and it is best to automate the procedure. In doing so I have generalised the problem: Two players A and B each have counters, the number to be chosen by whoever runs the program. The user may also determine the probability of A's winning on a given trial, and how many trials to carry the game to. I have also programmed a graphical representation of the play of the game: as each trial is finished, the probabilities that A has 0, 1, ... counters at the end of that stage are graphed.

The program naturally divides into three major stages. The first stage is simply to ask the user for some key parameters:

```
:PROGRAM:RUIN
:ClrHome
:Disp "PLEASE ENTER"
:Disp "THE FOLLOWING:"
:Disp "NUMBER OF"
:Disp "COUNTERS FOR"
:Input "A=",A
```

```
:Input "B=",B
:A+B+1→T
:ClrHome
:Disp "ENTER THE"
:Disp "PROBABILITY"
:Disp "THAT A WINS"
:Disp "A GIVEN GAME:"
:Input "P=",P
:1−P→Q .
```

The second stage of the program is to generate the matrices [A] and [B]:

```
:{1,T}→dim([A])
:Fill(0,[A])
:1→[A](1,A+1)
:{T,T}→dim([B])
:Fill(0,[B])
:1→[B](1,1)
:1→[B](T,T)
:For(I,2,T−1)
:Q→[B](I,I−1)
:P→[B](I,I+1)
:End .
```

The final stage of the program is to carry out the iteration, successively replacing [A] by new lists of probabilities and graphing them as the program proceeds a user-defined number N of steps:

```
:ClrHome
:Disp "HOW MANY TRIALS?"
:Input "N=",N
:.5→Xmin
:T+.5→Xmax
:1→Xscl
:−.1→Ymin
:1.1→Ymax
:seq(X,X,1,T)→L₁
:For(K,1,N)
:[A]*[B]→[A]
:seq([A](1,X),X,1,T)→L₂
:Text(1,5,K)
:Plot1(xyLine,L₁,L₂)
:DispGraph
:End
```

This part of the program is fairly unremarkable: The user is asked for a number of iterations, then an initial [A] is generated and a window for the graph is chosen. An xyLine is to be graphed, using 1, 2, ... for the x-coordinates. The current incarnation of the matrix [A] is converted into a list of y-coordinates and then a statistical plot is generated. The new command, Text(1,5,K) is a DRAW command that places the value of K in the picture in the first row, starting in the 5th column. It's purpose is simply to show the user how far along the iteration has proceeded. The redrawing is fairly rapid and I would suggest slowing it down by placing all the window assignment commands and the command generating L_1 inside the last For(loop or inserting a Pause command, but if one is going to iterate the procedure a couple hundred times these might not be welcome suggestions, especially the latter.

One should add a final minor clean-up stage deleting unnecessary variables:

```
:DelVar A
:DelVar B
:DelVar P
:DelVar Q
:DelVar T
:DelVar I
:DelVar K
:DelVar N
:DelVar [A]
:DelVar [B]
```

If one is done with the graph, one can clear the list L_1 as well. There is still one more thing to be done with L_2 and that is to compare the probabilities that A ends up with no counters with the probability that he ends up with all the counters. These probabilities are $L_2(1)$ and $L_2(T)$, respectively, the latter being obtained by entering $L_2(\dim(L_2))$ if one has deleted T.

10 Exercise. Run the above program for Huyghens's values A $= 12$, B $= 12$, P $= 5/14$. Choose 100 when prompted for a value of N. Compare $L_2(1)$ with $L_2(T)$ and compare their ratio with $282429536481/244140625$. Repeat the exercise with 100 replaced by 200 when prompted for a value of N.

11 Exercise. Run RUIN using the values:
i. 12 for A, 12 for B, $1/2$ for P, and 200 for N;
ii. 2 for A, 10 for B, $1/2$ for P, and 100 for N.

If you run the RUIN program for a variety of choices of A, B, and P, and suitably large N in each case, you should notice the following: The entries x_0, x_{n-1} of the matrix [A] each steadily increase to some limiting values and the quantities $x_1, x_2, \ldots, x_{n-2}$ each decrease to 0, so that the sequence

$$[A], [A][B], [A][B]^2, \ldots \tag{11}$$

tends to a limit of the form

$$[x\ 0\ 0\ \ldots\ 0\ y], \tag{12}$$

the exact values of x, y depending on the initial choices of A, B, and P. Using some advanced mathematics it can be shown that for matrices [B] arising from the Gambler's Ruin problem, the sequence (11) will converge to a limit. That the limit has the form (12) follows from this algebraically, either by the argument of de Moivre given back in Chapter 3, or, more simply by solving a set of simultaneous linear equations.

Suppose $[C] = [\,x_0\ x_1\ \ldots\ x_{n-1}\,]$ is the limit of the sequence (11). Then

$$[C][B] = (\lim_{n\to\infty} [A][B]^n)[B] = \lim_{n\to\infty} [A][B]^{n+1} = [C]$$

and we can try to determine $[C]$ by solving $[C][B] = [C]$, i.e.,

$$[\,x_0\ x_1\ \ldots\ x_{n-1}\,][B] = [\,x_0\ x_1\ \ldots\ x_{n-1}\,]. \tag{13}$$

Since $[B]$ has the form,

$$\begin{bmatrix}
1 & 0 & 0 & 0 & \cdots & 0 & 0 & 0 \\
q & 0 & p & 0 & \cdots & 0 & 0 & 0 \\
0 & q & 0 & p & \cdots & 0 & 0 & 0 \\
\vdots & & & & & & & \\
0 & 0 & 0 & 0 & \cdots & p & 0 & 0 \\
0 & 0 & 0 & 0 & \cdots & 0 & p & 0 \\
0 & 0 & 0 & 0 & \cdots & q & 0 & p \\
0 & 0 & 0 & 0 & \cdots & 0 & 0 & 1
\end{bmatrix},$$

the matrix equation (13) reduces to the system

$$
\begin{aligned}
1x_0 + qx_1 &&&&&= x_0 \\
qx_2 &&&&&= x_1 \\
px_1 &+ qx_3 &&&&= x_2 \\
\vdots &&&&& \\
px_{n-4} &&+ qx_{n-2} &&&= x_{n-3} \\
&px_{n-3} &&&&= x_{n-2} \\
&&px_{n-2} + x_{n-1} &&&= x_{n-1}.
\end{aligned}
$$

Subtracting the right from the left yields

$$
\begin{aligned}
qx_1 &&&= 0 \\
-x_1 + qx_2 &&&= 0 \\
px_1 - x_2 + qx_3 &&&= 0 \\
\vdots &&& \\
px_{n-4} - x_{n-3} + qx_{n-2} &= 0 \\
px_{n-3} - x_{n-2} &= 0 \\
px_{n-2} &= 0.
\end{aligned}
$$

The first equation yields $x_1 = 0$, the second then yields $x_2 = 0$, and so on down to $x_{n-2} = 0$. The variables x_0, x_{n-1} have disappeared, thus allowing arbitrary choices $x_0 = x, x_{n-1} = y$ for them. Hence the solution [C] is of the form (12) as promised.

For the problem at hand, starting with some distribution of the counters between the two players, the sum of the elements of each row matrix $[C][B]^n$ will be 1 and there will be the additional probabilistic constraint on the solution (12) that $x + y = 1$. Any further determination of [C] must take the actual values assumed for A, B, P into account. One can either use the existence of the limit and solve for it algebraically as was done in Chapter 3, or carry out the multiplications, e.g., by running the RUIN program, to obtain a good numerical approximation.

In elementary courses on Finite Mathematics, the examples given of Markov chains often have the special property that the limit [C] of the sequence (11) does not depend on the initial value matrix [A]. For example, in courses for business majors we offer problems like the following.

12 Example. Two companies X and Y control the market for a certain commodity. At present X owns 25% of the market and Y 75%. Experience teaches us that over the period of a year, X will keep 30% of its customers and lose 70% of them to Y, while Y will keep 40% of its customers and lose 60% to X. Given that this is an immutable law of human nature, what shares of the market will the two companies own in the long run?

Here, we think of the average customer as being in one of two possible states, X or Y, according to which product he purchases. At the outset, the probability of being in state X is .25 and that of being in state Y is .75, i.e.,

$$[A] = [.25 \ .75].$$

The transition matrix is

$$[B] = \begin{bmatrix} .3 & .7 \\ .6 & .4 \end{bmatrix}.$$

And we can solve

$$[x \ y][B] = [x \ y]$$

by setting up the equations,

$$.3x + .6y = x$$
$$.7x + .4y = y$$
$$x + y = 1,$$

i.e.,

$$-.7x + .6y = 0$$
$$.7x - .6y = 0$$
$$x + y = 1,$$

which amounts to solving

$$-.7x + .6y = 0$$
$$x + y = 1.$$

A little algebra yields $x = 6/13, y = 7/13$.

Grabbing our calculators and entering [A], [B] as above, we can calculate the matrices [A], [A][B], [A][B]^2, ... and watch the results quickly converge to

$$[.4615384615 \ .5384615385].$$

In fact, one gets exactly these digits when calculating [A][B]^{18}, the final digits differ for [A][B]^{19}, but revert for [A][B]^{20}. And if one follows the calculation of [A][B]^{22} with a ▶Frac command, the screen reads

[[6/13 7/13]].

The behaviour of the preceding example is special. Another special behaviour occurs when the transition matrix is merely a rotation of states: For each state s_i there is a state s_j for which the transition probability is $p_{ij} = 1$, all other values p_{ik} for $k \neq j$ thus being 0. For such a case there could fail to be any limit at all.

13 Example. Let the transition matrix [B] be

$$\begin{bmatrix} 0 & 0 & 1 \\ 1 & 0 & 0 \\ 0 & 1 & 0 \end{bmatrix}.$$

For [A] $= [\, x_0 \ x_1 \ x_2 \,]$ the sequence [A], [A][B], [A][B]^2, ... is periodic:

$$[\, x_0 \ x_1 \ x_2 \,], \quad [\, x_1 \ x_2 \ x_0 \,], \quad [\, x_2 \ x_0 \ x_1 \,], \quad [\, x_0 \ x_1 \ x_2 \,], \quad \dots$$

The general behaviour of Markov chains is rather complicated, with discussions of "transient states", "persistent states", periodicity and non-periodicity, etc. I refer the reader with a strong mathematical background to Chapters XV and XVI of the first volume of Feller's classic text on probability[14] for more information.

3 Randomness and Simulations

We have only encountered one application of random number generation in this book thus far. This was in the program PETE simulating the toss of a coin in illustrating Buffon's experimental approach to the Petersburg Problem. The sequences of random numbers, perhaps better termed random sequences

---

[14] Feller, *op. cit.*, "XV Markov Chains", pp. 338 - 379, and "XVI Algebraic Treatment of Finite Markov Chains", pp. 380 - 396.

of numbers, used by the calculator are not truly random. They are sequences generated by simple rules. In his book on applications of number theory, Manfred R. Schroeder begins his discussion of random number generators with the words

> In contemporary computation there is an almost unquenchable thirst for random numbers. One particularly intemperate class of customers is comprised of the diverse *Monte Carlo* methods. Or one may want to study a problem (or control a process) that depends on several parameters which one doesn't know how to choose. In such cases random choices are often preferred, or simply convenient. Finally, in system analysis (including biological systems) random "noise" is often a preferred test signal. And, of course, random numbers are useful—to say the least—in cryptography.
>
> In using arithmetical methods for generating "random" numbers great care must be exercised to avoid falling into deterministic traps. Such algorithms never produce truly random events (such as the clicks of a Geiger counter near a radioactive source), but give only *pseudo*random effects.[15]

Like much of early probability theory, Monte Carlo methods, the use of computers to generate random sequences of numbers to simulate complex processes, were conceived in gaming. Stanislaw M. Ulam (1909 - 1984) describes this in his autobiography:

> The idea for what was later called the Monte Carlo method occurred to me when I was playing solitaire during my illness. I noticed that it may be much more practical to get an idea of the probability of the successful outcome of a solitaire game (like Canfield or some other where the skill of the player is not important) by laying down the cards, or experimenting with the process and merely noticing what proportion comes out successfully, rather than to try to compute all the combinatorial possibilities which are an exponentially increasing number so great that, except in very elementary cases, there is no way to estimate it...In a sufficiently complicated problem, actual sampling is better than an examination of all the chains of possibilities.
>
> . . .
>
> The idea was to try out thousands of such possibilities and, at each stage, to select by chance, by means of a "random number" with suitable probability, the fate or kind of event, to follow it in a line, so to speak, instead of considering all branches. After examining the possible histories of only a few thousand, one will have a good sample and an approximate answer to the problem. All one needed was to have the means of producing such sample histories. It so happened that com-

---

[15] M.R. Schroeder, *Number Theory in Science and Communication: With Applications in Cryptography, Physics, Biology, Digital Information and Computing*, Springer-Verlag, Berlin, 1984, p. 271.

puting machines were coming into existence and here was something suitable for machine calculation.[16]

One of Ulam's colleagues was John von Neumann, a mathematician who consulted for the military in Los Alamos, where the atomic bomb was designed, as well as on the East Coast at Aberdeen Proving Grounds in Maryland, where, in part, the first American all purpose electronic computer was being designed as part of the war effort. Ulam consulted with von Neumann on the matter:

> The Monte Carlo method came into concrete form with its attendant rudiments of a theory after I proposed the possibilities of such probabilistic schemes to Johnny in 1946 during one of our conversations... After this conversation we developed together the mathematics of the method. It seems to me that the name Monte Carlo contributed very much to the popularization of this procedure. It was named Monte Carlo because of the element of chance, the production of random numbers with which to play the suitable games.
>
> Johnny saw at once its great scope even though in the first hour of our discussion he evinced a certain skepticism. But when I became more persuasive, quoting statistical estimates of how many computations were needed to obtain rough results with this or that probability, he agreed, eventually becoming quite inventive in finding marvelous technical tricks to facilitate or speed up these techniques.[17]

Many of the most outstanding physicists of the Second World War period worked in Los Alamos on the A-bomb project and stayed on after the end of the war. One of these, the Greek-American Nick Metropolis (1915 - 1999), got involved with Ulam and von Neumann and wrote:

> John von Neumann saw the relevance of Ulam's suggestion and, on March 11, 1947, sent a handwritten letter to Robert Richtmyer, the Theoretical Division leader (see "Stan Ulam, John von Neumann, and the Monte Carlo Method"). His letter included a detailed outline of a possible statistical approach to solving the problem of neutron diffusion in fissionable material.
>
> Johnny's interest in the method was contagious and inspiring. His seemingly relaxed attitude belied an intense interest and a well-disguised impatient drive. His talents were so obvious and his cooperative spirit so stimulating that he garnered the interest of many of us. It was at that time that I suggested an obvious name for the statistical method—a suggestion not unrelated to the fact that Stan had an uncle who would borrow money from relatives because he "just had to go to Monte Carlo." The name seems to have endured.[18]

---

[16] S.M. Ulam, *Adventures of a Mathematician*, Charles Scribner's Sons, New York, 1976, pp. 196 - 198.

[17] *Ibid.*, p. 199.

[18] N. Metropolis, "The beginning of the Monte Carlo Method", *Los Alamos Science* 15 (1987), pp. 125 - 130; here, p. 127. This special issue of *Los Alamos Science*

John von Neumann (1903 - 1957) rivals Jakob Bernoulli in the quantity of aliases. Orig-inally Neumann János, he became Margittai Neumann János in 1913 when his uxorious father was elevated to hereditary nobility in his native Hungary and chose to honour his wife Margit by using the name of the village Margitta in his title. János germanised his name to Johann Neumann von Margitta when he went to study in Zürich in 1923 and this became Johann von Neumann after moving to Germany in 1926. A final americanisation of his name to John von Neumann took place after moving to the United States in 1930, where his friends all called him Johnny.

The postcard pictured was issued on 27 December 1982, one day before von Neumann's birthday and 25 years after his death. The occasion was a computer conference organised by the Johann Neumann Computer Society of Hungary. One sees two versions of his name in the cachet and a nice portrait of him on the imprinted stamp.

In 1992 Hungary issued a postage stamp depicting von Neumann, with a personal com-puter in the background, as one of two stamps honouring famous Hungarian-Americans, the other stamp bearing the likeness of Theodore von Kármán. Other philatelic recogni-tion of von Neumann was granted by Portugal in 2000 when that country paired him with fellow computer pioneer Alan Turing on one of a set of stamps celebrating the accom-plishments of the 20th century—the same set in which John Maynard Keynes (p. 59) and Andrei Nikolaevich Kolmogorov (p. 317) appear. When the United States Postal Service decided to break with its long-standing tradition somewhat late in the game in 2005 and honour scientists on stamps, one of those so honoured was von Neumann.

Italian physicist Enrico Fermi (1901 - 1954) had actually used, but never published, Monte Carlo methods in physical calculations for some years before they were more fully developed by Ulam, von Neumann, and their colleagues.
The stamp pictured was issued in 1967 on the occasion of the 25th anniversary of the first controlled nuclear reaction under Fermi's guidance in Chicago in 1942.

Our use here of random number generators will be much simpler than their originating use in studying complex physical processes. But even so we share the problem: how do we obtain a seemingly random sequence of numbers? Von Neumann himself proposed a "middle-square digits" algorithm. To produce a sequence of n-digit numbers, start with an n-digit number, square it, take the middle n-digits to form a new number, square it, etc. This is not a precise description as the square of an n-digit number could have $2n - 1$ or $2n$ digits and only one of these will have an unambiguously determined middle n digits. If $n = 3$, for example, some squares will have 6 digits and it is not clear which 3 interior digits are meant by the "middle 3". We could decide in advance that, for 6-digit numbers, we drop the leading digit. Thus, starting with 131 for example, we would generate

$$131^2 = 17161, \quad 716^2 = 512656, \quad 265^2 = 70225, \quad 022^2 = 484 = 00484$$
$$048^2 = 2304, \quad 230^2 = 52900, \quad 290^2 = \dots,$$

and we see that from here on every 3-digit number generated will end in 0. Whatever randomness is exhibited by the sequence

$$131, \quad 716, \quad 265, \quad 22, \quad 48, \quad 230$$

does not continue full strength beyond this point.

A better method was introduced by Derrick Henry Lehmer (1905 - 1991), an American number theorist who got involved in computing on the ENIAC in Aberdeen at the end of the Second World War and became one of the pioneers of computing in Number Theory. One of his contributions was the *linear congruential generator*, as his algorithm is called.

dedicated to the memory of Stanislaw Ulam included two articles on the origin of the Monte Carlo Method. Roger Eckhardt's "Stan Ulam, John von Neumann, and the Monte Carlo Method", cited above by Metropolis, followed Metropolis's article on pp. 131 - 136 and offered facsimile reproductions of parts of von Neumann's letter to Richtmyer mentioned.

Let a, b, c, m be positive integers and consider the sequence

$$x_0 = c$$
$$x_{n+1} = \text{remainder of } ax_n + b \text{ after division by } m.$$

In Number Theory it is proven that if m is a power of a prime number, then a, b can be chosen so that the sequence

$$x_0, x_1, x_2, \ldots$$

is a periodic sequence,

$$x_0, x_1, \ldots, x_{m-1}, x_0, x_1, \ldots, x_{m-1}, \ldots,$$

with $x_0, x_1, \ldots, x_{m-1}$ being a reshuffling of all the integers $0, 1, \ldots, m - 1$. With the right choice, there is no easily discernible pattern to the partial subsequence $x_0, x_1, \ldots, x_{m-1}$. If d is a divisor much smaller than m, the sequence of remainders of $x_0, x_1, \ldots, x_{m-1}$ after division by d is a pseudorandom sequence of numbers taken from $\{0, 1, \ldots, d - 1\}$.

For computational simplicity, m is usually taken to be a large power of 2. The rules governing the choices of a and b result from number theoretic considerations and cannot be gone into here, but when m is a power of 2, the rule is that a and b must both be odd, with a having a remainder of 1 after division by 4. The number c is arbitrary and is called the *seed* of the random number generator. It determines the starting point of the random sequence to be generated. If one wants reproducible results one would always choose the same seed before running a simulation; if one wanted variety, one would vary the seed. For example, if a teacher in a classroom wanted his/her students to come up with the same results in following a demonstration with the calculator, then he/she would instruct them to use a particular seed. If, however, the teacher wanted to simulate multiple sampling, he/she would assign his/her students different seeds.

As I say, the choice of a, b, m is subject to various rules I cannot go into here, but I can illustrate the process with an example of the generation of a random number sequence on the calculator. I will take $m = 256$ and $d = 10$. The method described requires us to find the remainders of numbers after dividing by m and d. The *TI-83-Plus* does not have a built-in remainder function, so we have to calculate it. If n is a positive integer, we can write

$$\frac{n}{m} = k + \frac{h}{m},$$

where k, h are non-negative integers and $0 \leq h/m < 1$. k is called the *greatest integer* in n/m or the *integral part* of n/m and h/m is the *fractional part*. The number h is the remainder and can be solved for as follows:

$$h = m \cdot \frac{n}{m} - mk = n - mk$$

or

$$h = m \cdot \frac{h}{m}.$$

On the calculator, the greatest integer is represented by the function int(, the integral part by iPart([19], and the fractional part by the function fPart(. All of these are found in the NUM menu accessed by pressing the MATH button. Thus we can calculate the remainder of X after dividing by, say 256, by entering[20]

$X-256\,$int$(X/256)$, $X-256\,$iPart$(X/256)$,   or $256\,$fPart$(X/256)$.

Thus, go into the equation editor and enter

$Y_1 = 256\,$fPart$(X/256)$
$Y_2 = 10\,$fPart$(X/10)$.

Now store these values in A, B, C:

$17 \to A$
$23 \to B$
$0 \to C$.

Press the MODE button, move the cursor over Seq and press ENTER and QUIT to enter the sequential mode. Then press Y= and enter the following

nMin$=1$
$u(n) = Y_1(Au(n-1)+B)$
$u(n$Min$)=C$.

Now generate a sequence

$u(1,100) \to L_1$,

and take the remainders on dividing by 10:

$Y_2(L_1) \to L_1$.

You should now have a list of digits with no obvious pattern to them. It is seemingly random. A minimal test on randomness is to see what percent of the digits are 0's, 1's, ..., 9's. Entering

sum$(L_1 = K)$

for each value of K from 0 to 9 will result in *Table 3*, below.

14 Exercise. Write a program that does all of the above, including defining Y_1, Y_2, u and creating *Table 3* in the form of a 2×10 matrix.

---

[19] For negative x, int(x) and iPart(x) differ, whence the two implementations. For positive x, the functions agree. In this discussion we are only concerned with positive values of x.

[20] Because of the rounding error, $X-A*$int$(X/256)$ and $A*$fPart(X/A) need not actually be equal. For example, for $X = 22$ and $A = 21$, the latter is too small by an error of 10^{-12}.

0	1	2	3	4	5	6	7	8	9
12	10	11	12	9	8	8	9	10	11

Table 3. Uniform Distribution of the Digits

The sequence u(0), u(1), ... is not completely random and will fail more complex tests of randomness. There are various ways of modifying the procedure to introduce greater randomness into the generated sequences. I refer the reader to Schroeder's book cited above for more on the matter.

The *TI-83 Plus* uses one of these more sophisticated random number generating algorithms. According to the TI-BASIC Developer website[21], Texas Instruments calculators use an implementation of an algorithm devised by the Canadian mathematician Pierre L'Ecuyer[22] (*b.* 1950), who combined a pair of Lehmer's linear congruential generators to obtain random number generators satisfying stringent theoretical criteria. The result is known as *L'Ecuyer's algorithm*. The implementation of it presented at the aforementioned web site is given by a pair of programs written for the *TI-89 Titanium*, which has a more generous stock of variable names, allowing more readable programs. Where the web site's program uses mod1, mod2 for divisors and mod(x,mod1), mod(y,mod2) for the remainders of x, y modulo mod1 and mod2, respectively, we use A, B, A∗fPart(x/A) and B∗fPart(y/B), respectively. Two multipliers they name mult1 and mult2 we call C and D. And their seed1, seed2, we label less mnemonically S, T, respectively. The first program, which seeds the generator with a pair of seeds, assumes a number serving as a single seed stored in the variable N. These programs, rewritten for the *TI-83 Plus*, read as follows:

```
PROGRAM:RANDSEED
:2147483563→A
:2147483399→B
:40014→C
:40692→D
:abs(int(N))→N
:If N=0
:Then
:12345→S
:67890→T
:Else
:A∗fPart(C∗N/A)→S
```

```
PROGRAM:RAND
:A∗fPart(S∗C/A)→S
:B∗fPart(T∗D/B)→T
:(S−T)/A→R
:If R < 0
:R+1→R
:R→R
```

---

[21] I am hesitant to give the URL as URLs seem to change frequently. However, at the time of writing, the address is www.tibasicdev.wikidot.com/home. Type "random number generator" in the search box.

[22] Pierre L'Ecuyer, "Efficient and portable combined random number generators", *Communications of the ACM* 31 (1988), pp. 742 - 749 and 774.

:B∗fPart(N/B)→T
:End

Neither program deletes the variables A, B, C, D, S, T at the end; these are used to create new S,T seed pairs with each run of RAND.

We obtain random numbers in the unit interval as follows. First plant a seed N. For the following, I stored 0 in the variable N. Then, run the program RANDSEED. When Done appears on screen, you can run RAND and iterate it by repeatedly pressing ENTER. The readout should be

.9435974025
.9083188612
.1466941786
.7731603432

and so forth. There are no Done's. You may have noticed some programs finish with a Done appearing on screen and some do not. Basically, if the program ends by producing a value, it displays the value on screen and the flashing cursor appears on the next line without any Done; if the last executed command does something else, expect to see Done. The seemingly pointless final command R→R of RAND is redundant from a computational point of view and is there only to guarantee that the final executed command produces a value.

15 Exercise. Delete the command R→R from RAND and run RAND several times in succession. Sometimes R will appear onscreen without Done and sometimes Done will appear without R. Explain why. If you replace R→R by Disp R, what do you think will happen?

As stated, these programs are already built into the calculator. Going to the PRB menu accessed by the MATH button, one finds a number of functions beginning with the syllable rand:

1: rand
5: randInt(
6: randNorm(
7: randBin(,

and the MATH menu accessed via the MATRX button has

6: randM(.

The most important of these are the first two, rand and randInt(.

rand does double duty executing the internal versions of both RANDSEED and RAND. To seed the random number generator, one simply stores the seed in rand:

0→rand
17→rand ,

etc. Any real number can be used as a seed, but only the absolute value of the integral part matters. The command rand itself generates a random real number in the interval $[0, 1]$. A bit more accurately phrased, a random 14 digit number representing a real number between 0 and 1 is generated (10 digits of which are displayed). If one chooses a positive integer $n < 999$, rand(n) will produce a list of n such random reals.

The default seed is 0. Coming straight out of the box, entering rand will yield .9435974025 on the calculator. Entering rand again will yield .908318861. Each successive entry produces the next item in the list. Setting the seed to 0 starts the process over with .9435974025.

Try it yourself:

16 Exercise. Store 0 in rand and then run rand several times in succession and compare the sequence of numbers obtained with that generated by iterating RAND above. Store 5 in N and in rand and run RAND and rand several times in succession and compare the values.

Once one has seeded the random number generator, the function of greatest interest to us here is randInt(. Its syntax is

randInt(*lower, upper*[,*numtrials*]) ,

where *lower* is a lower bound, *upper* is an upper bound, and *numtrials* the quantity of random numbers to be produced. *lower* and *upper* must be integers, preferably with *lower* < *upper*, though it seems not to matter. If the third variable is not present, randInt(randomly produces an integer between *lower* and *upper*, inclusive. Thus[23]:

    0→rand
    randInt(0,9)

produces 9. If a third value is entered, a random list of digits of the prescribed length is produced:

    0→rand
    randInt(0,9,5)

produces {9 9 1 5 4}. The output is a list even if the third value is 1:

    0→rand
    randInt(0,9,1)

produces {9}.

The numbers produced by rand(and randInt(are *uniformly distributed*. This means that all values in the ranges of these functions are equally likely. If you split the $[0, 1]$ interval into subintervals of equal length, generate a long list of random reals in the interval using rand(, and then graph their histogram using

---

[23] One need not seed the random number generator every time one uses it. I have done so in these examples for the sake of obtaining fixed reproducible results. In repeating these on your calculator, first do the reseeding as illustrated here, then perhaps choose a different seed or repeat using no new seed.

a variant of INTVLFRQ, it will be very flat. And, using FREQLIST to gather the results of, say, randInt(0,9,100) and making an xyPlot using these frequencies as y-values and $\{0,1,\ldots,9\}$ as x-values should result in a close approximation to a horizontal line. The functions randNorm(and randBin(are analogues to rand and randInt(yielding random reals and random integers with approximately normal and binomial distributions, respectively. See the manual for details.

The matrix function randM(takes the dimension of a matrix as input, its syntax being

randM(*rows,columns*),

and produces a *rows*×*columns* matrix of random integral entries n from the range $-9 \le n \le 9$. Its use might add a little spice in a classroom when students get tired of always seeing the matrices

$$\begin{bmatrix} 1 & 2 \\ 3 & 4 \end{bmatrix}, \quad \begin{bmatrix} 1 & 2 & 3 \\ 4 & 5 & 6 \end{bmatrix}, \quad \begin{bmatrix} 1 & 2 & 3 \\ 4 & 5 & 6 \\ 7 & 8 & 9 \end{bmatrix}, \ldots$$

Note that the LIST menus do not include a comparable function—randInt(already does the trick. In fact, randM(is itself redundant. Try this[24]:

```
0→rand
randInt(−9,9,9)→L₁
0→rand
randM(3,3)→[A] .
```

How do L_1 and [A] compare?

We now have a grasp of the calculator's random number generation capabilities, and it is time to apply them to simulate some random processes. We have done so in the text only once, in Chapter 2 in the program PETE where we used randInt(0,1) to simulate a single toss of a coin. The game of *treize* of Chapter 3 struck me as one worthy of a good calculator simulation. Unlike the Petersburg Problem, where I wrote two programs, PETE to simulate the coin tossing and BUFFON to iterate PETE 2048 times, with *treize* I have collected everything into a single program to make it ever so slightly more efficient. The key randomising element of the program, however, should be explained outside the confines of the main program, so I shall first present it as a standalone, RANDPERM (for "random permutation"), and then incorporate it into a game simulation program TREIZE.

The task RANDPERM has is to simulate the shuffling of a deck of cards. We can represent the deck of cards by a list of successive numbers beginning at 1: $\{1, 2, \ldots, n\}$. A shuffled deck would be represented by some random permutation of the list. For example, if we input

$\{1,2,3,4,5\}$,

the shuffling might produce

---

[24] Definitely do the reseeding in this example.

{2 1 5 4 3}.

There is a more-or-less obvious strategy for doing this. Pick a random number from $\{1, 2, \ldots, n\}$ to be first in the reshuffled deck. Then pick another
random number from $\{1, 2, \ldots, n\}$. If it is not the same as the first number,
choose it to be the second card in the reshuffled deck; if it agrees with the
first, ignore it and try again. This translates fairly directly into the following
program, which assumes the size of the deck of cards has already been stored
in the variable N:

```
PROGRAM:RNDPRM
:seq(X,X,1,N)→∟NUMBS
:N→dim(∟PERM)
:Fill(0,∟PERM)
:randInt(1,N)→J
:∟NUMBS(J)→Y
:Y→∟PERM(1)
:For(I,2,N)
:Repeat max(∟PERM=Y)=0
:randInt(1,N)→J
:∟NUMBS(J)→Y
:End
:Y→∟PERM(I)
:End .
```

And, of course, this would be followed by DelVar commands for the variables
X, Y, I, J.

Mindful of the 16 trials de Morgan's pupil went through tossing a coin
before heads came up in one of his simulations of the Petersburg Problem, it
struck me that this could be an inefficient program, the more so the larger the
values stored in N as many values will repeat before all the distinct values have
appeared. Storing 13 in N, running the following simple iteration program took
roughly 7 minutes, 20 seconds:

```
PROGRAM:RNDTEST1
:For(M,1,100)
:prgmRNDPRM
:End .
```

A slightly different approach produced a more efficient program than RND-
PRM. The idea behind this one is to keep track of those numbers that haven't
been chosen and only randomly pick from these:

```
PROGRAM:RANDPERM
:seq(X,X,1,N)→∟NUMBS
:N→dim(∟AUX)
:N→dim(∟PERM)
:Fill(0,∟AUX)
:For(I,N,1,−1)
```

```
:randInt(1,I)→J
:ʟNUMBS(J)→ʟPERM(N−I+1)
:1→ʟAUX(J)                    When J is picked, ʟAUX(J) is 1
:SortA(ʟAUX,ʟNUMBS)          This puts the unchosen numbers.
:End                          at the beginning of ʟNUMBS .
```

Of course, one could now add the desired DelVar commands. Storing 13 in N and running the following program took only 3 minutes, 20 seconds—less than half the time iterating RNDPRM:

```
PROGRAM:RNDTEST2
:For(M,1,100)
:prgmRANDPERM
:End .
```

The TREIZE program is little more than a more sophisticated version of RNDTEST2: It starts with a request for the size N of a card deck and the number T of games to play. Then, with ever so slightly more suggestively named lists, incorporates a T-fold iteration of RANDPERM, keeping track in a list ʟTALLY of the numbers of games with no matches and at least one match, respectively. The program ends deleting all the unnecessary variables and displaying first the list ʟTALLY and then the proportion of games in which there was at least one match.

```
PROGRAM:TREIZE
:ClrHome
:Disp "PLEASE ENTER"
:Disp "SIZE OF DECK:"
:Input "N=",N
:Disp "HOW MANY TRIALS?"
:Input "T=",T
:seq(X,X,1,N)→ʟDECK1
:{0,0}→ʟTALLY
:N→dim(ʟAUX)
:N→dim(ʟDECK2)
:For(K,1,T)
:ʟDECK1→ʟNUMBS
:Fill(0,ʟAUX)
:For(I,N,1,−1)
:randInt(1,I)→J
:ʟNUMBS(J)→ʟDECK2(N−I+1)
:1→ʟAUX(J)
:SortA(ʟAUX,ʟNUMBS)
:End
:max(ʟDECK1=ʟDECK2)+1→I
:ʟTALLY(I)+1→ʟTALLY(I)
:End
:DelVar I
:DelVar J
```

```
:DelVar K
:DelVar T
:DelVar ∟AUX
:DelVar ∟DECK1
:DelVar ∟DECK2
:DelVar ∟NUMBS
:ClrHome
:Disp ∟TALLY
:Disp ∟TALLY(2)/sum(∟TALLY) .
```

In a classroom, one could use the program to illustrate the Central Limit Theorem. Each student is to enter the program in his calculator, and run it using 13 for N and, say, 100 for the number T of games of *treize* to be played. In class one would ask the students to start running the programs and 10 or 15 minutes later the teacher could tally the results: graph the histogram of second values of ∟TALLY and calculate the proportion ∟TALLY(2)/sum(∟TALLY) (= ∟TALLY(2)/T if one hasn't deleted the variable T) of successes and see how normal the distribution looks, calculate a mean and compare it with the computed probability $.6321205588 \approx 1 - 1/e$, and calculate the standard deviation of the distribution. Or, one can send the students home with the instruction to perform the task for $T = 1000$ trials, warning them not to sit and watch their calculators, but to start the calculator running the program before going off to do their French homework or television viewing.

In such a demonstration, it is important that the students all have different samples. One way of guaranteeing this is to assign them different seeds by, say, randomly choosing five digit numbers. For 30 students, one can obtain these seeds quickly (in a second or two) by entering

randInt(10000,99999,30)→L₁ .

Of course, the numbers $1, 2, \ldots, 30$ are already distinct and there is no need for five digit seeds chosen at random, but it is a nice touch and should give the students a greater illusion of the randomness of the sampling.

Perhaps it would be easier to illustrate the Central Limit Theorem in class by writing a Monte Carlo simulation on the calculator. Suppose as in Example 13 of Chapter 6 we take $\{0, 1, \ldots, 20\}$ as our population with $f(i) = 1/21$ as distribution function, and we wish to consider the distribution of averages of 10-element samples of the population. Now the number of such samples is $_{21}C_{10} = 352716$, too large a number to permit enumeration of all the possibilities. Now, the Central Limit Theorem itself, as stated in section 3 of Chapter 6, concerns sampling with replacement, i.e., the successive choices are independent. Thus, we first illustrate the Theorem itself by randomly choosing some number N of sequences of length 10 of elements from the population and determining the distribution of the averages of these sequences. This distribution should be close to normal for N "large". Now, if we do this, there will be the possibility of the same sequence being chosen twice. However, there are $21^{10} = 16679880978201$

such sequences and N will be small compared to this, say ≤ 1000. There will likely be no overlap[25]. Thus we can safely ignore this possibility.

A second point is that the maximum length of a list on the *TI-83 Plus* is 999. For illustrative purposes one might want to restrict N to lying below this number as the larger N is, the longer it takes the calculator to perform all the calculations involved. But for $N \geq 1000$, or for the purposes of graphing the distribution, one will have to group the data in intervals. Thus our program will not list the actual averages, but will create a list of midpoints of appropriate intervals along with a frequency list of how many values are in the given intervals.

A third point is that taking the sums instead of the averages results merely in a change in scale and will not affect the overall shape of the statistical plot.

With these things in mind, I present the following program:

```
PROGRAM:CENTRAL1
:ClrHome
:Disp "HOW MANY TRIALS?"
:Input "N=",N
:ClrHome
:Disp "HOW MANY"
:Disp "INTERVALS?"
:Input "I=",M .
```

I tend to use I as a counter in programs, which is a good mnemonic for "interval". Hence I use I here in the display but not for the variable in the program. N is a good generic name for the numbers of things, but it is already used here for the number of trials. Hence I use M instead.

Now, the smallest sum is $0 + 0 + \ldots + 0 = 0$ and the largest is $20 + 20 + \ldots + 20 = 200$. Thus we have the range of possible sums of the given samples.

```
:200/M→L
:L/2→H
:seq(LX,X,0,M)→LENDPT
:seq(LENDPT(X)+H,X,1,M)→LMIDS
:DelVar LENDPT .
```

This portion of the program sets up the list ∟MIDS of midpoints of the intervals for the graphing.

The next few lines set up a few more lists, ∟SAMP and ∟FREQ. The elements of ∟FREQ are set to 0 and will be incremented as appropriate during the running of the program. Each element of ∟SAMP will be overwritten in the loop to follow and no values need be specified.

```
:10→dim(LSAMP)
:M→dim(LFREQ)
:Fill(0,LFREQ) .
```

---

[25] Indeed, using Feller's estimate (10) cited on page 74 in the discussion of the Birthday Problem, the probability of two sequences being the same is calculated to be approximately .000000030006 for $N = 1000$.

We now come to the loop which takes N samples, sums each one, and increments the appropriate elements of ⌞FREQ in the process.

```
:For(I,1,N)
:randInt(0,20,10)→⌞SAMP
:sum(⌞SAMP)→X
:int(X/L)→K
:If K≠M
:Then
:⌞FREQ(K+1)+1→⌞FREQ(K+1)
:Else
:⌞FREQ(M)+1→⌞FREQ(M)
:End
:End
:DelVar ⌞SAMP .
```

Next, the program sets up the xyLine plot of the data.

```
:−H→Xmin
:200+H→Xmax
:L→Xscl
:max(⌞FREQ)→ F
:F/5→Y
:−Y→Ymin
:F+Y→Ymax
:DelVar F
:DelVar Y
:DelVar H
:PlotsOff
:FnOff
:ClrDraw
:Plot1(xyLine,⌞MIDS,⌞FREQ) .
```

And, finally, we define the bell curve approximating the data and graph the two distributions for comparison:

```
:1-Var Stats ⌞MIDS,⌞FREQ
:"L*N*normalpdf(X,x̄,σx)"→Y₁
:DispGraph .
```

The use of 1-Var Stats is a convenience, not a necessity, as the values $\bar{x}, \sigma x$ can be calculated directly, replacing the 3 lines above by:

```
:sum(⌞MIDS*⌞FREQ)/N→ B
:√(sum((⌞MIDS−B)²/N))→C
:"L*N*normalpdf(X,B,C)"→Y₁
:DispGraph .
```

17 Exercise. Run CENTRAL1 several times with various choices of values for N and M, e.g.,

i. N = 30, M = 10
ii. N = 60, M = 20
iii. N = 600, M = 20

Run each set of N, M values several times. (Because of the randomising element, the graphs will differ from run to run, even with common values of N, M.)

The problem of sampling without replacement is more complicated. First note that the distributions of 10-element combinations from $\{0, 1, \ldots, 20\}$ and 10-element permutations therefrom will have the same means and variances. For, each combination occurs 10! times among the permutations and the probability of an event $\overline{X} = x$ for a given mean x will be the same whether taking permutations or combinations. The distributions of the means will thus be the same, even though there are only 352716 combinations as opposed to $_{21}P_{10} = 1279935820800$ permutations. Again, the size of $_{21}P_{10}$ pretty much allows us to ignore the possibility of two identical sequences of draws when sampling only a thousand or so times. Thus, what I am saying is, in effect, that our next program will proceed as before. The differences will be that i. we have to make sure that we only generate sequences with no repetitions, ii. because the extreme sums are now $0 + 1 + \ldots + 9 = 45$ and $11 + 12 + \ldots + 20 = 155$, the range of possible sums differs, and iii. for technical reasons we have to store $\{0, 1, \ldots, 20\}$ into a list.

The new program will first store the population in a list, but otherwise will start as before:

```
PROGRAM:CENTRAL2
:seq(X,X,0,20)→∟POP
:ClrHome
:Disp "HOW MANY TRIALS?"
:Input "N=",N
:ClrHome
:Disp "HOW MANY"
:Disp "INTERVALS?"
:Input "I=",M .
```

Because of the different range, two commands from the next batch are slightly different:

```
:(155−45)/M→L
:L/2→H
:seq(45+LX,X,0,M)→∟ENDPT
:seq(∟ENDPT(X)+H,X,1,M)→∟MIDS
:DelVar ∟ENDPT .
```

The generation of randomly chosen sequences of length 10 from ∟POP will be handled analogously to the way it was done in RANDPERM and TREIZE. We will introduce a list ∟AUX as in these programs and ∟SAMP will now behave like the lists ∟PERM of RANDPERM and ∟DECK2 of ∟TREIZE. Thus our setting up commands are

```
:10→dim(∟SAMP)
:21→dim(∟AUX)
:M→dim(∟FREQ)
:Fill(0,∟FREQ).
```

Next comes the big loop. It begins by setting the elements of ∟AUX to 0 and copying ∟POP into a new list ∟NUMBS. Both of these commands are designed to restore these lists at the beginning of each execution of the loop. The next element of the loop is the subloop generating a permutation. The outer loop then incorporates the looped step from the program INTVLFRQ.

```
:For(I,1,N)
:Fill(0,∟AUX)
:∟POP→∟NUMBS
:For(J,1,10)
:randInt(1,22−J)→K
:∟NUMBS(K)→∟SAMP(J)
:1→∟AUX(K)
:SortA(∟AUX,∟NUMBS)
:End
:sum(∟SAMP)→X
:int((X−45)/L)→K
:If K≠M
:Then
:∟FREQ(K+1)+1→∟FREQ(K+1)
:Else
:∟FREQ(M)+1→∟FREQ(M)
:End
:End
:DelVar I
:DelVar J
:DelVar K
:DelVar ∟SAMP
:DelVar NUMBS.
```

The rest is now pretty much the same as before, taking into consideration the different range:

```
:45−H→Xmin
:155+H→Xmax
etc.
```

18 Exercise. Repeat Exercise 17 using program CENTRAL2 in place of CENTRAL1.

A good in-class demonstration of the Central Limit Theorem in action is afforded by the *Galton board*, or *quincunx*, of Francis Galton (1822 - 1911). A cousin of Charles Darwin, Galton contributed to various sciences—exploration, meteorology, genetics, and, most relevant to our purposes, statistics. One of his

lesser contributions to this last field was the Galton board, an inclined board
with a storage area for lead shot or similar pellets we will call *balls* at the top,
a lattice of *pins* below that, and rectangular compartments at the bottom for
collecting the balls after they have worked their way through the lattice. The
top storage area is tapered at the bottom so as to allow only one ball to escape
at a time. The lattice of pins consists of staggered rows of evenly spaced pins.
Traditionally the staggering is such that each triple of pairwise adjacent points
forms an equilateral triangle with sides slightly larger than the diameter of the
descending ball. The basic quincuncial pattern from which Galton derived his
name for the Galton board is illustrated in *Figure 1*, below.

Fig. 1. The quincunx

One pin of the top row lies directly below the exit hole through which the
balls pass to roll down the board. A ball rolling down will hit it and be deflected
left or right with equal probability before rolling down, encountering another
pin to be deflected, etc. When the ball rolls through a pair of pins in the last
row, it rolls straight down and enters one of the compartments, the width of
which being equal to the distance between pins on a given row.

When I was young, Chicago's Museum of Science and Industry had a huge
wall-sized Galton board in which ping pong balls would drop down, bounce
their way to the bottom, and pile up in a heap closely approximating the bell
curve painted on the glass of the exhibit. When I last visited the Museum a
decade or two ago, the exhibit had been replaced by a much smaller board
with flippers replacing pins so that the probability of bouncing left or right
could uniformly be given some value other than 1/2. The mathematics itself
was not deemed sufficiently interesting in itself to justify the exhibit, so it
was presented as a health warning. One entered the quantity of cigarettes one
smoked per day and, perhaps, one's current age, and when all the balls had
passed through the maze, the heap at the bottom represented the probability
distribution for surviving to various ages, or for getting lung cancer after that
many years—I'm afraid I've never had a good memory for facts and cannot
recall exactly what the final pile represented, only that it was supposed to
frighten impressionable youths away from smoking.

I confess I don't see an immediate connexion between the angles of the
flippers and smoking or between said flippers and cancer and imagine the pro-
pagandistic message could be conveyed more convincingly with a few graphs
representing clinical data. The purely mathematical illustration of the Central
Limit Theorem, the bell shape as the outcome of successive actions of chance,
is, however, most viscerally achieved. If I turn now to discuss a calculator sim-

ulation of the Galton board as a replacement for an actual board, I do so with some sense of guilt.

I mentioned the wall-size board for a ping-pong-ball-based display to make the point that a board large enough for the entire class to see will not be very portable. If your school has a math lab, it would be feasible to place one there, perhaps with other probabilistic devices for a one-period "field trip" for the class, but otherwise a small portable board is called for. The construction of such begins to require some skill. An animation on the calculator, however, is fairly easy and, while students watching such an animation without reading the program cannot tell if the demonstration is "truly random" or carefully worked out, i.e., faked, the program will offer a visual illustration, if not a convincing demonstration, of the Central Limit Theorem. Also, it allows for experimentation that would be very difficult for a hardware-based device utilising real balls, unless the device were fashioned by a skilled artisan.

As I say, programming an animation on the calculator is an easy task. The result, unfortunately, is not very impressive. This is due in large part to the smallness of the screen and the speed of the calculator. One can represent dropping balls by moving points, the movement achieved by turning on and off successive points in the paths the balls must follow. The default sizes of points are 1-by-1 dots, 3-by-3 boxes, and 3-by-3 plus signs. The dots are difficult to see, especially when they are moving, so visibility seemingly demands a higher dimension than 1-by-1. But 3-by-3 boxes effectively cut the room for manouvre to one-third the space available. The sum of the number of rows and the height of the highest column of balls at the bottom is now at most 21—one-third of the 63 rows of pixels in the screen. When these numbers are limited the heap of balls does not always present the most convincing bell-shaped silhouette. On top of that, while moving, the balls are still not very visible. I tried slowing them down by pausing via a do-nothing loop,

```
:For(K,1,25)
:End ,
```

between turning a point on and turning it off. The result allowed one to see the balls if one watched carefully, but the process was slow. Finally, I tried not turning the points off until the ball had completed its journey. The result was rather pleasing and splendidly visible even with 1-by-1 points representing the balls.

I offer two programs, GALTON1 and GALTON2, to simulate the workings of the Galton board. The first will forego the animation and simply give a graphical representation of the outcome, drawing the predicted bell curve and superimposing a histogram over it. The second will replace this with the promised animation.

The key randomising element of both programs is the choice at encountering a pin of whether the ball should go left or right to pass it. We can do this in either of two ways: iterate the application of randInt(0,1) as many times (R) as there are rows of pins, or make all the choices up front by generating a list ∟DIRS of directions taken using randInt(0,1,R). The former might feel a little more faithful to the actual process, but I have, for better or worse, chosen the latter approach.

GALTON1 begins by clearing the screens and asking the reader for the numbers R of rows of pins and B of balls. For technical reasons, the program requests an even number of rows.

```
PROGRAM:GALTON1
:ClrHome
:AxesOff
:PlotsOff
:FnOff
:ClrDraw
:Disp "HOW MANY", "ROWS OF PINS?"
:Disp "(CHOOSE AN EVEN","NUMBER.)"
:Input "R=",R
:ClrHome
:Disp "HOW MANY BALLS?"
:Input "B=",B .
```

If one chooses r for the number of rows, the ball can be diverted left or right up to r times, thus yielding $2r + 1$ horizontal positions. The program creates a list ∟HIGHT acting as a tally for the numbers of balls accumulating in these positions:

```
:2R+1→dim(∟HIGHT)
:Fill(0,∟HIGHT) .
```

This ends the preparation part of the program. The next step is to generate the moves, keep track of positions of the balls, and tally the number of balls finishing in each possible position.

```
:For (I,1,B)
:2*randInt(0,1,R)−1→∟DIRS
:cumSum(∟DIRS)→∟DEVS
:∟DEVS(R)→X
:R+1+X→N
:∟HIGHT(N)+1→∟HIGHT(N)
:End .
```

A few words of explanation: randInt(0,1,R) is a list of 0's (left turns) and 1's (right turns); multiplying by 2 and subtracting 1 results in a list of −1's (left) and 1's (right). This is simply a trick to avoid the use of a multiline If-Then-Else construction. Taking the cumulative sum determines the total deviation left or right from the centre the ball is at a given row during its descent. X is the last horizontal position before the ball continues straight down and collects at the bottom of the board. Now ∟DEVS is the list of deviations from the centre of action, which should be at position R + 1 among the positions. The absolute coordinate of the ball's final position is thus R + 1 plus its final deviation from this position. Hence, the definition of N.

We are now ready to perform the graphing. Were one to examine the list ⌐HIGHT, one would notice that the second, fourth, sixth,... values are all 0. The reason is that the balls drop down and only move half a ball-radius to the left or right in passing from one row to another. With an even number of rows, when the ball finally exits the lattice of pins, it will not be at one of these half-way positions. Thus, we extract the R + 1 values not guaranteed[26] to be 0 and use them to construct a histogram:

```
:seq(X,X,0,R)→L₁
:seq(⌐HIGHT(2X+1),X,0,R)→L₂
:Plot1(Histogram,L₁,L₂)
:−.5→Xmin
:R+.5→Xmax
:1→Xscl
:max(⌐HIGHT)+1→Ymax
:−Ymax/4 →Ymin
:DispGraph .
```

The strange value for Ymin is to allow enough space on the bottom of the screen to view the entire histogram when using the TRACE function of the calculator.

Each ball contributes 1 square unit to the area of the histogram, thus yielding an area of B. The Central Limit Theorem predicts $R/2$ as mean and $R/4$ as variance. Thus we associate the following bell curve with the distribution:

```
:"B/((√(2π)√(R/4))e^(−.5(X−R/2)²/(R/4))"→Y₁
:DispGraph .
```

And, of course, one would finish with the usual cleaning up commands deleting I, X, N, ⌐HIGHT, ⌐DIRS, and ⌐DEVS. The variables R, B should not be deleted nor the lists L_1, L_2 cleared until one is finished admiring the graph.

19 Exercise. Run the program with various values of R and B. You might start with R being 10, 20, 30 and B being 50, 100, 150, 200.

An actual, physical Galton board would yield a more convincing in-class display than such a calculator simulation, but the calculator is more flexible. It would for example require greater skill and more tools to create an actual Galton board with variable probabilities.

20 Project. Rewrite the program so that, instead of asking for a number R of rows, it asks the user for a list ⌐PROBS (of even length) of probabilities of being deflected left. The program should then define R to be the dimension of ⌐PROBS and use rand(R) to generate a list L_1 of numbers representing the directions to be taken. ⌐DIRS is defined by

$$\text{⌐DIRS}(I) = \begin{cases} -1, & L_1(I) < \text{⌐PROBS}(I) \\ 1, & \text{otherwise.} \end{cases}$$

---

[26] For N close to 0 or R, ⌐HIGHT(N) will likely be 0 whether N is even or not.

[*N.B.* i. Just as I used doubling and subtracting 1 to avoid an If-Then-Else construction in GALTON1, you can avoid such here by the command

$2*\text{int}(L_1/\llcorner\text{PROBS})-1\rightarrow\llcorner\text{DIRS}$.

This assumes, of course, the user of the program never chooses 0 for the probability of a left turn. If you wish to allow for this possibility, you can use a test:

$2*(\llcorner\text{PROBS}\leq L_1)-1\rightarrow\llcorner\text{DIRS}$.

ii. Don't forget to recalculate the mean and variance of the overall distribution to determine the predictive bell curve.]

While, mathematically speaking, GALTON1 is about as faithful a calculator simulation of the Galton board as one could hope for, the connexion is not immediately obvious to the senses. There are two problems. First, the histogram can be a bit irregular and not quickly recognisably bell-shaped. This is offset partially by drawing the bell curve that supposedly predicts the shape of the histogram. Choosing values R and B carefully should further correct for this. Bear in mind here that on the calculator a histogram can have at most 46 columns, whence R cannot exceed 45. Further, the average number of balls in a column is B/R and if this ratio is too small, even small deviations of the histogram from the bell curve will stand out.

The second problem, of course, is that one simply does not *see* how the histogram was formed. To this end, we turn to animation.

The animated version of our simulation begins with the same setting up commands as before.

```
PROGRAM:GALTON2
:ClrHome
:AxesOff
:PlotsOff
:FnOff
:ClrDraw
:Disp "HOW MANY", "ROWS OF PINS?"
:Disp "(CHOOSE AN EVEN","NUMBER.)"
:Input "R=",R
:ClrHome
:Disp "HOW MANY BALLS?"
:Input "B=",B
:2R+1→dim(∟HIGHT)
:Fill(0,∟HIGHT) .
```

After this the program differs. First, we have to set up a suitable window for the graphics. The calculator screen is 95 pixels by 63 pixels. Now, we imagine the ball starting in the top centre pixel, dropping down through the rows deviating left and right, and then dropping straight down until it meets the pile of accumulated balls. A convenient window would thus be

```
:−47→Xmin
:47→Xmax
:1→Xscl
:−62→Ymin
:0→Ymax
```

And we could represent the pins of a row by turning on every other dot in the row:

```
:For(I,1,R)
:If 2>2int(I/2)
:Then
:For(J,−46,46,2)
:Pt-On(J,−I)
:End
:Else
:For(J,−47,47,2)
:Pt-On(J,−I)
:End
:End
:End .
```

In this program we are graphing the individual balls, which must each be done after the trajectory of the ball has been calculated and before work on the next ball is started. We will thus be building up the heap of balls as we go along and will not graph this at the end of the program as before. Thus, except for the variable-deleting clean up commands at the end, the rest of the program is one big loop for balls numbered 1 to B. The computations begin as before:

```
:For (I,1,B)
:2∗randInt(0,1,R)−1→∟DIRS
:cumSum(∟DIRS)→∟DEVS
:∟DEVS(R)→X
:R+1+X→ N
:∟HIGHT(N)+1→∟HIGHT(N) .
```

There is no End command because this is only part of the loop, namely the computational part. There is next the graphical part:

```
:Pt-On(0,0)
:For(J,1,R)
:Pt-On(∟DEVS(J),−J)
:End
:For(J,R+1,62−∟HIGHT(N))
:Pt-On(X,−J)
:End
:Pt-On(X,−62+∟HIGHT(N)−1)
:Pt-Off(0,0)
:For(J,1,R)
```

```
:Pt-Off(ᴸDEVS(J),−J)
:End
:For(J,R+1,62−ᴸHIGHT(N))
:Pt-Off(X,−J)
:End
:End.
```

The last End command closes the big loop. One may now delete the unneeded variables.

I have not graphed the bell curve this time round. There are two reasons for this. The draw commands are intended to be applied to a graph, either an existing one or a new one, and the results can be erased when a new element is graphed. Thus, one should graph the curve first. Unfortunately, the Pt-Off(commands erasing the paths of falling balls eat away at the curve and ruin the æsthetics of the representation. One could program one's way around this by only turning off points not lying too close to the curve. I decided this was way too much work. One work around is to append a few lines to the end of GALTON2, or to create a third program that calls GALTON2:

```
:StorePic 0
:"B/((√(2π)√(R/4))e^(−.5(X/2)²/(R/4))−62"→Y₁
:RecallPic 0
:DispGraph.
```

Some explanation is called for here. The commands StorePic and RecallPic are draw commands and are to be found in the STO menu accessed via the DRAW button. The respective syntaxes of the two commands are

StorePic *n* and RecallPic *n*.

The former creates a snapshot of the screen and stores a picture named Pic*n*; the second recalls the picture and draws it over the existing picture in the graphics screen. In the case of the above lines, the pre-existing picture in the graphics screen will be the bell curve and Pic0 will be superimposed on it. Once again, the area under the histogram is B, thus yielding that factor in the expression. The expected mean and variance of the distribution are again $R/2$ and $R/4$, respectively, but algebraic transformations have been applied: a horizontal translation $R/2$ to the left placing the centre at 0 instead of at the mean (whence the disappearance of the subtracted $R/2$ in the exponent), a horizontal stretch factor of 2 (whence the division of X by 2), and a downward vertical translation of 62 pixels (whence the final subtraction of 62).

If you now run the program a few times with various values of R and B, you will notice two little flaws in the design. The greater horizontal spread of the screen is mostly unused as the columns are bunched narrowly together in the centre. And the vertical space is easily exhausted. *Figure 2*, below, shows a screen capture of one run with R = 20 and B = 200. The tallest column actually invades the space occupied by the rows of pins, but the extra height is not displayed.

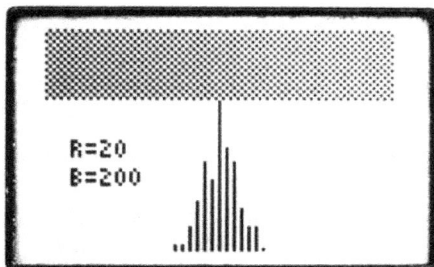

Fig. 2. A Run of GALTON2

21 Project. Rewrite GALTON2 interchanging dimensions so that the balls originate from the vertical centre on the left, pass through R columns of pins, and then move horizontally to the right to accumulate there. Note that you cannot graph the rotated bell curve by storing an expression in Y_1, but must use parametric function graphing by storing expressions in the variables X_{1T} and Y_{1T}.

After running GALTON1 and GALTON2, there are variables still around which will no longer be needed, especially if one hasn't added all the DelVar commands for the disposable variables at the ends of the programs. To this end, I offer a simple clean up program for use after one has finished admiring the pretty pictures:

```
PROGRAM:GLTCLNUP
:DelVar I
:DelVar J
:DelVar X
:DelVar R
:DelVar B
:DelVar ∟DIRS
:DelVar ∟DEVS
:DelVar ∟HIGHT
:DelVar Pic0
:DelVar Y₁
:ClrList L₁,L₂
:PlotsOff
:AxesOn
:−4.7→Xmin
:4.7→Xmax
:1→Xscl
:−3.1→Ymin
:3.1→Ymax
:1→Yscl.
```

The PlotsOff command is necessary because the lists L_1 and L_2 have been cleared and leaving Plot1 on will result in an INVALID DIMENSIONS error when one graphs something anew. The AxesOn command restores the display of axes. The windows variables Xmin, Xmax, etc. cannot be deleted, but they can be restored to some more commonly used setting. I have chosen the ZDecimal setting. A single line,

    :ZDecimal

could have been used in place of the 6 explicit commands given. However, this command enters the graphics screen and the program will quit with the calculator in the graphics screen uselessly displaying the axes. This command could be followed up with a simple command, say

    :ClrHome

which will bring one back to the home screen. This approach, however, leaves one with an unappealing flash as the axes pop into and out of sight quickly. The ClrHome command placed at the end of the program might still be a nice touch as it removes the last request for a value of B and the user's choice from sight.

B

Some Mathematical Extras

1 Preliminaries and Notations

ϕ

When I was a college student back in the 1960's, all mathematics textbooks began with an unreadable chapter of preliminary information, consisting of explanations of notations the reader was not assumed to have seen in prior courses. Every book explained that $\{a, b, c\}$ was a set containing exactly the elements a, b, c, that $X \cap Y$ was the intersection of two sets X and Y, etc. For the most part such a practice is not necessary, it being best to explain notation when it occurs, provided the explanation is not so involved as to interrupt the flow of the exposition. For these exceptions, the practice now is to defer the explanations to an appropriate appendix.

Looking back, I find a few notations and concepts in this book that could be given such appendicial treatment. The first of these, occurring as early as page 5 where "the divine proportion ϕ" is mentioned in passing, is this number. As I say, it was mentioned in passing and its nature and meaning are of no significance to anything in this book and nothing is lost if the reader has no idea what it is. The additional reference to the "other geometric topics" informs the reader that whatever ϕ is, it is geometric in nature, hence mathematical, and hence indicative of Pacioli's mathematical credentials—the point its citation was made in support of. There may, however, be some benefit to be derived from saying a few words about ϕ here.

The benefit to discussing ϕ has nothing to do with probability. It does however give me an excuse to include an additional illustration, an example of the recognition by the postal authorities of some nations of their ability to promote science education. The divine proportion ϕ, also called the "golden ratio", is the ratio of length to width of a rectangle with the curious reproductive property that, if you remove from one end of the rectangle the square with sides equal to the rectangle's width, the rectangle that is left over has the same proportions as the original one. (*Cf. Figure 1*, below.) Choosing 1 for the width and ϕ for the length, we have

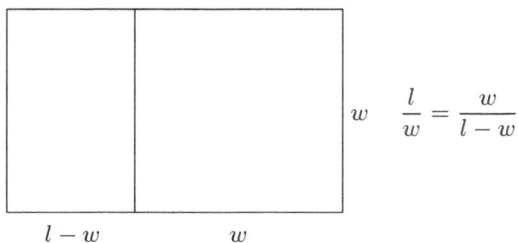

$$w \qquad \frac{l}{w} = \frac{w}{l-w}$$

$l - w$ \qquad w

Fig. 1.

$$\frac{\phi}{1} = \frac{1}{\phi - 1},$$

i.e., $\phi^2 - \phi = 1$. The quadratic formula yields

$$x = \frac{1 \pm \sqrt{5}}{2}$$

as the solutions to $x^2 - x - 1 = 0$. Since ϕ must be positive, this yields

$$\phi = \frac{1 + \sqrt{5}}{2} \approx 1.618033989.$$

A rectangle with dimensions in divine proportion was considered by the classical Greeks to have the most pleasing shape of all rectangles and it frequently appeared in Greek architecture. Its modern application in art includes the use of the inscribed spiral approximately illustrated in the stamp displayed on the next page: If one rotates the rectangle 90° counterclockwise, the "best" place to position the eyes in a portrait is within that part of the spiral inscribed in the rectangle consisting of the squares of sides 1 and 2. Those not involved in the production of art masterpieces may also encounter the "golden rectangle" every day. My widescreen monitor, for example, has two sets of resolutions—the traditional 4 : 3 ratio of old American televisions,

$$640 \times 480, 800 \times 600, 1024 \times 768, 1280 \times 960, 1344 \times 1008, 1600 \times 1200,$$

and the widescreen,

$$800 \times 500, 960 \times 600, 1024 \times 640, 1280 \times 800, 1344 \times 840, 1600 \times 1000, 1920 \times 1200,$$

with the aspect ratio 1.6 : 1, approximately ϕ.

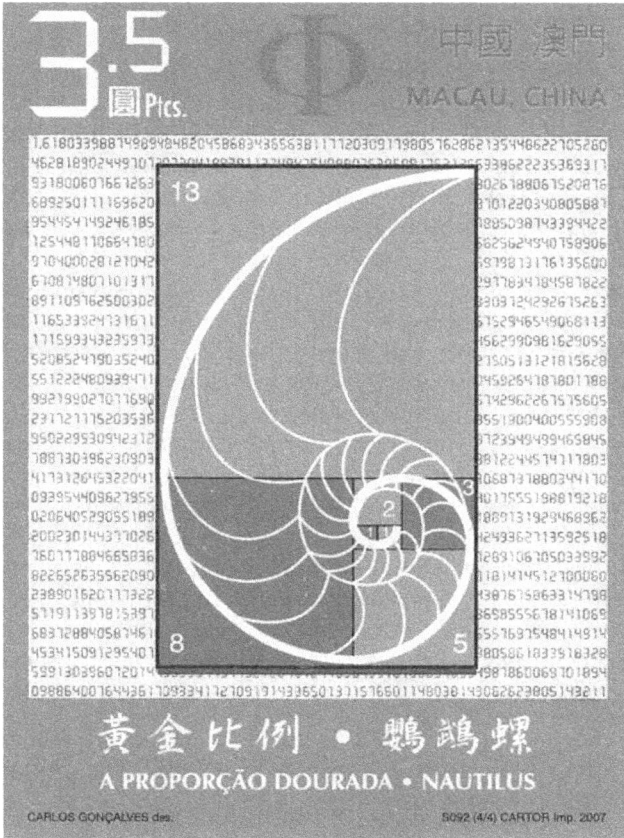

One of four stamps devoted to the divine proportion or golden ratio issued by Macao in 2007 in its annual set of science stamps. The symbol "Φ" denoting the golden ratio ϕ tops the image. The picture actually demonstrates the close relation between ϕ and the sequence $1, 1, 2, 3, 5, 8, 13, \ldots$ of Fibonacci numbers introduced by Leonardo of Pisa. These numbers, each one after the second being the sum of the two previous ones, bear to their individual predecessors a ratio that approximates the golden ratio more and more closely the farther out in the sequence one goes. For example,

$$5/3 = 1.\overline{6}, \ 8/5 = 1.6, \ 13/8 = 1.625, \ (13+8)/13 = 21/13 \approx 1.61538, \ldots,$$

thus illustrating the close relation between these numbers and ϕ. Surrounding the central image is the golden ratio itself, $\phi = 1.61803398\ldots$ I have upped the contrast in this portion of the image (and enlarged the whole as much as I did) to render these digits more visible. Other stamps in the set picture Leonardo's Rabbit Problem which gave rise to the Fibonacci numbers, geometric shapes of Roger Penrose related trigonometrically to the golden rectangle and used to tile the plane, and the spirals of the sunflower which, like the Nautilus shell in the picture, exhibit the Fibonacci numbers in nature.

Rounding

Slightly more relevant is the *greatest integer function* [x] introduced on page 163 and used on page 164 in the program BERNULLI, at various points in Appendix A, on page 356 in the program DIGITS, most importantly on page 367, again in BERNULLI, and on page 386 in connexion with random number generators. The greatest integer function belongs to a suite of functions and calculator implementations all related to integral rounding, i.e., rounding to an adjacent integer.

In elementary or middle school we learned how to round numbers to a specified number of significant digits. There were three versions of rounding—rounding down, rounding up, and general rounding. The greatest integer function rounds down to the greatest integer less than or equal to its argument. The function that rounds up to the least integer greater than or equal to its argument has no special name or notation in mathematical practice, but does in computer science where it is called *ceiling* and is denoted by $\lceil x \rceil$. As one might guess, the greatest integer in x is called *floor* by computer scientists who denote it by $\lfloor x \rfloor$. The *nearest integer* to x, i.e., the result of rounding x to the nearest integer is occasionally used in Number Theory, where it is denoted $\langle x \rangle$.

There is some ambiguity in ordinary rounding if one wants only to eliminate the last displayed digit when it is a 5. The choice is arbitrary and can vary from project to project. In my youth we were taught to round up, thus

$$\langle 3.5 \rangle = 4, \quad \langle 4.5 \rangle = 5;$$

in numerical analysis, however, one rounds up if the digit preceding the 5 is odd and down if it is even, thus

$$\langle 3.5 \rangle = 4, \quad \langle 4.5 \rangle = 4.$$

In involved computations this generally reduces the odds of errors accumulating and increases the odds of their partially cancelling out. The *TI-83 Plus* has a built-in rounding function that operates according to the first, simpler rule.

Another ambiguity occurs where negative numbers occur as arguments. For positive real numbers x, the greatest integer in x is calculated by simply deleting the digits to the right of the decimal point:

$$[3.5] = 3, \quad [4.167] = 4.$$

For negative numbers doing this would lose the order-theoretic property,

$$[x] \leq x,$$

i.e., defining $[-3.5] = -3$ would yield

$$-3.5 \leq -3, \text{ not } -3 \leq -3.5.$$

Order-theoretically we should have $[-3.5] = -4$, as $-4 \leq -3.5 < -3$. The function assigning -3 to -3.5 should also be useful and for this reason the calculator offers two implementations, int(and iPart(, respectively denoting the greatest integer (order-theoretic function) and the *integral part* (the truncating function) which agree on non-negative values of x, but do not agree in general for negative x.

The best way to sort things out is to graph the various functions, a task for which the calculator can be used, but not passively. The calculator plots graphs by plotting points and connecting the dots. When a function has a discontinuous jump the calculator will connect the last point plotted before the jump with the first point following it, thus resulting in a nearly vertical line segment as an element of the graph. One gets around this by pressing the MODE button, scrolling down to the line reading

Connected Dot,

and positioning the cursor over Dot before pressing the Enter button and then exiting to the home screen. If one now graphs $y = [x]$ by entering

Y$_1$=int(X)

in the equation editor, one will get something that looks like *Figure 2*, below, which will be hard to distinguish from the graph of

Y$_2$=−int(1−X).

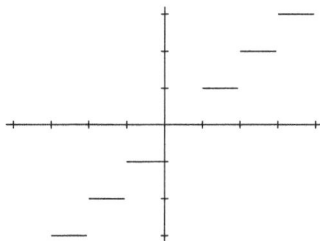

Fig. 2.

Looking very carefully at the screen when graphing the two functions separately, one will see that the left-most pixels of the horizontal pieces lie directly above or below the tick marks on the x-axis in the graph of Y$_1$, while it is the right-most pixels having this property in the graph of Y$_2$. The difference between the two also manifests itself when one returns to the Connected mode: The nearly vertical pieces of the graph of Y$_2$ are 1 pixel to the right of the vertical pieces in the graph of Y$_1$, thus yielding vertical lines of double thickness when graphing both functions.

The best graphical representation for such functions is done on paper, using filled circles for the endpoints on the graphs and empty circles for those not on the graphs, as in *Figure 3*, below.

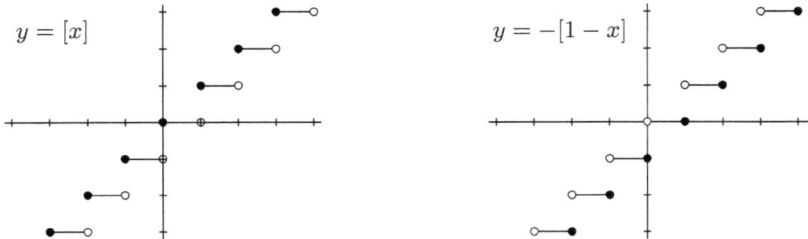

$y = [x]$

$y = -[1 - x]$

Fig. 3.

We can get something of the same effect on the *TI-83 Plus* by using the DRAW function to fill in those points on the graph at the points of discontinuity. To this end, I present the following program:

```
PROGRAM:GRAFINT
:FnOff
:ClrDraw
:Dot
:"int(X)"→Y₁
:ZDecimal
:For(I,−3,3)
:Circle(I,Y₁(I),.1)
:End
```

[I note that one can enter Dot either from the CATALOG or by pressing the MODE button, positioning the cursor above Dot, and then pressing the ENTER button as before. The same sort of thing can be done with ZDecimal. The Circle(command is accessed via the DRAW button or, again, via the CATALOG.]

1 Exercise. Graph the following functions
i. $y = -[-x] \ (= \lceil x \rceil)$
ii. $y = -[1 - x]$.
Do each of these by hand on paper and on the calculator via modifications of GRAFINT.

The *TI-83 Plus* has its own built-in rounding function round(, with the syntax

round(*value*[,#*decimals*]) .

It will take a number *value* as input and output the number rounded to the given number #*decimals* of decimals. The default, when #*decimals* is not specified, is 10. When #*decimals* is 0, the result is an integer:

```
round(3.2,0) yields 3
round(3.5,0) yields 4
round(−3.2,0) yields −3
round(−3.5,0) yields −4
etc.
```

Imitating GRAFINT, enter and run the following program

```
PROGRAM:GRAFRND
:FnOff
:ClrDraw
:Dot
:"round(X,0)"→Y₁
:ZDecimal
:For(I,−4,3)
:Circle(I+.5,Y₁(I+.5),.1)
:End
```

2 Exercise. Graph $y = [x + .5]$ and compare it to the result of running GRAFRND. Which of the two should you take to be $y = \langle x \rangle$?

Our major use of the greatest integer function has been to test if a number x is an integer:
$$x \text{ is an integer iff } x = [x].$$
Put differently, for positive integers m, n,

$$m \text{ divides } n \quad \text{iff} \quad n/m \text{ is an integer}$$
$$\text{iff} \quad n/m = [n/m]$$
$$\text{iff} \quad n = m[n/m].$$

Or, recalling the Division Algorithm, by which for positive integers m, n there are unique positive integers q, r satisfying

$$n = mq + r, \quad 0 \le r < m,$$

we have a means of finding the quotient q of n on dividing by m:

$$q = [n/m];$$

and the remainder r after such division:

$$r = n - m[n/m].$$

The *TI-83 Plus* only deals exactly with positive integers without too many digits. It will store a number up to 14 digits before rounding and displaying 10 digits. Thus the next remarks only hold on the calculator for such small numbers: If m is a power of 10, the digits of the quotient $[n/m]$ are the first few digits of n and the digits of the remainder $n - m[n/m]$ are the digits of the tail:

$$[123456789/10^4] = 12345$$
$$123456789 - 10^4 \cdot 12345 = 6789.$$

More generally, if we write n in base 10,

$$n = a_{k-1}a_{k-2}\ldots a_1 a_0, \quad 0 \le a_i \le 9,$$

i.e.,

$$n = a_{k-1}10^{k-1} + a_{k-2}10^{k-2} + \ldots + a_1 10 + a_0,$$

then

$$[n/10^i] = a_{k-1}a_{k-2}\ldots a_{i+1}a_i$$
$$[n/10^{i+1}] = a_{k-1}a_{k-2}\ldots a_{i+1}$$
$$10[n/10^{i+1}] = a_{k-1}a_{k-2}\ldots a_{i+1}0,$$

and

$$[n/10^i] - 10[n/10^{i+1}] = a_i.$$

That is, we can easily pick a specified digit from the decimal expansion of n using the greatest integer function. Given that this is trivial when we are given the number in its usual decimal representation, this may not be very impressive. But the trick works for any base. If b is fixed,

$$[n/b^i] - b[n/b^{i+1}] = c_i,$$

where the representation of n in base b is $c_{l-1}c_{l-2}\ldots c_1 c_0$.

If one adds to this the fact that the number of base b digits of a number n is $[\log_b n] + 1 = [(\log n)/(\log b)] + 1$, we can readily use lists to represent numbers in various bases and perform routine calculations with them.[1] A practical application of this is to overcome the limitations of integer arithmetic on the *TI-83 Plus*.[2] For example, von Neumann's middle-square algorithm for generating random numbers wouldn't begin to be useful unless the numbers being multiplied had a lot of digits—we saw how quickly randomness disappeared with 3-digit numbers. 10 digits make more sense, but we lose half of the middle digits we want if we multiply them directly. Representing a 10-digit number as a 2-digit number in base 100000, however, makes such multiplication easy.

Choose a 10-digit number at random (the first 10 numerical digits of the serial number of some nearby appliance, a sequence of 10 random digits generated by your calculator, etc.) and store it in the variable X:

    2585003983→X.

Turn it into a list of two 5-digit numbers:

    2→dim(L₁)
    int(X/100000)→L₁(1)
    X−100000L₁(1)→L₁(2).

---

[1] We already did the representation, but not the calculations, with the program DIGITS on page 356, back in Appendix A.

[2] The *TI-89 Titanium* treats integers and rational numbers exactly when the calculator is in EXACT mode, and integers are not limited to 14 digits. On the *TI-83 Plus*, if we want multiple precision, we have to program it ourselves.

Multiplication of X by itself, in base 100000, before performing the carries, looks like

	25850	03983
	25850	03983
	102960550	15864289
668222500	102960550	
668222500	205921100	15864289

In terms of lists, this multiplication is given by

$\{L_1(1)^2, 2*L_1(1)*L_1(2), L_1(2)^2\} \to L_2$.

One now performs the carrying operations:

$int(L_2(3)/100000) \to Y$
$L_2(3) - 100000Y \to L_2(3)$
$L_2(2) + Y \to L_2(2)$

$int(L_2(2)/100000) \to Y$
$L_2(2) - 100000Y \to L_2(2)$
$L_2(1) + Y \to L_2(1)$

$int(L_2(1)/100000) \to Y$
$L_2(1) - 100000Y \to L_2(1)$
$augment(\{Y\}, L_2) \to L_2$.

The calculator now displays the list,

$\{6682 \ 24559 \ 21258 \ 64289\}$,

representing the 20-digit number 06682245592125864289, the central 10 digits of which are given by

$seq(L_2(I), I, 2, 3) \to L_1$,

representing the number

$100000L_1(1) + L_1(2) \to X$,

or, eliminating the unnecessary explanatory step,

$100000L_2(2) + L_2(3) \to X$.

3 Project. Write a random number generation program MIDSQRNG for the calculator which, given a 10-digit number X, will square X, take the middle 10 digits of the result and store it in X. Make a list of 20 or 30 random 10-digit numbers by starting with some 10-digit number stored in X and iterating the application of MIDSQRNG.

4 Project. Another method of random number generation is the *shift-register generator*. Suppose we have a 10-digit number to start with, say 2455921258 as we ended up with above. Shift the digits 4 places to the right, with no wrapping: 0000245592. Add the two numbers with no carries in any column:

$$2455921258$$
$$0000245592$$
$$\overline{2455166740}$$

Now shift left 5 digits getting 6674000000, and add this, *sans* carries, to the previous sum:

$$2455166740$$
$$6674000000$$
$$\overline{8029166740}$$

The result, 8029166740, is the new 10-digit number. Write a program SHIFTRNG for the *TI-83 Plus* to automate the process and produce a list of 10-digit random numbers.

Δ, E, and Σ

Two complementary bits of notation are the symbols Δ and Σ. These are the Greek D for *Difference* and S for *Summation*, and one comes across them near the beginning of one's study of the more advanced parts of mathematics, certainly no later than in introductory Calculus. In the present book we were introduced to Δ on page 40 in connexion with tables; we encountered Σ in passing in the illustration on page 65, and began to use it in earnest on page 80 when it became no longer feasible always to write "$+\ldots+$" to represent sums.

The letter Δ, when prefixed to an independent variable, denotes an arbitrary increment, say h:

$$\Delta x = x + h - x = h.$$

When prefixed to a dependent variable $y = f(x)$, Δx denotes the induced increment:

$$\Delta y = f(x + \Delta x) - f(x).$$

The student's first introduction to the use of the notation could well be in an Analytic Geometry course, where the *difference quotient*,

$$\frac{\Delta y}{\Delta x} = \frac{f(x + \Delta x) - f(x)}{\Delta x},$$

is the slope of the line connecting the points $\langle x, f(x) \rangle$ and $\langle x + \Delta x, f(x + \Delta x) \rangle$. And, in the Calculus course, one learns that making the increment Δx get smaller and smaller, this ratio $\Delta y / \Delta x$ approaches the slope of the line tangent to the curve $y = f(x)$ at $\langle x, f(x) \rangle$.

The fashion has recently been to use the letter h in this context in place of Δx. Before the days of computerised typesetting, this would have been advantageous to whoever was typing the manuscript to be submitted. Today, with difference quotients introduced for practice free of any motivation in College Algebra texts, it has the pædagogical advantage of not introducing any notational complication where it is not yet needed.

We have already seen, on page 133 above, another use of the Δ-notation, namely, in tables. Tables of functions will list in adjacent rows or columns pairs of values of the independent (x_i) and dependent variables (y_i). Generally the successive x_i's are a fixed distance h apart,

$$\Delta x_i = x_{i+1} - x_i = h.$$

As a calculational aid in *interpolation*, the table might include an extra row or column labelled Δ of the differences $\Delta y_i = y_{i+1} - y_i$. When using a table one invariably finds that the argument x for which one wants a corresponding value $f(x)$ is not to be found in the table, but lies strictly between two entries x_i and x_{i+1}. One would estimate the value of $f(x)$ by assuming the changes in y-values and x-values proportional, i.e., one would solve the proportion

$$\frac{y - y_i}{x - x_i} = \frac{\Delta y_i}{\Delta x_i}.$$

Equivalently, one would assume $\langle x, f(x) \rangle$ to lie on the straight line connecting $\langle x_i, f(x_i) \rangle$ to $\langle x_{i+1}, f(x_{i+1}) \rangle$:

$$y = f(x_i) + \frac{\Delta y_i}{\Delta x_i}(x - x_i).$$

In some tables yet another row or column labelled Δ^2 would list the *second differences*,

$$\Delta^2 y_i = \Delta y_{i+1} - \Delta y_i.$$

Taken with the first differences, the second differences can be used to find a good quadratic curve connecting the points $\langle x_i, f(x_i) \rangle$ and $\langle x_{i+1}, f(x_{i+1}) \rangle$, allowing greater accuracy in interpolating a value for $f(x)$ for x in the interval $[x_i, x_{i+1}]$. $\Delta^3, \Delta^4, \ldots$ allow even further refinement.

Tables, for centuries one of the most important aids to calculation around, may strike one today as obsolete. This or that table may indeed be obsolete, replaced by a button on one's calculator, but there are always newer tables. Indeed, many of the functions on one's calculator are computed via small tables of numbers allowing for rapid computation of the said functions via simple operations. While not the same in practice, it was this principle that led Charles Babbage (1791 - 1871) to conceive the modern computer: Any continuous function on a finite interval can be approximated as closely as desired by a polynomial and, given the value and differences of a polynomial at the initial point, its value so many units (i.e., multiples of Δx) away can be calculated using only additions—and additions, unlike multiplications, are easily performed mechanically.

To illustrate the abstract principle behind, if not the mechanical workings of, Babbage's *difference engine*, as he called his initially envisioned device, let us begin with the polynomial $p(x) = x^5 - 2x^3 - 3x$ and find the successive differences at $x = 0$, using $\Delta x = 1$. It happens that the n-th difference of an n-th degree polynomial is constant, all further differences being 0. Thus we need to find $p(0), \Delta p(0), \Delta^2 p(0), \Delta^3 p(0), \Delta^4 p(0)$ and $\Delta^5 p(0)$. For this we need 6 successive values of p, starting with $p(0)$:

$Y_1 = X^5 - 2X^3 - 3X$
$\text{seq}(Y_1(X),X,0,5) \to L_1$.

$L_1(1)$ is $Y_1(0)$, i.e., $p(0)$, and is the first of the 6 values sought. Let us create a list to store these in:

$6 \to \dim(\llcorner\text{DIFFS})$
$L_1(1) \to \llcorner\text{DIFFS}(1)$.

Now the list of differences of elements of L_1 can all be calculated in a single stroke. The 7th item in the OPS menu accessed via the LIST button is ΔList(, which performs this calculation:

$\Delta\text{List}(L_1) \to L_1$.

L_1 is now the list $\{\Delta p(0), \Delta p(1), \ldots, \Delta p(4)\}$ and is shorter by one element than L_1. It's first element is the second one needed in $\llcorner\text{DIFFS}$:

$L_1(1) \to \llcorner\text{DIFFS}(2)$.

We can now repeat the step:

$\Delta\text{List}(L_1) \to L_1$
$L_1(1) \to \llcorner\text{DIFFS}(3)$
$\Delta\text{List}(L_1) \to L_1$
$L_1(1) \to \llcorner\text{DIFFS}(4)$
$\Delta\text{List}(L_1) \to L_1$
$L_1(1) \to \llcorner\text{DIFFS}(5)$
$\Delta\text{List}(L_1) \to L_1$
$L_1(1) \to \llcorner\text{DIFFS}(6)$.

This is a bit repetitive. One can render the process less painful by doing the first differencing operation and subsequent storing instruction on a single line,

$\Delta\text{List}(L_1) \to L_1 : L_1(1) \to \llcorner\text{DIFFS}(2)$,

then hitting ENTRY (i.e., 2nd ENTER) and editing the argument of $\llcorner\text{DIFFS}$ before pressing ENTER.

Another alternative, of course, is to write a program to perform the iteration using a For(loop.

Once one has the list $\llcorner\text{DIFFS}$, if we assume, as is the case with our polynomial, that the last differences are constant, we can generate a table of values, differences, second differences, etc., using only additions. The table will be given in the form of a matrix:

$$[A] = \begin{bmatrix} p(0) & p(1) & p(2) & \ldots & p(n) \\ \Delta p(0) & \Delta p(1) & \Delta p(2) & \ldots & \Delta p(n) \\ \Delta^2 p(0) & \Delta^2 p(1) & \Delta^2 p(2) & \ldots & \Delta^2 p(n) \\ \vdots & & & \\ \Delta^5 p(0) & \Delta^5 p(1) & \Delta^5 p(2) & \ldots & \Delta^5 p(n) \end{bmatrix}.$$

Note that the dimension of the matrix is $6 \times (n+1)$. The construction of [A] is simple, but tiresome unless we automate the entire process. Our program will assume ∟DIFFS given and some value of n stored in the variable N.

```
PROGRAM:BABBAGE
:dim(∟DIFFS)→D
:{D,N+1}→dim([A])
:For(I,1,D)
:∟DIFFS(I)→[A](I,1)
:End .
```

The second line creates the matrix and the For(loop fills in the first column. Because the last differences are supposed to be constant, we can quickly fill in the bottom row as well:

```
:For(J,2,N+1)
:[A](D,1)→[A](D,J)
:End .
```

Now we work our way up filling each row from left to right: From

$$\Delta^{i+1} p(j) = \Delta^{i}(j+1) - \Delta^{i} p(j),$$

we have

$$\Delta^{i} p(j+1) = \Delta^{i+1} p(j) + \Delta^{i} p(j).$$

Now, for $i+1, j+1$ stored in I,J, respectively, [A](I,J) is supposed to be $\Delta^{i} p(j)$. This leads to

```
:For(I,D−1,1,−1)
:For(J,2,21)
:[A](I,J−1)+[A](I+1,J−1)→[A](I,J)
:End
:End .
```

Except for the clean-up commands deleting the no longer needed variables D, N, I, J, and ∟DIFFS, one is finished.

5 Exercise. i. Store 20 in N and run the program. For any integer $j \in \{0, 1, \ldots, 20\}$, store j in J and calculate $Y_1(J)$ and compare the result to [A](1,J). ii. Transpose [A] using the transpose function in the MATH submenu accessed via the MATRX button:

$$[A]^{T} → [B] .$$

This turns the rows of [A] into columns of [B]. The first column of [B] now lists the values $p(0), p(1), \ldots, p(20)$. Convert these columns into lists using the command

Matr▶list([B],L$_1$,L$_2$,L$_3$,L$_4$,L$_5$,L$_6$)

and compare ΔList(L$_i$) with L$_{i+1}$ for several values of $i < 6$.

Babbage's difference engine was designed to generate and print tables automatically following such a procedure. He never completed his machine, having around 1830 conceived the grander design of a general purpose computer. He did, however, return to the difference engine, producing a new design in 1851. Almost a century and a half later, using these plans and technology as close to that of Babbage's day as possible, the London Science Museum produced a working model of the machine, putting it on exhibit in 1991 in celebration of Babbage's 200th birthday.

The existence of the function ΔList(on the calculator points to a once abstract and now quite concrete aspect of Δ: it is not just a notation, handy for abbreviation, but a function converting one list to another on the calculator and one numerical function to another in theory. Given a fixed difference for the independent variable, say $\Delta x = h$, one defines $\Delta_h(f)$, usually written Δf, to be the function $g(x) = f(x + h) - f(x)$, i.e.,

$$\Delta_h f(x) = f(x + h) - f(x).$$

For convenience, one usually takes h to be 1. In theory, this is no restriction, for one can take x to be ht for a new independent variable t and apply one's theoretical results to

$$f^*(t) = f(ht).$$

With $h = 1$, one writes Δf instead of $\Delta_1 f$.

Functions of functions are often called *functionals*, especially if they map functions to numbers, or *operators*, if they map functions to functions. Thus we would say that Δ is an operator.

Now, operators can be manipulated algebraically. The justification for such manipulation is not difficult, but is relatively modern and is given in a course called Modern Algebra, or Abstract Algebra. Justification generally comes after practice, and de Moivre, Lagrange, Laplace made some clever uses of formal manipulation of operators not backed up by rigorous justification. We shall illustrate this with a simple example.[3]

First, note that we can add operators just as we add functions:

$$(F + G)(f) = F(f) + G(f),$$

where

$$(f + g)(x) = f(x) + g(x).$$

We can similarly multiply operators, though it is more common to treat composition as multiplication:

$$(FG)(f) = F(G(f)).$$

Thus, for example,

$$\Delta^2 = \Delta\Delta, \quad \Delta^3 = \Delta\Delta^2, \quad \ldots$$

---

[3] Another simple example using the operator E is given on page 102 in Chapter 3. And in section 5, below, we will give a truly impressive example.

Great Britain's philatelic tributes to Charles Babbage. The stamp on the left, which was issued in 1991 in a set of 4 stamps honouring scientific achievements, represents everything wrong with the British Post's silly practice of honouring its scientists with meaningless abstractions. What about Babbage or his work is represented here? Does an empty head filled with numbers tell us he was a mathematician or an accountant? Do the numbers tell us he was involved in computation? As the founder of the computer, or as one of those lightning calculators who can compute the cube root of a multidigit number in his head in seconds? The design may be of mild interest to any art majors one shows it to, but if offers little for the scientist or mathematics student. I will grant, however, that the profile does bear a resemblance to a well-known portrait of the 36 or 40 year old Babbage (Babbage's age at the time of the portrait being variously reported).

The stamp on the right was issued in 2010 in celebration of the 350th anniversary of the founding of the Royal Society. This extremely attractive set of 10 stamps does its honoree's right. The image on each stamp features in split screen both a recognisable portrait of the scientist so honoured and the work for which he or she is known. Admittedly, the latter representation is not always perfect, but it is never far off the mark. In Babbage's case, the depicted gears may not immediately bring computers to mind, but the picture serves to remind us that i. from Pascal's adding machine well into the 20th century, additions in mechanical computing were performed by gears, not electrical circuitry, and ii. Babbage's machines were designed as massive and complex arrays of gears. More to the point, the gears depicted are from the addition carriage of Babbage's Difference Engine No. 2, the one that resides in the London Science Museum. The portrait on the stamp, incidentally, is from the obituary of Babbage published in the *Illustrated London News* on 4 November 1871 and is based on a photograph taken at the Fourth International Statistical Congress held in London in July 1860. Babbage had more than a passing interest in statistics, having organised the Statistical Section of the British Association for the Advancement of Science in 1833, founded the Statistical Society of London in 1834, and participated in Quetelet's organisation of the First International Statistical Congress in Brussels in 1853. Additionally, he authored several publications of statistical concern.

Each real number r induces a multiplicative operator we also denote by r:

$$(rf)(x) = r \cdot f(x).$$

In particular, $1f = f$.

We are now in position to perform some mathematical sleight of hand. Define the operator E by

$$E = 1 + \Delta,$$

i.e.,

$$E(f) = (1 + \Delta)f,$$

i.e.,

$$(Ef)(x) = ((1 + \Delta)f)(x) = (1f)(x) + (\Delta f)(x)$$
$$= f(x) + f(x+1) - f(x) = f(x+1).$$

Similarly,

$$E^2 f(x) = f(x+2), \quad E^3 f(x) = f(x+3), \quad \cdots$$

and generally

$$E^n f(x) = f(x+n).$$

But $E = 1 + \Delta$, whence

$$E^n = (1 + \Delta)^n = 1 + \frac{n}{1}\Delta + \frac{n(n-1)}{2 \cdot 1}\Delta^2 + \ldots + \frac{n(n-1) \cdots 1}{n!}\Delta^n,$$

i.e.,

$$f(n) = E^n f(0) = f(0) + \frac{n}{1}\Delta f(0) + \frac{n(n-1)}{2 \cdot 1}\Delta^2 f(0) + \ldots + \frac{n(n-1) \cdots 1}{n!}\Delta^n f(0).$$

Thus, for positive integral x

$$f(x) = f(0) + \frac{x}{1}\Delta f(0) + \frac{x(x-1)}{2 \cdot 1}\Delta^2 f(0) + \ldots$$
$$+ \frac{x(x-1) \cdots (x-n+1)}{n!}\Delta^n f(0) + \ldots,$$

the infinite series having 0's for all terms with $n > x$. If we stop the series at the n-th term, we have an n-th degree polynomial

$$p(x) = f(0) + \frac{x}{1}\Delta f(0) + \frac{x(x-1)}{2 \cdot 1}\Delta^2 f(0) + \ldots$$
$$+ \frac{x(x-1) \cdots (x-n+1)}{n!}\Delta^n f(0) \quad (1)$$

which agrees with f for $x = 0, 1, \ldots, n$ and which can be used to interpolate f generally quite well in the interval $[0, 1]$. (1) is called the *Newton Forward Difference Formula* after Isaac Newton, who made use of it alongside his *Backward Difference Formula* and *Central Difference Formula*, which place the base (here: 0) at the right end and center of the points at which $f(x)$ is sampled to produce the differences.

Formula (1) had to be split and spread over two lines. The right-hand side of the equation is the sum of several similar looking items of the form

$$\frac{x(x-1)\cdots(x-k+1)}{k!}\Delta^k f(0), \tag{2}$$

and the whole of (1) can be written more compactly if one introduces a short-hand notation for taking a sum. Before the 16th century's introduction of much of our familiar algebraic symbolism, the first letter or letters of the appropriate Latin word were commonly used as such symbols: r for *radix* to denote a root, *æq* for *æqualitas* to denote equality, and some shortened form of *summa* to denote a sum. Leibniz, for example, used what is often described as an elongated "S" and which was, in fact, the italic form of one of the two typographical representations of the letter "S" then in use. Thus he would have abbreviated the sum of terms (2) as

$$\int \frac{x(x-1)\cdots(x-k+1)}{k!}\Delta^k f(0).$$

However, Leibniz also used $\int$ for integration, which in those days was considered a sum of infinitely many infinitesimal elements.[4] Later Leonhard Euler introduced the Greek Σ for finite sums and infinite series, distinguishing such from integrals. According to Florian Cajori[5] Euler's Σ did not immediately catch on. Cajori cites the use of S and $\bar{a}$ for the sum of an infinite series $a_0 + a_1 + a_2 + \ldots$ And, closer to home, one finds $[\alpha, \alpha]$ for the sum $a^2 + (a')^2 + (a'')^2 + \ldots$ and $[\alpha, \beta]$ for $ab + a'b' + a''b'' + \ldots$ in Gauss's *Theoria combinationis*. For the most part, however, Euler's Σ eventually caught on and the practice of listing the range of indices with Σ developed.

Today we write

$$\sum_{i=a}^{b} c_i \quad \text{or} \quad \Sigma_{i=a}^{b}\, c_i$$

for $c_a + c_{a+1} + \ldots + c_b$, where $a \le b$ are integers. The index can be varied:

$$\sum_{i=a}^{b} c_i = \sum_{j=a}^{b} c_j = \sum_{k=a}^{b} c_k.$$

And there are variants:

$$\sum_{a \le i \le b} c_i, \quad \sum_{i \in I} c_i, \quad \sum_i c_i, \quad \sum c_i,$$

---

[4] We discuss integration as a limit of sums in section 2, below.

[5] Florian Cajori, *A History of Mathematical Notations*, two volumes, Open Court Publishing Company, La Salle (Illinois), 1928 and 1929; here: volume 2, article 438, p. 61. A single volume reprint by Dover Publications, Mineola (New York), was published in 1993.

the latter two presupposing a given range, e.g., $I = \{a, a+1, \ldots, b\}$, for the indices.

One tends to first encounter the notation in one's mathematical training—certainly no later than in the Calculus course—when one starts learning formulæ for special sums:

$$\sum_{i=1}^{n} i = \frac{n(n+1)}{2}$$

$$\sum_{j=0}^{n} r^j = \frac{r^{n+1} - 1}{r - 1}$$

$$\sum_{k=1}^{n} k^2 = \frac{(n+1)n(2n+1)}{6},$$

etc. Thus I do not imagine I am introducing anything new here. We discuss Σ here because of its special relation to Δ.

The various Newtonian Difference Formulæ are useful in interpolation and perform one of the basic tasks of the Calculus of Finite Differences. Another is the solution of *difference equations*, equations involving Δ or E as on page 102, above. Another basic task of this calculus is in finding closed expressions for simple sums. For those who know the Calculus, I am referring to a discrete analogue of integration. And, just as in the Calculus, where integration is an operation inverse to differentiation, summation is accomplished by an operation inverse to taking differences. In the Calculus, this is a deep result and a proof is generally absent from the beginning course. In the Calculus of Finite Differences, however, it is a fairly simple matter. One considers a *telescoping sum*:

$$(a_1 - a_0) + (a_2 - a_1) + \ldots + (a_{n+1} - a_n) =$$
$$- a_0 + (a_1 - a_1) + (a_2 - a_2) + \ldots + (a_n - a_n) + a_{n+1} = a_{n+1} - a_0,$$

i.e.,

$$\Delta a_0 + \Delta a_1 + \ldots + \Delta a_n = a_{n+1} - a_0.$$

Thus, if $f(x) = \Delta F(x)$, one has

$$f(0) + f(1) + \ldots + f(n) = F(n+1) - F(0).$$

That is, one can find a sum,

$$\sum_{k=a}^{b} f(x),$$

by finding a function F which differences to f and observing

$$\sum_{k=a}^{b} f(x) = F(b+1) - F(a).$$

It is important to note the offset: one evaluates F at $b+1$, not at b.

6 Lemma. *Let f be a function defined on all integers. There is a function F, unique up to an additive constant, such that* $\Delta F = f$.

Proof. If f were only defined on non-negative integers, we could prove existence by setting

$$F(x) = \sum_{k=0}^{x-1} f(k).$$

For f defined on all integers, the question of where to set the lower limit of the summation arises. Formally, we could try

$$F(x) = \sum_{k=-\infty}^{x-1} f(k).$$

However, this is an infinite series and might not converge. With a little fiddling one can show that, for any integer a, one can choose

$$F(x) = \begin{cases} \displaystyle\sum_{k=a}^{x-1} f(k), & x > a \\ 0, & x = a \\ \displaystyle-\sum_{k=x}^{a-1} f(k), & x < a. \end{cases} \tag{3}$$

To establish uniqueness up to a constant, let $\Delta F = \Delta G = f$. Then $0 = \Delta F - \Delta G = \Delta(F - G)$, i.e.,

$$(F - G)(x + 1) - (F - G)(x) = 0,$$

whence

$$(F - G)(x + 1) = (F - G)(x),$$

and a simple induction shows $F - G$ to be constant: $F(x) = G(x) + C$ for some constant C. $\qquad\square$

We let $\Delta^{-1} f$ ambiguously denote any function F, like (3), for which $\Delta F = f$.

7 Remark. The *TI-89 Titanium* has a built-in version of Δ^{-1}, namely the function $\Sigma(\,)$ in its Calculus menu accessed via the MATH button. Its syntax is

$\Sigma(\text{expr},\text{var},\text{low},\text{high})$,

where expr is an expression defining a function of an independent variable var and low, high are the limits of summation. This function is quite handy. If, for example, one enters

$\Sigma(\text{k}/2\text{\textasciicircum k},\text{k},1,\text{n})$, then, provided no value has been stored in n, the answer portion of the screen will exhibit

$$\sum_{k=1}^{n} \left(\frac{k}{2^k} \right)$$

on the left and on the right one line below will give the answer,

$$2-(n+2)2^{-n} \text{ or } 2.-(n+2.)(.5)^n,$$

according to whether the calculator is in EXACT or APPROXIMATE mode, which result agrees with that of Lemma 1 of Chapter 2. The *TI-83 Plus* will not do this for us, but it can find the sums for any chosen value of n greater than the lower limit of summation. Entering

sum(seq(X/2^X,X,1,N))

will yield the sum

$$\sum_{k=1}^{n} \frac{k}{2^k}$$

for the given value of $n \geq 1$; and

cumSum(seq(X/2^X,X,1,N))

will yield the list

$$\left\{ \sum_{k=1}^{1} \frac{k}{2^k} \quad \sum_{k=1}^{2} \frac{k}{2^k} \quad \cdots \quad \sum_{k=1}^{n} \frac{k}{2^k} \right\}$$

i.e.,

$$\{.5 \ 1 \ 1.375 \ \ldots \}.$$

(*Exercise.* Try this for $n = 25$.)

The sums of finite arithmetic and geometric progressions are arrived at via simple *ad hoc* tricks and were known in ancient times. For an arithmetic progression $a, a + d, a + 2d, \ldots, a + nd$, one places a copy of the progression in reverse order immediately below the original and adds the matching terms:

a	$a + d$	$a + 2d$	$\ldots$	$a + nd$
$a + nd$	$a + (n-1)d$	$a + (n-2)d$	$\ldots$	a
$2a + nd$	$2a + nd$	$2a + nd$	$\ldots$	$2a + nd$

Then one sums the bottom row:

$$(2a + nd) + (2a + nd) + \ldots + (2a + nd) = (n+1)(2a + nd),$$

and, this being the sum of two copies of the arithmetic progression, one simply divides by 2:

$$\sum_{k=0}^{n} (a + kd) = \frac{(n+1)(2a + nd)}{2}.$$

We have already discussed the trick for summing the geometric progression (page 104, above) and I need not repeat it here.

To find the sums of more complicated functions f, like $f(x) = x/2^x$, requires a calculus of its own if one doesn't have the *TI-89 Titanium* to perform the task for one. To this end, one makes a table of differences of familiar functions and observes how differencing interacts with various methods of combining functions. A simple, but useful, table would be given by

$f(x)$	C	x	$x^{(n)}$	$1/x^{|n|}$	r^x
$\Delta f(x)$	0	1	$nx^{(n-1)}$	$-n/x^{|n+1|}$	$(r-1)r^x$,

where C is any constant, $n \neq 0, r \neq 1$, and we define

$$x^{(n)} = x(x-1)\cdots(x-n+1), \quad x^{|n|} = x(x+1)\cdots(x+n-1).$$

The first rule for differencing combined functions is the linearity of Δ:

8 Lemma. *Let f, g be functions and a, b constants.*

$$\Delta(af + bg) = a\Delta f + b\Delta g.$$

9 Corollary. *Let f, g be functions and a, b constants.*

$$\Delta^{-1}(af + bg) = a\Delta^{-1}f + b\Delta^{-1}g. \tag{4}$$

Proof. Observe

$$\Delta\left(a\Delta^{-1}f + b\Delta^{-1}g\right) = a\Delta\Delta^{-1}f + b\Delta\Delta^{-1}g = af + bg. \qquad \square$$

[Equation (4), indeed any equation $\Delta^{-1}f = G$ is to be interpreted as saying that $\Delta G = f$ and any F satisfying $\Delta F = f$ is of the form $G + C$ for some constant C.]

Our earlier table and the Corollary yield another simple table:

$f(x)$	C	$x^{(n)}$	$1/x^{|n|}$	r^x
$\Delta^{-1}f(x)$	Cx	$x^{(n+1)}/(n+1)$	$-1/(n+1) \cdot 1/x^{|n+1|}$	$r^x/(r-1)$.

To see the last one for example, observe

$$\Delta^{-1}r^x = \Delta^{-1}\frac{r-1}{r-1}r^x = \frac{1}{r-1}\Delta^{-1}(r-1)r^x = \frac{1}{r-1}r^x.$$

Using this, we can sum a geometric progression:

$$\sum_{k=0}^{n} ar^k = \Delta^{-1}ar^k\Big|_0^{n+1} = a\left(\Delta^{-1}r^k\Big|_0^{n+1}\right)$$

$$= a\left(\frac{1}{r-1}r^k\Big|_0^{n+1}\right) = \frac{a}{r-1}\left(r^{n+1} - r^0\right)$$

$$= a\frac{r^{n+1} - 1}{r-1}.$$

We can sum the arithmetic progression even more simply:

$$\sum_{0}^{n}(a+kd) = \Delta^{-1}(a+kd)\Big|_{0}^{n+1} = \left(\Delta^{-1}a + d\Delta^{-1}k\right)\Big|_{0}^{n+1}$$

$$= \left(ak + d\frac{k^{(2)}}{2}\right)\Big|_{0}^{n+1}$$

$$= a(n+1) + d\frac{(n+1)n}{2} - \left(a\cdot 0 + d\frac{0\cdot(-1)}{2}\right)$$

$$= (n+1)\left(a + \frac{dn}{2}\right) = \frac{(n+1)(2a+nd)}{2},$$

as before.

Other familiar sums are easily obtained:

$$\sum_{k=0}^{n}x^2 = \sum_{k=0}^{n}\left(x(x-1)+x\right) = \sum_{k=0}^{n}\left(x^{(2)}+x\right)$$

$$= \Delta^{-1}\left(x^{(2)}+x^{(1)}\right)\Big|_{0}^{n+1} = \left(\frac{x^{(3)}}{3}+\frac{x^{(2)}}{2}\right)\Big|_{0}^{n+1}$$

$$= \frac{(n+1)(n)(n-1)}{3} + \frac{(n+1)n}{2} - 0 = (n+1)n\left(\frac{n-1}{3}+\frac{1}{2}\right)$$

$$= (n+1)n\frac{2n-2+3}{6} = \frac{(n+1)n(2n+1)}{6},$$

which the reader may recognise.

Another combination is multiplication, which is messier:

$$\Delta fg(n) = f(n+1)g(n+1) - f(n)g(n)$$
$$= f(n+1)g(n+1) - f(n)g(n+1) + f(n)g(n+1) - f(n)g(n)$$
$$= \Delta f(n)\cdot g(n+1) + f(n)\Delta g(n).$$

From this one gets

$$f(n)\Delta g(n) = \Delta fg(n) - \Delta f(n)\cdot g(n+1).$$

Thus

$$\Delta^{-1}\left(f(n)\Delta g(n)\right) = \Delta^{-1}\Delta fg(n) - \Delta^{-1}\left(\Delta f(n)\cdot g(n+1)\right)$$
$$= f(n)g(n) - \Delta^{-1}\left(\Delta f(n)\cdot g(n+1)\right),$$

and we have the *summation by parts* formula, also called *Abel's Summation Formula* after Niels Henrik Abel (1802 - 1829) who published it in 1826. Let us apply it to the product

$$h(x) = x\cdot\frac{1}{2^x}.$$

Here we take $f(x) = x, \Delta g(x) = 1/2^x$. Then $\Delta f(x) = 1$ and

$$g(x) = \Delta^{-1}\left(\frac{1}{2^x}\right) = \Delta^{-1}\left(\frac{1}{2}\right)^x = \left(\frac{1}{2}\right)^x \Bigg/ \left(\frac{1}{2} - 1\right) = -2\left(\frac{1}{2}\right)^x.$$

Thus

$$\Delta^{-1}h(x) = x\left(-2\left(\frac{1}{2}\right)^x\right) - \Delta^{-1}\left(1\cdot(-2)\left(\frac{1}{2}\right)^{x+1}\right)$$

$$= \frac{-2x}{2^x} + \Delta^{-1}\left(\frac{1}{2}\right)^x = \frac{-2x}{2^x} + \frac{-2}{2^x}$$

$$= \frac{-2x - 2}{2^x}.$$

Hence

$$\sum_{k=1}^{n}\frac{k}{2^k} = \frac{-2k-2}{2^k}\bigg|_{1}^{n+1} = \frac{-2(n+1)-2}{2^{n+1}} - \frac{-2-2}{2}$$

$$= \frac{-2n-4}{2^{n+1}} + 2$$

$$= 2 - \frac{n+2}{2^n}, \tag{5}$$

as before.

10 Exercise. Use Abel's formula to show:

i. $\displaystyle\sum_{k=0}^{n}k^2 = \frac{(n+1)n(2n+1)}{6}.$

ii. $\displaystyle\sum_{k=0}^{n}k^3 = \left(\frac{n(n+1)}{2}\right)^2.$

There is much more fun to be had along these lines, but we have derived all the sums used in this book, which already is a task not necessary for our present purposes. We could, as in Lemma 1, Chapter 2, simply have stated the formulæ as needed and used induction to verify them. If there is any real justification in my having gone as far into these matters as I have, it is the historically close relation between the development of Probability theory and that of the Calculus of Finite Differences. For, as with our discussion of Gambler's Ruin, methods of this calculus were called upon to calculate probabilities. For now, however, it is time to move on.

Approximations and Limits

It often happens that one wants to use a number that cannot be pinned down exactly. In such cases one resorts to an approximation and writes $a \approx b$ to indicate that the approximating value b is close enough for the purpose at hand. Thus, when I was a student, we wrote

$$\pi \approx \frac{22}{7} \quad \text{or} \quad \pi \approx 3.14$$

and

$$\sqrt{2} \approx 1.414.$$

With the advent of calculators one supplied many more digits in decimal approximations, currently

$$\pi \approx 3.141592654 \quad \text{and} \quad \sqrt{2} \approx 1.414213562.$$

The extra precision is generally uncalled for in practical applications and working by hand one would not use it. Deciding how much accuracy is needed requires serious thought, however, and it is simpler to use all the digits the calculator supplies.

Mathematically, $\approx$ is not a well-behaved relation.

11 Definition. *A* binary relation $\equiv$ on a set X *is an* equivalence relation *if it satisfies the following properties for all* $x, y, z \in X$
i. $x \equiv x$ *(reflexivity)*
ii. $x \equiv y \Rightarrow y \equiv x$ *(symmetry)*
iii. $x \equiv y \ \& \ y \equiv z \Rightarrow x \equiv z$ *(transitivity).*
An equivalence relation $\equiv$ *is a* congruence *with respect to a family* $\mathcal{F}$ *of functions on* X *if for each function* $f \in \mathcal{F}$ *and all* $x_0, \ldots, x_{n-1} \in X, y_0, \ldots, y_{n-1} \in X$ *(assuming* f *is n-ary),*

$$x_0 \equiv y_0 \ \& \ x_1 \equiv y_1 \ \& \ \ldots \ \& \ x_{n-1} \equiv y_{n-1} \Rightarrow$$
$$f(x_0, x_1, \ldots, x_{n-1}) = f(y_0, y_1, \ldots, y_{n-1}).$$

We cannot really begin to discuss whether or not $\approx$ is an equivalence relation or a congruence because it is too vaguely defined. The truth or falsity of a given assertion $a \approx b$ depends on the purpose at hand and the level of precision required. For example, for the purposes of measuring wealth, a single dollar does not constitute a significant difference for the middle or upper classes:

$$500000 \approx 500001 \approx 500002 \approx \ldots \approx 999999 \approx 1000000.$$

If transitivity held, we could conclude $500000 \approx 1000000$ and, indeed, to a pauper to whom \$1 could mean the difference between eating and not eating, \$500000 and \$1000000 may look the same, both being approximately $+\infty$. To a man possessing \$500000, the sums assuredly are not the same.[6]

Similarly, $\approx$ cannot be said to be a congruence with respect to addition. In dollars, 500000 and 500001 are barely discernible, thus $500000 \approx 500001$. Yet

---

[6] Recall the Weber-Fechner Law mentioned in Chapter 2, section 2 (page 40, above), in connexion with the Petersburg Problem.

adding -500000 yields 0 and 1, respectively. We would not say, e.g., for the purposes of buying a snack, that \$0 and \$1 are the same. Thus

$$a \approx b \nRightarrow a + -500000 \approx b + -500000.$$

If you insist on a positive summand, note that there is not much difference between $-40°C$ and $-42°C$ with respect to the properties of water, $-40 \approx -42$ to the Antarctic scientist setting the thermostat of the refrigeration unit in which he preserves his ice cores. But if we add $41°$, the temperatures become $1°C$ and $-1°C$, the difference between liquid and solid.

There really isn't much we can do to sharpen the notion of approximate equality of two numbers. It is vague, clear only in some individual cases where we have a purpose in mind allowing us to say more definitively, "This particular a is sufficiently close to b for the purpose at hand".

When faced with a number that is hard to pin down, however, we can always find a better approximation and a better, and a better, ... That is, we can use a sequence of approximations, and when we do this, the notion $\approx$ of approximate equality can be sharpened—in several different ways.

The basic notion of approximation of a number by a sequence is that of the number being the limit of the sequence. The definition is logically complex.

12 Definition. *Let $a : a_0, a_1, a_2, \ldots$ be a sequence of real numbers. A number L is said to be the* limit *of the sequence, written*

$$\lim_{n \to \infty} a_n = L,$$

if, for every $\epsilon > 0$, no matter how small, we can find a number n_0 sufficiently large so that for all $n > n_0$, $|a_n - L| < \epsilon$.

Basically, the definition says that L is the limit of the sequence if the elements of the sequence eventually get as close to L as we want—and it stays close.

Consider, for example, the sequence $a_n = 1 - 1/2^n$. Its elements are

$$0, \frac{1}{2}, \frac{3}{4}, \frac{7}{8}, \frac{15}{16}, \ldots,$$

all approaching 1. They also approach 2, but do not get arbitrarily close to 2, staying at least one unit away. For 1, however, the sequence gets as close as we want to the number. The sequence

$$b : 0, 1, \frac{1}{2}, \frac{3}{2}, \frac{3}{4}, \frac{7}{4}, \frac{7}{8}, \frac{15}{8}, \ldots$$

also gets as close as we want to 1, the sequence $b_0, b_2, b_4, \ldots$ getting closer at each step. However, the numbers $b_1, b_3, b_5, \ldots$ keep moving away from 1, approaching ever nearer to 2. b would not be said to have 1 as a limit, and this is reflected in the definition by the requirement that $|a_n - L| < \epsilon$ for *all* large n.

In the early days of the Calculus, one did not use the formal definition of limit given above. It came out only in the second decade of the 19th century. Instead, mathematicians reasoned about the value of a_N for an *infinite* integer N. L was the limit of the sequence a just in case a_N was infinitesimally close to L for infinite N. The modern practice is to use $\approx$ in this case to mean infinitesimal closeness: $a_N \approx L$ means $L - a_n$ is infinitesimal or 0. Accepting that the sum and product of infinitesimals are infinitesimal and the product of an infinitesimal with an ordinary real number is infinitesimal, it is easy to use the alternate definition to prove basic properties of limits such as:

1. limits are unique: $\lim a_n = L_1$ & $\lim a_n = L_2 \Rightarrow L_1 = L_2$
2. $\lim(a_n + b_n) = \lim a_n + \lim b_n$
3. $\lim(a_n \cdot b_n) = \lim a_n \cdot \lim b_n$
4. $\lim(ca_n) = c \cdot \lim a_n$,

etc., provided the limits on the right exist.

Let us prove the second of these assertions as an example. Assume $\lim a_n = L_1, \lim b_n = L_2$ and let N be an infinite integer. Thus

$$a_N \approx L_1, \quad b_N \approx L_2.$$

Let $\eta_1 = L_1 - a_N, \eta_2 = L_2 - b_2$ be infinitesimal or 0 and note that

$$a_N + b_N = (L_1 - \eta_1) + (L_2 - \eta_2) = (L_1 + L_2) - (\eta_1 + \eta_2)$$

and $\eta_1 + \eta_2$ is infinitesimal or 0, whence $a_N + b_N \approx L_1 + L_2$, i.e.,

$$\lim_{n \to \infty} (a_n + b_n) = L_1 + L_2 = \lim_{n \to \infty} a_n + \lim_{n \to \infty} b_n.$$

From some time in the 19th century until the latter half of the 20th century, the use of infinitesimals was increasingly regarded as merely heuristic. With the development of nonstandard analysis under Curt Schmieden, Detlef Laugwitz, and Abraham Robinson, such arguments are now accepted as rigorous.

The relation $\approx$ extends nicely to the comparison of two sequences:

$$a \approx b \text{ iff } \lim_{n \to \infty} a_n = \lim_{n \to \infty} b_n$$

iff for some real number L and any infinite integer N,

$$a_N \approx L \text{ and } b_N \approx L,$$

or even

$$a \approx b \text{ iff for all infinite integers } N, a_N \approx b_N$$

$$\text{iff for all infinite integers } N, a_N - b_N \approx 0$$

$$\text{iff } \lim_{n \to \infty} a_n - b_n = 0.$$

These two notions do not agree, for it is possible for two sequences to approach each other arbitrarily closely and yet not share a common limit because the limit doesn't exist. A simple example is given by the pair,

$$a_n = n, \quad b_n = n + \frac{1}{n}.$$

If sequences a, b grow large without bound, one might want to consider them to be close to each other if their difference is small relative to the sequences themselves. For example, we could define

$$a \equiv b \text{ iff the differences } |a_n - b_n| \text{ are bounded.}$$

A famous pair of sequences so related is given by

$$a_n = \sum_{k=1}^{n} \frac{1}{k}, \quad b_n = \ln n,$$

for $n = 1, 2, 3, \ldots$ For this pair, the difference is not only bounded, but has a limit known as the *Euler-Mascheroni constant* $C = .5772156649\ldots$

The "Big O" notation used on page 226, above, can be used to generalise this. One writes

$$a_n = O(c_n)$$

to mean that there is some bound B such that $|a_n| < B|c_n|$ for all n. And we write

$$a_n = b_n \, O(c_n).$$

to mean

$$a_n - b_n = O(c_n).$$

Thus, from what we said above,

$$\sum_{k=1}^{n} \frac{1}{k} = \ln n \, O(1).$$

The $O(\cdot)$ notation offers a handy refinement of $\approx$ should one want to express how rapidly two sequences approach one another.

Again, if a, b grow large without bound, we can consider them to be close to each other if their ratios are bounded:

$$\frac{a_n}{b_n} = O(1) \quad \text{and} \quad \frac{b_n}{a_n} = O(1).$$

If the ratios have 1 as a limit, we call the sequences *asymptotic* and write

$$a \sim b \text{ (or, } a_n \sim b_n\text{)} \quad \text{iff} \quad \frac{a_N}{b_N} \approx 1 \quad \text{iff} \quad \lim_{n \to \infty} \frac{a_n}{b_n} = 1.$$

These latter relations are equivalence relations, but not necessarily congruences, and one must be careful in performing computations with them. If a, b are sequences that grow large without bound, and r is any nonzero number, we have

$$\lim_{n \to \infty} \frac{r a_n}{r b_n} = \lim_{n \to \infty} \frac{a_n}{b_n} = 1, \quad \text{if } a_n \sim b_n,$$

i.e., $a \sim b \Rightarrow ra \sim rb$. Similarly,

$$\lim_{n \to \infty} \frac{a_n + r}{b_n + r} = \lim_{n \to \infty} \frac{\frac{a_n}{b_n} + \frac{r}{b_n}}{\frac{b_n}{b_n} + \frac{r}{b_n}}$$

$$= \frac{\lim_{n \to \infty} \frac{a_n}{b_n} + \lim_{n \to \infty} \frac{r}{b_n}}{1 + \lim_{n \to \infty} \frac{r}{b_n}}$$

$$= \frac{1 + 0}{1 + 0}, \quad \text{if } a \sim b,$$

i.e., $a \sim b \Rightarrow a + r \sim b + r$. In the first of these arguments we can replace r by any sequence c of nonzero numbers:

$$a \sim b \Rightarrow ac \sim bc.$$

The second argument, however, only holds for those sequences c for which $\lim_{n \to \infty} c_n/b_n = 0$, which assumption was automatically true for constant sequences $c_n = r$ because of the assumption $b_n \to \infty$. I hasten to add that I have been careful in dealing with asymptotic sequences in the text.[7]

Derivatives

We have almost exhausted the list of unexplained notions and notations used in this book that some readers may not already be familiar with. There remain only the concepts of limit, derivative, and integral and their associated notations. That of integral requires more space and is given its own section following the present one.

In the immediately preceding subsection we considered limits of sequences as $n \to \infty$. In the Calculus one also considers the limit of values $f(x)$ of a function as the variable x approaches a given value a. Informally the notion is neither more nor less intuitive than that of the limit of a sequence. Formally, the definition is slightly more complicated.

13 Definition. *Let f be a function defined for all $x \neq a$ in an interval around a. A number L is the* limit *of $f(x)$ as x approaches a, written*

$$\lim_{x \to a} f(x) = L,$$

just in case, for every $\epsilon > 0$, no matter how small, there is a $\delta > 0$ such that, for all x,

$$0 < |x - a| < \delta \implies |f(x) - L| < \epsilon.$$

---

[7] *Cf.*, e.g., pages 230*ff*.

The condition that $|f(x) - L| < \epsilon$ for whatever small value of ϵ is chosen is analogous to the condition that $|a_n - L| < \epsilon$. And the existence of δ such that this condition holds for $0 < |x - a| < \delta$ is analogous to the existence of n_0 such that the corresponding desired condition holds for $n > n_0$; both assert that x, n, is in a neighbourhood of the element a, ∞, respectively, being approached by x or n. That $|x - a| > 0$, i.e., that $x \neq a$, is required because one is not interested in the value of f at a, but in the values of f at x near(er and nearer to) a. It could be that one hasn't defined f at a and wants to use the limit to determine the value to use. A number of limits encountered in first year Calculus are of this form, the most common being the derivative.

14 Definition. *Let f be a function defined in an interval around a. The derivative of f at a, written $f'(a)$, is the limit*

$$\lim_{h \to 0} \frac{f(a + h) - f(a)}{h},$$

provided this limit exists.

The process of finding the derivative of a function is called *differentiation* and, but for the taking of the quotient and then the limit, is analogous to *differencing*, i.e., applying Δ.

The interest in this limit is made clear by drawing a simple diagram like *Figure 4*, below. The fraction given is the slope of the secant line connecting

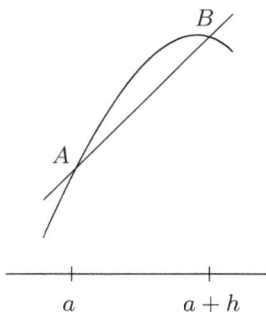

Fig. 4.

the points $A = \langle a, f(a) \rangle$ and $B = \langle a+h, f(a+h) \rangle$. As $h \to 0$ and B approaches A, the line rotates around A and approaches the line tangent to the curve at A, the limit of the ratio approaching the slope of this tangent.[8]

In this case, the ratio

$$g(h) = \frac{f(a + h) - f(a)}{h} \tag{6}$$

---

[8] Perhaps it would be a good programming exercise to write a program that will produce a graphic animation of the procedure on the calculator.

is undefined at $h = 0$, arithmetically because division by 0 is undefined and geometrically because one needs two points to determine a secant.

The use of the letter "h" in (6) is relatively modern, the old standby having been Δx. Writing the corresponding increment in g as Δy, (6) can be written

$$\frac{f(a + \Delta x) - f(a)}{\Delta x} = \frac{\Delta y}{\Delta x},$$

and consequently

$$f'(a) = \lim_{\Delta x \to 0} \frac{\Delta y}{\Delta x} = \frac{dy}{dx}\bigg|_{x=a} = \frac{dy}{dx}(a).$$

Other notations include

$$\frac{df(a)}{dx}, \quad \frac{df}{dx}(a), \quad \frac{d}{dx}f(a).$$

One also emphasises the fact that differentiation is an operation on functions by writing Df for f'.

When f is a function of two or more variables, say $f(x, y)$, one can imagine fixing each variable but one and differentiating with respect to the single unfixed variable. When this happens one replaces d by ∂ and has several *partial derivatives*—in the case of a two-variable function:

$$\frac{\partial f}{\partial x} \quad \text{and} \quad \frac{\partial f}{\partial y}.$$

The notations f_x and f_y are also used.

The chief computational algorithm from which the Calculus derives its name is the determination of the derivative of a function. The derivatives of a few simple functions are determined directly by taking limits, e.g.,

$$\frac{dx^n}{dx} = nx^{n-1}, \quad \frac{d\sin x}{dx} = \cos x, \quad \frac{de^x}{dx} = e^x, \quad \frac{d\ln x}{dx} = \frac{1}{x},$$

and the derivatives of more complex expressions are then determined by simple rules, e.g.,

$$\frac{d}{dx}(f + g) = \frac{df}{dx} + \frac{dg}{dx}, \quad \frac{d}{dx}(f \cdot g) = g\frac{df}{dx} + f\frac{dg}{dx}.$$

The algorithm is a simple deterministic one, easy to program in any computer language that handles strings. I recall assigning such a program in LOGO to beginning programming students around 1980. Writing such a program on the *TI-83 Plus* with its limited access to strings might be a major undertaking, but I note that such has been done for the *TI-89 Titanium*. On that calculator one can enter an expression for f and the calculator will find an expression for f'.

The *TI-83 Plus* does not find an expression for the derivative when given an expression for f, but, if also supplied with a value for a, can approximate $f'(a)$. The syntax is nDeriv(*expr*, *variable*, *value*)

and the function nDeriv(is to be found in the MATH menu accessed via the MATH button. For example, on entering

nDeriv(sin(X),X,$\pi/4$),

one obtains .7071066633, correct to six decimal places, the exact value being $\cos(\pi/4) = \sqrt{2}/2 \approx .7071067812$.

In the present book none of this is really much needed. There are a few places where I have included some calculations that should only be fully accessible to the reader with the background in the Calculus. Most of these calculations, however, concern polynomials, where the rule for finding the derivative is very simple: if

$$p(x) = a_n x^n + a_{n-1}x^{n-1} + \ldots + a_1 x + a_0,$$

then

$$p'(x) = na_n x^{n-1} + (n-1)a_{n-1}x^{n-2} + \ldots + a_1.$$

One application of differentiation is in determining where a maximum or minimum occurs. We applied it implicitly on page 303 (Chapter 6, section 4) in minimising expression (39) of that page and will use it again to maximise an expression in section 7, below (page 476). If the maximum or minimum of a function f occurs at some point c in an open interval on which f is differentiable, then $f'(c) = 0$. The reason is very simple: the tangent to a curve at a maximum or minimum must be horizontal, whence its slope at that point, $\langle c, f(c)\rangle$, must be 0: $f'(c) = 0$.

Likewise, for a function $f(x,y)$ of two variables, if the maximum or minimum occurs at some point $\langle a, b\rangle$ and the partial derivatives $\partial f/\partial x, \partial f/\partial y$ exist at all points $\langle x, y\rangle$ in some open disc around $\langle a, b\rangle$, then

$$\frac{\partial f}{\partial x}(a,b) = \frac{d}{dx}f(x,b)\bigg|_{x=a} = 0$$

$$\frac{\partial f}{\partial y}(a,b) = \frac{d}{dy}f(a,y)\bigg|_{y=b} = 0,$$

where we fix b and a, respectively, and apply our observation to $g(x) = f(x,b)$ and $h(y) = f(a,y)$.

Expression (39) of page 303 can be rewritten:

$$f(a,b) = \sum_{i=0}^{n-1}(y_i - a - bx_i)^2 = \sum_{i=0}^{n-1}(y_i^2 + a^2 + b^2 x_i^2 - 2ay_i - 2bx_i y_i + 2abx_i)$$

$$= \left(\sum_{i=0}^{n-1} 1\right)a^2 + \left(\sum_{i=0}^{n-1}(-2y_i + 2bx_i)\right)a + \sum_{i=0}^{n-1}(y_i^2 + b^2 x_i^2 - 2bx_i y_i).$$

Thus f is exhibited as a polynomial in the variable a, with b held constant. Differentiating with respect to a yields

$$\frac{\partial f}{\partial a} = 2 \left(\sum_{i=0}^{n-1} 1 \right) a + \sum_{i=0}^{n-1} (-2y_i + 2bx_i) + 0$$

$$= 2 \sum_{i=0}^{n-1} (a - y_i + bx_i) = 2 \sum_{i=0}^{n-1} (y_i - a - bx_i)(-1),$$

as on page 303. Similarly, collecting the b-terms and differentiating with respect to b will yield the second equation solved on that page to find a, b minimising the sum of squares.

2 $\int_a^b f(x)dx$

The reader with no background in the Calculus may have found it a bit daunting when, in flipping through the pages of this book, he saw so many integrals. There is no cause for alarm. Except for a few scattered spots where small points are made, a knowledge of the Calculus is not necessary. All one really needs are i. the knowledge that the *definite integral*

$$\int_a^b f(x)dx \tag{7}$$

represents the area under the curve $y = f(x)$ between the *limits* $x = a$ and $x = b$ as in *Figure 5*, below, and ii. the ability to find this area approximately

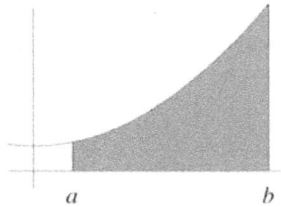

Fig. 5.

on the calculator. The first of these matters is almost self-explanatory, there being a slight subtlety only when the curve $y = f(x)$ dips below the x-axis. The explanation for this, the formal definition of (7), and an explanation of how the calculator estimates this area are by-and-large one and the same discussion. We will get to that shortly. First I will explain how to find the value of the integral on the *TI-83 Plus*.

Given a function $y = f(x)$ and the endpoints a and b (called the *limits of integration*) of the interval over which one wants to find the integral, say

$$f(x) = x^2 + 2, \quad a = 1, \quad b = 5,$$

the first step is to graph the function over the interval to get some idea what one is looking for. Thus, one enters f in the equation editor,

$Y_1=X^2+2$,

and chooses a window, say

Xmin=0
Xmax=6
Xscl=1
Ymin=-5
Ymax=40
Yscl=5 .

The choice of the window is fairly easy: Xmin and Xmax should be slightly to the left and right of a, b, respectively; Ymin and Ymax smaller and larger, respectively, than the minimum and maximum values of f on the graphed interval, and Ymin should also be smaller than 0 so that the x-axis will be seen. In the present case, values are easy to find. In general, one can make any sort of guess, graph the function, and if one doesn't like what one sees, press the ZOOM button and select

0: ZoomFit .

This will now show the complete graph on the Xmin-Xmax interval, but may now hide the x-axis, so a visit to the WINDOW menu again to lower Ymin and maybe raise Ymax to convenient values may be necessary. Xscl and Yscl determine the spacing of the tick marks on the x-axis and the y-axis or, as in the present case, the side of the screen and are chosen to taste. The values should be large enough to show some separation of the ticks.

One can now use the draw command Shade(, the syntax of which is

Shade(*lowerfunc,upperfunc,*[*Xleft,Xright,pattern,patresl*]) .

Here one would enter

Shade(0,Y_1(X),1,5) ,

choosing the default pattern and pattern resolution. The result is that the area in question has been shaded in.

This last step brings us no closer to actually finding the area and is, in any event, redundant. A ClrDraw command removes the shading and restores the simple graph. To find the area, press the CALC button and select option

7: $\int$f(x) dx .

You get a prompt asking for the lower limit a. You will also see the cursor flashing and can move it along the curve using the left- and right-arrow keys until the lower limit is as closely reached as possible. More simply, you can just enter 1. Just below the top line of the screen a blackened right-pointing triangle ($\blacktriangleright$) appears, indicating that the lower limit has been chosen. Its coordinates X=1, Y=3 appear at the bottom of the screen and directly above them is a prompt for the upper limit. Enter 5. A quick animation now shades in the region whose area is being calculated and the value appears at the bottom line of the screen:

$\int f(x)\,dx{=}49.333333$.

In the present simple example, the answer is exact. The area is $148/3$, not 49.333333, but the difference is a matter of representation. To see this, escape to the home screen and press the ANS button. One sees a slightly more precise 49.33333333, but pressing the MATH button and choosing the first option,

1: ►Frac,

one sees the fraction $148/3$ on the screen.

15 Example. Find the area under the curve

$$y = \frac{1}{\sqrt{2\pi}}e^{-x^2/2}$$

between $x = 0$ and $x = 1$ by the above method. Using the window

Xmin$=-.1$
Xmax$=1.1$
Xscl$=.1$
Ymin$=-.1$
Ymax$=.5$
Yscl$=.1$,

my calculator yields

$\int f(x)\,dx{=}.34134475$.

Returning to the home screen again and pressing the ANS button gives $.3413447461$ with slightly greater precision. The integral is just the normal probability integral, which we can calculate by entering

normalcdf(0,1).

The result, $.3413447399$, does not agree with the value calculated by the integration technique, underlying the fact that integration on the calculator is achieved by approximation and is not generally exact.

The graphical approach is perhaps the simplest to use. In any event, it tells the user what information it needs when it needs it. There is another way to perform the integration, one that is more useful in programming. This is to use the function fnInt(, entry 9 in the MATH menu accessed via the MATH button. The function's syntax is

fnInt(*expression,variable,lowerlimit,upperlimit*[,*tolerance*]).

In the case of Example 15, one would enter

fnInt(Y_1(X),X,0,1)

and get $.341447461$ as before.

The optional variable *tolerance* is used to control the accuracy of the calculation and fnInt(assumes the value 10^{-5} as default. The graphical version

$\int f(x)\,dx$ is the same as fnInt(when the tolerance is set at 10^{-3}. Such approximations are often calculated iteratively, with one choosing to stop when successive
iterates agree to a previously determined number of digits or when the difference between successive values of the iterates drops below a given level. The
greater the tolerance allowed in the difference, the less accurate the estimate,
but the lesser effort expended in obtaining an estimate. It is a trade-off and
one must decide in a given instance if anything is to be gained by increasing
precision.

16 Example. While one might no longer use the normal curve to approximate
binomial probabilities which are easily and accurately calculated quickly these
days, this was the approach in the past and I justify this example on historic
grounds[9]. A coin is tossed 100 times. What is the probability of its landing
heads between 60 and 75 times, inclusive? Here we have $p = q = \frac{1}{2}$ as the
common probability of success or failure, $n = 100$ trials, and thus $\mu = np = 100 \cdot \frac{1}{2} = 50, \sigma^2 = npq = 100 \cdot \frac{1}{2} \cdot \frac{1}{2} = 25$. The normal approximation to the
binomial expansion is

$$y = \frac{1}{5\sqrt{2\pi}}e^{-1/2((x-50)/5)^2}. \tag{8}$$

As discussed in Chapter 4, and to be formally clarified shortly, the actual
probability in question,

$$\sum_{i=60}^{75} \binom{100}{i}\left(\frac{1}{2}\right)^i \left(\frac{1}{2}\right)^{n-i}, \tag{9}$$

is approximated by the integral,

$$\int_{59.5}^{75.5} y(x)dx, \tag{10}$$

which we can estimate on our calculators as follows. A little experimentation
yields the window

Xmin=59
Xmax=76
Xscl=1
Ymin=−.001
Ymax=.016
Yscl=.001.

Integrating graphically using the limits 59.5 and 75.5 yields

$\int f(x)\,dx = .02871639$.

Again, this only approximates the normal area,

---

[9] Perhaps more honestly expressed, since this is the way I was taught, I take the
approach on nostalgic grounds

normalcdf(59.5,75.5,50,5),

which is .0287163227. In each case, however, the displayed precision is completely bogus. The sought for probability is given by entering

binomcdf(100,.5,75)−binomcdf(100,.5,59),

which yields .0284438759. Neither estimate of the probability is correct beyond the third place. The error is impressively small, but not so small one would carry out the computation to 10 decimals were one doing it by hand.

Finding areas and volumes goes back to the earliest civilisations. Good approximations and some exact results can be found in all of them, the Greeks even having known how to calculate the area of any polygonal figure as well as that of the circle by the time of Euclid. And Archimedes found the areas trapped by parabolic segments. Further progress had to await the 16th and 17th centuries, with formal definitions and rigorous justifications given in the 19th century and abstract generalisations in the 20th. The definition of the definite integral (7) in the standard undergraduate Calculus course is a kind of limit of approximations to the area under a curve by a lot of rectangles. These rectangles can be inscribed in the region in question as in *Figure 6.a*, below; they can circumscribe the region as in *Figure 6.c*; or they can be crossed by the curve as in *Figure 6.b*. The sum of the areas of the rectangles in *Figure 6.a* yields

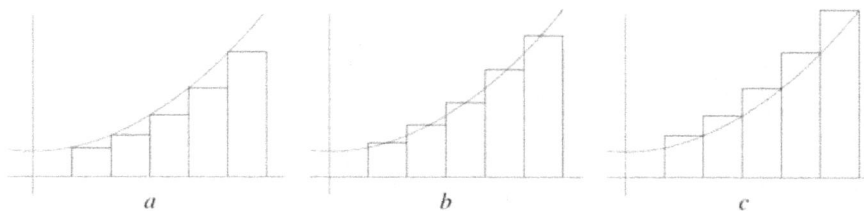

Fig. 6.

a lower bound for the area A, the sum of the areas of the rectangles in *Figure 6.c* generally yields an upper bound on A, and the areas of *Figure 6.b* generally yield a closer approximation to A than the other two, though one will usually not know if it is higher or lower than A. Moreover, unless f is monotone increasing or decreasing, it is not always easy to determine the points in an interval where the function attains a minimum (the point at which the value of the function gives the height of an inscribed rectangle) or maximum (for determining the height of a circumscribing rectangle). Thus, although the sums of *Figures 6.a* and *6.c* are nice to have, one often uses the third type to estimate the area numerically. Such sums, as approximations to area, are termed *Riemann sums* after Bernhard Riemann (1826 - 1866) who first used them in such generality and proved their limits to exist under certain conditions. The sums of *Figures*

6.a and *6.c* given by inscribed and circumscribed rectangles are termed *lower sums* and *upper sums*, respectively.

The formal definition is as follows.

Given a function $y = f(x)$ and an interval $[a,b]$ over which one wishes to perform the integration, one approximates the area

$$A = \int_a^b f(x)dx \tag{11}$$

by partitioning the interval $[a,b]$ into n subintervals with endpoints $a = x_0 < x_1 < \ldots < x_{n-1} < x_n = b$, chooses an element x_i^* in each subinterval $[x_i, x_{i+1}]$, and uses the rectangle of height $f(x_i^*)$ and base $x_{i+1} - x_i$ to approximate the area under f over the interval $[x_i, x_{i+1}]$. Thus A itself will be approximated by the sums

$$A^* = \sum_{i=0}^{n-1} f(x_i^*)(x_{i+1} - x_i). \tag{12}$$

For conceptual simplicity one could choose all the bases to be identical:

$$x_{i+1} - x_i = \frac{b-a}{n}.$$

In practice, one would save oneself some work by choosing smaller intervals where the graph of f is steep and fewer, larger intervals where f is fairly flat. There is any number of ways of choosing x_i^*. The simplest choices would be the left endpoints, right endpoints, or midpoints[10] of the intervals:

$$x_i^* = x_i, \quad x_i^* = x_{i+1}, \quad \text{or} \quad x_i^* = \frac{x_i + x_{i+1}}{2}.$$

For lower sums, one would choose x_i^* to be the point where $f(x)$ is minimum for $x \in [x_i, x_{i+1}]$; and for upper sums x_i^* would be where the maximum occurs. For a monotone function, as in *Figure 6*, above, x_i^* would be one of the endpoints of the interval. For non-monotone functions, there are techniques from the Calculus which allow one to find x_i^* more or less exactly. If a rough approximation will do, the *TI-83 Plus* has built in routines for finding where the minimum and maximum occur. For the maximum, for example, one finds on the MATH menu accessed by the MATH button the item

7: fMax(,

---

[10] Note that, because

$$\binom{100}{i}\left(\frac{1}{2}\right)^i\left(\frac{1}{2}\right)^{n-i} \quad \text{and} \quad \frac{1}{5\sqrt{2\pi}}e^{-\frac{1}{2}((i-50)/5)^2}$$

are approximately equal, the sum (9) is basically a Riemann sum for the function $y(x)$ of (8) using intervals of width 1 centered at $i = 60, 61, \ldots, 76$, thus over the interval $[59.5, 75.5]$. This explains the limits of integration in (10). This slight shift in the limits of integration is called the *continuity correction* and was introduced by Augustus de Morgan.

with syntax

fMax(*expression,variable,lower,upper*[,*tolerance*]) .

Considering again the function $f(x) = x^2 + 2$ on the interval $[1, 5]$, which is increasing and thus has a maximum at $x = 5$, we would enter

fMax(X^2+2,X,1,5)

and obtain 4.999994726, which is very close to 5.

The formal definition of the area (11) trapped between a curve $y = f(x)$ and the x-axis over the interval $[a, b]$ is that it is the limit of the sums (12) as the sizes $x_{i+1} - x_i$ of the intervals approach 0. When the limit exists, f is called *integrable* and the limit is called the *definite integral* of f (over the interval $[a, b]$). Most functions one can enter in the equation editor are integrable—the exceptions occur when somewhere in the interval the evaluation of the expression involves dividing by 0, taking the logarithm of 0 or a negative number, etc.

The first rigorous treatment of the integral as the limit of a sum was given by Augustin Louis Cauchy (1789 - 1857), whose treatment restricted the choice of x_i^* to the end-points of the partitioning subintervals and consequently presupposed the continuity of the function being integrated. The stamp pictured here, issued by France in 1989 to celebrate Cauchy's 200th birthday, is a philatelic tribute of the best kind. It features Cauchy's portrait and three representations of his mathematical work, including his most famous contribution, the Cauchy Integral Formula of Complex Analysis (upper right) and, more relevant here, a reference to his treatment of the integral as the limit of sums of areas of approximating rectangles (left).
Cauchy was a prolific mathematician, contributing to all areas of mathematics, including Probability. His work in this field, however, is not central like his work in other fields of mathematics and we have not considered it in the present book.

One should probably work through an example or two by hand to get a feel for the process. Now, if one chooses too small a number n of intervals the estimate is likely to be very rough, and if one chooses n too large the computation can quickly become tedious. However, one can automate the process on the calculator easily enough. Choose a function—let's stick with $f(x) = x^2 + 2$—and enter it into the calculator:

Y_1=X^2+2 .

Georg Friedrich Bernhard Riemann
1826   1866

German mathematician who made
important contributions to analysis
and differential geometry

CARTE
POSTALA

Georg Friedrich Bernhard Riemann (1826 - 1866) led a short but productive mathematical life. Riemann figures prominently in discussions of types of mathematical minds, his heavily geometric intuitive style often compared with or contrasted to the more formal logical style of Karl Weierstrass (1815 - 1897). One of his important contributions to the field was the Riemann integral, which sharpened the definition of Cauchy's integral and made it more amenable to study, e.g., Riemann himself providing examples of integrable functions with discontinuities at dense sets of points. The notion of integral has been generalised to larger classes of functions, most importantly by Henri Léon Lebesgue (1875 - 1941), whose integral formed the basis of Measure Theory on which Andrei Nikolaevich Kolmogorov based his modern treatment of Probability Theory.

So far as I know there are no stamps or banknotes depicting Riemann's portrait. However, post offices can also commemorate subjects, sometimes at a local level, via special cancellations—as we saw for Gauss on page 128, above. Here we see such a cancellation featuring Riemann's portrait on a Romanian postal card. The date on the cancellation—20 July—suggests the card is meant to commemorate the 40th anniversary of Riemann's death.

Let us also continue with 1 and 5 as values of a and b, respectively.

We can subdivide the interval $[a, b]$ into n equal subintervals of length $\Delta = (b - a)/n$ after choosing n, say $n = 5$:

$(5-1)/5 \rightarrow D$.

This gives us endpoints

$$x_0 = 1, \quad x_1 = x_0 + \Delta = 1 + \frac{4}{5}, \quad \ldots,$$

$$x_4 = x_3 + \Delta = 1 + 4\Delta = 1 + 4 \cdot \frac{4}{5},$$

$$x_5 = x_4 + \Delta = 1 + 5\Delta = 1 + 5 \cdot \frac{4}{5} = 5.$$

Now we know that $x_i = 1 + i\Delta$, but we don't necessarily need to make a list of these points, but rather of the points x_i^* in the intervals which we will use to determing the heights of the intervals. We collect these $n = 5$ points into a list: $\{x_0^*, x_1^*, \ldots, x_4^*\}$. To automate the process we need a rule. We can choose

left endpoint:	1+ID
right endpoint:	1+(I+1)D
midpoint:	1+(2I+1)D/2
randomly chosen point:	1+(I+rand(1))D
point where minimum occurs:	fMin(Y₁(X),X,1+ID,1+(I+1)D)
point where maximum occurs:	fMax(Y₁(X),X,1+ID,1+(I+1)D).

Finding the lower sum or upper sum using fMin or fMax, respectively, is only going to approximate the lower or upper sum and can be time consuming as fMin and fMax are calculated iteratively. And, since our chosen function is strictly increasing, we can accurately calculate the lower and upper sums using the left and right endpoints, respectively. So let us use, say, the left endpoint rule to calculate the list $\{x_0, x_1, \ldots, x_4\}$:

seq(1+ID,I,0,4)→L₁ .

Finally, let us calculate the sum of the areas of the rectangles:

$$\sum_{i=0}^{4} f(x_i^*)(x_{i+1} - x_i) = \sum_{i=0}^{4} f(x_i)\Delta.$$

The values $f(x_0), f(x_1), \ldots, f(x_4)$ are collected together in the list $Y_1(L_1)$. Δ is constant, so we can perform the summation either by multiplying this list by Δ and summing it,

sum(Y₁(L₁)∗D) ,

or appealing to the distributive law and summing the list before multiplying:

sum(Y₁(L₁))∗D .

The result, 40.16, is too low, as was to be expected, but is not a bad approximation. If we similarly use right endpoints to find an upper sum,

seq(1+ID,D,1,5)→L₁
sum(Y₁(L₁))∗D ,

we get 59.36, which is too high, but still of the right order of magnitude. The mean of these two figures is 49.76, which is quite close to the actual value of $49.\overline{3}$. Basing the sum on midpoints yields 49.12, which begins to seem phenomenally close.

17 Project. (For those who have completed Chapter 6 and section 3 of Appendix A.) Analyse the results of estimating the area 200 times using a randomly chosen point $x_i^* \in [x_i, x_{i+1}]$ for our usual $f(x) = x^2 + 2, a = 1, b = 5, n = 5$. What value would you expect for the mean of the distribution?

The fundamental result is that the sums (12) approach a limit (11) as the differences $x_{i+1} - x_i$ approach 0. One can verify this by carrying out the computations for larger and larger values of n. A nice way of doing this is to double the number of intervals at each step, as I do in the following program. This program will both perform the successive calculations and depict the curve and rectangles approximating the area under the curve (à la the program BELL2 on page 137, above—but with the curve drawn first and the rectangles second this time).

```
PROGRAM:AREAS
:PlotsOff
:FnOff
:ClrDraw
:"X²+2"→Y₁
:1→L
:5→R
:.5→Xmin
:5.5→Xmax
:1→Xscl
:−5→Ymin
:40→Ymax
:5→Yscl .
```

As my goal is illustration, in this case a simple succession of calculations that could be projected onscreen in class, I am writing the program for a fixed function and interval, namely our usual $f(x) = x^2 + 2$ on $[1,5]$. A more general program assuming a function expression stored in Y_1 and left and right endpoints stored in L and R, respectively, could easily be written. The choice of values for the WINDOW parameters must be different from that used above. I suggest replacing the lines after the ClrDraw command by

```
:(R−L)/8→E
:L−E→Xmin
:R+E→Xmax
:E→Xscl
:fMax(Y₁(X),X,L,R)→T
:Y₁(T)→M
:M/8→F
:−F→Ymin
:M+F→Ymax
:F→Yscl
:FnOn 1 .
```

In the original program, the command assigning the expression to Y_1 turned the function on. As this new version assumes an expression already stored in Y_1, and the general FnOff command has been applied, the graph of Y_1 must explicitly be turned on, whence the FnOn 1 command.

Once all of this has been settled, the rest is a loop.

```
:For(J,1,7)
:ClrDraw
:2^J→N
:(R−L)/N→D .
```

This will start with 2 subintervals, doubling the number at each pass until $2^7 = 128$ subintervals are given. There being only 95 pixels horizontally, there will be no visual difference in the display with more doublings than this, though, of course the numerical estimate can continue to increase in accuracy beyond this point. The ClrDraw command is obviously included to keep each graph from cluttering up its successor. For a monotone increasing function like $f(x) = x^2+2$, however, one might want to omit this command and see a nicer animation revealing the cumulative effect on the lower sums of the interval splitting.

```
:seq(1+DX,X,0,N−1)→L₁
:Y₁(L₁)→L₂
:sum(L₂)*D→A .
```

This evaluates the sum of the areas of the rectangles. Note that sum(L_2)*D utilises 1 multiplication by D, whereas sum(L_2*D) utilises N such multiplications and is thus less efficient.

The graphing of the rectangles is done this time via DRAW commands (as opposed to via histograms as in BELL2).

```
:For(I,1,N)
:Line(L₁(I),0,L₁(I),L₂(I))
:Line(L₁(I),L₂(I),L₁(I)+D,L₂(I))
:Line(L₁(I)+D,0,L₁(I)+D,L₂(I))
:End
:Text(0,2,"AREA=",A)
:DispGraph .
```

The Line(commands draw the sides of the rectangles, the first doing the vertical lines from $\langle x_i, 0\rangle$ to $\langle x_i, f(x_i)\rangle$, the second the horizontal lines from $\langle x_i, f(x_i)\rangle$ to $\langle x_{i+1}, f(x_i)\rangle$, and the third the vertical lines from $\langle x_{i+1}, f(x_i)\rangle$ to $\langle x_{i+1}, 0\rangle$. Note that x_{i+1} is given by $x_i + \Delta$ (L₁(I)+D) and not directly as x_{i+1} (L₁(I+1)). This is to avoid having to add a separate line in the program handling the case $i = n$. The Text(command is similar to the Output command allowing the exact placement of text, but it places the text in the graphics screen instead of the home screen. One might prefer to place the message at the bottom of the screen by using the command

```
:Text(57,2,"AREA=",A) .
```

It now remains only to end the loop and clean up unwanted variables. The loop ending looks like this:

```
:If J<7
:Pause
:End .
```

The Pause command pauses execution of the program to allow one to admire the pretty picture before pressing ENTER and performing the next iteration. There is no need to do this in the last iteration, whence the conditional. The conditional itself is a simple If command and needs no End command to close it off, whence the given End command closes the outer loop.

A last step is to add clean up commands deleting no longer needed variables and clearing the lists L_1, L_2.

I mentioned earlier that a complication arises when f dips below the x-axis. What happens is that when f is negative, the terms in the sums (12) are negative. The overall result is that the integral (11) does not give the area trapped between the curve $y = f(x)$ and the x-axis, but the *signed* area, regions above the x-axis having positive area and regions below the x-axis having negative area. This is actually quite useful in applications.

We have been discussing the definite integral over *finite* intervals. Yet in probability theory the intervals are often unbounded. Basically, one defines

$$\int_{-\infty}^{\infty} f(x)dx = \int_{-\infty}^{0} f(x)dx + \int_{-0}^{\infty} f(x)dx$$

$$\int_{-\infty}^{b} f(x)dx = \lim_{a \to -\infty} \int_a^b f(x)dx \qquad (13)$$

$$\int_{-a}^{\infty} f(x)dx = \lim_{b \to +\infty} \int_a^b f(x)dx. \qquad (14)$$

If these integrals exist, the area outside the bounded interval will be so small as to be negligible for some large value of the variable bound. A moderately large choice of a in (13) or b in (14) will generally result in an error smaller than that given by the approximation yielded by fnInt(.

For some functions f the integral (11) can be determined exactly by the methods of the Calculus, while others must be calculated approximately. The use of rectangles, while conceptually simple, easily visualised, and even theoretically useful, is not all that efficient. The program AREAS, after 7 iterations— 128 subintervals—, yields 48.958984, too small by over 1/3. Approximating the area by trapezia (specifically, by the quadrilaterals with sides connecting the points $\langle x_i, 0 \rangle$, $\langle x_i, f \langle x_i \rangle \rangle$, $\langle x_{i+1}, f \langle x_{i+1} \rangle \rangle$ and $\langle x_{i+1}, 0 \rangle$) yields 49.333984 with 128 intervals. In fact, 8 intervals already yield an estimate of 49.5 with an error less than 1/5.

If one does the algebra, one sees that the use of trapezia, called the *Trapezium Rule*[11], assumes the form

---

[11] *Trapezoid Rule* in American usage, as trapezia are commonly called trapezoids in the United States.

$$\int_a^b f(x)dx \approx f(x_0)\frac{b-a}{2n} + f(x_1)\frac{b-a}{n} + \ldots + f(x_{n-1})\frac{b-a}{n} + f(x_n)\frac{b-a}{2n},$$

the constant coefficient being $(b-a)/(2n)$ for the first and last terms and $(b-a)/n$ for the remaining ones. This, as well as (12), has the form

$$\sum_{i=0}^m w_i f(x_i), \tag{15}$$

for some points $x_0 < x_1 < \ldots < x_m$ and weights $w_0, w_1, \ldots, w_m$. The calculator estimates the area through an iterative algorithm called the *Gauss-Kronrod method* after Carl Friedrich Gauss who developed the original method and Aleksandr Semenovich Kronrod (1921 - 1986) who improved its efficiency. This method makes a succession of choices of points and weights, stopping the iteration when the difference of two successive sums falls below a preassigned tolerance level. Any discussion of this algorithm must lie beyond the scope of this book, other than, of course, to make the observation that it is more efficient than the simple methods described here.

3 Footnote to Bernoulli's Proof

Bernoulli's Theorem, when written in a more convenient modern form, reads:

18 Theorem. *Let r, s be positive integers, $t = r + s$, and $p = r/t$. If for any positive value N we let $m_N =$ the number of successes in N Bernoulli trials with probability p of success on a given trial, then, for any $\epsilon, \delta > 0$, a number N_0 can be chosen so that*

$$P\left(\left|\frac{m_N}{N} - p\right| < \epsilon\right) > 1 - \delta,$$

for $N > N_0$.

Jakob Bernoulli only proved his theorem for N of the form nt. A proof for N of the more general form $nt + k$, with $0 < k < t$, can be given along the same lines, but is a notational mess. Fortunately, the general result follows from the special case fairly simply.

Let $\epsilon, \delta > 0$ be given, $N = nt + k$ with $0 < k < t$. Let m_N denote the number of successes in the first nt trials and k_n denote the number of successes in the last k trials. Thus $m_n + k_n$ is the number of successes in $N = nt + k$ trials.

Now a standard trick of the Calculus in what is called an $\epsilon/2$-argument is to add and subtract a term to change a difference into a sum of two simpler differences:

$$\frac{m_n + k_n}{nt + k} - p = \frac{m_n + k_n}{nt + k} - \frac{m_n}{nt} + \frac{m_n}{nt} - p$$

$$= \frac{m_n}{nt+k} - \frac{m_n}{nt} + \frac{m_n}{nt} - p + \frac{k_n}{nt+k},$$

whence

$$\left| \frac{m_n + k_n}{nt+k} - p \right| \leq \left| \frac{m_n}{nt+k} - \frac{m_n}{nt} \right| + \left| \frac{m_n}{nt} - p \right| + \frac{k_n}{nt+k}$$

$$\leq \frac{m_n|nt - nt - k|}{nt(nt+k)} + \left| \frac{m_n}{nt} - p \right| + \frac{k_n}{nt+k}$$

$$< \frac{k \cdot m_n}{nt \cdot nt} + \left| \frac{m_n}{nt} - p \right| + \frac{1}{n}, \tag{16}$$

this last since

$$nt + k > nk + k = (n+1)k > nk \geq nk_n.$$

Dividing the first and last of these by $n(nt+k)$ yields,

$$\frac{k_n}{nt+k} < \frac{1}{n}.$$

Continuing from (16),

$$\left| \frac{m_n + k_n}{nt+k} - p \right| \leq \frac{1}{n} \cdot \frac{k}{t} \cdot \frac{m_n}{nt} + \left| \frac{m_n}{nt} - p \right| + \frac{1}{n}$$

$$< \frac{1}{n} \cdot 1 \cdot 1 + \left| \frac{m_n}{nt} - p \right| + \frac{1}{n}$$

$$< \left| \frac{m_n}{nt} - p \right| + \frac{2}{n}.$$

Now choose n_0 so large that $\epsilon > 2/n$ for $n > n_0$ and let $N_0 > n_0t$ be so large that, for $nt > N_0$,

$$P\left(\left| \frac{m_n}{nt} - p \right| < \epsilon - \frac{2}{n} \right) > 1 - \delta.$$

But

$$\left| \frac{m_n}{nt} - p \right| < \epsilon - \frac{2}{n} \Rightarrow \left| \frac{m_n + k_n}{nt+k} - p \right| < \epsilon - \frac{2}{n} + \frac{2}{n} = \epsilon,$$

whence

$$P\left(\left| \frac{m_n + k_n}{nt+k} - p \right| < \epsilon \right) \geq P\left(\left| \frac{m_n}{nt} - p \right| < \epsilon - \frac{2}{n} \right) \geq 1 - \delta.$$

Thus, Bernoulli's restricting his consideration to N of the form nt is theoretically unimportant. The estimate on N_0 is larger, however, as $|m_n/(nt) - p|$ must now be less than $\epsilon - 2/n$ and not just less than ϵ.

One more point we might address is the difference between Bernoulli's motivational description of why his theorem should be true (Chapter 5, section 2) and his ultimate proof of this theorem (Chapter 5, section 3). The description and our numerical experience via the program BERNULLI suggest that each sequence of tails $q_i, q_{i+t}, q_{i+2t}, \ldots$ tends exponentially to 0, i.e., for some positive constant $c < 1$ and all k, $q_{i+kt} > cq_{i+(k+1)t}$. If one compares the terms of the sum q_{i+kt} with those of $q_{i+(k+1)t}$, one finds indeed that there is a $c < 1$ such that each term of q_{i+kt} is greater than c times the corresponding term of $q_{i+(k+1)t}$. Unfortunately, not every term of $q_{i+(k+1)t}$ is such a corresponding term. $q_{i+(k+1)t}$ contains t additional summands and one has to make an additional comparison of inner to outer terms of these tails. After some thought, one realises that the comparison of q_{i+kt} with $q_{i+(k+1)t}$ is not needed and Bernoulli's final proof or his nephew's proof results.

4 The Integral $\int_{-\infty}^{\infty} e^{-x^2} dx$

[This section presupposes a knowledge of the Calculus.]

The function $f(x) = e^{-x^2}$ has no elementary anti-derivative on any interval, as first proven by Joseph Liouville (1809 - 1882). This means that there is no expression $F(x)$ built up from $x, e^x, \ln x, \sin x, \cos x$, etc., for which

$$\int_{\alpha}^{\beta} f(x) dx = F(\beta) - F(\alpha)$$

for all $\alpha < \beta$, whence these areas must be evaluated numerically. For the intervals $(-\infty, \infty)$ and $[0, \infty)$, however, the areas under the curve $y = f(x)$ can be evaluated in closed form by various tricks.

As we quoted from Thomas Galloway on page 126, above, the substitution $e^{-t^2} = z$ converts the integral

$$A = \int_0^{\infty} e^{-t^2} dt \tag{17}$$

into

$$\frac{1}{2} \int_0^1 \frac{dz}{\sqrt{\ln 1/z}},$$

which Leonhard Euler had already shown to equal $\sqrt{\pi}/2$. As this looks to be as difficult to handle as the original, I don't find this an intuitively appealing approach.

The first to evaluate (17) was Pierre Simon de Laplace, who was a master at integrating transformed functions. For reasons not immediately evident he evaluated the double integral,

$$I = \int_0^{\infty} \int_0^{\infty} e^{-s(1+u^2)} ds\, du.$$

Using the indicated order of integration one has

$$I = \int_0^{\infty} \left. \frac{e^{-s(1+u^2)}}{-(1+u^2)} \right|_0^{\infty} du$$

$$= \int_0^\infty \frac{du}{1+u^2} = \arctan u \Big|_0^\infty = \frac{\pi}{2}.$$

Reversing the order of integration yields

$$I = \int_0^\infty \int_0^\infty e^{-s(1+u^2)}du\,ds$$

$$= \int_0^\infty \int_0^\infty e^{-s-su^2}du\,ds$$

$$= \int_0^\infty e^{-s}\int_0^\infty e^{-su^2}du\,ds$$

$$= \int_0^\infty e^{-s}\int_0^\infty e^{-v^2}s^{-1/2}dv\,ds, \text{ using } v = s^{1/2}u, dv = s^{1/2}du$$

$$= \int_0^\infty e^{-s}s^{-1/2}\left(\int_0^\infty e^{-v^2}dv\right)ds$$

$$= A \int_0^\infty e^{-s}s^{-1/2}ds$$

$$= A \int_0^\infty e^{-t^2}t^{-1}2t\,dt, \text{ using } s = t^2, ds = 2tdt$$

$$= 2A^2.$$

Equating the two expressions for I,

$$2A^2 = \frac{\pi}{2},$$

whence

$$\int_0^\infty e^{-x^2}dx = A = \sqrt{\frac{\pi}{4}} = \frac{\sqrt{\pi}}{2}.$$

The preferred modern proof is just a tad more intuitive, but ultimately is similar. One begins by squaring the integral:

$$A^2 = \left(\int_0^\infty e^{-x^2}dx\right)^2 = \left(\int_0^\infty e^{-x^2}dx\right)\left(\int_0^\infty e^{-y^2}dy\right)$$

$$= \int_0^\infty \int_0^\infty e^{-x^2-y^2}dx\,dy.$$

A change to polar coordinates yields

$$A^2 = \int_0^{\pi/2}\int_0^\infty e^{-r^2}r\,dr\,d\theta.$$

Substituting $u = r^2$, we have $du = 2rdr$ and

$$\int_0^\infty e^{-r^2}rdr = \int_0^\infty \frac{1}{2}e^{-u}du = -\frac{1}{2}e^{-u}\Big|_0^\infty = 0 - \left(-\frac{1}{2}\right) = \frac{1}{2},$$

whence

$$A^2 = \int_0^{\pi/2}\frac{1}{2}d\theta = \frac{1}{2}\theta\Big|_0^{\pi/2} = \frac{\pi}{4}.$$

Thus, again, we have $A = \sqrt{\pi}/2$.

5 Stirling's and Wallis's Formulæ

A standard result of Calculus is Taylor's Theorem expressing a function near a point a as an infinite series

$$f(x) = f(a) + (x-a)f'(a) + (x-a)^2 \frac{f''(a)}{2} + \ldots$$
$$= f(a) + \sum_{k=1}^{\infty} (x-a)^k \frac{f^{(k)}(a)}{k!}, \tag{18}$$

where $f^{(k)}$ denotes the k-th derivative of f. This, of course, does not hold for all functions f, all a, or all x near a. The obvious failures occur, of course, when the derivatives do not exist or when the series does not converge. As first noted by Cauchy, the function,

$$f(x) = \begin{cases} e^{-1/x^2}, & x \neq 0 \\ 0, & x = 0, \end{cases}$$

has all the necessary derivatives at $a = 0$ and the series converges everywhere, but the two sides of (18) do not agree for any $x \neq 0$. But the equation does hold for many functions and one can work with the *Taylor series* as the infinite series of (18) is called. One example is the familiar expansion of the exponential function,

$$e^x = 1 + \frac{x}{1} + \frac{x^2}{2!} + \frac{x^3}{3!} + \ldots = \sum_{k=0}^{\infty} \frac{x^k}{k!}. \tag{19}$$

Combining this with the operators E, Δ, Σ considered in section 1, we note

$$E(f)(x) = f(x+1) = f(x) + \sum_{k=1}^{\infty} \frac{f^{(k)}(x)}{k!}, \tag{20}$$

where we have $a = x, x+1-a = 1$. Writing D for the differentiation operation, D^k for the k-fold iteration: $D^k f = f^{(k)}$, (20) reads

$$E(f)(x) = f(x) + \sum_{k=1}^{\infty} \frac{D^k}{k!} f(x) = \left(1 + \sum_{k=1}^{\infty} \frac{D^k}{k!}\right)(f)(x)$$
$$= e^D(f)(x).$$

That is,

$$E = e^D.$$

Now, $E = 1 + \Delta$, whence $\Delta = E - 1$ and

$$\Delta^{-1} = \frac{1}{E-1} = \frac{1}{e^D - 1}$$
$$= \frac{1}{D} \cdot \frac{D}{e^D - 1}. \tag{21}$$

However,

$$\frac{e^x - 1}{x} = 1 + \frac{x}{2} + \frac{x^2}{3!} + \frac{x^3}{4!} + \ldots$$

and long division (as in the box below) yields

$$
\begin{array}{r}
1 - x/2 + x^2/12 + \quad 0 \quad - x^4/720 + \ldots \\
\hline
1 + x/2 + x^2/3! + x^3/4! + \ldots \,\big|\, 1 \\
1 + x/2 + x^2/3! + x^3/4! + x^4/5! + \ldots \\
\hline
- x/2 - x^2/2 - x^3/4! - x^4/5! + \ldots \\
- x/2 - x^2/4 - x^3/12 - x^4/48 + \ldots \\
\hline
x^2/12 + x^3/24 + x^4/80 + \ldots \\
x^2/12 + x^3/24 + x^4/72 + \ldots \\
\hline
- x^4/720 + \ldots \\
\vdots
\end{array}
$$

$$\frac{x}{e^x - 1} = \frac{B_0}{0!} + \frac{B_1 x}{1!} + \frac{B_2 x^2}{2!} + \ldots \tag{22}$$

for some sequence $B_0, B_1, B_2, \ldots$ of numbers, called the *Bernoulli numbers* after Jakob Bernoulli who first introduced them in the combinatorial part of his *Ars conjectandi*.

Combining (21) and (22) we have

$$\Delta^{-1} = \frac{1}{D} \cdot \left(B_0 + B_1 D + \frac{B_2 D^2}{2!} + \ldots \right)$$

$$= B_0 D^{-1} + B_1 + \frac{B_2}{2!} D + \ldots$$

Thus

$$\Delta^{-1} f(x) = \int f(x)\, dx + B_1 f(x) + \frac{B_2}{2!} f'(x) + \ldots$$

and we have the

19 Theorem (Euler-Maclaurin Summation Formula). *Let f be given, and $n > 1$.*

$$\sum_{k=1}^{n-1} f(x) = \int_1^n f(x)\, dx + B_1 f(n) + \frac{B_2}{2!} f'(n) + \ldots - \left(B_1 f(1) + \frac{B_2}{2!} f'(1) + \ldots \right).$$

The Euler-Maclaurin Summation Formula was discovered independently by Leonhard Euler, one of the greatest mathematicians of all time, and Colin Maclaurin (1698 - 1746), a Scottish mathematician best known for his more convincing demonstration of Taylor's Theorem than Taylor's (1685 - 1731). Rigorous proofs of these results, together with the conditions necessary for their validity, came much later than the results and their widespread use. Joseph Louis Lagrange, whose approach to Euler-Maclaurin I have followed here, gave the first proof of Taylor's Theorem asserting, under certain conditions, the Taylor series of a function to converge to the function itself by providing an estimate for the difference,

$$f(x) - \sum_{k=0}^{n} \frac{f^k(a)}{k!}(x-a)^k.$$

And a rigorous formulation and proof of the Euler-Maclaurin Theorem would first be given by Simeon Denis Poisson and more definitively in 1834 by Carl Gustav Jacobi (1804 - 1851).

If one accepts the validity of the Euler-Maclaurin Summation Formula for $f(x) = \ln x$, de Moivre's Theorem follows as an easy corollary:

$$\ln n! = \sum_{k=1}^{n} \ln k = \ln n + \sum_{k=1}^{n-1} \ln k$$

$$= \ln n + \int_{1}^{n} \ln x \, dx + B_1 \ln n + \frac{B_2}{2!} \cdot \frac{1}{n} + \frac{B_3}{3!}\left(\frac{-1}{n^2}\right) + \ldots$$

$$- \left(B_1 \ln 1 + \frac{B_2}{2!} \cdot 1 + \frac{B_3}{3!}(-1) + \ldots\right)$$

$$\approx (x \ln x - x)\Big|_1^n + (1 + B_1)\ln n - \left(\frac{B_2}{2!} - \frac{B_3}{3!} + \ldots\right),$$

the omitted terms being close to 0 for large n,

$$\approx n \ln n - n + 1 + (1 + B_1)\ln n - \left(\frac{B_2}{2!} - \frac{B_3}{3!} + \ldots\right)$$

$$\approx \left(n + \frac{1}{2}\right)\ln n - n - \left(-1 + \frac{1}{12} - 0 + \ldots\right),$$

using the values of $B_1, B_2, \ldots$ given by the long division in the box above. Exponentiating,

$$n! \sim A n^{n+1/2} e^{-n},$$

for some constant A. But for the determination of A, this is Stirling's formula. From here one can continue as in Chapter 6, section 1, pages 230 - 234, using A in place of $\sqrt{2\pi}$:

$$\binom{2m}{m}\left(\frac{1}{2}\right)^{2m} \sim \frac{A(2m)^{2m+1/2}e^{-2m}}{Am^{m+1/2}e^{-m}Am^{m+1/2}e^{-m}} \cdot \frac{1}{2^{2m}}$$

$$\sim \frac{2^{2m+1/2}m^{2m+1/2}}{A2^{2m}m^{2m+1}}$$

$$\sim \frac{\sqrt{2}}{A\sqrt{m}}.$$

The derivation ends, for $n = 2m, \mu = np = m, \sigma^2 = npq = n/4$, and $x = \mu \pm l$,

$$\binom{n}{x} \sim \frac{1}{\sqrt{A}\sigma}e^{-1/2((\mu-x)/\sigma)^2},$$

and one can use Laplace's evaluation of $\int_{-\infty}^{\infty} e^{-x^2}dx$ and the probabilistic assumption,

$$\int_{-\infty}^{\infty} \frac{1}{\sqrt{A}\sigma}e^{-1/2((\mu-x)/\sigma)^2}dx = 1,$$

to conclude $A = \sqrt{2\pi}$ as determined by Stirling. I leave the details to the energetic reader.

That de Moivre could have proven his result in this way is a point often made in histories of the subject. In textbooks, however, a more rigoroous approach is preferred. The proof in Feller's book[12] stems from the observation that

$$\ln n! = \sum_{k=1}^{n} \ln n = \sum_{k=2}^{n} \ln n$$

is a Riemann sum for $\int_1^n \ln x \, dx$, but uses trapezia to approximate the area under the curve $y = \ln x$ over the intervals $[1, n]$ and $[3/2, n]$. The former is a straightforward application of the Trapezium Rule using the subintervals $[1, 2], [2, 3], \ldots, [n - 1, n]$ of $[1, n]$. The result is

$$\sum_{k=1}^{n-1} \frac{1}{2}\big(\ln k + \ln(k+1)\big) = \frac{1}{2}\ln 1 + \sum_{k=2}^{n-1} \ln k + \frac{1}{2}\ln n < \int_1^n \ln x \, dx,$$

i.e.,

$$\ln n! = \sum_{k=1}^{n} \ln k < \int_1^n \ln x \, dx + \frac{1}{2}\ln n. \tag{23}$$

The second area calculation uses trapezia with top lines tangent to the curve $y = \ln x$ at $k = 2, 3, \ldots, n - 1$ over the intervals $[3/2, 5/2], [5/2, 7/2], \ldots, [(n - 2)/2, (n - 1)/2]$ and a single rectangle of height $\ln n$ over the interval $[(n - 1)/2, n]$. The result yields

$$\int_{3/2}^{n} \ln x \, dx < \sum_{k=2}^{n-1} \ln k + \frac{1}{2}\ln n = \sum_{k=1}^{n-1} \ln k + \frac{1}{2}\ln n,$$

i.e.,

[12] William Feller, *An Introduction to Probability Theory and Its Applications*, 2nd ed., John Wiley & Sons, Inc., New York, 1957, pp. 50 - 53.

$$\int_{3/2}^{n} \ln x \, dx + \frac{1}{2} \ln n < \ln n!. \tag{24}$$

In the Calculus one learns that $\int \ln x \, dx = x \ln x - x$, whence (23) and (24) yield

$$n \ln n - n - \left(\frac{3}{2} \ln \frac{3}{2} - \frac{3}{2}\right) + \frac{1}{2} \ln n < \ln n! < n \ln n - n - 0 + \frac{1}{2} \ln n,$$

i.e.,

$$\left(n + \frac{1}{2}\right) \ln n - n - \left(\frac{3}{2} \ln \frac{3}{2} - \frac{3}{2}\right) < \ln n! < \left(n + \frac{1}{2}\right) \ln n - n,$$

whence

$$0 < \left(n + \frac{1}{2}\right) \ln n - n - \ln n! < \frac{3}{2} \ln \frac{3}{2} - \frac{3}{2}.$$

Now the sequence $a_n = (n + 1/2) \ln n - n - \ln n!$ is bounded and, the terms being the sums of more and more differences between larger and smaller areas, is monotone increasing and hence has a limit K. Thus

$$e^K = \lim_{n \to \infty} \exp\left(\left(n + \frac{1}{2}\right) \ln n - n - \ln n!\right)$$

$$= \lim_{n \to \infty} \frac{\exp((n + \frac{1}{2}) \ln n - n)}{e^{\ln n!}}$$

$$= \lim_{n \to \infty} \frac{n^{n + \frac{1}{2}} e^{-n}}{n!},$$

i.e.,

$$n! \sim e^{-K} n^{n + \frac{1}{2}} e^{-n},$$

and we will again have Stirling's formula once we find the value of $A = e^{-K}$.

The determination of A is a sort of footnote to the rest of the proof. As mentioned earlier, de Moivre himself originally settled for a close numerical approximation. Stirling provided the value $\sqrt{2\pi}$, which de Moivre conjectured Stirling had obtained by appeal to Wallis's formula (introduced on page 231, above). Indeed, today the main methods of deriving Stirling's formula in probability texts end with the determination of the constant A by appeal to Laplace's calculation of the probability integral, as I've done here. Feller[13] refers to the integral, while Arne Fisher (1887 - 1944) provides an example of a textbook[14] in which Wallis's formula is used at the end of yet another derivation[15]. Before

---

[13] *Ibid.*, p. 169.

[14] Arne Fisher, *The Mathematical Theory of Probabilities and Its Application to Frequency Curves and Statistical Methods*, 2nd. ed., The Macmillan Company, New York, 1922.

[15] *Ibid.*, pp. 92 - 95. His presentation, incidentally, is based on a simple proof published by Ernesto Cesáro (1859 - 1906) in 1884.

deriving Stirling's formula, Fisher presents a proof of Wallis's formula[16] which is simple enough to present here.

One begins by considering the integrals

$$J_n = \int_0^{\pi/2} \sin^n x \, dx.$$

For $n = 0, 1$, one calculates directly:

$$J_0 = \int_0^{\pi/2} 1 dx = x \Big|_0^{\pi/2} = \frac{\pi}{2} - 0 = \frac{\pi}{2}$$

$$J_1 = \int_0^{\pi/2} \sin x \, dx = -\cos x \Big|_0^{\pi/2} = -0 - (-1) = 1.$$

For larger n, one integrates by parts:

$$J_n = -\cos x \sin^{n-1} x \Big|_0^{\pi/2} + \int_0^{\pi/2} \cos x \cdot (n-1) \sin^{n-2} x \cos x \, dx$$

$$= -0 \cdot 1 - (-1 \cdot 0) + (n-1) \int_0^{\pi/2} \sin^{n-2} x \cos^2 x \, dx$$

$$= (n-1) \int_0^{\pi/2} \sin^{n-2} x (1 - \sin^2 x) dx$$

$$= (n-1) \int_0^{\pi/2} \sin^{n-2} x - \sin^n x \, dx$$

$$= (n-1) J_{n-2} - (n-1) J_n,$$

from which it follows that

$$n J_n = (n-1) J_{n-2},$$

i.e.,

$$J_n = \frac{n-1}{n} J_{n-2}. \tag{25}$$

Expanding (25) for $n = 2m, 2m-1$, respectively, gives us two interesting formulæ:

$$J_{2m} = \frac{(2m-1)(2m-3) \cdots 1}{2m(2m-2) \cdots 2} \cdot \frac{\pi}{2}$$

$$J_{2m-1} = \frac{(2m-2)(2m-4) \cdots 2}{(2m-1)(2m-3) \cdots 3}.$$

The connexion with Wallis's formula is easy to see:

$$\frac{J_{2m}}{J_{2m-1}} = \frac{\dfrac{(2m-1)(2m-3) \cdots 1}{2m(2m-2) \cdots 2} \cdot \dfrac{\pi}{2}}{\dfrac{(2m-2)(2m-4) \cdots 2}{(2m-1)(2m-3) \cdots 3}}$$

---

[16] *Ibid.*, pp. 90 - 92.

$$= \frac{(2m-1)(2m-3)\cdots 1 \cdot (2m-1)(2m-3)\cdots 3}{2m(2m-2)\cdots 2 \cdot (2m-2)(2m-4)\cdots 2} \cdot \frac{\pi}{2},$$

and

$$\lim_{m \to \infty} \frac{J_{2m}}{J_{2m-1}} = \frac{\pi}{2} \cdot \frac{3 \cdot 3 \cdot 5 \cdot 5 \cdot 7 \cdot 7 \cdot 9 \cdot 9 \cdots}{2 \cdot 2 \cdot 4 \cdot 4 \cdot 6 \cdot 6 \cdot 8 \cdot 8 \cdots},$$

and Wallis's formula will be proven once one shows

$$\lim_{m \to \infty} \frac{J_{2m}}{J_{2m-1}} = 1. \tag{26}$$

Unfortunately, Fisher's justification of (26) is a bit weak:

The difference of the ordinates of these graphs, namely:

$$(\sin x - 1) \sin^{2m-1} x$$

is evidently decreasing with increasing values of the positive integer n, since $\sin x$ lies between 0 and 1 and $\sin^{2m-1} x$ approaches the value 0 except for certain values of x The larger we select m the less is the difference of the two areas and the ratio will therefore approach 1.[17]

However, (26) is not too difficult to establish.

$$J_{n+1} = \int_0^{\pi/2} \sin x \cdot \sin^n x \, dx < \int_0^{\pi/2} 1 \cdot \sin^n x \, dx = J_n,$$

since $0 \leq \sin x \leq 1$ with equality only at $x = \pi/2$ on $[0, \pi/2]$. Thus the sequence $J_0, J_1, J_2, \ldots$ is strictly decreasing. In particular

$$\frac{J_{2m}}{J_{2m-1}} < 1 \quad \text{and} \quad \frac{J_{2m-1}}{J_{2m-2}} < 1 = \frac{J_{2m}}{J_{2m}}.$$

But the second of these yields

$$\frac{J_{2m}}{J_{2m-1}} > \frac{J_{2m}}{J_{2m-2}} = \frac{2m-1}{2m} \quad \text{by (25)}.$$

Thus J_{2m}/J_{2m-1} is squeezed between $(2m-1)/(2m)$ and 1 which both have 1 as a limit as $m \to \infty$. This establishes (26) and therewith Wallis's formula.

6 Expectations: Mean and Variance

Our first goal is to prove a couple of results from Chapter 4 on binomial probability. The first of these is the characterisation of the expected number μ of successes as $\mu = np$, where n is the number of trials and p the probability of success on a given trial. Write $q = 1 - p$ for the probability of failure on a given trial. By definition,

---

[17] *Ibid.*, p. 92.

$$\mu = 0\binom{n}{0}p^0 q^n + 1\binom{n}{1}p^1 q^n + \ldots + n\binom{n}{n}p^n q^0$$

$$= \sum_{k=0}^{n} k\binom{n}{k}p^k q^{n-k}.$$

I offer two direct proofs that $\mu = np$, a simple one for those who know some Calculus, and a slightly more involved one that avoids the use of the Calculus.

Proof using Calculus. Let

$$f(x) = (x+q)^n = \sum_{k=0}^{n}\binom{n}{k}x^k q^{n-k}.$$

Differentiation yields

$$f'(x) = n(x+q)^{n-1} = \sum_{k=0}^{n}\binom{n}{k}k x^{k-1} q^{n-k}$$

$$= \frac{1}{x}\sum_{k=0}^{n} k\binom{n}{k}x^k q^{n-k}.$$

Setting $x = p, q = 1 - p$, we have

$$n = n \cdot 1 = n(p+q)^{n-1} = \frac{1}{p}\sum_{k=0}^{n} k\binom{n}{k}p^k q^{n-k},$$

whence

$$np = \sum_{k=0}^{n} k\binom{n}{k}p^k q^{n-k} = \mu.$$

Proof without Calculus. Write $n = m + 1, k = j + 1$ and observe

$$k\binom{n}{k} = k \cdot \frac{n!}{k!(n-k)!}$$

$$= \frac{n(n-1)!}{(k-1)!(n-k)!}$$

$$= n \cdot \frac{(n-1)!}{(k-1)!(n-1-(k-1))!}$$

$$= n\binom{n-1}{k-1}$$

$$= n\binom{m}{j}. \tag{27}$$

Thus

$$\mu = \sum_{k=0}^{n} k\binom{n}{k}p^k q^{n-k} = \sum_{k=1}^{n} k\binom{n}{k}p^k q^{n-k}$$

$$= \sum_{k=1}^{n} n \binom{m}{k-1} p^k q^{n-k}, \text{ by (27)}$$

$$= n \sum_{j=0}^{m} \binom{m}{j} p^{j+1} q^{m-j}$$

$$= pn \sum_{j=0}^{m} \binom{m}{j} p^j q^{m-j}$$

$$= pn(p+q)^m = pn \cdot 1 = pn.$$

A simpler, more conceptual proof is had as follows: Let e_n denote the expected number of successes in n trials. Note that

$$e_1 = 1 \cdot p + 0 \cdot q = 1p = p.$$

Assume, by way of induction, that $e_n = np$ and note that we expect np successes in n trials and p successes in the $(n+1)$-th trial, whence we should expect $np + p = (n+1)p$ successes in $n+1$ trials, i.e., $e_{n+1} = (n+1)p$. This proof is ultimately correct, though it may really require an additional argument (Corollary 23, below).

20 Exercise. Apply the identity

$$\binom{n+1}{k} = \binom{n}{k} + \binom{n}{k-1},$$

for $k \geq 1$, to

$$e_{n+1} = \sum_{k=0}^{n+1} k \binom{n+1}{k} p^k q^{n+1-k} = \sum_{k=1}^{n+1} k \binom{n+1}{k} p^k q^{n+1-k}$$

to show directly that $e_{n+1} = e_n + p$.

The characterisation of the variance,

$$\sigma^2 = \sum_{k=0}^{n} (k-\mu)^2 \binom{n}{k} p^k q^{n-k},$$

as $\sigma^2 = npq$ can also be given proofs using and not using, respectively, the Calculus. Both proofs computationally reduce to the result for the mean. Again, the proof by appeal to the Calculus is slightly simpler.

Proof using Calculus. For f as before, a second differentiation yields

$$f''(x) = n(n-1)(x+q)^{n-2} = \sum_{k=0}^{n} k(k-1) \binom{n}{k} x^{k-2} q^{n-k}$$

$$= \frac{1}{x^2} \sum_{k=0}^{n} k^2 \binom{n}{k} x^k q^{n-k} - \frac{1}{x^2} \sum_{k=0}^{n} k \binom{n}{k} x^k q^{n-k},$$

whence for $x = p, q = 1 - p,$

$$n(n-1) = \frac{1}{p^2} \sum_{k=0}^{n} k^2 \binom{n}{k} p^k q^{n-k} - \frac{1}{p^2} \sum_{k=0}^{n} k \binom{n}{k} x^k q^{n-k},$$

i.e.,

$$n(n-1)p^2 = \sum_{k=0}^{n} k^2 \binom{n}{k} p^k q^{n-k} - \mu,$$

$$\sum_{k=0}^{n} k^2 \binom{n}{k} p^k q^{n-k} = p^2 n^2 - p^2 n + \mu = \mu^2 - p\mu + \mu$$

$$= \mu^2 + q\mu = \mu^2 + npq. \tag{28}$$

But

$$\sigma^2 = \sum_{k=0}^{n} (k-\mu)^2 \binom{n}{k} p^n q^{n-k}$$

$$= \sum_{k=0}^{n} k^2 \binom{n}{k} p^k q^{n-k} - 2\mu \sum_{k=0}^{n} k \binom{n}{k} p^k q^{n-k} + \mu^2 \sum_{k=0}^{n} \binom{n}{k} p^k q^{n-k}$$

$$= \sum_{k=0}^{n} k^2 \binom{n}{k} p^k q^{n-k} - 2\mu^2 + \mu^2$$

$$= \sum_{k=0}^{n} k^2 \binom{n}{k} p^k q^{n-k} - \mu^2 \tag{29}$$

$$= \mu^2 + npq - \mu^2 = npq, \text{ by (28)}.$$

Proof without Calculus. The proof without appeal to the Calculus is scarcely any more revealing. First note that the proof of (29) did not require any use of the Calculus, so we assume that equation and need merely to give a Calculus-free derivation of (28):

$$\sum_{k=0}^{n} k^2 \binom{n}{k} p^k q^{n-k} = \sum_{k=1}^{n} k \cdot n \binom{n-1}{k-1} p^k q^{n-k}, \text{ by (27)}$$

$$= n \sum_{j=0}^{n-1} (j+1) \binom{n-1}{j} p^{j+1} q^{n-j-1}$$

$$= np \left[\sum_{j=0}^{n-1} j \binom{n-1}{j} p^j q^{n-1-j} + \sum_{j=0}^{n-1} \binom{n-1}{j} p^j q^{n-1-j} \right]$$

$$= np[(n-1)p + 1], \text{ by the result for means}$$

$$= (np)^2 - (np)p + np$$
$$= \mu^2 - \mu p + \mu = \mu^2 + \mu q$$
$$= \mu^2 + npq.$$

We can take a more general view of the situation and consider the problem of expected values of arbitrary random variables. Let us use the notation of Chapter 6, section 4: Our sample space is $\Omega = \{O_0, O_1, \ldots, O_{m-1}\}$, f is a distribution function on Ω, and $x_0, x_1, \ldots, x_{m-1}$ are the values of the random variable X: $X(O_i) = x_i$. The expected value of X is then defined to be

$$E(X) = x_0 f(O_0) + x_1 f(O_1) + \ldots + x_{m-1} f(O_{m-1}). \tag{30}$$

If several outcomes $O_{i_0}, O_{i_1}, \ldots, O_{i_{k-1}}$ have the same value x under X, the summands can be collected:

$$x f(O_{i_0}) + x f(O_{i_1}) + \ldots + x f(O_{i_{k-1}}) = x(f(O_{i_0}) + f(O_{i_1}) + \ldots + f(O_{i_{k-1}}))$$
$$= x P(\{O_{i_0}, O_{i_1}, \ldots, O_{i_{k-1}}\}),$$

thus yielding the following lemma.

21 Lemma. *Suppose Ω is partitioned into the union of mutually disjoint events $A_0, A_1, \ldots, A_{k-1}$, i.e.,*

$$\Omega = A_0 \cup A_1 \cup \ldots \cup A_{k-1}, \quad A_i \cap A_j = \quad \text{for } i \neq j.$$

Suppose further that X is a random variable constant on each set A_i: There are $x_0, x_1, \ldots, x_{k-1}$ such that $X(O) = x_i$ for all $O \in A_i$. Then

$$E(X) = x_0 P(A_0) + x_1 P(A_1) + \ldots + x_{k-1} P(A_{k-1}). \tag{31}$$

In particular, if $x_0, x_1, \ldots, x_{k-1}$ is a listing without repetition of all possible values of X, then (31) assumes the form

$$E(X) = x_0 P(X = x_0) + x_1 P(X = x_1) + \ldots + x_{k-1} P(X = x_{k-1}). \tag{32}$$

We turn now to matters of a bit more substance.

22 Lemma (Linearity of E). *Let X, Y be random variables and α, β real numbers. Then:*

$$E(\alpha X + \beta Y) = \alpha E(X) + \beta E(Y).$$

Proof. Letting y_i denote the value of $Y(O_i)$, we have

$$E(\alpha X + \beta Y) = \sum_{i=0}^{m-1} (\alpha x_i + \beta y_i) f(O_i)$$
$$= \alpha \sum_{i=0}^{m-1} x_i f(O_i) + \beta \sum_{i=0}^{m-1} y_i f(O_i)$$
$$= \alpha E(X) + \beta E(Y). \qquad \square$$

23 Corollary. *Let $X_0, X_1, \ldots, X_{n-1}$ be random variables and $\alpha_0, \alpha_1, \ldots, \alpha_{n-1}$ real numbers. Then:*

$$E(\alpha_0 X_0 + \alpha_1 X_1 + \ldots + \alpha_{n-1} X_{n-1}) = \alpha_0 E(X_0) + \alpha_1 E(X_1) + \ldots + \alpha_{n-1} E(X_{n-1}).$$

In particular,

$$E(X_0 + X_1 + \ldots + X_{n-1}) = E(X_0) + E(X_1) + \ldots + E(X_{n-1}).$$

The Corollary is an easy induction on $n \geq 2$.

Now suppose we repeat n Bernoulli trials with fixed probability p of success on a given trial. Ω is the set of sequences of outcomes of the n trials. Define X_i for $i = 0, 1, \ldots, n - 1$ by

$$X_i(O) = \begin{cases} 1, & \text{the } (i+1)\text{-th element of } O \text{ is a success} \\ 0, & \text{otherwise.} \end{cases}$$

We also let $S_n = X_0 + X_1 + \ldots + X_{n-1}$. Note that S_n is the number of successes of a given sequence of trials. By the Corollary,

$$E(S_n) = E(X_0) + E(X_1) + \ldots + E(X_{n-1}).$$

But $E(X_i) = 1 \cdot p + 0 \cdot (1 - p) = p$, whence

$$E(S_n) = p + p + \ldots + p = np,$$

as before.

If Y is constant, say $Y(O) = b$, we have by (32)

$$E(Y) = b \cdot P(\Omega) = b \cdot 1 = b.$$

Together with linearity, this yields the following:

24 Corollary (Change of Variables). *Let $Y = aX + b$. Then*

$$E(Y) = aE(X) + b.$$

The variance of a random variable X is the expected value of the squares of the deviations from the expected value:

$$\sigma^2 = E\big((X - E(X))^2\big) = E\big((X - \mu)^2\big),$$

where $\mu = E(X)$. A little algebra yields the following:

25 Lemma. $\sigma^2(X) = E(X^2 - \mu^2) = E(X^2) - \mu^2$.

Proof. Observe

$$\begin{aligned}
\sigma^2(X) &= E\big((X - \mu)^2\big) = E\big(X^2 - 2\mu X + \mu^2\big) \\
&= E\big(X^2\big) - 2\mu E(X) + E\big(\mu^2\big), \text{ by linearity} \\
&= E\big(X^2\big) - 2\mu^2 + \mu^2 = E\big(X^2\big) - \mu^2. \qquad \square
\end{aligned}$$

This is essentially formula (7) of Appendix A on page 361, above, derived now with a bit less notation.

26 Corollary (Change of Variables). *Let* $Y = aX + b$. *Then*

$$\sigma^2(Y) = a^2 \sigma^2(X).$$

Proof. Simply perform the calculation:

$$
\begin{aligned}
\sigma^2(Y) &= E(Y^2) - (E(Y))^2 \\
&= E((aX + b)^2) - (E(ax + b))^2 \\
&= E(a^2 X^2 + 2abX + b^2) - (aE(X) + b)^2 \\
&= a^2 E(X^2) + 2abE(X) + b^2 - (a\mu + b)^2 \\
&= a^2 E(X^2) + 2ab\mu + b^2 - (a^2\mu^2 + 2ab\mu + b^2) \\
&= a^2 (E(X^2) - \mu^2) \\
&= a^2 \sigma^2(X). \qquad\qquad\qquad\qquad\qquad\qquad\qquad\qquad \square
\end{aligned}
$$

This is just a slight abstract generalisation of our earlier switch to standard scores. If a distribution has an average μ and variance σ^2, the substitution

$$Y = \frac{X - \mu}{\sigma}$$

will have mean

$$E(Y) = \frac{1}{\sigma} E(X) - \frac{\mu}{\sigma} = \frac{\mu}{\sigma} - \frac{\mu}{\sigma} = 0$$

and variance

$$\sigma^2(Y) = \frac{1}{\sigma^2} \cdot \sigma^2(X) = \frac{\sigma^2}{\sigma^2} = 1.$$

To present an abstract calculation of variance in the binomial case to accompany that of the mean, we must first have a look at $E(XY)$. By Lemma 25, the expectation of a product is not generally the product of the expectations. This is the case, however, when the random variables are independent.

27 Lemma. *If* X, Y *are independent, then*

$$E(XY) = E(X) \cdot E(Y).$$

Proof. We can write

$$E(X) = \sum_{i=0}^{m-1} x_i P(X = x_i) \quad \text{and} \quad E(Y) = \sum_{j=0}^{n-1} y_j P(Y = y_j).$$

Then

$$
\begin{aligned}
E(X) \cdot E(Y) &= \left(\sum_{i=0}^{m-1} x_i P(X = x_i) \right) \left(\sum_{j=0}^{n-1} y_j P(Y = y_j) \right) \\
&= \sum_{i=0}^{m-1} \sum_{j=0}^{n-1} x_i y_j P(X = x_i) P(Y = y_j)
\end{aligned}
$$

$$= \sum_{i=0}^{m-1} \sum_{j=0}^{n-1} x_i y_j P(X = x_i \,\&\, Y = y_j), \text{ by independence}$$

$$= E(XY), \text{ by Lemma 21.} \qquad \square$$

28 Corollary (Bienaymé Identity). *Let $X_0, X_1, \ldots, X_{n-1}$ be independent. Then:*

$$\sigma^2 \left(\sum_{i=0}^{n-1} X_i \right) = \sum_{i=0}^{n-1} \sigma^2 (X_i).$$

Proof. Write $\mu_i = E(X_i)$ and $\mu = \sum \mu_i = E(\sum X_i)$. Observe

$$\sigma^2 = E\left(\left(\sum X_i - \sum \mu_i \right)^2 \right)$$

$$= E\left(\left(\sum X_i \right)^2 - 2 \left(\sum X_i \right) \left(\sum \mu_i \right) + \left(\sum \mu_i \right)^2 \right)$$

$$= E\left(\left(\sum X_i \right)^2 \right) - 2 \left(\sum E(X_i) \right) \left(\sum \mu_i \right) + \left(\sum \mu_i \right)^2$$

$$= E\left(\left(\sum X_i \right)^2 \right) - 2 \left(\sum \mu_i \right) \left(\sum \mu_i \right) + \left(\sum \mu_i \right)^2$$

$$= E\left(\left(\sum X_i \right)^2 \right) - \left(\sum \mu_i \right)^2$$

$$= E\left(\sum X_i^2 + \sum_{i \neq j} X_i X_j \right) - \left(\sum \mu_i^2 + \sum_{i \neq j} \mu_i \mu_j \right)$$

$$= E\left(\sum X_i^2 \right) + \sum_{i \neq j} E(X_i X_j) - \sum \mu_i^2 - \sum_{i \neq j} \mu_i \mu_j$$

$$= \sum E(X_i^2) - \sum \mu_i^2 + \sum_{i \neq j} E(X_i) E(X_j) - \sum_{i \neq j} \mu_i \mu_j$$

using independence of X_i, X_j,

$$= \sum \left(E(X_i^2) - \mu_i^2 \right) + \sum_{i \neq j} \mu_i \mu_j - \sum_{i \neq j} \mu_i \mu_j$$

$$= \sum \sigma^2(X_i). \qquad \square$$

The application to the binomial case is simple. Let n Bernoulli trials with probability p of success on a single trial be given. Define X_i for $i = 0, 1, \ldots, n-1$ as before and note

$$\sigma^2(X_i) = E(X_i^2) - E(X_i)^2$$
$$= E(X_i) - E(X_i)^2, \text{ since } X_i^2 = X_i$$
$$= p - p^2 = p(1 - p) = pq,$$

whence

$$\sigma^2(S_n) = \sigma^2\left(\sum X_i\right) = \sum \sigma^2(X_i) = \sum pq = npq,$$

as before.

We can use this sort of calculation to explain the difference between the variance of the population and the variance of the sample introduced on page 362 in Appendix A, above. To this end, imagine taking n successive measurements from a population P consisting of N individuals where N is a very large number, much larger than n. We choose these n individuals in succession, thus obtaining a sequence $I_0 I_1 \ldots I_{n-1}$ of individuals as an outcome. For $i = 0, 1, \ldots, n-1$, we let $X_i(I_0 I_1 \ldots I_{n-1})$ be the measurement associated with individual I_i. For example, we might be trying to estimate the average height of a student in a school of 2500 students by sampling 25 students and measuring their heights; X_i will then be the height of the i-th student measured. If N is sufficiently larger than n, the probabilities of obtaining various values will not change significantly from choice to choice and we can liken the process to drawing balls from an urn with replacement. In particular, the random variables $X_0, X_1, \ldots, X_{n-1}$ can be treated as if they were independent.

For each of the variables X_i, the expected value $E(X_i)$ will be the mean μ of the values of the measurements of the entire population. To see this, let m_I denote the measurement associated with individual I and let

$$\Omega = \{I_0 I_1 \ldots I_{n-1} \mid I_0, I_1, \ldots, I_{n-1} \in P\}, \quad A_I = \{I_0 I_1 \ldots I_{n-1} \in \Omega \mid I_i = I\}.$$

The A_I's partition Ω and $X_i(I_0 I_1 \ldots I_{n-1}) = m_I$ for all $I_0, I_1, \ldots, I_{n-1} \in A_I$. Thus

$$E(X_i) = \sum_{I \in P} m_I P(A_I).$$

But A_I has N^{n-1} elements and Ω has N^n elements, whence $P(A_I) = N^{n-1}/N^n = 1/N$ and

$$E(X_i) = \sum_{I \in P} m_I \cdot \frac{1}{N} = \frac{\sum m_I}{N}$$

is the average of the values m_I over P.

In a similar fashion $\sigma^2(X_i)$ is σ^2, the variance of the population P. Thus we may write $E(X_i) = \mu, \sigma^2(X_i) = \sigma^2$.

Given the measurements $m_0, m_1, \ldots, m_{n-1}$ of a sample S, we usually take their mean $(m_0 + m_1 + \ldots + m_{n-1})/n$ to estimate the mean of the population. Thus we consider

$$\overline{X} = \frac{X_0 + X_1 + \ldots + X_{n-1}}{n}$$

and evaluate

$$E(\overline{X}) = E\left(\frac{1}{n}\sum_{i=0}^{n-1} X_i\right) = \frac{1}{n}\sum_{i=0}^{n-1} E(X_i) = \frac{1}{n}\sum_{i=0}^{n-1} \mu = \frac{1}{n} \cdot n\mu = \mu.$$

For the variance, Bienaymé's Identity tells us

$$\sigma^2(S_n) = \sum \sigma^2(X_i) = n\sigma^2$$

for $S_n = X_0 + X_1 + \ldots + X_{n-1}$, whence by Change of Variables[18],

$$\sigma^2(\overline{X}) = \sigma^2\left(\frac{S_n}{n}\right) = \frac{1}{n^2}\sigma^2(S_n) = \frac{n\sigma^2}{n^2} = \frac{\sigma^2}{n}.$$

Note too that Lemma 25 yields

$$E(X_i^2) = \sigma^2 + \mu^2, \quad E(\overline{X}^2) = \frac{\sigma^2}{n} + \mu^2. \tag{33}$$

Now simply combine all of this in another calculation. For each i,

$$
\begin{aligned}
E((X_i - \overline{X})^2) &= E\left(X_i^2 - 2X_i\overline{X} + \overline{X}^2\right) \\
&= E(X_i^2) - 2E(X_i\overline{X}) + E(\overline{X}^2) \\
&= E(X_i^2) - \frac{2}{n} \cdot E\left(X_i^2 + \sum_{j\neq i} X_i X_j\right) + E(\overline{X}^2) \\
&= \frac{n-2}{n}E(X_i^2) - \frac{2}{n}\sum_{j\neq i} E(X_i)E(X_j) + E(\overline{X}^2) \\
&= \frac{n-2}{n}(\sigma^2 + \mu^2) - \frac{2}{n}\sum_{j\neq i}\mu^2 + \left(\frac{\sigma^2}{n} + \mu^2\right) \\
&= \frac{n-1}{n}\sigma^2 + \frac{n-2}{n}\mu^2 - \frac{2}{n}(n-1)\mu^2 + \mu^2 \\
&= \frac{n-1}{n}\sigma^2 + \frac{n-2-2n+2+n}{n}\mu^2 = \frac{n-1}{n}\sigma^2.
\end{aligned}
$$

Thus

$$\sum_{i=0}^{n-1} E((X_i - \overline{X})^2) = n \cdot E((X_i - \overline{X})^2) = n\frac{n-1}{n}\sigma^2 = (n-1)\sigma^2$$

and

$$E\left(\frac{1}{n-1}\sum_{i=0}^{n-1}(X_i - \overline{X})^2\right) = \frac{1}{n-1}E\left(\sum_{i=0}^{n-1}(X_i - \overline{X})^2\right) = \frac{1}{n-1}(n-1)\sigma^2 = \sigma^2.$$

Hence the mean of the sample is a good estimate of the mean of the population and the *variance of the sample*,

$$S^2 = \frac{\sum_S (m_i - \overline{m})^2}{n-1},$$

---

[18] Note: $\sigma(\overline{X}) = \sigma/\sqrt{n}$ is called the *standard error*. Note that this calculation assumes the variables $X_0, X_1, \ldots, X_{n-1}$ to be independent.

is a good estimate of the *variance of the population*,

$$\sigma^2 = \frac{\sum_P (m_i - \overline{m})^2}{N}.$$

The estimate one might expect to use here

$$\sigma_S^2 = \frac{\sum_S (m_i - \overline{m})^2}{n},$$

is too small by a factor of $(n-1)/n$ and is considered a *biased estimator*, while S^2 is an *unbiased estimator*. If n is large, the factor is near 1 and either estimator is usable, but if n is small, the practice is to use S^2 as the variance of a sample to estimate the variance and with it the standard deviation of a population. Thus one finds both estimates on the *TI-83 Plus* when one runs 1-Var Stats.

7 Bošković, Gauss, and Hagen Revisited

The present section collects a few footnotes to Chapter 6, section 5, on Gauss, the Error Function, and the Method of Least Squares. It begins with the Calculus-free discussion of Bošković's method, then presents Gauss's and Hagen's derivations of the error curve. Both of these assume the Calculus and are not meant for all readers. And Hagen's derivation uses infinitesimals and infinitely large numbers in a non-rigorous, not totally convincing way. Finally, I follow this up with a loosely related justification of the use of the arithmetic mean for linear measurements due to the astronomer Schiaparelli.

Bošković's Method

Recall the central problem of Bošković's estimation method. Given a sequence of observed pairs of observations of x- and y-values, say $\langle x_0, y_0 \rangle, \langle x_1, y_1 \rangle, \ldots,$ $\langle x_{n-1}, y_{n-1} \rangle$, one wishes to find a, b such that

$$\sum_{i=0}^{n-1} (y_i - a - bx_i) = 0$$

$$\sum_{i=0}^{n-1} |y_i - a - bx_i| \text{ is a minimum.}$$

The first condition affirms the so-called centre of gravity $\langle \overline{x}, \overline{y} \rangle$ of the points $\langle x_i, y_i \rangle$ to lie on the line $y = a + bx$, and converts the second problem to that of minimising

$$\sum_{i=0}^{n-1} |(y_i - \overline{y}) - b(x_i - \overline{x})|.$$

Writing a_i for $y_i - \bar{y}$, b_i for $x_i - \bar{x}$, and the more traditional variable x for b, we find ourselves interested in minimising the expression

$$y = \sum_{i=0}^{n-1} |a_i - b_i x|. \tag{34}$$

If we separate out those terms for which $b_i = 0$, and write

$$y = \sum_{b_i \neq 0} |a_i - b_i x| + \sum_{b_i = 0} |a_i|,$$

we see that y is minimised when

$$\sum_{b_i \neq 0} |a_i - b_i x|$$

is minimised and we need only consider the sums (34) for which no b_i is 0.

I remarked in Chapter 6, but did not make use of the fact, that (34) is minimised at one of the points

$$c_i = \frac{a_i}{b_i},$$

preferring instead to use the familiar technique of finding a minimum via graphing. The fact that some c_i minimised the expression in question was used only as an aid in choosing the window. A direct, non-graphical approach using lists proceeds as follows. As we will be using six lists, we could get by with $L_1, \ldots,$ L_6, but it might be better to use some more mnemonic names. First, create two lists ∟X and ∟Y in which to store the x_i- and y_i-values, respectively. Next store the deviations in two new lists:

∟X−mean(∟X)→∟B
∟Y−mean(∟Y)→∟A .

Then create the list of values $c_i = a_i/b_i$

∟A/∟B→∟C ,

and for the values $Y_1(c_i)$

Y_1(∟C)→∟YC ,

where one has already entered

Y_1=sum(abs(∟A−X∗∟B))

in the equation editor. Using the data stored in L_1 in our discussion of Bošković's method in Chapter 6, section 4and L_2 in ∟X and ∟Y, respectively, inspection reveals a minimum value of .11748 in ∟YC shared by the entries .28453 and .28382 in ∟C. One can see this more clearly by sorting. Now, we will want to use these lists again later, so let us first copy the lists into some new ones and sort those:

⌊C→⌊C2
⌊YC→⌊YC2 .

Then we can sort these new lists:

SortA(⌊C2,⌊YC2)

will sort the c_i's in ascending order and carry along the sums $Y_1(c_i)$ as one can see by entering

SetUpEditor ⌊C2,⌊YC2 .

And

SortA(⌊YC2,⌊C2)

sorts by values $Y_1(c_i)$, the minimum occurring first in the new ⌊YC2, the corresponding c_i being the first element of the new ⌊C2— or, in this case, the first two elements: any x between .28382 and .28453 will minimise (34).

Examining the two lists after the first sort reveals that the function decreases to a minimum and increases thereafter, as we saw in graphing the function in Chapter 6.

So we see how we can easily implement Bošković's method of estimation on the calculator, and how we can get a feel for how the method works through further exploration. But the question remains: does it always work? That is, how do we know that the minimum of the expression (34) exists and occurs at one or more of the ratios c_i?

Let $y = f(x)$ be the function defined by (34). Assume the c_i's given in increasing order:

$$c_0 \leq c_1 \leq \ldots \leq c_{n-1}$$

and assume each b_i positive (if b_i is negative, replace a_i, b_i by $-a_i, -b_i$, respectively). Now, for

$$x \leq c_0, \qquad \text{we have } f(x) = \sum_{i=0}^{n-1}(a_i - b_i x)$$

$$c_j \leq x \leq c_{j+1}, \text{ we have } f(x) = \sum_{i=0}^{j-1}(b_i x - a_i) + \sum_{i=j}^{n-1}(a_i - b_i x)$$

$$c_{n-1} \leq x, \qquad \text{we have } f(x) = \sum_{i=0}^{n-1}(b_i x - a_i).$$

Now on each of the intervals,

$$(-\infty, c_0), \ (c_0, c_1), \ \ldots, \ (c_{n-2}, c_{n-1}), \ (c_{n-1}, \infty),$$

f is linear, and the slope is either positive, negative, or 0. If it is either positive or negative, the minimum cannot occur in the open interval; while if the slope is 0, the function is constant on the interval and assumes the same value at both endpoints. Thus the minimum can only occur at one of the points $c_0, c_1, \ldots, c_{n-1}$.

29 Exercise. Define the linear functions $f_i(x)$ by

$$f_0(x) = \sum_{i=0}^{n-1}(a_i - b_i x)$$

$$f_j(x) = \sum_{i=0}^{j-1}(b_i x - a_i) + \sum_{i=j}^{n-1}(a_i - b_i x)$$

$$f_n(x) = \sum_{i=0}^{n-1}(b_i x - a_i).$$

Let m_i be the slope of the line $y = f_i(x)$ and show

$$m_0 \le m_1 \le \ldots \le m_n.$$

Bošković's own approach was also elementary, but more geometrically motivated, a bit more involved, but, where calculation is done by hand, less computationally intensive. It still requires n divisions, plus an extra division by 2, plus some additions, but not the multiplications involved in calculating $Y_1(1)$, $Y_1(2)$, ...

One begins by observing that the point $\langle \overline{x}, \overline{y} \rangle$ lies at the point of intersection of all the lines

$$l_i : y = \overline{y} + c_i(x - \overline{x}).$$

Assume we have listed the pairs $\langle x_i, y_i \rangle$ in such a way that

$$c_0 \le c_1 \le \ldots \le c_{n-1},$$

and consider the configuration of lines as in *Figure 7*.

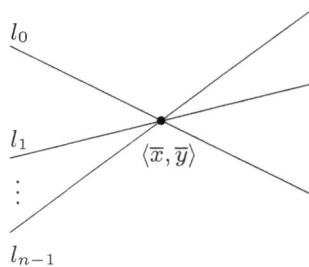

Fig. 7.

We assume no $x_i = \overline{x}$, so none of these lines is vertical and imagine an initially vertical line passing through $\langle \overline{x}, \overline{y} \rangle$ rotating clockwise around that point. In doing so the equation of the line takes on the form

$$l_c : y = \overline{y} + c(x - \overline{x}),$$

as c decreases from $+\infty$ to $-\infty$. As it does so, the vertical distances between points on the line l_c and a line l_i decrease until l_c crosses the line l_i and increase thereafter. So long as $c > c_{n-1}$ all these distances are decreasing, and so long as $c < c_0$, they are all increasing.

We look more closely at these differences.

$$l_c(x_i) - y_i = l_c(x_i) - l_i(x_i)$$
$$= \bar{y} + c(x_i - \bar{x}) - (\bar{y} + c_i(x_i - \bar{x}))$$
$$= (c - c_i)(x_i - \bar{x}),$$

whence the total of the absolute distances is

$$v_c = \sum_{i=0}^{n-1} |l_c(x_i) - y_i| = \sum_{i=0}^{n-1} \left(|c - c_i| \cdot |x_i - \bar{x}| \right).$$

For $c_j \leq c \leq c_{j+1}$,

$$v_c = \sum_{i=0}^{j} (c - c_i)|x_i - \bar{x}| + \sum_{i=j+1}^{n-1} (c_i - c)|x_i - \bar{x}|$$
$$= \sum_{i=0}^{j} (c - c_i)|x_i - \bar{x}| - \sum_{i=j+1}^{n-1} (c - c_i)|x_i - \bar{x}|.$$

Suppose now, l_c rotating clockwise, c moves to c' in the same interval: $c_j \leq c' < c \leq c_{j+1}$. Then

$$v_c - v_{c'} = \sum_{i=0}^{j} (c - c_i - (c' - c_i))|x_i - \bar{x}| - \sum_{i=j+1}^{n-1} (c - c_i - (c' - c_i))|x_i - \bar{x}|$$
$$= \sum_{i=0}^{j} (c - c')|x_i - \bar{x}| - \sum_{i=j+1}^{n-1} (c - c')|x_i - \bar{x}|$$
$$= (c - c') \left(\sum_{i=0}^{j} |x_i - \bar{x}| - \sum_{i=j+1}^{n-1} |x_i - \bar{x}| \right),$$

i.e.,

$$v_c > v_{c'} \text{ if } \sum_{i=j+1}^{n-1} |x_i - \bar{x}| < \sum_{i=0}^{j} |x_i - \bar{x}|$$

$$v_c = v_{c'} \text{ if } \sum_{i=j+1}^{n-1} |x_i - \bar{x}| = \sum_{i=0}^{j} |x_i - \bar{x}|$$

$$v_c < v_{c'} \text{ if } \sum_{i=j+1}^{n-1} |x_i - \bar{x}| > \sum_{i=0}^{j} |x_i - \bar{x}|.$$

So v_c ceases its decrease when

$$\sum_{i=j+1}^{n-1} |x_i - \overline{x}| \geq \sum_{i=0}^{j} |x_i - \overline{x}|,$$

i.e., when

$$\sum_{i=j+1}^{n-1} |x_i - \overline{x}| \geq \frac{1}{2}\sum_{i=0}^{n-1} |x_i - \overline{x}|. \tag{35}$$

We can now simply implement Bošković's algorithm on the calculator. Assume ∟X, ∟Y, ∟A, ∟B, and ∟C are as before. [*Note.* ∟YC is not needed.] Eliminate all entries for which $x_i = \overline{x}$, at which ∟B is 0. This never occurs for the data of our ongoing example. Copy ∟B (the values of $x_i - \overline{x}$) and ∟C into new lists, taking absolute values in the first case:

    abs(∟B)→∟DEV
    ∟C→∟C2 .

Sort the lists in descending order using ∟C2 as the key:

    SortD(∟C2,∟DEV) .

Find the cumulative sums of the absolute deviations:

    cumSum(∟DEV)→∟SUMS

and test where (35) is true:

    ∟SUMS≥(∟SUMS(dim(∟SUMS))/2)→∟TEST .

The first 1 in ∟TEST gives the index i of the c_i minimising the sum of the absolute deviations. If one wants to be clever:

    sum(1−∟TEST)+1→I
    ∟C2(I)→X

solves the minimisation problem.

Gauss's Derivation of the Error Distribution

Gauss's derivation of the error distribution makes essential use of first-year Calculus, whence the following is only for readers who have taken such a course.

Let $\varphi(x)$ be the error function. It is assumed that

i. φ assumes its maximum value at $x = 0$ and decreases monotonically to 0 as $|x|$ gets large without bound;

ii. φ is even: $\varphi(-x) = \varphi(x)$; and

iii. $\varphi(x)$ has a continuous derivative.

If $x_0, x_1, \ldots, x_{n-1}$ are values that can possibly turn up in observations, then

$$\Omega(\theta) = \varphi(x_0 - \theta)\varphi(x_1 - \theta)\cdots\varphi(x_{n-1} - \theta)$$

represents the distribution function for the conditional probability that θ is the true value given that the values $x_0, x_1, \ldots, x_{n-1}$ have been observed in n independent observations. The value of θ maximising Ω satisfies

$$\frac{d\Omega(\theta)}{d\theta} = 0,$$

i.e., by logarithmic differentiation,

$$\frac{d\ln\Omega(\theta)}{d\theta} = \frac{\Omega'(\theta)}{\Omega(\theta)} = \sum_{i-0}^{n-1} \frac{-\varphi'(x_i - \theta)}{\varphi(x_i - \theta)} = 0$$

Gauss makes the assumption that $\Omega(\theta)$ is maximised for the mean $\overline{x} = (x_0 + \ldots + x_{n-1})/n$ of the observed values, i.e.,

$$\sum_{i=0}^{n-1} \frac{\varphi'(x_i - \overline{x})}{\varphi(x_i - \overline{x})} = 0. \tag{36}$$

This is assumed to hold for all possible sets of observations $x_0, x_1, \ldots, x_{n-1}$. Gauss now makes a careful choice of values for such observations. Let x_0 be arbitrary and, for some real number r, let $x_1 = x_2 = \ldots = x_{n-1} = x_0 - r$. Thus

$$\overline{x} = \frac{1}{n}\sum_{i=0}^{n-1} x_i = \frac{1}{n}\left(x_0 + \sum_{i=1}^{n-1}(x_0 - r)\right) = x_0 + \frac{n-1}{n}(-r)$$

$$x_0 - \overline{x} = \frac{n-1}{n}r$$

$$x_i - \overline{x} = -r + \frac{n-1}{n}r = -\frac{r}{n}, \quad \text{for } i \neq 0.$$

By (36)

$$\frac{\varphi'\left(\dfrac{n-1}{n}r\right)}{\varphi\left(\dfrac{n-1}{n}r\right)} + \sum_{i=1}^{n-1}\frac{\varphi'\left(-\dfrac{r}{n}\right)}{\varphi\left(-\dfrac{r}{n}\right)} = 0,$$

i.e.,

$$\frac{\varphi'\left(\dfrac{n-1}{n}r\right)}{\varphi\left(\dfrac{n-1}{n}r\right)} = -(n-1)\frac{\varphi'\left(-\dfrac{r}{n}\right)}{\varphi\left(-\dfrac{r}{n}\right)}. \tag{37}$$

To simplify notation, let

$$\psi(x) = \frac{\varphi'(x)}{\varphi(x)}, \quad R = \frac{r}{n}.$$

Thus (37) reads

$$\psi((n-1)R) = -(n-1)\psi(-R). \tag{38}$$

Because φ is even, φ' is odd, whence so is ψ: $\psi(-x) = -\psi(x)$. Thus (38) becomes

$$\psi((n-1)R) = -(n-1)(-\psi(R)) = (n-1)\psi(R).$$

Now r was arbitrary, whence R is arbitrary. And n is an arbitrary integer greater than 1. Hence, for all x and all positive integers k, we have

$$\psi(kx) = k\psi(x).$$

Let p, q be any positive integers and observe:

$$\psi\left(\frac{p}{q}\right) = p\psi\left(\frac{1}{q}\right) = p \cdot \frac{1}{q}\psi(1), \quad \text{since } \psi(1) = q\,\psi\left(\frac{1}{q}\right).$$

Thus $\psi(x) = x\psi(1)$ for rational x. The same holds for general x by continuity.[19]
    Thus there is a constant C_1 such that

$$\frac{\varphi'(x)}{\varphi(x)} = \psi(x) = C_1 x.$$

Integration yields

$$\ln\varphi(x) = \frac{C_1}{2}x^2 + C_2,$$

whence

$$\varphi(x) = e^{C_2} \cdot e^{\frac{1}{2}C_1 x^2}.$$

Because this decreases as x gets larger, Gauss determines one can write

$$\frac{1}{2}C_1 = -h^2$$

for some h, and, because the integral of φ over the real line must equal 1, he calculates

$$e^{C_2} = \frac{h}{\sqrt{\pi}}.$$

Thus, Gauss concludes that, for some h,

$$\varphi(x) = \frac{h}{\sqrt{\pi}}e^{-h^2 x^2}.$$

Hypothesis of Elementary Errors

The derivation of the normal error curve given in Gotthilf Hagen's probability textbook for civil engineers included a simplified proof of de Moivre's Theorem and, according to Anders Hald[20], the proof soon became popular. The

---

[19] Note that the continuity of $\psi = \varphi'/\varphi$ requires φ' to be continuous.
[20] Anders Hald, *A History of Parametric Statistical Inference from Bernoulli to Fisher, 1713 - 1935*, Springer-Verlag, New York, 2007, p. 107.

popularity of the proof alone offers justification on historical grounds for its inclusion here. However, the detailed proof of de Moivre's Theorem, as outlined in Chapter 6, was a bit heavy-going and I think the heuristic approach, however inadequate in terms of rigour, would also make it welcome.

Hagen derives the normal curve for errors by taking the Hypothesis of Elementary Errors literally. Recall his statement of the Hypothesis:

> The error in the result of a measurement is the algebraic sum of an infinitely large number of elementary errors which are all equally large, and each of which can be positive or negative with equal ease.

Now forget for the moment the "infinitely large number" and think of some large number n of elementary errors, each of which contributes either positively or negatively to the total error. If the common size of the elementary errors is ϵ and it can be positive or negative, the overall error is

$$\varepsilon = \sum_{i=0}^{n-1}(-1)^{\delta_i}\epsilon = \left(\sum_{i=0}^{n-1}(-1)^{\delta_i}\right)\epsilon,$$

where δ_i is 0 or 1 according as the elementary error i is positive or negative.

What are the possible values of ε? Well, if all the elementary errors are positive,

$$\varepsilon = \left(\sum_{i=0}^{n-1}1\right)\epsilon = n\epsilon\,;$$

if $n-1$ are positive and 1 is negative, then

$$\varepsilon = \left(\sum_{i=0}^{n-2}1 + -1\right)\epsilon = (n - 1 - 1)\epsilon = (n-2)\epsilon\,;$$

if $n-2$ are positive and 2 are negative, then

$$\varepsilon = \left(\sum_{i=0}^{n-3}1 + -1 + -1\right)\epsilon = (n - 2 - 1 - 1)\epsilon = (n-4)\epsilon\,;$$

and so on. Thus the possible values of the coefficients of ϵ are

$$-n, -n+2, -n+4, \ldots, n-4, n-2, n,$$

and, if we follow Hagen's lead in assuming n to be even, these values are

$$-n, -n+2, \ldots, -2, 0, 2, \ldots, n-2, n,$$

with a central term 0.

The next question is: what are the probabilities of the possible accumulated errors $\varepsilon = 2k\epsilon$ for $2k = -n, -n+2, \ldots, n-2, n$? Fix k, let a be the number of positive elementary errors among the n elementary errors, and b the number of negative such errors. We can work out using simple algebra that

k	$-n$	$-n+2$	...	-2	0	2	...	$n-2$	n
a	0	1	...	$\frac{n-2}{2}$	$\frac{n}{2}$	$\frac{n+2}{2}$	...	$n-1$	n

Table 1

$$a = \frac{n+2k}{2}, \qquad b = \frac{n-2k}{2},$$

or simply refer to *Table 1*, below. Now the probability of a positives in n elementary errors, with a fixed probability of $\frac{1}{2}$ of an elementary error being positive ("positive or negative [occur] with equal ease") is clearly a binomial probability:

$$\binom{n}{a}\left(\frac{1}{2}\right)^a \left(\frac{1}{2}\right)^{n-a} = \binom{n}{a}\frac{1}{2^n}.$$

The distribution is binomial and we can appeal to de Moivre's Theorem in this special case to conclude that the error distribution, if n is allowed to become infinite, is normal, and if n is merely a finite, but large, number, is approximately normal. For, de Moivre's Theorem says just that: in the limit the binomial distribution becomes the normal distribution.

Now Hagen began his discussion of the Hypothesis of Elementary Errors on page 27 of his book. He states the hypothesis formally on page 34, and only then begins his discussion of binomial probability on that page. He hasn't got to de Moivre's Theorem yet and derives it in §§14 - 15 with a simple argument, one that hardly amounts to a proof, but which is suggestive.

We now consider the binomial distribution graphed as follows: Write $n = 2m$ and plot the binomial coefficients above the shifted interval, with $_nC_k$ placed above $x = -m + k$. Consider the polygonal line connecting these points. [To visualise this, choose an even number and store it in the variable N on your calculator, then successively enter

N/2→M
seq(X,X,−M,M)→L₁
binompdf(N,1/2)→L₂ .

Then press the STAT PLOT button, choose a plot, turn it on, choose the second (polygonal line) icon, enter L₁ for Xlist, L₂ for Ylist, and the small square for the Mark. Finally, graph the data using ZoomStat.[21]]

Hagen next derives two formulæ involving binomial coefficients that we have seen before: For $k \geq 0$,

$$\binom{2m}{m+k+1} = \frac{(2m)!}{(m+k+1)!(m-k-1)!}$$
$$= \frac{m-k}{m+k+1} \cdot \frac{(2m)!}{(m+k)!(m-k)!}$$

---

[21] I would have suggested using one of the programs BINGRAPH or BELL1 of Chapter 4, but I would assume the reader has had to delete the occasional program from the calculator to clear up memory and these programs are eminently deleteable.

$$= \frac{m-k}{m+k+1}\binom{2m}{m+k}, \tag{39}$$

whence

$$\binom{2m}{m+k} = \frac{m}{m+1} \cdot \frac{m-1}{m+2} \cdots \frac{m-k-1}{m+k}\binom{2m}{m}, \tag{40}$$

for $k > 0$. Letting

$$y_k = \binom{2m}{m+k},$$

for $k \geq 0$, (39), (40) read

$$y_{k+1} = \frac{m-k}{m+k+1}y_k \tag{41}$$

$$y_k = \frac{m}{m+1} \cdot \frac{m-1}{m+2} \cdots \frac{m-k-1}{m+k}y_0, \tag{42}$$

respectively. Now the slope of the segment connecting the points $\langle k, y_k \rangle$ and $\langle k+1, y_{k+1} \rangle$ is

$$\frac{y_{k+1} - y_k}{k+1-k} = y_{k+1} - y_k = \left(\frac{m-k}{m+k+1} - 1\right)y_k$$
$$= \frac{m-k-m-k-1}{m+k+1}y_k$$
$$= -\frac{2k+1}{m+k+1}y_k. \tag{43}$$

Hagen now applies some infinitary sleight of hand.

Coordinating our notation, let us write $x_{-m}, x_{-m+1}, \ldots, x_m$ for the abscissas $-m, -m+1, \ldots, m-1, m$ of our plotted points. Hagen now multiplies the x_i's and y_i's by an infinitesimal scaling factor dx, thus obtaining

$$Y_k = y_k dx, \quad X_k = x_k dx = k dx$$
$$\Delta Y_k = (y_{k+1} - y_k)dx, \quad \Delta X_k = (x_{k+1} - x_k)dx = dx,$$

and

$$\frac{\Delta Y_k}{\Delta X_k} = \frac{(y_{k+1} - y_k)dx}{(x_{k+1} - x_k)dx} = \frac{y_{k+1} - y_k}{x_{k+1} - x_k}$$
$$= -\frac{2k+1}{m+k+1}y_k, \text{ by (43).} \tag{44}$$

Simultaneously, he lets $n = 2m$ and m be infinite.

Hagen doesn't say explicitly how n and dx relate. It is convenient to assume $\sqrt{n} \cdot dx$ is infinite. One way to achieve this is to take $dx = 1/\sqrt[4]{n}$ since then

$$\sqrt{n} \cdot dx = \sqrt{n} \cdot \frac{1}{\sqrt[4]{n}} = \sqrt[4]{n}$$

is infinite. Now, every real number x lies between kdx and $(k+1)dx$ for some k with $|k| < \sqrt{n}$. For such k, he argues we can replace

$$\frac{2k+1}{m+k+1} \quad \text{by} \quad \frac{2k}{m}, \quad \text{i.e., by} \quad \frac{2x_k}{m}.$$

Thus

$$\frac{\Delta y_k}{\Delta x_k} = -\frac{2x_k}{m} y_k, \tag{45}$$

i.e.,

$$\frac{dy}{dx} = -\frac{2x}{m} y$$

$$\frac{dy}{y} = -\frac{2x}{m} dx.$$

Integration yields

$$\ln y = -\frac{2x^2}{2m} + C_0 = -\frac{x^2}{m} + C_0,$$

for some constant C_0, whence

$$y = e^{-x^2/m + C_0} = Ce^{-x^2/m},$$

for some constant C. For $x = 0$, we have $y(0) = C$. But

$$y(0) = y_0 = \binom{2m}{m},$$

and Hagen applies Wallis's formula (*Cf.* pp. 231 - 231.) to conclude

$$C = y_0 = \sqrt{\frac{2}{\pi n}} \cdot 2^n.$$

Now in the above we dealt with the binomial coefficients and not the probabilities, i.e., we did not divide by 2^n. If we do so now, we get the distribution function

$$\varphi(x) = \sqrt{\frac{2}{\pi n}} e^{-x^2/m}. \tag{46}$$

We are dealing with binomial probabilities, for which $\sigma = \sqrt{npq} = \sqrt{n/4} = \sqrt{n}/2$, i.e., $\sqrt{n} = 2\sigma, m = n/2 = 4\sigma^2/2 = 2\sigma^2$, turning (46) into

$$\varphi(x) = \frac{1}{\sigma\sqrt{2\pi}} e^{-\frac{1}{2}x^2/\sigma^2},$$

as was to be shown. Hagen is a little less detailed than this, specifically in the step from (44) to (45) and I cannot say I entirely trust it or my reconstruction thereof. I do know that Hagen's proof appeared in subsequent texts. For example, the 1879 edition of Czuber's translation of Meyer's text[22] presents first

---

[22] Anton Meyer and Emanuel Czuber, *Vorlesungen über Wahrscheinlichkeitsrechnung*, Verlag von B.G. Teubner, Leipzig, 1879; here: pp. 245 - 247.

Hagen's and then Gauss's derivation of the error curve, but this treatment of Hagen's proof is also sketchy.

The story of the Hypothesis of Elementary Errors does not end with Hagen's proof and its subsequent expositions. Hagen's teacher Friedrich Wilhelm Bessel derived the error function in the more general case in which the probability distributions of the errors varied in successive trials.[23] And a couple of decades later, George Biddell Airy (1801 - 1882) made a contribution to perspective in his book on the Method of Least Squares[24]:

> In formulating his Central Limit Theorem Laplace envisioned a large number of observations being made in sequence under the condition that the errors arising from these repeated observations are all governed by the same probability law. However, the hypothesis of Laplace's Central Limit Theorem could also be interpreted from the point of view of the hypothesis of elementary errors and viewed as asserting that an error of observation is the sum of numerous very small elementary errors which arise from independent sources, where all elementary errors, regardless of their sources, are governed by the same law of error $f(x)$ which satisfies $f(x) = f(-x)$. The mathematical structure of Laplace's Central Limit Theorem does not depend on whether the hypothesis is viewed from the point of view adopted by Laplace or from the point of view of the hypothesis of elementary errors. It is in G.B. Airy's derivation of the normal law of error that we first find this marriage of Laplace's Central Limit Theorem and the hypothesis of elementary errors. Airy interpreted Laplace's Central Limit Theorem from the hypothesis of elementary errors point of view and a number of writers subsequently attributed this point of view to Laplace.[25]

Schiaparelli on the Arithmetic Mean

Two reasons Gauss's star shines so brightly next to Laplace's in the probabilistic heavens are that the problem of errors was one of immediate practical importance in a way the more general Central Limit Theorem was not and that Gauss's approach via functional equations was easier to emulate than Laplace's more advanced approach. The error curve can be derived by relatively elementary means and a vast literature developed. I said earlier (page 262, above)

---

[23] Friedrich Wilhelm Bessel, "Untersuchungen über die Wahrscheinlichkeit der Beobachtungsfehler", *Astronomische Nachrichten* 15 (1838), pp. 368 - 404.

[24] George Biddell Airy, *On the Algebraical and Numerical Theory of Errors of Observations and the Combination of Observations*, Macmillan and Co., Ltd., Cambridge and London, 1861. The second edition of the book (1875) is available online at Google Books. Airy gives a nicely motivated sketch of Laplace's proof for the reader familiar with Fourier series.

[25] William J. Adams, *The Life and Times of the Central Limit Theorem*, 2nd edition, American Mathematical Society, Providence (RI), 2009, p. 42. The first edition was published by Kaedmon Pub. Co., New York, 1974.

Sir George Airy
Pioneer of the Greenwich Meridian

"As to the need of a Prime Meridian, no practical man wants such a thing."

England's seventh Astronomer Royal, for the years from 1835 to 1881, George Biddell Airy suffers the misfortune of being better known for his failures than his successes. His oppositions included the Prime Meridian, Babbage's calculating engine ("On Sept. 15th Mr. Goulburn, Chancellor of the Exchequer, asked my opinion on the utility of Babbage's calculating machine, and the propriety of expending further sums of money on it. I replied, entering fully into the matter, and giving my opinion that it was worthless." (*Autobiography of George Biddell Airy, K.C.B.,...*, Cambridge University Press, Cambridge, 1896, p. 152.)), and, most famously, his refusal to depart from observatory routine to search for the planet Neptune, the existence and position of which had been predicted by John Couch Adams (1819 - 1892). He was the archetypal scientific bureaucrat, beautifully described by America's premier female astronomer Maria Mitchell (1818 - 1889) in her memoirs (Phebe Mitchell Kendall, ed., *Maria Mitchell; Life, Letters, and Journals*, Lee and Shepard Publishers, Boston, 1896, pp. 96 - 97.):

> He is naturally a despot and his position increases this tendency.
> ... Near the throne of the astronomical autocrat is another proof of his system, in a case of portfolios. These contain the daily bills, letters, and papers, as they come in and are answered in order. When a portfolio is full, the papers are removed and sewed together. Each year's accumulation is bound, and the bound volumes of Mr. Airy's time nearly cover one side of his private room.

Despite his opinion of the Prime Meridian, in 1884 at a conference in Washington, D.C., it was officially decided, in honour of the excellent transit instrument Airy had installed at Greenwich Observatory, to position the Prime Meridian there. In 1984, England and Tonga both issued stamps celebrating the anniversary of this decision. The four English stamps featured the Prime Meridian ($0°$ longitude) on a globe and a map, Greenwich Observatory, and Airy's transit instrument, respectively. The stamp of Tonga, pictured here, includes Airy's portrait and a map showing the International Date Line ($180°$ longitude) as it zigzags to avoid slicing Tonga and other Pacific island nations in half. Another portrait of Airy appears on a stamp of Nicaragua issued in 1994 in a set honouring famous scientists.

that the Central Limit Theorem was subject to intense investigation for more than a century after its appearance. More precisely, I should have emphasised that for the first several decades the main focus of such investigation was the theory of errors. Every author had his own assumptions from which he derived the error curve by some argument using only results of elementary Calculus.[26] Among these investigations one also finds justifications of the use of the arith-

---

[26] *Cf.* Emanuel Czuber, *Theorie der Beobachtungsfehler*, Verlag von B.G. Teubner, Leipzig, 1891, for an extensive collection of such examples. Or, peek ahead to

metical mean to combine the discordant results of multiple observations. Two such justifications were supplied by the astronomers Johann Franz Encke and Giovanni Schiaparelli (1835 - 1910). I am rather taken with the latter's proof and present it here.[27]

ITALIA   € 0,65

GIOVANNI VIRGINIO SCHIAPARELLI 1835 - 1910

I.P.Z.S. S.p.A · ROMA · 2010   S. ISOLA

The astronomer Giovanni Schiaparelli is best known for his map of Mars on which he depicted lines he called *canali*, which translates most properly as "channels" but more popularly as "canals". Philatelic depictions of Schiaparelli, of which there are several, always show Mars and its canals and usually also display his portrait. The earliest of these stamps was issued by Hungary in 1974 as the high value of a set celebrating results of research on Mars. Pictured here is a stamp issued by his native Italy in 2010 on the occasion of the 100th anniversary of his death.

Schiaparelli makes 4 assumptions about the choice x of a proper approximation to the "true value" of a quantity x approximated by n independent measurements yielding values $a_0, \ldots, a_{n-1}$. First is the purely mathematical assumption that there is an analytic function $F(x_0, \ldots, x_{n-1})$ determining x:

$$x = F(a_0, \ldots, a_{n-1}).$$

Following this are three assumptions about the nature of the measurements:

1. The value of x must be independent of the unit of measurement; i.e., a change of scale of the a_i's merely changes the scale of x. Mathematically, this reads

$$F(ta_0, ta_1, \ldots, ta_{n-1}) = tF(a_0, a_1, \ldots, a_{n-1}). \tag{47}$$

2. The value of x must be independent of the position of the measuring instrument; i.e., a translation of the ruler merely translates the measure:

$$F(a_0 + t, a_1 + t, \ldots, a_{n-1} + t) = F(a_0, a_1, \ldots, a_{n-1}) + t. \tag{48}$$

3. A small change in the value of a given measurement has the same effect as the identical change in any of the other measurements:

Maxwell's derivation of the error curve in Physics in the next section for a single example.

[27] In this, I follow the exposition of Czuber (*ibid.*, pp. 31 - 33).

$$F(a_0, \ldots, a_{i-1}, a_i + t, a_{i+1}, \ldots, a_{n-1}) = F(a_0, \ldots, a_{j-1}, a_j + t, a_{j+1}, \ldots, a_{n-1}).$$
$$(49)$$

Under these assumptions, Schiaparelli concludes

$$F(a_0, \ldots, a_{n-1}) = \frac{a_0 + \ldots + a_{n-1}}{n}.$$

The argument is short and sweet. First, one differentiates (47) with respect to t. The left side is

$$\frac{\partial F}{\partial t}(ta_0, ta_1, \ldots, ta_{n-1}) = \sum_{i=0}^{n-1} \frac{\partial F}{\partial x_i}(ta_0, ta_1, \ldots, ta_{n-1}) \cdot \frac{\partial(ta_i)}{\partial t}$$

$$= \sum_{i=0}^{n-1} \frac{\partial F}{\partial x_i}(ta_0, ta_1, \ldots, ta_{n-1})a_i,$$

while the right side is

$$\frac{\partial}{\partial t}(tF(a_0, a_1, \ldots, a_{n-1})) = F(a_0, a_1, \ldots, a_{n-1}) = x.$$

Thus, letting $t = 1$,

$$\sum_{i=0}^{n-1} \frac{\partial F}{\partial x_i}(a_0, a_1, \ldots, a_{n-1})a_i = x,$$

which we may write

$$\frac{\partial F}{\partial a_0}a_0 + \frac{\partial F}{\partial a_1}a_1 + \ldots + \frac{\partial F}{\partial a_{n-1}}a_{n-1} = x. \tag{50}$$

The second step, as reported by Czuber, is to expand $F(a_0 + t, a_1 + t, \ldots, a_{n-1} + t)$ into a power series in t:

$$F(a_0 + t, \ldots, a_{n-1} + t) = F(a_0, \ldots, a_{n-1}) + \left(\frac{\partial F}{\partial a_0} + \ldots + \frac{\partial F}{\partial a_{n-1}}\right)t + \ldots,$$

i.e.,

$$F(a_0, \ldots, a_{n-1}) + t = F(a_0, \ldots, a_{n-1}) + \left(\frac{\partial F}{\partial a_0} + \ldots + \frac{\partial F}{\partial a_{n-1}}\right)t + \ldots$$

Equating coefficients of the linear terms yields

$$1 = \frac{\partial F}{\partial a_0} + \ldots + \frac{\partial F}{\partial a_{n-1}}. \tag{51}$$

One can, in fact, weaken the assumption on the analyticity of F to differentiability by differentiating (48) to conclude

$$1 = \frac{\partial}{\partial t}(F(a_0, \ldots, a_{n-1}) + t) = \sum_{i=0}^{n-1} \frac{\partial F}{\partial x_i}(a_0 + t, \ldots, a_{n-1} + t)) \cdot \frac{\partial(a_i + t)}{\partial t}$$

$$= \sum_{i=0}^{n-1} \frac{\partial F}{\partial x_i}(a_0 + t, \ldots, a_{n-1} + t),$$

whence, for $t = 0$,

$$1 = \sum_{i=0}^{n-1} \frac{\partial F}{\partial x_i}(a_0, \ldots, a_{n-1}) = \frac{\partial F}{\partial a_0} + \ldots + \frac{\partial F}{\partial a_{n-1}}.$$

The third step is to apply (49):

$$\frac{\partial F}{\partial a_i} = \lim_{t \to 0} \frac{F(a_0, \ldots, a_{i-1}, a_i + t, a_{i+1}, \ldots, a_{n-1}) - F(a_0, \ldots, a_{n-1})}{t}$$

$$= \lim_{t \to 0} \frac{F(a_0, \ldots, a_{j-1}, a_j + t, a_{j+1}, \ldots, a_{n-1}) - F(a_0, \ldots, a_{n-1})}{t}$$

$$= \frac{\partial F}{\partial a_j}. \tag{52}$$

The rest is child's play: By (51) and (52),

$$1 = \frac{\partial F}{\partial a_0} + \frac{\partial F}{\partial a_1} + \ldots + \frac{\partial F}{\partial a_{n-1}} = \frac{\partial F}{\partial a_0} + \frac{\partial F}{\partial a_0} + \ldots + \frac{\partial F}{\partial a_0} = n \cdot \frac{\partial F}{\partial a_0},$$

i.e.,

$$\frac{\partial F}{\partial a_0} = \frac{1}{n}.$$

But with (50) and (52) this yields

$$x = \sum_{i=0}^{n-1} \frac{\partial F}{\partial a_i} a_i = \sum_{i=0}^{n-1} \frac{\partial F}{\partial a_0} a_i = \sum_{i=0}^{n-1} \frac{1}{n} a_i = \frac{a_0 + \ldots + a_{n-1}}{n}.$$

As I say, I like the proof. As to the result, it is only as convincing as the assumptions made. I wonder if it is reasonable to assume the function combining disparate measurements into a single estimate be analytic, or even just differentiable. Continuity would be a lot easier to believe. And, as to the numbered assumptions, I find 1 and 2 (i.e., formulæ (47) and (48)) readily acceptable, but do not wholeheartedly embrace the third (formula (49)). It is, for example, not a property of the often useful median, a continuous but not universally differentiable choice for F.

It was not uncommon in this stage of the development of the theory of errors for A to derive a formula on the basis of some assumptions and B to criticise these assumptions and find a new derivation on the basis of his own favoured assumptions. What seems notable is the robustness of some of the results, e.g., the frequent derivation of the bell curve under disparate assumptions on the nature of errors. Another example of this latter is taken up in the next section.

8 Maxwell and the Error Curve

In the mid-19th century, physicists began to apply probabilistic methods in their field. The ground for such application had been laid with the early atomism of the Greeks and a particulate view of matter had been growing for a couple of centuries. In 1738 Daniel Bernoulli

> first derived an exact expression for the pressure exerted by a gas in terms of the motions of its assumed constituent particles or atoms, and thus initiated the modern interpretation of the properties of bodies in terms of the motions of their constituent particles.[28]

Real progress on the molecular theory of gases began just over a century later. First was a paper published in 1856 by August Carl Krönig in which he set up a "very primitive model for the gas enclosed in a rectangular container"[29]. The following year Rudolph Clausius gave a better foundation for a mathematical model of gases, even introducing the statistical concept of the mean distance between collisions, but he didn't really apply probabilistic methods. This was first done a couple of years later by James Clerk Maxwell (1831 - 1879):

> The recognition of the necessity of combining the methods of statistics with those of dynamics in order to arrive at a valid kinetic theory of gases, that is, the foundation of... molecular mechanics, we owe to Maxwell. The velocity distribution... first published by him in 1859, is known as a Maxwellian distribution. Much of the subsequent development of molecular mechanics centers around the matter of the satisfactory derivation of the law expressing the distribution. The original proof assumed that the component velocities were independent, which would seem to require demonstration. Further, Maxwell's derivation contemplated only simple particles and thus could be valid only for monatomic gases.[30]

I present here Maxwell's original derivation of the Maxwellian distribution in his own words[31], together with some annotations. There is more of probabilistic interest in this paper, e.g., the calculation of the probability that a gas particle will proceed a given distance before a collision occurs, but of particular interest

---

[28] Lynde Phelps Wheeler, *Josiah Willard Gibbs; The History of a Great Mind*, Archon Books, 1970, p. 146. This is a reprint of the third edition. The original editions (1951, 1952, 1962) were published by Yale University Press, New Haven.

[29] Ivo Schneider, ed., *Die Entwicklung der Wahrscheinlichkeitstheorie von der Anfängen bis 1933; Einführungen und Texte*, Wissenschaftliche Buchgesellschaft, Darmstadt, 1989, pp. 299 - 300.

[30] Wheeler, *op. cit.*, p. 149.

[31] J.C. Maxwell, "Illustrations of the dynamical theory of gases.—Part I. On the motions and collisions of perfectly elastic spheres", *Philosophical Magazine* 19 (1860), pp. 19 - 32; here: pp. 22 - 23. The paper, presented to a meeting of the British Association in Aberdeen on 21 September 1859, was published in two parts in successive volumes of *Philosophical Magazine*.

Maxwell has several philatelic appearances to his credit, mostly for his rôle in the development of radio: He shares with Heinrich Hertz a Mexican stamp issued in 1967 on the occasion of an international conference on telecommunications held in Mexico City; one of his equations and a radio tower grace a stamp in a set issued by Nicaragua in 1971 celebrating famous laws of mathematics and physics; and his portrait and an equation appear on the stamp of San Marino pictured here commemorating 100 years of radio. More recently, in 2010 a lovely stamp featuring his portrait was issued on a small souvenir sheet of Mali honouring him as the father of electrodynamics; a second stamp on the sheet features a radio telescope named after him. His work in molecular mechanics has yet to attract philatelic attention.

here is his derivation of the Maxwellian distribution and the presence, in a new context, of the error function as a distribution.

For those familiar with vector calculus, I first offer a word or two about notation. The modern usage is to represent the velocity of a particle by a vector $\mathbf{v}$, the components in the x-, y-, and z-directions by v_x, v_y, and v_z. Maxwell writes x, y and z in place of these, making $\mathbf{v} = \langle x, y, z \rangle$. He gives no notation for $\mathbf{v}$ itself, but considers its absolute value, which he denotes by $v = \sqrt{x^2 + y^2 + z^2}$. Hence, when he refers to "velocity", it is what we refer to today as "speed", the magnitude but not the direction of the velocity. With this in mind, let us read what Maxwell has to say:

> If a great many equal spherical particles were in motion in a perfectly elastic vessel, collisions would take place among the particles, and their velocities would be altered at every collision; so that after a certain time the *vis viva* will be divided among the particles according to some regular law, the average number of particles whose velocity lies between certain limits being ascertainable, though the velocity of each particle changes at every collision.
>
> Prop. IV. To find the average number of particles whose velocities lie between given limits, after a great number of collisions among a great number of equal particles.
>
> Let N be the whole number of particles. Let x, y, z be the components of the velocity of each particle in three rectangular directions, and let the number of particles for which x lies between x and $x + dx$ be N$f(x)dx$, where $f(x)$ is a function of x to be determined.

The number of particles for which y lies between y and $y + dy$ will be $\mathrm{N}f(y)dy$; and the number for which z lies between z and $z + dz$ will be $\mathrm{N}f(z)dz$, where f always stands for the same function.

Here $f(x)$ can be thought of as the probability distribution for the probability that the velocity of a particle in the x-direction is x. The probability that the velocity lies between x and $x + dx$ will be approximately $f(x)dx$, whence the number of such particles will be $\mathrm{N}f(x)dx$. Maxwell makes the simplifying assumption that the same distribution works in all three dimensions.

Now the existence of the velocity x does not in any way affect that of the velocities y or z, since these are all at right angles to each other and independent, so that the number of particles whose velocity lies between x and $x+dx$, and also between y and $y+dy$, and also between z and $z + dz$, is
$$\mathrm{N}f(x)f(y)f(z)dx\,dy\,dz.$$
If we suppose the N particles to start from the origin at the same instant, then this will be the number in the element of volume $(dx\,dy\,dz)$ after unit of time, and the number referred to unit of volume will be
$$\mathrm{N}f(x)f(y)f(z).$$

Maxwell makes here another simplifying assumption, that the velocities in the three directions are independent, whence the probabilities are independent. This makes the probability that all three conditions are met the product of the three probabilities, i.e., $f(x)dx \cdot f(y)dy \cdot f(z)dz$. The number of such particles is thus this probability multiplied by N.

But the directions of the coordinates are perfectly arbitrary, and therefore this number must depend on the distance from the origin alone, that is[32]
$$f(x)f(y)f(z) = \phi(x^2 + y^2 + z^2). \tag{53}$$
Solving this functional equation, we find
$$f(x) = \mathrm{C}e^{Ax^2}, \quad \phi(r^2) = \mathrm{C}^3 e^{Ar^2}. \tag{54}$$

Here, of course, r is defined by $r^2 = x^2 + y^2 + z^2$, and it is easily verified that f, ϕ as defined by (54) satisfy the equation (53) in question. That it is the general solution is not immediately obvious. The basic result on which the solution rests is probably not covered in the standard Calculus course, but is an early result in the first post-Calculus Analysis course. It is the characterisation of exponentiations as the only continuous non-constant solutions to the functional equation
$$g(x + y) = g(x)g(y). \tag{55}$$
I digress from Maxwell's exposition to present the details.

---

[32] These equations are not numbered in Maxwell's paper. I number them here for the sake of the ensuing discussion.

To solve the functional equation (53), first note that

$$f(x)f(0)f(0) = \phi(x^2 + 0 + 0) = \phi(x^2),$$

whence, letting

$$C = f(0), \tag{56}$$

we have

$$\phi(x^2) = C^2 f(x).$$

Similarly, $\phi(y^2) = C^2 f(y)$. Thus

$$\phi(x^2)\phi(y^2) = C^2 f(x) \cdot C^2 f(y) = C^3 f(x)f(y)f(0) = C^3 \phi(x^2 + y^2).$$

A change of variables, u, w for x^2, y^2, respectively, and division by $\phi(0)^2$ converts this to[33]

$$\frac{\phi(u + w)}{\phi(0)} = \frac{\phi(u)\phi(w)}{\phi(0)^2} = \frac{\phi(u)}{\phi(0)} \cdot \frac{\phi(w)}{\phi(0)}.$$

Defining $g(u) = \phi(u)/\phi(0)$, this last becomes the more readable

$$g(u + w) = g(u)g(w),$$

i.e., (55) written with different variables. Taking logarithms, we can simplify this even further: for $h(u) = \ln(g(u))$,

$$h(u + w) = h(u) + h(w). \tag{57}$$

The solutions to (57) among continuous functions are all of the form $h(x) = Ax$ for $A = h(1)$. To see this, note that, for any positive integers m, n,

$$h(m) = h(1 + \ldots + 1) = h(1) + \ldots + h(1) = mh(1)$$

and

$$h(1) = h\left(n \cdot \frac{1}{n}\right) = nh\left(\frac{1}{n}\right), \quad \text{whence } h\left(\frac{1}{n}\right) = \frac{1}{n} \cdot h(1).$$

It follows that

$$h\left(\frac{m}{n}\right) = mh\left(\frac{1}{n}\right) = \frac{m}{n}h(1),$$

i.e.,

$$h(u) = uh(1) = Au, \tag{58}$$

for positive rational u, where we take $A = h(1)$. For $u = 0$, (57) quickly yields $h(0) = 0$, and (58) holds for all nonnegative rational values of u. By continuity, the identity holds for all nonnegative real u.

---

[33] Note that $\phi(0) = 0$ entails $C = 0$, whence $\phi(r^2) = C^2 f(r) = 0$ for all r. In particular, for any x, let $r = \sqrt{3}x$ and note $\phi(r^2) = \phi(3x^2) = f(x)^3$, whence $f(x) = 0$. But a probability distribution cannot be identically 0. Hence $\phi(0) \neq 0$ and the division is permissible.

30 Remark. Instead of coupling the arithmetic argument and continuity to extend (58) to all nonnegative reals, we can assume the continuous differentiability of g, whence that of h, and apply L'Hôpital's Rule:

$$h(u + \Delta u) = h(u) + h(\Delta u),$$

whence

$$\frac{h(u + \Delta u) - h(u)}{\Delta u} = \frac{h(\Delta u)}{\Delta u}.$$

Taking the limit as $\Delta u \to 0$,

$$h'(u) = \lim_{\Delta u \to 0} \frac{h(\Delta u)}{\Delta u} = \lim_{w \to 0} \frac{h(w)}{w} = \lim_{w \to 0} \frac{h'(w)}{1} = \frac{h'(0)}{1}.$$

Letting $A = h'(0)$ and integrating, we have

$$h(u) = Au + K,$$

for some constant K. But $h(0) = 0 = A0 + K$, whence $K = 0$.

Now $g(u) = e^{h(u)} = e^{Au}$ and $\phi(u) = \phi(0) \cdot g(u)$, whence $\phi(u) = C^3 e^{Au}$. Returning to our original variable x the square of which was u, this yields

$$\phi(x^2) = C^3 e^{Ax^2}, \tag{59}$$

and

$$f(x) = \frac{\phi(x^2)}{C^2} = C e^{Ax^2}. \tag{60}$$

The rest of Maxwell's derivation, given (59) and (60), is fairly simple:

If we make A positive, the number of particles will increase with the velocity, and we should find the whole number of particles infinite. We therefore make A negative and equal to $-\dfrac{1}{\alpha^2}$, so that the number between x and $x + dx$ is

$$NCe^{-\frac{x^2}{\alpha^2}} dx.$$

Integrating from $x = -\infty$ to $x = +\infty$, we find the whole number of particles,

$$NC\sqrt{\pi}\alpha = N, \quad \therefore C = \frac{1}{\alpha\sqrt{\pi}},$$

$f(x)$ is therefore

$$\frac{1}{\alpha\sqrt{\pi}} e^{-\frac{x^2}{\alpha^2}}.$$

Whence we may draw the following conclusions:—
1st. The number of particles whose velocity, resolved in a certain direction, lies between x and $x + dx$ is

$$N\frac{1}{\alpha\sqrt{\pi}} e^{-\frac{x^2}{\alpha^2}} dx. \tag{61}$$

2nd. The number whose actual velocity lies between v and $v + dv$ is

$$\text{N}\frac{4}{\alpha^3\sqrt{\pi}}v^2 e^{-\frac{v^2}{\alpha^2}}\,dv. \tag{62}$$

3rd. To find the mean value of v, add the velocities of all the particles together and divide by the number of particles; the result is

$$\text{mean velocity} = \frac{2\alpha}{\sqrt{\pi}}. \tag{63}$$

4th. To find the mean value of v^2, add all the values together and divide by N,

$$\text{mean value of } v^2 = \frac{3}{2}\alpha^2. \tag{64}$$

This is greater than the square of the mean velocity, as it ought to be. It appears from this proposition that the velocities are distributed among the particles according to the same law as the errors are distributed among the observations in the theory of the "method of least squares."

The value of C is, as usual, determined by integration:

$$1 = \int_{-\infty}^{\infty} \text{C}e^{-x^2/\alpha^2}\,dx = \text{C}\int_{-\infty}^{\infty} e^{-x^2/\alpha^2}\,dx$$
$$= \text{C}\int_{-\infty}^{\infty} e^{-u^2}\alpha\,du, \text{ where } u = x/\alpha$$
$$= \text{C}\alpha\sqrt{\pi},$$

yielding $\text{C} = 1/(\alpha\sqrt{\pi})$. Equation (62) may at first seem to come from nowhere, one expecting

$$\text{N}f(x)f(y)f(z)\,dx\,dy\,dz = \text{N}\frac{1}{\alpha^3\pi\sqrt{\pi}}e^{-v^2/\alpha^2}\,dx\,dy\,dz.$$

However, Maxwell is not considering here the number of particles in the cube-shaped region of space anchored at the point $\langle x, y, z\rangle$ and extending to the diagonally opposite point $\langle x+dx, y+dy, z+dz\rangle$; he is considering the spherical shell trapped between two spheres of radii v and $v + dv$, respectively. The volume of such is

$$\frac{4}{3}\pi\left((v+dv)^3 - v^3\right) = \frac{4}{3}\pi\left(3v^2 dv + 3v(dv)^2 + (dv)^3\right).$$

Now, dv is taken to be very small, whence $(dv)^2$ will be very very small and $(dv)^3$ very very very small. Hence the terms involving $(dv)^2$ and $(dv)^3$ can be ignored and the volume in question is taken to be

$$\frac{4}{3}\pi \cdot 3v^2 dv = 4\pi v^2 dv.$$

Multiplying by $f(x)f(y)f(z)$ yields the distribution

$$\left(\frac{1}{\alpha\sqrt{\pi}}\right)^3 e^{-x^2/\alpha^2 - y^2/\alpha^2 - z^2/\alpha^2} \cdot 4\pi v^2 dv = \frac{4}{\alpha^3\sqrt{\pi}} v^2 e^{-v^2/\alpha^2} dv.$$

Multiplying by N yields (62).

Equation (63) is obtained by evaluating[34] the mean

$$\int_0^\infty vg(v)dv$$

of the distribution

$$g(v) = \frac{4}{\alpha^3\sqrt{\pi}} v^2 e^{-v^2/\alpha^2}$$

as in (62). And (64) is the mean of the squares:

$$\int_0^\infty v^2 g(v)dv.$$

Of particular relevance to our present interests is $f(x)$, which is just the normal distribution with $\mu = 0, \sigma = \alpha/\sqrt{2}$. Or, as Maxwell points out, f is the error distribution (i.e., $\mu = 0$) with $\sigma = \alpha/\sqrt{2}$. Indeed, it is generally accepted that Maxwell got his idea for the proof from a review, "Quetelet on probabilities" published by John Herschel in 1850 in the *Edinburgh Review* and reprinted in a collection of Herschel's essays published in 1857. Herschel stated

---

[34] For those familiar with the Calculus, this is an easy exercise in the use of substitution ($u = v^2/\alpha^2$) and integration by parts. Ignoring the constant term, owners of a *TI-89 Titanium* might try entering

∫(v^3*e^(−(v/a)^2),v,0,∞) ,

after applying a DelVar a command to make sure a is taken to be an arbitrary constant and not some value stored in the variable a, and, of course, after putting the calculator into EXACT mode. The result is undef. With a change of variable, $u = v/\alpha$, however, one has

$$\int_0^\infty v^3 e^{-(v/\alpha)^2} dv = \int_0^\infty \alpha^3 u^3 e^{-u^2} \alpha du = \alpha^4 \int_0^\infty u^3 e^{-u^2} du,$$

and one can enter

(a^4)∫(u^3*e^(−u^2),u,0,∞)

to get

$$\frac{a^4}{2}.$$

Whereas differentiation of terms is a simple deterministic algorithm, symbolic integration is a partial, nondeterministic algorithm much more difficult to program, and, depending on the hardware and software, one may still have to be clever.

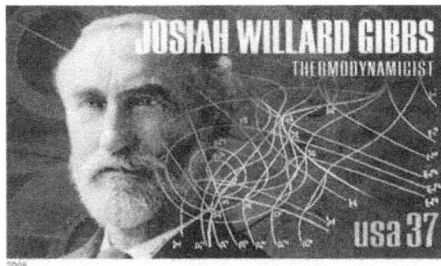

The two most important figures in statistical mechanics are Ludwig Boltzmann (1844 - 1906) and Josiah Willard Gibbs (1839 - 1903). In his biography of Gibbs, Wheeler (*op. cit.*, p. 150) describes their contributions as follows:

> The first of the numerous attempts which have been made to put the Maxwellian distribution upon a firmer logical basis was made by Ludwig Boltzmann. He extended the application to include polyatomic molecules, confirmed the result that the distribution was unaffected by molecular collisions, and showed further than any assumed initial distribution tended to approach the Maxwellian under the influence of these encounters... later development of Boltzmann's gave an entirely new slant on the problem. Up to this time attention had been centered on the statistics of the individual particles constituting a system of particles. In this later work Boltzmann introduced the idea of studying the statistics of the whole system of particles as a unit... This point of view has dominated the subsequent development of the subject and is that adopted by Gibbs in his work.
>
> Gibbs recognized that this point of view made possible the creation of a science of statistical mechanics which could be independent of all considerations of molecular structure and dependent solely on the laws of mechanics and of statistics.

The stamp featuring Boltzmann was issued by Austria in 1971 on the occasion of the 75th anniversary of his death. The stamp depicting Gibbs is one of four in an unprecedented set of commemoratives honouring American scientists as scientists issued by the United States Postal Service in 2005. A second stamp of the set of probabilistic interest featured a portrait of John von Neumann.

Suppose a ball is dropped from a great height given the intention that it should fall on a given mark. Fall as it may, its deviation from the mark is error and the probability of that error is the unknown function of its square i.e. the sum of the deviations in any two rectangular directions. Now, the probability of any deviation depending solely on its magnitude, and not on its direction, it follows that the probability of each of these rectangular deviations must be the same function of its square. And since the observed oblique deviation is equivalent to the

David Hilbert (1862 - 1943) was one of the leading mathematicians in the world in the decades from 1890 to 1920. For the International Congress of Mathematicians held in Paris in 1900 he prepared a list of 24 problems for the new century, 10 of which he presented at the Congress and 23 of which were discussed in the resulting paper. The text of the 6th problem of the published list in the translation of Mary Winston Newson that appeared in the *Bulletin of the American Mathematical Society* in 1902 begins:

6. Mathematical Treatment of the Axioms of Physics.
The investigations of the foundations of geometry suggest the problem: *To treat in the same manner, by means of axioms, those physical sciences in which mathematics plays an important part; in the first rank are the theory of probabilities and mechanics.*
As to the axioms of the theory of probabilities, it seems to me desirable that their logical investigation should be accompanied by a rigorous and satisfactory development of the method of mean values in mathematical physics, and in particular in the kinetic theory of gases.
Important investigations by physicists on the foundations of mechanics are at hand: I refer to the writings of Mach, Hertz, Boltzmann and Volkmann. It is therefore very desirable that the discussion of the foundations of mechanics be taken up by mathematicians also. Thus Boltzmann's work on the principles of mechanics suggests the problem of developing mathematically the limiting processes, there merely indicated, which lead from the atomistic view to the laws of motion of continua. Conversely one might try to derive the laws of the motion of rigid bodies by a limiting process from a system of axioms depending upon the idea of continuously varying conditions of a material filling all space continuously, these conditions being defined by parameters. For the question as to the equivalence of different systems of axioms is always of great theoretical interest.

The first few decades of the 20th century saw several attempts at axiomatising Probability Theory, the most successful of which was the measure theoretic treatment of Andrei Nikolaevich Kolmogorov of 1933.
The stamp shown is from a rather large set of stamps issued by the Democratic Republic of the Congo in 2001 celebrating Personalities of the 20th Century.

two rectangular ones, supposed concurrent, and which are essentially independent of one another, and is, therefore, a compound event of which they are the simple independent constituents, therefore the probability will be the product of their separate probabilities. The form of the unknown function comes to be determined from this condition, viz., that the product of such functions of two independent elements is equal to the same function of their sum. But it is shown in every work of algebra that this property is the peculiar characteristic of, and belongs only to, the exponential function. This, then, is the function of the square of the error, which expresses the probability of committing that error.[35]

While, mathematically speaking, the kernel of Maxwell's proof is already in this remark of Herschel's, Maxwell must be credited with the realisation that it applied to velocities and for introducing probabilistic and statistical methods into Physics.

Hagen's, Herschel's, and Maxwell's derivations suffered from reliance on special assumptions, as well as a lack of rigour. Criticisms and attempts at better founding and generalising their results followed. The physical question was treated, following a second proof by Maxwell in 1867, by a number of papers, beginning in 1868, by Ludwig Boltzmann. The error distribution was considered by a number of authors who derived the error curve on the basis of a number of different sets of assumptions. A nice summary of them is given by Part I, section 6, of Emanuel Czuber's book on the theory of observational errors[36].

9 Population Growth

Aside from the occasional population reducing disaster like the Black Death which killed an estimated 30 to 60 percent of the European population in the 14th century, population generally grows, a fact which must have been recognised early on and which would account for municipalities and eventually nations taking periodic censuses. Indeed, the United States Constitution included provision for the taking of a census every decade, as has been done since 1790. The purpose of the American census at that time was to determine representation in Congress, not for any scientific purpose although many of the Founding Fathers were themselves amateur scientists. Nonetheless, American population data provided fodder for the first, crude scientific study of population growth by Thomas Robert Malthus (1766 - 1834) in his *Essay on the Principle of Population* (1798) and the census data were referred to in the more sophisticated "Mathematical investigations on the law of population growth" (1845) by Pierre François Verhulst (1804 - 1849).

The second chapter of the first edition of Malthus begins

---

[35] Quoted from the note "Maxwell distribution of molecular velocities" by David Forfar posted on the web site of the James Clerk Maxwell Foundation.
[36] Czuber, *op. cit.*, pp. 99 - 113.

I said that population, when unchecked, increased in a geometrical ratio; and subsistence for man in an arithmetical ratio.
Let us examine whether this proposition be just.
I think it will be allowed, that no state has hitherto existed (at least that we have any account of) where the manners were so pure and simple, and the means of subsistence so abundant, that no check whatever has existed to early marriages; among the lower classes, from a fear of not providing well for their families; or among the higher classes, from a fear of lowering their condition in life. Consequently in no state that we have yet known, has the power of population been left to exert itself with perfect freedom.
. . .
In the United States of America, where the means of subsistence have been more ample, the manners of the people more pure, and consequently the checks to early marriages fewer, than in any of the modern states of Europe, the population has been found to double itself in twenty-five years.
This ratio of increase, though short of the utmost power of population, yet as the result of actual experience, we will take as our rule; and say, That population, when unchecked, goes on doubling itself every twenty-five years, or increases in a geometrical ratio.[37]

Writing about half a century later, Verhulst again referred to America, but he now used the results of the 6 censuses that had been made:

The United States provide us with an example of such a rapid rate of population growth. According to the U.S. official censuses, the counts were. . .[38]

There follows a table of the populations of the United States for the years $1790, 1800, \ldots, 1840$.[39] Verhulst extended the table by linear interpolation to include values for 5-year periods and then for the the years $1815, 1820, \ldots, 1840$ calculated the ratios r of the populations for the given years to those 25 years earlier[40], obtaining values

$$2.147, 2.087, 2.120, 2.052, 2.076, 2.021.$$

From this we see an evident doubling every 25 years.

---

[37] T.R. Malthus, *Parallel Chapters from the First and Second Editions of An Essay on the Principle of Population*, Macmillan and Co., New York, 1895, pp. 9 - 10.

[38] P.F. Verhulst, "La loi d'accroissment de la population", *Nouveaux mémoires de l' académie royale des sciences et belles-lettres de Bruxelles* 18 (1845), pp. 1 - 38. I quote the truncated English translation (p. 70) in: H.A. David and A.W.F. Edwards, eds., *Annotated Readings in the History of Statistics*, Springer-Verlag New York, Inc., New York, 2001.

[39] Verhulst's values for the populations disagree with those I found online at the web site of the U.S. Census Bureau. Both sets of values can be found in Appendix D.

[40] His table of these values, with populations rounded to the nearest thousand, is reproduced in Appendix D.

The calculator's regression capabilities come in handy here. In the CALC menu accessed by pressing the STAT button, one finds 10 curve-fitting applications designed to give the function of a specific form that "best fits" plotted data. If one has collected data in the form of a simple table of xy-values, $\langle x_0, y_0 \rangle, \langle x_1, y_1 \rangle, \ldots, \langle x_{n-1}, y_{n-1} \rangle$, one can store the x- and y-values in two lists, L_1, L_2, and use STAT PLOT to graph the points. For theoretical purposes, one likes to have a function $y = f(x)$ the graph of which comes as close to the plotted points as possible. *Regression* is the process of finding such a function f of a predetermined form. In *linear regression*, one tries to fit a straight line to the data. That is, one assumes f to be of the form $f(x) = ax + b$ and tries to find a, b such that the line $y = ax + b$ is the closest possible fit to the data. The problem is not well-defined until one says how one measures closeness. In practice one assumes the x_i's as given and minimises the sum of the squares of the deviations of the actual values y_i from the predicted values $f(x_i) = ax_i + b$. That is, one finds a, b such that $\Sigma(y_i - ax_i - b)^2$ is minimal.

With a little Calculus this is, conceptually, a simple matter. Thinking of the x_i's and y_i's as constants, $\Sigma(y_i - ax_i - b)^2$ is a function of a, b. To find where it is minimum, differentiate with respect to b, a, respectively:

$$\frac{\partial}{\partial b} \sum (y_i - ax_i - b)^2 = 2 \sum (y_i - ax_i - b)(-1) \tag{65}$$

$$\frac{\partial}{\partial a} \sum (y_i - ax_i - b)^2 = 2 \sum (y_i - ax_i - b)(-x_i). \tag{66}$$

Setting these equal to 0 yields, for (65)

$$\sum (y_i - ax_i - b) = 0$$

$$\sum y_i - a \sum x_i - \sum b = 0,$$

and, dividing by n,

$$\bar{y} - a\bar{x} - b = 0,$$

i.e.,

$$a\bar{x} + b = \bar{y}; \tag{67}$$

and for (66),

$$\sum x_i(y_i - ax_i - b) = 0$$

$$\sum x_i y_i - a \sum x_i^2 - b \sum x_i = 0,$$

i.e.,

$$a \sum x_i^2 + b \sum x_i = \sum x_i y_i. \tag{68}$$

Equations (67) and (68) form a system of two linear equations in two unknowns a and b, the coefficients of which are readily calculated. Indeed, if $x_0, x_1, \ldots, x_{n-1}$ and $y_0, y_1, \ldots, y_{n-1}$ have been entered as L_1 and L_2, respectively, the coefficients $\bar{x}$ and $\bar{y}$ in (67) are just mean(L_1) and mean(L_2), respectively. The coefficients $\Sigma x_i^2, \Sigma x_i, \Sigma x_i y_i$, as well as $\bar{x}$ and $\bar{y}$, are all quickly calculated by entering

2-Var Stats L_1, L_2.

This command stores the results in the variables Σx^2, Σx, Σxy, $\bar{x}$, $\bar{y}$ listed in the XY and Σ submenus of the Statistics... submenu accessed by pressing the VARS button. The usual methods of solving simultaneous linear equations now apply.

Now all of this has been built into the calculator. Provided one has entered the x's and y's into lists L_1 and L_2, one can press the STAT button, choose the CALC menu, and then select 4: LinReg(ax+b) to enter

LinReg(ax+b) L_1, L_2.

The home screen will display[41]

LinReg
  y=ax+b
  a=*value of a*
  b=*value of b*.

If, instead, one enters

LinReg(ax+b) L_1, L_2, Y_1,

the function $ax + b$ with the given values of a, b of a,b will be entered as Y_1 in the equation editor. If one has already graphed the statistical plot of the pairs $\langle x_i, y_i \rangle$, one can again press the GRAPH button and see the plotted points with the line $y = ax + b$ now passing among them.

The calculator has a number of regressions built-in, including two versions of linear regression,

4: LinReg(ax+b)
8: LinReg (a+bx),

the pair accommodating those who prefer b or a, respectively, to be the constant coefficient; quadratic regression

5: QuadReg,

yielding the "best" quadratic function $y = ax^2 + bx + c$ fitting the data; and so on. The two regressions most relevant for population data are *exponential regression*,

0: ExpReg,

which fits an exponential model $y = ab^x$ to the data, and *logistic regression*,

A: Logistic,

which fits a *logistic* curve,

$$y = \frac{c}{1 + ae^{-bx}}$$

to the data.

---

[41] If one has set DiagnosticsOn on the calculator, extra information that doesn't concern us here will also appear.

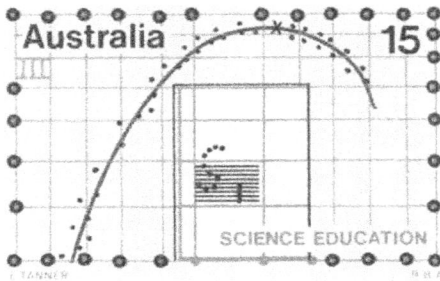

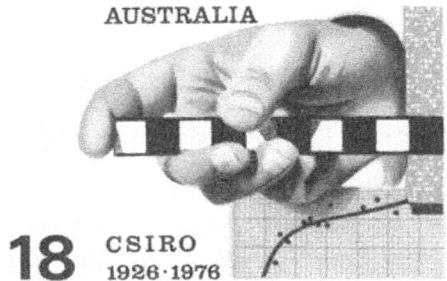

Above are two Australian stamps illustrating curve fitting. The stamp on the left celebrating science education is one of four definitives celebrating education issued in 1974. The stamp on the right is a commemorative issued in 1976 celebrating the 50th anniversary of the founding of the Commonwealth Science and Industrial Research Organisation, Australia's national science agency. The other elements of the design are the easily recognisable measuring stick, and, between it and the hand, the now obsolete punch tape for a computer.

The stamp on the right, issued by Japan in 1981 in connexion with a pharmacology congress, features a logistic curve.
It should be noted that not every "S"-shaped curve is a logistic curve. For example, the cumulative normal distribution has such an "S"-shape.

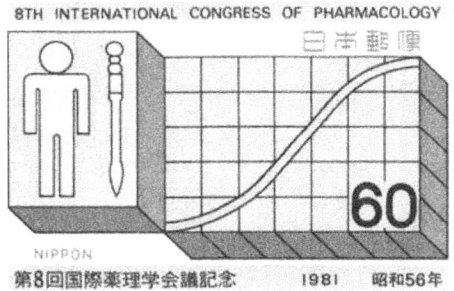

In theory, exponential regression can be accomplished in either of two ways. One can repeat the above analysis involving differentiating

$$\sum (y_i - ab^{x_i})^2$$

with respect to a, b, and setting up some messy equation in a, b to solve, or one can perform a linear regression on the logarithms of the y_i's to obtain an equation

$$\ln y = A + Bx,$$

whence

$$y = e^{A+Bx} = e^A e^{Bx} = a \cdot b^x,$$

where $a = e^A, b = e^B$. Or, of course, one can bypass this by using the built-in exponential regression feature of the calculator.

31 Project. Store the U.S. Census Table of Appendix D in two lists ∟YEARS and ∟CNSUS, the first listing the years $1790, 1800, \ldots, 2000$, and the second the corresponding population figures. Note that ∟YEARS can be generated quickly by entering

seq(X,X,1790,2000,10)→∟YEARS,

but ⌐CNSUS has to be copied one item at a time. This is possibly most simply done by entering

    SetUpEditor ⌐YEARS,⌐CNSUS ,

and entering the values in the List editor. The presence of the list ⌐YEARS alongside ⌐CNSUS makes it easier to check you haven't skipped any numbers in creating ⌐CNSUS. We will use the lists ⌐YEARS and ⌐CNSUS throughout the rest of this section.

Now copy ⌐YEARS and ⌐CNSUS to L_1, L_2 by entering

    ⌐YEARS→L_1
    ⌐CNSUS→L_2 ,

truncate these to the years available to Verhulst (1790 to 1840),

    6→dim(L_1)
    6→dim(L_2) ,

and enter STAT PLOT to prepare to graph the data: Make sure that all the plots other than Plot1 are off and do the same for any functions, enter Plot1 and choose the first icon, the dotted line, as Type, enter L_1 for XList and L_2 for YList, and choose the open box or small plus sign as Mark. Press the Zoom button and graph the data by entering ZoomStat. To see how well an exponential curve fits the data, go back to the home screen and enter

    ExpReg L_1,L_2,Y_1 .

The screen will display the values of a, b and, if you press the Y= button, you will see the equation of the exponential function under Y_1. For the purpose of this project, however, it is best to press the GRAPH button to see visually how well the curve fits the data. Even bearing in mind that the vertical scale is highly compressed, the fit is impressive.

Now copy ⌐YEARS and ⌐CNSUS into L_1 and L_2 again, but don't perform the truncation, and graph using ZoomStat again. The separation after 1870 is visually apparent. Entering

    ExpReg L_1,L_2,Y_1 .

again will yield an exponential function better fitting the data throughout the range of the x's, if not for the original years studied by Verhulst. Pressing GRAPH again will demonstrate visually that the population growth is not exponential in the long run.

Malthus had already emphasised that exponential growth would only occur if there were no checks to growth, such as the limitation of the means to produce food. Verhulst went farther, analysing the problem mathematically. Ignoring plagues, wars, conversion to a religion which encourages or discourages large family size, the numbers of births and deaths would be proportional to the population size P. Hence the increase in population would be proportional to the population. In terms of the Calculus, this yields a differential equation

$$\frac{dP}{dt} = kP, \tag{69}$$

where P = population, t = time, k = some constant. The solution to this equation is of the form

$$P(t) = a \cdot e^{kt}, \text{ for some constant } a$$
$$= ab^t, \text{ for } b = e^k.$$

Verhulst reasoned that population growth would continue exponentially until such time as all the good farm land was taken. Thereafter the growth would slow down.

> We will no longer insist on the geometric progression hypothesis, seeing that it holds only in quite exceptional circumstances, for example, when a territory, fertile and virtually unlimited in extent, is inhabited by a people with a very advanced civilization, such as the first settlers of the United States.[42]

> It is an observed fact that in all of Europe the ratio of the annual excess of births over deaths to the corresponding population size, and consequently the coefficient l/M [43], is constantly declining. But the annual growth, whose absolute value continually increases when there is a geometric progression, appears to follow a progression that is arithmetic at the very outside. This remark confirms the famous aphorism of Malthus, that *the population tends to grow in geometric progression whereas the production of subsistence follows an arithmetic progression at the very best*, since the population is obliged to adjust itself to its means of subsistence.

> We can make an infinity of hypotheses about the law of decline of the coefficient l/M. The simplest consists in regarding this decline as proportional to the growth of the population, from the moment when the difficulty of finding good land has made itself felt. The population at that special moment, from which we now measure time, will be called a *normal population* and designated N_0.[44]

Thus, Verhulst postulates exponential population growth up to the point where the easy increase in production of subsistence has ended. The population at this time is N_0, for which we shall write P_0 using the mnemonic "P" for "population". He now decides that there is a retardant element to growth proportional to $P(P - P_0)$ and thus he replaces (69) by

$$\frac{dP}{dt} = k_1 P - k_2 P(P - P_0). \tag{70}$$

For a slightly clearer explanation for the form of (70) I quote Martin Braun[45]:

---

[42] Verhulst, in David and Edwards, *op. cit.*, p. 71.
[43] l/M is our k above.
[44] Verhulst, in David and Edwards, *op. cit.*, p. 72.
[45] Martin Braun, *Differential Equations and Their Applications*, 3rd ed., Springer-Verlag New York, Inc., New York, 1983, p. 28.

The way out of our dilemma is to observe that linear models for population growth are satisfactory *as long as* the population is not too large. When the population gets extremely large though, these models cannot be very accurate, since they do not reflect the fact that individual members are now competing with each other for the limited living space, natural resources and food available. Thus, we must add a competition term to our linear differential equation. A suitable choice of a competition term is $-bp^2$, where b is a constant, since the statistical average of the number of encounters of two members per unit time is proportional to p^2. We consider, therefore, the modified equation

$$\frac{dp}{dt} = ap - bp^2.$$

Braun's and Verhulst's equations are the same, with $a = k_1 - k_2 P_0$, $b = k_2$. The solution to this equation has the form

$$P(t) = \frac{c}{1 + ae^{-bt}},$$

for some constants a, b, c the names of which I have chosen to agree, not with Braun or Verhulst, but the calculator.

Verhulst describes population growth by a pair of functions, an exponential one up to a point of saturation (P_0), followed by a logistic curve—so named by him. As Braun's remarks make clear, the logistic curve generally works well enough alone.

32 Project. The growth pattern of the U.S. population exhibited classic Verhulstian logistic behaviour. Copy ∟YEARS and ∟CNSUS to L_1 and L_2 again, truncate to 1940,

    16→dim(L₁)
    16→dim(L₂),

perform the regression

    Logistic L₁,L₂,Y₁,

and apply ZoomStat to graph the curve against the data. Notice how good the fit is.

Now copy ∟YEARS and ∟CNSUS to L_1 and L_2, and apply ZoomStat to compare how well the 1790 - 1940 logistic curve fits the data over the entire 1790 - 2000 period. What do you see?

There are various factors explaining the graph. First, the population for 1940 falls slightly below the curve—the Great Depression of the 1930's would have delayed some marriages as well as the having of children by already married couples. The high numbers for 1950 are readily explained by the willingness of those who had gone off to war to marry and settle down on returning home, along with a much improved economy and GI benefits making this feasible. The continued rapid growth can be accounted for in various ways: improved agriculture, an expanded welfare system, a breakdown in morality as a response to the Vietnam War. The subsistence level of the lower classes, as Malthus described them, was raised by the first two of these, and the third removed the social barriers discouraging extramarital teen pregnancies.

I end with the remark that population studies offer good examples of the difficulties of making long term forecasts in the face of large numbers of variables. The Malthusian model describes the growth of the American population well for the period 1790 - 1840 and the exponential curve based on that data still had predictive value for a couple of decades. Verhulst's logistic curve based on an extra century of data far underestimates all future population figures till today.

33 Project. As I write this, it is just over a week since I sent in my form for the 2010 census; as you read this, the results of the 2010 census ought to have been tabulated. Run the logistic regression for all the data from 1790 to 2000 and compare its prediction for 2010 with the actual result.

10 Daniel Bernoulli on Smallpox

I have relegated the discussion of Daniel Bernoulli's analysis of the efficacy of inoculation to this appendix in part so as not to interrupt the flow of the exposition in the main text with technical details, in part because more math needs to be introduced, and also in part to allow myself the opportunity to make a digressive comment or two that, in the context of the main text, would have digressed further from a digression from a digression...

Bernoulli's analysis is a fine example of applied mathematics, illustrative not only of numerical technique, but of the entire process, including the making of assumptions—guessing parameters and deciding what is appropriate.

Todhunter's description[46] of Bernoulli's mathematical theory uses the notions of the Calculus and proceeds by setting up and solving a simple *differential equation* of the sort often solved in the standard Calculus course. The solution is accessible to the student who has completed such a course and I present it here. For those unfamiliar with the Calculus, I first consider a discrete analogue that should illustrate the general features of Bernoulli's approach.

Let $\xi_0, \xi_1, \xi_2, \ldots$ be the numbers of people surviving to ages $0, 1, 2, \ldots$, respectively, assuming an initial population of 1300, and let $s_0, s_1, s_2, \ldots$ be the numbers of people of ages $0, 1, 2, \ldots$ who have not yet had smallpox. Suppose the probability that a person who has not yet had smallpox will be so afflicted in a given year is p (which Bernoulli took to be $1/8$), and the probability of dying given that one has smallpox is q (which Bernoulli again took to be $1/8$).

As a first approximation, we might take $s_{n+1} = s_n - ps_n$ since ps_n of those who have not yet had smallpox will acquire the illness during the course of the

---

[46] Isaac Todhunter, *A History of the Mathematical Theory of Probability From the Time of Pascal to That of Laplace*, Cambridge University Press, Cambridge, 1865 (reprinted by Chelsea Publishing Company, New York, 1949, 1965), pp. 225 - 226.

year. This is an overestimate because it doesn't take into account the additional loss of those who do not come down with the illness, but die of other causes. Nonetheless, it is an approximation that is easy to calculate:

$$s_0 = 1300, \quad s_1 = \frac{7}{8} \cdot 1300, \quad s_2 = \frac{7}{8} \cdot s_1 = \left(\frac{7}{8}\right)^2 \cdot 1300, \quad \ldots,$$

i.e.,

$$s_{n+1} = \left(\frac{7}{8}\right)^n \cdot 1300. \tag{71}$$

And, however good or bad an approximation it may be, it is a definite upper bound on what the correct answer should be.

To improve the estimate we will subtract from s_n, not only ps_n, but also the number of those of age n who have not had smallpox but who will die during the year of some other cause. To this end, we assume the probability of someone of age n who has never had smallpox dying of some other cause is the same probability r_n of someone of age n in the general population dying of some other cause. Now the probability of a person aged n years dying of smallpox is the product of the probabilities of never having had smallpox, of getting smallpox given that one has never had it, and of dying of smallpox given that one has it, i.e.,

$$\frac{s_n}{\xi_n} \cdot p \cdot q.$$

And the probability of dying during the year, given that one is n years of age, is

$$\frac{\xi_n - \xi_{n+1}}{\xi_n}.$$

Thus,

$$r_n = \frac{\xi_n - \xi_{n+1}}{\xi_n} - \frac{s_n}{\xi_n} \cdot p \cdot q,$$

and we take

$$s_{n+1} = s_n - ps_n - r_n s_n$$
$$= s_n - ps_n - s_n \left(\frac{\xi_n - \xi_{n+1}}{\xi_n} - \frac{s_n}{\xi_n} \cdot pq\right). \tag{72}$$

The sequence $\xi_1, \xi_2, \xi_3, \ldots$ was known empirically through Halley's table, and the initial value $\xi_0 = 1300$ was prefixed to this list by Bernoulli via extrapolation. With these and an initial value $s_0 = 1300$, we can use (72) to calculate $s_1, s_2, s_3, \ldots$ successively.

Halley's table, copied from the appendix to de Moivre's *A Treatise of the Annuities on Lives*, is given in Appendix D. After placing 1300 in front of his list of numbers of survivors to given ages, copy the list $\{1300, 1000, 855, \ldots, 19\}$ into a new list ∟HALLY. Then put the calculator into Seq mode, open the equation editor, and enter the following:

nMin=0
u(n)=u(n−1)(1−1/8−((⌊HALLY(n)−⌊HALLY(n+1))/⌊HALLY(n)−u(n−1)
  /⌊HALLY(n)*(1/8)*(1/8))
u(nMin)={1300} .

The second line merely expresses (72): u(n) is supposed to approximate s_n and
⌊HALLY(n+1) is ξ_n: ages begin at 0, while the indices of lists begin at 1.
  Having defined u, on the home screen enter

u(0,84)→⌊SN .

The list ⌊SN is the list $s_0, s_1, s_2, \ldots$ up to age 84 obtained from ⌊HALLY via
(72). While one is at it, create two more lists:

seq(n,n,0,84)→⌊AGE
seq(1300*(1/8)^n,n,0,84)→⌊GP .

⌊AGE is just the list of ages given by Halley and ⌊GP is the geometric progression (71) discussed earlier.
  To get a better view of the data, enter

SetUpEditor ⌊AGE,⌊HALLY,⌊GP,⌊SN

and then enter the List Editor to view a spreadsheet presentation of the data
through a small window.
  In interpreting the results, we should remind ourselves that these numbers
represent approximations to the number of people from a group of 1300 who
survive to a given age n without ever having smallpox. That s_{50} as approximated by ⌊SN(51) is .416718... means we should expect 0 or 1 such out of
1300, 4 out of 13000, around 42 out of 130000, and somewhere near 417 out of
1300000.
  How accurate are the figures in the table and how can we improve them? [47]
To be honest, there is no way short of record keeping of knowing how accurate
these numbers are. What one can do is refine the model and hope for greater
accuracy. By shortening the periods under consideration from 1 year to half a
year, from half a year to a month, from a month to a week, ..., we can hope
for a better solution. When one does this, choosing a very small difference dx
for the period, so that the differences $s_{n+1} - s_n$ and $\xi_{n+1} - \xi_n$ become very
small numbers ds and $d\xi$, respectively, equation (72),

$$s_{n+1} - s_n = -ps_n - s_n \left(\frac{\xi_n - \xi_{n+1}}{\xi_n} - \frac{s_n}{\xi_n} pq \right),$$

transforms into a differential equation,

$$ds = -ps\,dx - s \left(\frac{-d\xi}{\xi} - \frac{s}{\xi} pq\,dx \right)$$

$$= -ps\,dx + \frac{s}{\xi} d\xi + \frac{s^2}{\xi} pq\,dx. \qquad (73)$$

---

[47] From here until page 508 one needs a knowledge of the Calculus to fully understand
the calculations.

Bernoulli[48] makes the substitution $u = \xi/s$ for which

$$du = \frac{s\, d\xi - \xi\, ds}{s^2}.$$

Isolating the dx terms of (73) on one side of the equation and then multiplying by ξ/s^2 yields successively

$$ps\, dx - \frac{s^2}{\xi} pq\, dx = \frac{s\, d\xi}{\xi} - ds$$

$$\frac{\xi p\, dx}{s} - pq\, dx = \frac{s\, d\xi}{s^2} - \frac{\xi\, ds}{s^2} = du.$$

Thus

$$du = \frac{\xi}{s} p\, dx - pq\, dx = up\, dx - pq\, dx = (pu - pq)dx.$$

Setting $v = pu - pq$ and remembering that $p = q = 1/8$ are constants, we have $dv = p\, du = pv\, dx$, whence

$$\frac{dv}{v} = p\, dx,$$

and integration yields

$$\ln v = px + C, \text{ for some constant } C.$$

Thus

$$pu - pq = v = e^{px+C}$$

and

$$\frac{\xi}{s} = u = \frac{pq + e^{px+C}}{p},$$

from which a wee bit of algebra yields

$$s = \frac{p\xi}{e^{px+C} + qp} = \frac{\xi}{\frac{1}{p}e^{px+C} + q}.$$

When $x = 0$, $s = \xi_0$ ($= 1300$), whence $u = 1$ and $v = p - pq = e^C$. Thus $e^{px+C} = e^{px}e^C = e^{px}(p - pq)$ and

$$s = \frac{\xi}{\frac{1}{p}p(1 - q)e^{px} + q} = \frac{\xi}{(1 - q)e^{px} + q}.$$

More formally:

$$s(x) = \frac{\xi(x)}{(1 - q)e^{px} + q}. \tag{74}$$

---

[48] I follow Todhunter's presentation of Bernoulli's argument.

We only know $\xi(x)$ for integral values $0, 1, 2, \ldots, 84$ from Halley's table, i.e., from the list ⌐HALLY. Recalling $p = q = 1/8$, we have Bernoulli's estimates[49] for $s_0, s_1, \ldots, s_{84}$:

seq(⌐HALLY(n+1)/((7/8)e^(n/8)+1/8),n,0,84)→⌐BERN.

One can again use SetUpEditor to add a column for ⌐BERN and compare the results to ⌐SN. The results seem to be reasonably close, just how close may perhaps be better appreciated by graphing. Using the STAT PLOT button, for each plot choose the xyLine type by positioning the cursor over the second pictogram and hitting ENTER, for each plot enter AGE for Xlist, and choose the small dot for the Mark. For the Ylist choose HALLY in Plot1, SN in Plot2, and BERN in Plot3. Plot 1 is for later reference and I suggest turning Plot1 off, and Plot2 and Plot3 on before graphing the lines via ZoomStat (after, of course, turning off the graphs of any functions and, if necessary, entering a ClrDraw command).

34 Remark. The Bernoullis pioneered the study of differential equations and their applications in physics, where the results more closely approximated to the real world than did the discrete analyses. Which approach is more accurate in the social sciences where the data are discrete rather than continuous should not be immediately clear. Only the collection of data—in the present case impossible as we have effective vaccines against smallpox—would settle the matter. I do note that, without a good calculator or computer, the computation of, say, $s(51)$ via (74) by hand in a day dominated by logarithm tables would have been much less formidable than a computation using (72). Thus, just as de Moivre's normal distribution afforded a convenient approximation to the binomial one, Bernoulli's differential equation approach at least offers the same convenience here.

Bernoulli did not stop here. Had he done so I would not be discussing his work now. He next assumed one had eradicated smallpox, repeated his analysis, setting up and solving a new differential equation, and estimated what the mortality table would be in the absence of smallpox. The calculation is similar to that given for s. Let $z_0, z_1, z_2, \ldots$ now be the numbers of individuals from an initial stock of 1300 who survive to ages $0, 1, 2, \ldots$, respectively. Let $\xi_0, \xi_1, \xi_2, \ldots$ be as before. Let p, q each have the value $1/8$ as before, but they are now no longer the probabilities of getting smallpox if one hasn't had it and of dying of smallpox once one has it, but the corresponding probabilities had smallpox not been eradicated. With smallpox, the probability of dying within the year for a person aged n is

---

[49] As a partial check on the calculation, the reader might wish to look up Bernoulli's original table and compare his results with those of the calculator. I, personally, compared the extract of Bernoulli's table printed by Todhunter (*op. cit.*, p. 226) and note that, after rounding, the calculator agrees with Bernoulli for $n \leq 16$ except for $s(9)$ where Bernoulli rounds down to 237 and the *TI-83 Plus* rounds up to 238.

$$\frac{\xi_n - \xi_{n+1}}{\xi_n}.$$

Now of that, the probability of dying of smallpox is

$$\frac{s_n}{\xi_n} \cdot pq,$$

as established earlier. Thus, with small pox eradicated

$$z_{n+1} = z_n - z_n \left(\frac{\xi_n - \xi_{n+1}}{\xi_n} - \frac{s_n}{\xi_n} \cdot pq \right),$$

i.e.,

$$z_{n+1} = z_n \left(1 - \frac{\xi_n - \xi_{n+1}}{\xi_n} + \frac{s_n}{\xi_n} \cdot pq \right). \tag{75}$$

As before, we can use (75) and the initial value $z_0 = \xi_0 = 1300$ to generate the z's:

```
nMin=0
v(n)=v(n−1)(1−(∟HALLY(n)−∟HALLY(n+1))/∟HALLY(n)+∟SN(n)
    /(64∗∟HALLY(n))
v(nMin)={1300}.
```

And again one can generate a list ∟ZN of values of z:

```
v(0,84)→∟ZN,
```

and compare ∟HALLY with ∟ZN:

```
SetUpEditor ∟AGE,∟Hally,∟ZN.
```

Bernoulli, of course, refined this by taking smaller periods, turning (75) into

$$z_{n+1} - z_n = \frac{z_n}{\xi_n} (\xi_{n+1} - \xi_n + s_n pq)$$

and thus derived the differential equation

$$dz = \frac{z}{\xi} (d\xi + spq\, dx).$$

Using (74) to replace s by an expression involving ξ, x, and proceeding as before, he obtained

$$z = \frac{\xi e^{px}}{(1-q)e^{px} + q} = se^{px}. \tag{76}$$

Again we can make a list:

```
seq(∟BERN(n+1)∗e^(n/8),n,0,84)→∟BERN2,
```

open the List Editor, position the cursor over the empty name in the fourth column, type in ∟BERN2 and hit ENTER to reproduce the list ∟BERN2 and compare ∟ZN with ∟BERN2. The results are very close.

The close agreement between ∟ZN and ∟BERN2 may also be checked graphically. If one hasn't made any statistical plots since performing those described a couple of pages ago, simply replace SN as Ylist in Plot2 by ZN, and BERN in Plot3 by BERN2. After examining the picture, turn on Plot1 and turn off either Plot2 or Plot3—they are so close one only needs display one of them—and compare the graphs of the numbers of people living to various given ages in the presence and absence of smallpox. Bear in mind that it is the horizontal separation, not the vertical that is of significance here.

We have not reached the end of Bernoulli's ingenuity or his paper. Todhunter informs us

> 406. After discussing the subject of the mortality caused by the small-pox, Daniel Bernoulli proceeds to the subject of Inoculation. He admits that there is some danger in Inoculation, but finds on the whole that it is attended with large advantages. He concluded that it would lengthen the average duration of life by about three years. This was the part of the memoir which at the time of publication would be of the greatest practical importance; but that importance happily no longer exists.[50]

Todhunter does not relate the details of this analysis, but it is not too hard to imagine how it would go. It is a fairly simple matter to determine the life expectancy of a person of a given age. Let p_i denote the probability of living to age i. In the presence of smallpox,

$$p_i = \frac{\xi_i}{1300}, \tag{77}$$

while if smallpox were eliminated we would have

$$p_i = \frac{z_i}{1300}. \tag{78}$$

The probability of living to age i, but not to age $i + 1$, is $p_i - p_{i+1}$. Basing ourselves on Halley's tables which only go up to 84, we assume $p_{85} = 0$. The expected age of death is thus

$$E = 0(p_0 - p_1) + 1(p_1 - p_0) + \ldots + 84(p_{84} - p_{85})$$

$$= \sum_{i=0}^{84} i(p_i - p_{i+1}). \tag{79}$$

This is readily calculated in each case. In the presence of smallpox, enter

−ΔList(∟HALLY)/1300→∟PRBSP
∟HALLY(85)/1300→∟PRBSP(85).

ΔList(takes a list $\{a_0, a_1, \ldots, a_{k-1}\}$ of length k and produces the list of differences, $\{a_1 - a_0, a_2 - a_1, \ldots, a_{k-1} - a_{k-2}\}$ of length $k - 1$. Thus

ΔList(∟HALLY) is $\{\xi_1 - \xi_0, \xi_2 - \xi_1, \ldots, \xi_{84} - \xi_{85}\}$

---

[50] Todhunter, op. cit., p. 228.

and the first line of code yields $\{p_0, p_1, \ldots, p_{83}\}$. The second line places $p_{84} - p_{85} = p_{84} = \xi_{84}/1300$ at the end of the list. We now calculate E by entering

```
sum(∟AGE∗∟PRBSP)
```

to get 26.07153846.

For the case in which smallpox has been eradicated, one would enter in turn

```
−ΔList(∟BERN2)/1300→∟PRB
∟BERN2(85)/1300→∟PRB(85)
sum(∟AGE∗∟PRB),
```

resulting in 29.14812044 as life expectancy. The expected gain, should smallpox be eliminated, is thus approximately

$$29.15 - 26.07 = 3.08$$

or 3 years, 1 month.

Once again, Bernoulli did not stop here. If one is already n years of age, one's life expectancy would be calculated by replacing the denominator 1300 in (77) and (78), and setting the lower bound of summation in (79) to n. Thus, for example, for a healthy 20 year old, in the presence of smallpox one would enter

```
−ΔList(∟HALLY)/∟HALLY(21)→∟PRB20
∟HALLY(85)/∟HALLY(21)→∟PRB20(85)
seq(∟PRB20(n),n,21,85)→∟PRB20
seq(∟AGE(n),n,21,85)→∟AGE20
sum(∟AGE20∗∟PRB20)
```

to get 53.58528428, approximately 53 years, 7 months, as the age of death, or an additional 33 years, 7 months of life.

Replacing ∟HALLY by ∟BERN2 yields 53.89958148 or between 53 years, 10 months and 53 years, 11 months as the expected age of death should smallpox be eliminated entirely. This makes a less impressive average increase of only 3 to 4 months. However, one must remember that the probability of having already had smallpox by age 20 is $1 - (7/8)^{19}$, which is .9209.... Over 90% have already had smallpox and are immune.

The average gain x for a 20 year old who has never had smallpox is determined algebraically:

$$E = .3143 = .921(0) + .079x,$$

so $x = .3143/.079 = 3.9784$, or about a week short of 4 years.

Bernoulli's analysis is a bit more sophisticated than this in that he takes into account, which I haven't here, that inoculation can be fatal. I quote Pearson:

> Bernoulli makes use of the slender data available, and determines from ages one to 24 the gain in mean length of life due to inoculation. In order to do this he has to make hypotheses where more exact data fail. He assumes in each year from 1 to 24, 1/8 of those living have

smallpox, and that $1/8$ of those attacked die of it. On the basis of this he then calculates the mean life for each age of those who have not had the smallpox. He further supposes that inoculation kills 1 in 200 of the inoculated, and he now calculates the mean life of the inoculated. Comparing the results that the two hypotheses furnish, he determines the time that one can hope to live longer by being inoculated than if one awaited the smallpox. This is Bernoulli's measure of the advantage one procures by inoculation.[51]

We can factor the occasional fatality into our calculations. If 1 in 200 die as a result of inoculation, only 199 out of 200, or 99.5% of those inoculated, will reap the benefit of the extended lifespan. The average increase in life expectancy of those inoculated at age 20 is thus

$$.005(-53.8996) + .995(3.9784) = 3.689\ldots,$$

or just over 3 years, 8 months, not much less than the result obtained ignoring the danger of inoculation.

35 Project. Produce a list of such expected gains for a person of age n who gets inoculated at age n for all $n < 40$.

My presentation, not being based on Bernoulli's paper, but instead taking Todhunter's and Pearson's accounts of this paper as points of departure, I caution the reader not to assume its complete historical accuracy. Think of it as a first approximation to history. The reader desirous of seeing more exactly how Bernoulli treated the problem and who has access to a decent research library is referred to Bernoulli's own papers. The original *Essai d'une nouvelle analyse de mortalité causée par la petite Vérole, et les avantages de l'Inoculation pour la prévinir* appeared in *Histoire de l'Académie royale des sciences avec les mémoires de mathématique et physique* (1766). Some volumes of the *Mémoires* can be found online and I assume it is just a matter of time before they all become available. Until then, a more likely source is the reprint in Bernoulli's collected works: L.P. Bouckært, B.L. van der Wærden, and David Speiser, eds., *Die Werke von Daniel Bernoulli: Band 2: Analysis und Wahrscheinlichkeitsrechnung*, Birkhäuser Verlag, Basel, 1982. Also included in the volume is a paper, "Réflexions sur les avantages de l'inoculation", a short mathematics-free version published in 1760 in *Mercure de France*, a literary journal that still exists today. For those of us who prefer English, there is a translation in: L. Bradley, *Smallpox Inoculation: An Eighteenth Century Mathematical Controversy*, Adult Education Department, Nottingham, 1971. The translation was reprinted: Trevor A. Sibbett and Steven Haberman, eds., *History of Actuarial Science, vol. VIII, Multiple Decrement and Multiple State Models*, William Pickering, London, 1995.

---

[51] Karl Pearson, *The History of Statistics in the 17th & 18th Centuries against the changing background of intellectual, scientific and religious thought*, Charles Griffin & Co. Ltd., London, 1978, p. 546.

For the purposes of the present book, however, the presentation given suffices. It gives the flavour of Bernoulli's treatment and of mathematical modelling in general. It also makes clear that Bernoulli makes a lot of assumptions.

Bernoulli wrote his paper in 1760, but its publication in the *Mémoires* only occurred in 1766. D'Alembert had seen it far earlier and chose to criticise it publicly before its publication, an act that pleased Bernoulli not at all. He wrote to Euler in April of 1768:

> What do you say about the enormous platitudes of the great d'Alembert about the probabilities; as I find myself too frequently unjustly treated in his publications, I have decided already some time ago to read nothing anymore which comes from his pen; I have taken this decision on the occasion of a manuscript about inoculation which I sent to the Academy in Paris eight years ago and which was greatly appreciated because of the novelty of the analysis; it was, I dare say, like incorporating a new province into the body of mathematics; it seems that the success of this new analysis caused him pains of the heart; he has criticized it in a thousand ways all equally ridiculous, and after having it well-criticized, he pretends to be the first author of a theory which he did not only hear mentioned. He, however, knew that my manuscript could only appear after some seven or eight years, and he could only have knowledge about it in his capacity as a member of the Academy and in this respect my manuscript should have stayed sacred until it was made public.[52]

The ethical question aside, d'Alembert's paper was a genuine contribution. Some of his criticisms were spot-on, and "D'Alembert developed in 1761 an alternative method for dealing with competing risks of death, which is also applicable to non-infectious diseases".[53]

It is the mathematical criticism that should be considered here:

> While praising the ingenuity of the calculations, he says that the suppositions of the number of persons who take the smallpox at each age and the number of these who die appear absolutely gratuitous. He says that it is highly doubtful whether 1/8th part of those who have not yet had smallpox die at every age, and still more doubtful that a constant number at each age of those who have it die of it. Some physicians assert that in the first ten years of life, one is ten times more subject to smallpox than in other years. Further, according to the inoculators, infants who die before four years of age—almost half of those born—die of other maladies than smallpox. According to these hypotheses the greatest danger of having the smallpox lies between three or four years and 10 years, and the danger of dying of this disease begins at four and not at one year as Bernoulli supposed. D'Alembert does not believe in the

---

[52] Quoted in: Klaus Dietz and J.A.P. Heesterbeek, "Daniel Bernoulli's epidemiological model revisited", *Mathematical Biosciences* 180 (2002), pp. 1 - 21; here, p. 12.
[53] *Ibid.*, p. 1.

deathrate of those attacked being the same at all years... D'Alembert points out that Bernoulli's hypothesis leads to the consequence that in the 9th year of life, there die of smallpox alone 2/3 of the number who die of all other causes. There is no one, D'Alembert says, who will not believe this excessive.[54]

In the absence of careful records, Bernoulli had to guess at various parameters: 1/8 for p, 1/8 for q, 1/200 for the probability of dying as the result of inoculation. The particular values, based on data supplied by la Condamine are probably approximately correct *averages* and, though one would like to have some concrete evidence of the reliability of the data on which they are based, we might as well accept these figures. What seems crucial, however, is the criticism that none of these three values is likely to be constant, but to depend on age. p should not be 1/8, but some function p_n, q should be q_n, and 1/200 also a function of age. Indeed, with our modern knowledge of viral mutation and evolution, we know that these parameters not only vary with age, but with time: natural selection favouring those viruses which do not kill their host organisms.

Soon after its publication, Bernoulli's paper was followed by Lambert who tried to generalize the method of Bernoulli taking into account age-dependent parameters. Trembley and Duvillard also pursued the same objective.[55]

J. Trembley (1749 - 1811) merits a chapter of his own in Todhunter's history and I recommend the discussion of his treatment of the inoculation problem[56] for an alternate discrete approach to the calculation of s_n. He obtains his estimate of s_n in closed form:

$$s_n = \frac{\xi_n}{(1-q)(1-p)^{-n} + q}. \tag{80}$$

Using smaller and smaller periods of time, the factor $(1-p)^{-n}$ approaches e^{pn} and Bernoulli's

$$s = \frac{\xi}{(1-q)e^{px} + q} \tag{81}$$

results, without resort to differential equations. The derivation of (80), however, is a bit involved:

Trembley's methods are laborious and like many other attempts to bring high mathematical investigations into more elementary forms, would probably cost a student more trouble than if he were to set to

---

[54] Pearson, *op. cit.*, pp. 546 - 547.

[55] Dietz and Heesterbeek, *op. cit.*, p. 4. English translations of Lambert's paper, "The mortality of smallpox in children", and Duvillard's paper, "Analysis and tables of the influence of smallpox on mortality", can be found in Sibbett and Haberman, cited on page 512, above.

[56] Todhunter, *op. cit.*, pp. 423 - 425.

work to enlarge his mathematical knowledge and then study the original methods.[57]

I note that the results obtained from (80) are not quite equal to the results stored in the list ∟SN discussed earlier, but they are remarkably close and using a table of logarithms would be as easy to calculate by hand as by Bernoulli's (81).

Todhunter does not go into detail on Trembley's calculations when p and q are replaced by age-dependent parameters, other than to say that, "There is no difficulty in working this hypothesis by Trembley's method; the results are of course more complicated than those obtained on Daniel Bernoulli's simpler hypotheses".[58]

Charles Marie de la Condamine (1701 - 1774) is probably best known for leading the incident-filled expedition to the Spanish colony of Peru, then including modern day Ecuador, to measure the arc of the meridian at the equator while Pierre Louis Moreau de Maupertuis (1698 - 1759) led a sister expedition to Lappland to perform the corresponding measurement near the pole. The two men stimulated Bernoulli to look into the problem of inoculation. La Condamine had written in favour of it himself, he provided Bernoulli with data, and it was he who read Bernoulli's memoir to the Academy in Paris when Bernoulli sent it to him from Basel.

La Condamine has been honoured philatellically a number of times by Ecuador (1936, 1986, 1993) and once by France (1986), the stamps of 1936 celebrating the bicentenniel of the la Condamine expedition, those of 1986 the 250th anniversary of the two expeditions, and that of 1993 the 250th anniversary of la Condamine's subsequent exploration of the Amazon. Of all these, I have chosen to picture the French stamp of 1986 because of its double portrait of Maupertuis (left) and la Condamine (right). I note, however, that although French stamps are generally among the most beautiful, in the present instance the Ecuadorean stamps of 1936 and 1993 are more attractive.

Lambert's paper appeared as early as 1772; Trembley's paper and a follow-up presented to the Royal Academy in Paris in 1776 and 1804, respectively, appeared in print in 1799 and 1807; and *Analyse et tableaux de l'influence de la petite vérole sur la mortalité á chaque age...* by Emmanuel Étienne Duvillard

---

[57] *Ibid.*, p. 415.
[58] *Ibid.*, p. 425.

de Durand (1755 - 1832) was published in 1807. More recently, in 2002 Klaus Dietz and J.A.P. Heesterbeek further refined Bernoulli's model.[59]

Finally, I should mention the paper "Smallpox and the double decrement table, a piece of actuarial pre-history" by Raymond H. Daw[60]. This article reviews in more detail than I've given here the work of Bernoulli, d'Alembert, Trembley, and Lambert.

---

[59] Dietz and Heesterbeek, *op. cit.* I recommend this paper to the reader. It will not be completely accessible to the reader without a background in the Calculus and differential equations, but the historical titbits alone make the paper worth reading. It also offers a mini-case study in how mathematical modelling works as models are refined to encompass greater variability and generality.

[60] *Journal of the Institute of Actuaries* 106 (1979), pp. 299 - 318.

C

Philately

Modern postage stamps go back to the Penny Black, the first adhesive stamp issued by Great Britain in 1840. It was an engraved profile of Queen Victoria against a black background, the choice of subject being dictated by the familiarity of the image: counterfeits would be readily recognisable. In the spring of 1843 the canton of Zürich in Switzerland followed suit and issued its first postage stamp, and in the fall of that year Brazil joined in. In the United States various postmasters began in 1845 to issue *provisional* stamps for temporary use until regular issues were available. These came in 1847 with the 5 cent Benjamin Franklin and 10 cent George Washington *definitives* of 1 July. Most early definitives were either simple designs, portraits of current rulers, or, in the United States, Benjamin Franklin and dead presidents. If a scientist of any sort appeared on a stamp, it was not for his contribution to science: Benjamin Franklin and Thomas Jefferson on American stamps, Albert I of Monaco and Carlos I of Portugal on European stamps, were all scientists, but appeared on stamps because of their political standings, not their scientific ones.

About half a century after the first definitives appeared, *commemoratives*, stamps issued for short periods of time to celebrate specific current or historical events, made their debut. In 1888 the Australian state of New South Wales issued a set to mark its having been a crown colony for 100 years. And in 1892/93 various American countries commemorated the 400th anniversary of Columbus's first voyage to the New World. The simplest of these was a Paraguayan overprint in 1892 of a stamp only first issued without overprint in January 1893, the overprint reading

1492

12 DE OCTUBRE

1892,

and the unrelated stamp bearing a portrait of a former president. At the opposite extreme, the United States issued 16 stamps in 1893 in simultaneous celebration of Columbus's voyage of discovery and the holding of the World Columbian Exposition in Chicago from May through October of that year. The next commemorative issues of the United States were the 1898 stamps celebrating the Trans-Mississippi Exposition in Omaha and the 1901 Pan-American Exposition in Buffalo. The pace has picked up since then and most countries issue several commemorative sets per year.

Another practice that has developed is the printing of *souvenir sheets*. Stamps are usually printed in densely packed sheets, but some commemoratives may be issued in sheets containing fewer stamps with the illustration extending beyond the stamp or with the stamp margin containing a completely different illustration. Additionally, stamps and marginal designs may be imprinted on ready-to-mail *postal cards*, *envelopes*, and *aerogrammes*. We have seen examples of the postal cards in the text—the marginal illustration, called a cachet, on the postal card featuring Jakob Bernoulli on page 31 and the imprinted stamp featuring John von Neumann on page 383.

There are different types of stamp collectors. There are those who collect stamps of specific countries. They often prefer stamps that have been used, though the cancellations obscure the pictures, something I cannot imagine bearable to one who collects the miniature works of art used as postage by France and its former colonies. There are those who specialise in air mail stamps, *semi-postals* (stamps with a surcharge going to charity), etc. The most recent addition to the group is the *topical* collector, who collects stamps for the subject matter of the images. Popular topics are mushrooms, butterflies, scouting, space, Rotary International, and, closer to our interests here, Einstein, astronomy, and Nobel Prize winners. Topical collectors have been a bit indiscriminate in our buying habits and are thus ultimately responsible for the existence of many *undesirable* issues.

For over half a century it has been the habit of some nations to produce stamps far in excess of domestic postal needs to sell to foreign collectors. The famous early examples of this were Monaco and San Marino. The stamps of Monaco are quite beautiful and one is pleased to own them. And San Marino did produce a couple of the earliest issues devoted solely to science, deliberately depicting numerous scientists because they were scientists. In the mid-1960's other countries got into the act. First were Ecuador, Panama and Paraguay, then Equatorial Guinea, and by the end of the decade various Arab emirates (Ajman, Fujeira, Sharjah, etc.). The Arab oil embargo and subsequent increases in oil prices removed the necessity of philatelic sales to provide revenue and since 1973 all stamps from the Middle East have been non-exploitative. The Central and South American post offices have pulled back and are mostly issuing only appropriate stamps. However, excessive and not culturally relevant issues are today being produced in large numbers by certain African nations and members of the British Commonwealth.

There is a premium to stamps issued by a country honouring its own, be it a native son, a famous visitor, or just someone with a connexion to the country. Of the many stamps depicting Edmond Halley issued in the 1985/86 rush to produce Halley's Comet stamps, the one's I treasure most are those from Great Britain and St. Helena. And of all the many stamps issued honouring Albert Einstein, the ones from Germany, Switzerland, and the United States (places he lived and worked), Israel (a country he had close ties to), Sweden

A souvenir sheet is usually a small sheet containing only a few stamps, sometimes only one. The margin contains its own illustration, which may or may not extend the image of the stamp and may or may not be related in any way to the subject of the stamp's image. Pictured above is an example, not as extreme as some, of a souvenir sheet with a loose connexion among the topics pictured. The sheet itself was issued to celebrate the projected "Conquest of Mars", the stamp featuring an imaginary "Mars photon rocket" and the trial of Galileo. Having observed the heavens and made a number of discoveries, Galileo is a common subject of space-themed stamps. His trial is not so common a theme, but as it concerned his writings about the solar system it is perhaps not as irrelevant as it seems when viewing the stamp alone. The caveman and dinosaur in the margin, however, are a bit harder to relate to Mars. Today, one would point to the popularity of dinosaur stamps, and dismiss its presence as mere exploitation. At the time of issue in 1970, however, dinosaurs were not a common philatelic subject.

(the country from which he was awarded the Nobel Prize), and St. Tomé et Príncipe (where Eddington verified his general theory of relativity via astronomical observation), seem to have a legitimacy the stamps of the rest of the world do not.

There was a backlash against undesirable stamps. The chief American stamp catalogue refused for years to list them and collector organisations banned their use in competitions. For educative purposes, the difference between desirable, culturally appropriate and undesirable, printed-for-profit stamps is less important. Insofar as the stamps are used locally and intended as propaganda to inspire the local population, they remind us of Plato's remark, "What is hon-

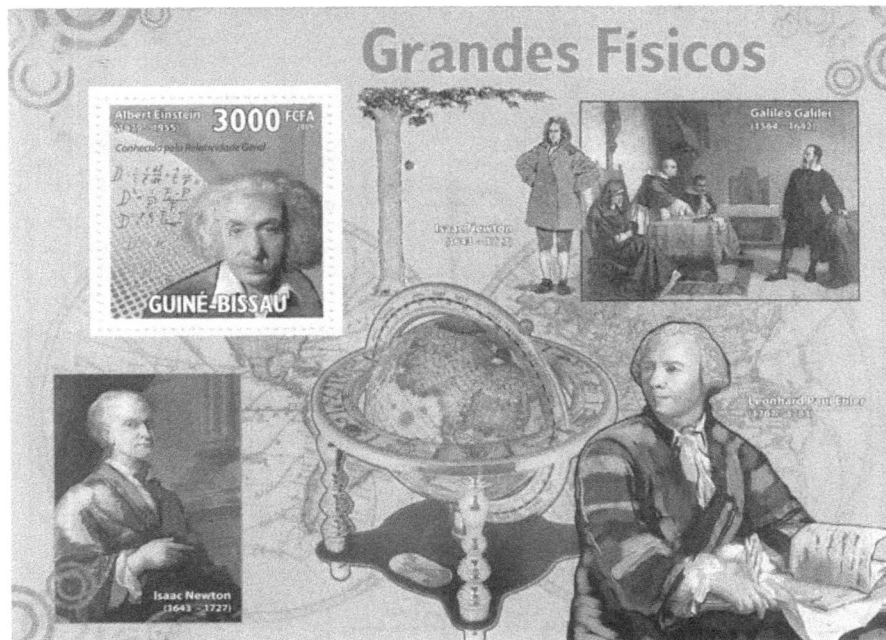

Here we have a nice example of a souvenir sheet issued in 2009 by the African nation of Gabon. While the main topic of the sheet has no Gabonese cultural significance, featuring portraits of white European males, none of whom visited the African continent, much less the territory now forming the nation of Gabon, the major elements of the design have a unity of topic—famous physicists—not always found in such sheets, as seen in the preceding illustration.

The present sheet sticks closely to its announced topic. The stamp pictures Albert Einstein and the margin contains portraits of Isaac Newton (lower left), Leonhard Euler (lower right), and a depiction of the same painting of the trial of Galileo by Cristiano Banti used in the stamp from the IPSA reproduced on page 217. The elements of the background—a map of the world, an antique globe, and a drawing of Newton standing, watching an apple fall from a tree—depart a bit from the topic, unless the antique nature of the map and globe is intended to set Einstein off chronologically from his predecessors.

oured in a country is cultivated there".[1] Besides, an "undesirable" stamp may

---

[1] Or words to that effect. This statement has stayed with me since my youth. I looked it up in Jowett's translation and found in book VIII of *The Republic*, "And what is honoured is cultivated, and that which has no honour is neglected". (Benjamin Jowett, *The Dialogues of Plato*, vol. I, Random House, New York, 1937, p. 808.) Every few years American politicians deplore the sad state of education in the United States and demand more mathematics and science for all students and propose or even legislate stop-gap measures. In the meantime, the message that mathematics is not valued in American society is broadcast across the land. Our banknotes depict political activists, our stamps have long celebrated sports figures and entertainers as such, and our scientists and mathematicians have been ignored or, like Benjamin Franklin, have been celebrated for some other reason. Even Al-

offer something that a "desirable" one does not. It would have been nice, on page 65, above, to have illustrated the early Chinese knowledge of the Binomial Theorem with a stamp of China; but I didn't have any, so I used the Liberian stamp from that country's Millennium issue celebrating Chinese achievements. From the point of view of local appropriateness, the early Ecuadorean (1936) and the first British issue (1982) honouring Darwin certainly beat my choice from St. Tomé et Príncipe. These stamps are much more attractive, feature Darwin, and hint at his work. However, the point of including a stamp of Darwin when I did was to illustrate evolution, the subject about which the Scopes trial was held, and the two direct references to evolution in the stamp of St. Tomé et Príncipe led to my preferring it over the more subtle hints from Ecuador and Great Britain.

Obtaining stamps for display in class—scanning them into the computer for projection from the computer, for printing enlargements for overhead transparencies, or for printing them on paper for distribution in class—can be a bit of a challenge. Stamp catalogues are massive multivolume affairs listing all stamps issued by country, often with only a single stamp from each set pictured. Finding those dealing with a particular subject by searching the catalogues can take quite some dedication. Fortunately, there are groups of collectors that maintain checklists and distribute newsletters. In the United States, affiliated with the American Topical Association are Study Units in mathematics, astronomy, and physics/chemistry. Also, there have been a number of books dedicated to stamps featuring scientific and mathematical topics. I cite a few here:

> W.J. Bishop and N.M. Matheson, *Medicine and Science in Postage Stamps*, Harvey & Blythe Limited, London, 1948. This, the oldest such book in my collection, has only 82 pages, 32 of which have black and white images of stamps set against a black background. Very few of the stamps featured depict persons of mathematical interest and none depict mathematical topics.

> R.W. Truman, *Science Stamps*, American Topical Association, Milwaukee, 1975. This is more up-to-date than the previous listing and has only a limited number of black and white images, some of which depict mathematicians, but again no mathematical topics. The text divides into four parts: 5 chapters on specific disciplines (physics, chemistry, natural history, medicine, and technology), 5 chapters on individual scientists (the Curies, Louis Pasteur, Alexander von Humboldt, Albert Schweitzer, and Leonardo da Vinci), 5 chapters on countries with good

---

bert Einstein, who appeared twice(!) on US postage stamps, was so honoured not so much for his scientific prowess as for his notoriety. I suspect that the recent decision on the part of the US Postal Service to start issuing stamps honouring scientists was a marketing strategy designed to capitalise on the popularity among collectors of stamps depicting Nobel Prize winners. That there are countries of the world that value intellectual achievement, including mathematics, and honour their scientists and mathematicians for their intellectual work, is easily demonstrated by examining the postage stamps issued by these countries.

records of honouring their scientists on stamps (Poland, France, Germany, Russia, and Italy), and several checklists of scientists by name, country, and discipline, the last including mathematics.

William L. Schaaf, *Mathematics and Science; An Adventure in Postage Stamps*, National Council of Teachers of Mathematics, Reston (Virginia), 1978. This is probably the first book devoted solely to mathematics on stamps. It features some text, numerous illustrations, some in colour, and a short checklist. The stamps include mathematical instruments, notations, and concepts as well as portraits.

Peter Schreiber, *Die Mathematik und ihre Geschichte im Spiegel der Philatelie* [*Mathematics and its history as reflected in philately*], B.G. Teubner Verlagsgesellschaft, Leipzig, 1980. The author is a math historian as well as stamp collector and the book is a brief history of mathematics with 16 pages of colour plates depicting various mathematical stamps. It also includes a checklist.

Hans Wussing and Horst Remane, *Wissenschaftsgeschichte en miniature* [*History of science in miniature*], VEB Deutscher Verlag der Wissenschaften, Berlin, 1989. This is another East German publication by authorities on the history of mathematics and topical collecting. The book is a collection of short one-column essays illustrated on adjacent columns by colour reproductions of stamps set against black backgrounds. Unlike Schaaf and Schreiber, Wussing and Remane cover all of science. The colour reproductions are of much higher quality than anything that had gone before.

Robin J. Wilson, *Stamping Through Mathematics*, Springer-Verlag New York, Inc., New York, 2001. The scope of this book is narrower historically than Schreiber's and scientifically than Wussing and Remane's, featuring short essays and beautiful colour reproductions, slightly larger than actual size, set against the customary black background. These are the sharpest and most accurate colours of any of the books cited here. Indeed, one could almost scan these images for classroom display.

It shows my age that I mention books before the Internet. I generally do not like to give references to web sites as they disappear or change addresses so often. However, the Internet can be a source of information on the topic. I refer specifically to the site Images of Mathematicians on Postage Stamps currently located at http://jeff560.tripod.com/stamps.html. It offers a checklist with links to images.

As for stamps featuring topics from probability, there are, in addition to those featured already in this book a number of stamps that might be added— and this brings us finally to the reason for including this appendix. While I could not find anything directly related to Probability proper, there are some pertinent to Statistics and the use of probabilistic concepts or methods in Physics.

From Statistics there are stamps depicting statisticians, stamps celebrating statistical congresses, stamps supporting the local census, and stamps featuring graphical data displays.

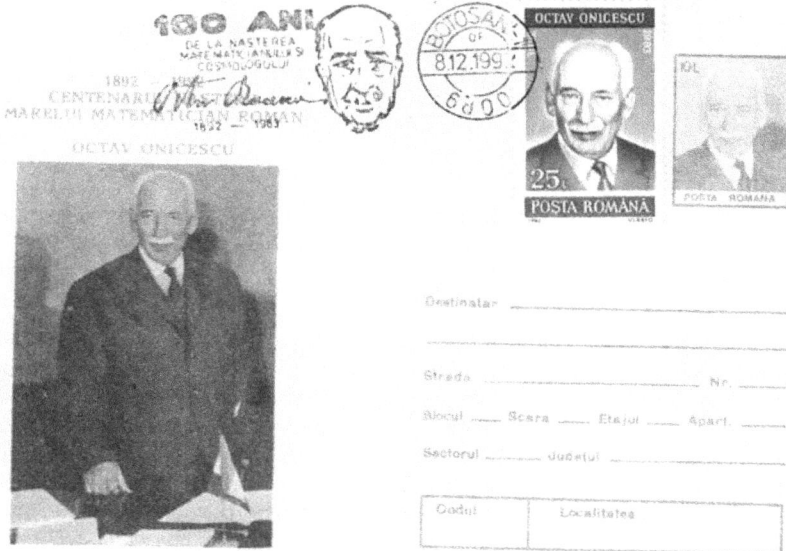

Octav Onicescu (1892 - 1983), who died one day short of his 91st birthday, exemplifies the best tradition of science in the public service often found by the first practitioners of a scientific discipline in a country, taking leading rôles in teaching the subject and founding societies dedicated to the field, and applying statistics to issues of national importance. Onicescu wrote a textbook in probability theory, as well as articles on scientific, cultural and philosophical matters, and involved himself with the Romanian Committee for the History and Philosophy of Science.

Pictured above is a quadruple homage to Onicescu issued in 1992 on the centenary of his birth: his portrait appears in the cachet, an imprinted stamp on the envelope, a stamp and a cancellation. To the right is an enlargement of the stamp.

Two generations of Indian statisticians are represented by Prasanta Chandra Mahalanobis (1893 - 1972) and Damodar Dharmananda Kosambi (1907 - 1966).

Mahalanobis was heavily involved in government planning, developing statistical tools for dealing with large-scale sample surveys.

> In his work as an applied statistician, Mahalanobis was very innovative, often introducing new concepts or methodologies or systematizations. His work on flood control combined innovative data analysis, understanding and modeling natural phenomena, and a systematization of the whole complex analysis which made his recommendations so readily acceptable to the government. The same gift for innovation and systematization are apparent in his work on large-scale sample surveys and planning. For him, theory grew out of a practical need and thus influenced subsequent practical work. He had nothing but contempt for irrelevant, poorly conceived abstractions which he would dismiss as "aerodynamics in a viscous fluid." (J.K. Ghosh, "Prasanta Chandra Mahalanobis", in: C.C. Heyde and E. Seneta, eds., *Statisticians of the Centuries*, Springer-Verlag New York, Inc., New York, 2001; here: pp. 437 - 438.)

The International Statistical Institute established the Mahalanobis Award in his honour in 2003.

Kosambi, the son of the nearly identically named Dharmananda Damodar Kosambi, a famous Indian philologist, had very broad interests and is in fact best known for his work in Indology, the study of Indian history, culture, etc. His work in the quantitative sciences began with his first research publication in 1930 on Physics, continued through various subjects in Analysis, and on into Statistics. He is particularly noted for his application of statistics in genetics.

The Indian stamp above left features the National Statistics Institute along with a cameo portrait of Mahalanobis and was issued in celebration of the centenary of his birth. The stamp on the right pictures Kosambi, a map of India, and a formula from genetics known as the *Kosambi map formula*.

विशेष आवरण Special Cover

पद्मविभूषण प्रो.सी.आर.राव, एफ.आर.एस.

Padma Vibhushan Prof. C.R. Rao, FRS

सांख्यिकी : भविष्य का विज्ञान

Statistics is the Science of the Future

Calyampudi Radhakrishna Rao (*1920) was a leading statistician of the 20th century whose *Linear Statistical Inference and Its Applications* is commonly cited. Eponymous terms bearing his name occur in most standard textbooks on statistics, economics, and engineering. He has been honoured many times: 32 honorary doctorates, the Wilks Medal of the American Statistical Association (1989), the Padma Vibhushan Award of the Indian government (2001), the Indian National Science Academy's Srinivasa Ramanujan Medal (2002), the American government's National Medal of Science (2002), the Desikottama Award of the University of Visva-Bharati (2002), and the first Mahalanobis Award of the International Statistical Institute (2003). The cachet reproduced above is from a special envelope issued by the Indian Post Office in December 2009 and refers explicitly to his Padma Vibhushan professorship.

Shown here are two stamps demonstrating official pride in statistical congresses. The Egyptian stamp (left) was issued in 1927 to commemorate the hosting of a statistical congress in Cairo. Pictured is Amenhotep, a scribe to Amenhotep III. The scribes were the record keepers and calculators of ancient Egypt and can be considered the statisticians of the day.

The Polish stamp (right) was issued in 1975 to celebrate the 40th World Statistical Congress, held in Warsaw, of the International Statistical Institute. It has more directly mathematical content, featuring a variant of a pie chart. The familiar pie charts divide a circle into sectors corresponding to various categories, the areas of the sectors being proportional to the measures applied to the given categories. In the chart pictured, the sectors are all of the same size, but the spokes radiating out from the centre have varying lengths proportional to these measures. They are direct descendants of the original pie charts used by Florence Nightingale. In hers the segments again shared a common angle, but could extend beyond or fall short of the circle, the areas of the resulting regions being proportional to the measures of the categories.

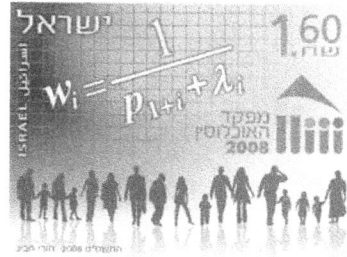

Stamps of statistical interest can be issued for a variety of reasons. There are stamps to remind people of the importance of the census, to celebrate economic growth, etc. And these stamps can include line graphs, histograms, stylised histograms, and such. Pictured here are two stamps honouring their respective country's central statistical bureau. The stamp on the left was issued by Austria in 1979 to celebrate the 150th anniversary of its bureau of statistics and features a couple of interesting histograms revealing the "age structure of the residential population". Here the vertical distance from the bottom indicates age and the horizontal distances from the center the number of people of a given age—left for men and right for women in side-by-side comparison. The big indentations around age 60 would be people born in the years immediately following the First World War and the great influenza epidemic. That the right-hand side is generally wider than the left—in the portions of the graph in which the extent of the right-hand side is visible—reflects the fact that women generally survive to an older age than men. It is a shame that the lower right portion of the graph is obscured by the icon for child pornography, as it would be nice to see the excess number of male over female births represented graphically. I have no explanation for the apparent "graph within a graph" given by the darkened patch around the center of the figure. My best guess is that the very dark region represents those who have never married, as the non-dark region begins only around age 20, but this is just a wild guess. The stamp on the right was issued by Israel in 2008 and I confess ignorance of the meaning of the formula.

Opposite bottom: A Bolivian stamp of 2011 celebrating the country's National Institute of Statistics features three statistical formulæ and a hypothetical graph.

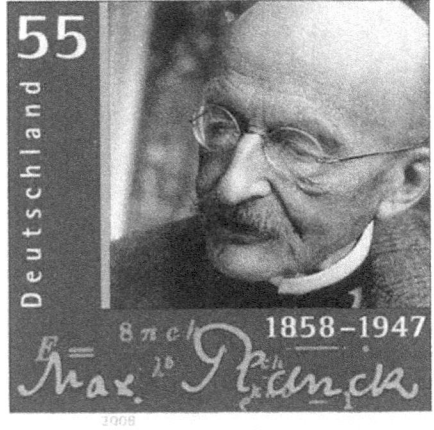

While some 19th century physicists may have embraced indeterminism, the application of probabilistic concepts in physics was statistical—a method of dealing with a large number of particles for which a deterministic approach would have been a computational nightmare. It was the 20th century development of quantum theory that transformed Probability from a useful tool into a central concept of modern Physics. This began in 1900 with the derivation of Planck's radiation law introducing quantum theory.

> Two kinds of probabilities can be found in the old quantum theory, namely, probabilities of energy states and *transition probabilities* between energy states. Einstein's crucial 1917 paper, 'On the quantum theory of radiation,' introduces both. (Jan von Plato, *Creating Modern Probability*, Cambridge University Press, Cambridge, 1994, p. 143.)

A Nobel Prize winner, Max Planck has appeared on a number of stamps. The two most attractive are that issued by West Berlin in 1953 as one of a set of 10 stamps celebrating famous Berliners, and one issued in 2008 by Germany on the occasion of his 150th birthday.

Albert Einstein and later S.N. Bose offered probabilistic derivations of Planck's radiation law. Out of Bose's work and Einstein's consideration of it arose the Bose-Einstein statistics, a new probabilistic model of ideal gases. Einstein, the "rock star" of science of the 20th century, has been honoured and exploited philatelically by postal services all around the globe. I have chosen for them stamps issued by India in 1979 and 1994 on the occasions of their 100th birthdays.

300

$$\Delta p \cdot \Delta q \sim h$$

Heisenbergsche
Unschärferelation

Deutschland

1,53 €

Werner Heisenberg
2001   **Physiker**   **1901 – 1976**

Direktor des
Max-Planck-Institutes für
Physik in München

1958 – 1970

The new quantum theory of the 1920's, i.e., quantum mechanics, cemented the rôle of indeterminacy in Physics. Every physicist dealt with a world of Probability and every physicist of the 20th century who appears on a stamp could be included here. As many of them are Nobel Prize winners and the Nobel Prize is a hot topic among stamp collectors, there is no shortage of possible candidates for display here. To represent them I choose a single stamp depicting Werner Heisenberg and his famous *uncertainty principle*.

There are several stamps featuring Werner Heisenberg issued in sets celebrating the Nobel Prize. The present one, issued in 2001 on the occasion of his 100th birthday by his native Germany, is particularly appropriate and attractive.

D

Tables

Except for the most theoretical treatises, books on probability and statistics used to come equipped with tables—tables of factorials, binomial coefficients, binomial and/or normal probabilities, and, after insurance and annuities became popular topics, tables giving the birth and death statistics in various locations, values of annuities, and the like. The modern calculator obviates the necessity of inclusion of those tables which can be calculated, but tables that must be tabulated by collecting data are still necessary. The first two tables resented here are Arbuthnot's Birth Tables for London for the years 1629 to 1710 and Halley's Mortality Tables for Breslau. In reproducing them I have altered the number and lengths of the columns, but have otherwise tried to reproduce their formats faithfully.

One thing to notice about Arbuthnot's table is that the columns are not labelled "Born", but "Christened". In his *Théorie analytique des probabilités* Laplace reports that the ratio of male to female births in London is 19 to 18, while the ratio of male to female baptisms in Paris is 25 to 24. Laplace performed a calculation and determined that the difference was significant and in need of an explanation. He came up with the following:

> Among the causes that might produce it, it seemed to me that baptisms of abandoned infants, which form part of the annual list of baptisms in Paris would have a noticeable influence on the ratio of baptisms of boys to those of girls; and that they would diminish this ratio, if, as it is natural to believe, parents in the surrounding countryside, finding some advantage in keeping male infants close to them, sent them to the hospice for abandoned infants in Paris in a ratio less that that of the births of the two sexes. This is what a reading of the register of this hospice made me see with very great probability.[1]

Halley's table is straightforward and elicits no comments from me.

---

[1] Cited in the original French and in English translation in: Jacqueline Stedall, *Mathematics Emerging: A Sourcebook 1540 - 1900*, Oxford Unifersity Press, Oxford, 2008, p. 188.

Arbuthnot's Table of Births

Christened.			Christened.		
Anno.	*Males.*	*Females.*	*Anno.*	*Males.*	*Females.*
1629	5218	4683	1670	6278	5719
30	4858	4457	71	6449	6061
31	4422	4102	72	6443	6120
32	4994	4590	73	6073	5822
33	5158	4839	74	6113	5738
34	5035	4820	75	6058	5717
35	5106	4928	76	6552	5847
36	4917	4605	77	6423	6203
37	4703	4457	78	6568	6033
38	5359	4952	79	6247	6041
39	5366	4784	80	6548	6299
40	5518	5332	81	6822	6533
41	5470	5200	82	6909	6744
42	5460	4910	83	7577	7158
43	4793	4617	84	7575	7127
44	4107	3997	85	7484	7246
45	4047	3919	86	7575	7119
46	3768	3395	87	7737	7214
47	3796	3536	88	7487	7101
48	3363	3181	89	7604	7167
49	3079	2746	90	7909	7302
50	2890	2722	91	7662	7392
51	3231	2840	92	7602	7316
52	3220	2908	93	7676	7483
53	3196	2959	94	6985	6647
54	3441	3179	95	7263	6713
55	3655	3349	96	7632	7229
56	3668	3382	97	8062	7767
57	3396	3289	98	8426	7626
58	3157	3013	99	7911	7452
59	3209	2781	1700	7578	7061
60	3724	3247	1701	8102	7514
61	4748	4107	1702	8031	7656
62	5216	4803	1703	7765	7683
63	5411	4881	1704	6113	5738
64	6041	5681	1705	8366	7779
65	5114	4858	1706	7952	7417
66	4678	4319	1707	8379	7687
67	5616	5322	1708	8239	7623
68	6073	5560	1709	7840	7380
69	6506	5829	1710	7640	7288

Halley's Mortality Table

Age	Living.	Age	Living.	Age	Living.	Age	Living.
1	1000	22	586	43	417	64	202
2	855	23	580	44	407	65	192
3	798	24	574	45	397	66	181
4	760	25	567	46	387	67	172
5	732	26	560	47	377	68	162
6	710	27	553	48	367	69	152
7	692	28	546	49	357	70	142
8	680	29	539	50	346	71	131
9	670	30	531	51	335	72	120
10	661	31	523	52	324	73	109
11	653	32	515	53	313	74	98
12	646	33	507	54	302	75	88
13	640	34	499	55	292	76	78
14	634	35	490	56	282	77	68
15	628	36	481	57	272	78	58
16	622	37	472	58	262	79	49
17	616	38	463	59	252	80	41
18	610	39	454	60	242	81	34
19	604	40	445	61	232	82	28
20	598	41	436	62	222	83	23
21	592	42	427	63	212	84	19

The next three tables contain data on the population of the United States. The first table contains figures posted online at the website of the U.S. Census Bureau and are the official figures. The second table are the figures used by Verhulst, as discussed in Appendix B, for the years 1790 to 1840. They are roughly the same as the official figures, but none exactly equals the official values. I do not know the reason for this discrepancy. The third table contains Verhulst's rounded figures for each decade, his interpolated figures for each additional half decade, and for the years 1815 to 1840 the ratio r of the population for that date to the population 25 years earlier, thus demonstrating the doubling of the population every 25 years.

US Census Figures

Year	Population	Year	Population	Year	Population
1790	3929214	1870	38558371	1950	151325798
1800	5308483	1880	50189209	1960	179323175
1810	7239881	1890	62979766	1970	203302031
1820	9638453	1900	76212168	1980	226542199
1830	12860702	1910	92228496	1990	248709873
1840	17063353	1920	106021537	2000	281421906
1850	23191876	1930	123202624		
1860	31443321	1940	132164569		

Source: U.S. Census Bureau

US Census according to Verhulst

Year	1790	1800	1810
Population	3929827	5305925	7239814

Year	1820	1830	1840
Population	9638151	12866020	17062566

Verhulst's Adjusted Census

Year	Population	Value of r
1790	3930000	
1795	4618000	
1800	5306000	
1805	6273000	
1810	7240000	
1815	8439000	2.147
1820	9638000	2.087
1825	11252000	2.120
1830	12866000	2.052
1835	14964000	2.076
1840	17063000	2.021

E

Recommended Reading

As the reader will have noticed, I have eschewed a bibliography in favour of footnoted citations. The present appendix is here, not to list every source cited in the text, but to cite the most important of these as well as some others.

I am neither a probability theorist nor a statistician, merely a mathematician who has occasionally taught elementary courses on these subjects. I am, consequently, not up to date on the textbooks available. Thus my purely mathematical recommendations are somewhat old fashioned, but have stood the test of time in that most of them are still in print. The first book is a fairly elementary classic that has been picked up for republication by Dover.

> Samuel Goldberg, *Probability: An Introduction*, Dover Publications, New York, 1987.

A full text running the gamut from providing elementary exercises to rigorous theory is given by the two volume textbook of William Feller. It is more expensive than the above. For elementary probability as discussed here, volume one suffices:

> William Feller, *An Introduction to Probability Theory and Its Applications*, 3rd edition, John Wiley & Sons, New York, 1968.

One final recommendation, apparently out of print, is Kolmogorov's classic. The book offers nothing by way of teaching one how to calculate probabilities, but instead presents a solid mathematical foundation for probability in terms of measure theory. It is a slim volume and, for one who has the proper mathematical background, offers a quick introduction to the measure theoretic treatment:

> A.N. Kolmogorov, *Foundations of the Theory of Probability*, 2nd English edition, Chelsea Publishing Company, New York, 1950.

Books on the history of Probability split into several categories: general history, history of statistics, history of the elementary theory, history of the Central Limit Theorem, and source books. The classic history of the subject is Todhunter's:

> Isaac Todhunter, *A History of the Mathematical Theory of Probability From the Time of Pascal to That of Laplace*, Cambridge University Press, Cambridge, 1865.

Todhunter offers more a summary of papers on the subject than a narrative history; it is a kind of technical annotated bibliography, extremely valuable as a reference, but hardly a good read. It can be downloaded for free on the Internet.

The following pair of books comes highly recommended and the claim has been made that they supersede Todhunter's work. They are, however, prohibitively expensive and I've not seen them:

> Anders Hald, *A History of Probability and Statistics and Their Applications before 1750*, John Wiley & Sons, New York, 1990.
>
> _____, *A History of Probability and Statistics and Their Applications from 1750 to 1830*, John Wiley & Sons, New York, 1998.

I have another of Hald's books and have made use of it in writing the present account and can readily believe the claim that Hald has surpassed Todhunter with this pair of books. However, I have noticed in the book I have that he leaves unexplained notation that the nonspecialist might not be familiar with and only recommend these books to the more advanced reader.

With respect to the history of Statistics, the books to consult are Karl Pearson's and Stephen Stigler's. Pearson's is much harder to come by, but is beautifully written, highly informative, and just a delight to read.

> Karl Pearson, *The History of Statistics in the 17th & 18th Centuries against the changing background of intellectual, scientific and religious thought*, Charles Griffin & Co. Ltd., London, 1978.
>
> Stephen M. Stigler, *The History of Statistics; The Measurement of Uncertainty before 1900*, Harvard University Press, Cambridge (Mass.), 1986.

The journal *Biometrika* ran a series of "Studies in the history of probability and statistics". The papers of this series were collected together in a volume edited by Egon Pearson and M.G. Kendall:

> E.S. Pearson and M. Kendall, eds., *Studies in the History of Statistics and Probability*, Charles Griffin & Co. Ltd., London, 1970.

I have not seen this book, which, I believe, is out of print and generally expensive in the antiquarian market, but I have read some of the papers in their original publication and found them illuminating.

Turning to the category of the history of elementary probability, one book stands out:

> F.N. David, *Games, Gods and Gambling; A History of Probability and Statistical Inference*, Charles Griffin & Co., Ltd., London, 1962 (reprinted by Dover Publications, Mineola (New York), 1998).

As I was in the final stages of writing the present book I discovered some books on the history of the Central Limit Theorem:

Emanuel Czuber, *Theorie der Beobachtungsfehler*, Verlag von B.G. Teubner, Leipzig, 1891 (reprinted by Nabu Press, 2010).

William J. Adams, *The Life and Times of the Central Limit Theorem*, 2nd edition, American Mathematical Society, Providence (RI), 2009.

Hans Fischer, *A History of the Central Limit Theorem: From Classical to Modern Probability Theory*, Springer-Verlag New York, Inc., New York, 2010.

Czuber's is the oldest of these and can also be downloaded for free from Google Books. It has a wealth of material about the error distribution and the Method of Least Squares, both mathematical and historical details presented. Adams's first edition was not very technical; the second edition becomes at least partially so through the inclusion of several important technical papers as appendices. The pre-appendicial material is pleasantly readable and informative. I have not seen the book by Fischer, it having appeared rather late and not being inexpensive, but the examination of pages freely displayed online convince me to put the book on my want list.

Of the various source books in the histories of Probability and Statistics, the most useful is the hard to get collection in German edited by Ivo Schneider:

Ivo Schneider, ed., *Die Entwicklung der Wahrscheinlichkeitstheorie von den Anfängen bis 1933; Einführungen und Texte*, Wissenschaftliche Buchgesellschaft, Darmstadt, 1989.

This book, with the introductory essays, itself constitutes a history of the theory and was the single most valuable resource I used in preparing the present work.

Two works that don't quite fit into any of these categories are:

Ian Hacking, *The Emergence of Probability; A Philosophical Study of Early Ideas About Probability*, Cambridge University Press, Cambridge, 1975.

Jan von Plato, *Creating Modern Probability: Its Mathematics, Physics and Philosophy in Historical Perspective*, Cambridge University Press, Cambridge, 1994.

Both authors are philosophers and touch on philosophical issues as well as historical ones. Hacking is particularly strong on the emergence of probabilistic concepts—the beginnings of probability and also of twentieth century developments like subjective approaches to probability. Von Plato deals with modern probability and has a lot more to say about developments in physics, as well as 20th century interpretations (frequentism, measure theory, subjectivism) of probability.

Index

www.ingramcontent.com/pod-product-compliance
Lightning Source LLC
Chambersburg PA
CBHW031350210326
41599CB00019B/2718